Science and the Soul

facts and philosophy

translated into

Natural Dialectic

Copyright © Michael Pitman 2017
ISBN 978-0-9930067-2-2

A catalogue record for this book is available from the British Library.

The right of Michael Pitman to be identified as the Author of this Work has been asserted by him in accordance with the Copyright, Designs and Patents Act 1988.

All Rights Reserved. Apart from any use expressly permitted under UK copyright law no part of this publication may be reproduced, stored in an alternative retrieval system to the one purchased or transmitted in any form or by any means electronic, mechanical, photocopying, recording or otherwise without the prior permission in writing of the Author.

Published by Merops Press

website: www.philosophyandscience.co.uk
also www.cosmicconnections.co.uk
www.scienceandthesoul.co.uk
www.michaelpitmanbooks.co.uk

Acknowledgements: Suzanne, my wife,
Françoise, my daughter,
Dave Gant, Rick Pulford
and Beth Churchard.

TOTAL CONTENTS

Summary:

Part 0: Setting the Scene

    Book 0   A Systematic Start
- Polar Perspectives
- Natural Dialectic's ABC
- Information
- Information

    Book 1   The Stage before a Play
- Energy (Physics)

Part 1: Playtime

    Book 2   A Player's Mind
- Psychology: Consciousness
- Psychology: Subconsciousness

    Book 3   A Player's Body
- Biology

    Book 4   The Whole Cast
- Community

Contents

Illustrations 12
Preface 14
Book 0: ABC and Information 18
Chapter 0: Anti-parallel Perspectives 18
    The Light in Your Eye 18
    Experience 18
    World-view 19
    Opposite Directions of Mind 19
    Two Sorts of Implement 20
    Two Pillars: a Dialogue of Faith 20
    Scientific Delusions 23
    Religious Delusions 25
    A Complementary Course 26

Chapter 1: Natural Dialectic's ABC 28
    Natural Dialectic 28

Chapter 2: First Principles 41
    First Principles 41
        Nothing 41
        Something 44
    Two Existential Principles 45
        Causality 47
        Polarity 50
    The Basic Existential Dipole. 52
        Information 54
        Energy 55
    Informed Energy 56

Chapter 3: Hierarchical Perspective 59
    Hierarchical, Triplex Construction of the
        Cosmic Pyramid 59
    Subtendence 61
    Transcendence 64
    The Cosmological Axis 66

Chapter 4: Truth, Appearance and Reality 68

Truth, Appearance and Reality 68
Two Value Systems 71
Rights and Wrongs 73
From Science to Conscience 74
Is There an Absolute Morality? 75

Chapter 5: The Immaterial Element 76

Information, Messages, Arrangement 76
Information is Immaterial 79
Information is Hierarchical 80
Orderly Creation 83
The God Delusion? 84
ID, IC, IE - Call it What You Will 86
Alignment with Truth 89
(Sat) Potential or Transcendent Information 91
   Top Teleology 92

Chapter 6: Mind, Machines and Mind Machines 94

(Raj) Active Information 96
   Purposely Down to Earth 96
   Getting Your Way - Pragmatics 97
   What Do You Mean? 97
(Tam) Passive Information 98
   Information's Infrastructure - Code 99
   The Lowest, Physical Level 100
Music 101
Machines 101
Mind Machines 103
Universal Authorship 104

Book 1: Physics 105

Chapter 7: Lady Luck and Lord Deliberate 105

The Order of Invariance 107
Precondition/ Potential 107
Cosmo-logic 109
Lack of Cosmo-logic 109
Getting to Grips with Lady Luck 111
Are You Certain? 113

Towards a Theory of Physic and Metaphysic 118
Can Mathematics Help Us? 119
Punting on the Cosmic Stream 121
Shots in the Dark 122
The Chance-Killer 124
A Culture of Doubt 125
The Matrix 127

Chapter 8: Principles of a Unified Theory 128
of Matter 128

The Principles of a Unified Theory of Matter 128
Infinity *133*
Unity and Unification *137*
    Holy Grails 140
    Snip, Snip 143
    Split Continuity 144
    Boxing the Infinite 147

Chapter 9: Nothing 150

The Nature of Nothing *150*
    The Order of Nothing 154
    How Does Nothing Physical Work? 156
    Old Vacuums 157
    New Vacuums 159
    Dialectical Vacuum 160
    Is Space a Waste? 161

Chapter 10: Time 162

Time *162*
Species of Time 163
    Super-time 165
    Psycho-time 165
    Archetypal Time 166
    Bio-time 166
    Material Times 166
The Geometry of Time 167
    AC-time 168
    DC-time 168
Grades, Principles and Times 169
How Long has Time Hung Round? 171

Chapter 11: Energy 174
    Matter's Holy Ghost 174
    Grit in the Ghost 177
    Alpha Points. 179
    How Fit? 180

Chapter 12: Magnificent Mythology 183
    Fit.. 183
    Pass The Paracetamol 184
    Labours of an Empty Womb 185
    Order Below a Physical Start 190
    You Ain't Seen Nothin' Yet....................................... 193
    Approaching a Magnificent Mythology 195
    Atheism's Last Refuge 196
    Points Omega... 198

Book 2: Psychology 200

Active Information (Psychological).............................. 200

Chapter 13: Consciousness.. 200
    Psyche and Psychology 200
    Soul to Sell?... 202
    The Neurological Delusion.. 202
    Does Brain Originate or Mediate?............................. 203
    Consciousness.. 204
    Open Up the Brainbox! 205
    Build Yourself a Brain! 210
    Towards a Unified Theory of Science and Psychology.. 215

Chapter 14: Five States of Mind..................................... 219
    Top-down, Hierarchical Psychology............................ 219
    First State of (Super-)Consciousness - 221
        The Psychology of Transcendence 221
    Second and Third States of Consciousness - 221
        The Psychology of Waking Normalcy................... 221
    Norms and Loops 223
    Upper, Metaphysical Loop 224
    Lower, Physical Loop... 225
    Quality of Mind 226
    Quality of Information.. 227

The Ascent of Man 228

Passive Information (Psychological))) 229

Chapter 15:   Subconsciousness 229

A Black Box 229
Sleepy Head 230
The Fourth State of Consciousness - 232
   The Psychology of Dreaming 232
The Fifth State of Consciousness - 233
   The Psychology of Deep Sleep 233
The Sixth State of Consciousness - 234
   The 'Psychology' of Physic 234
Frozen Time 234
Psychosomatic Linkage 236
Synchromesh 1 - Awareness and Memory 239
The Personal Mnemone 244

Chapter 16:   Archetype 246

The Typical Mnemone 246
H. archetypalis, the Image of Man 248
Signal Translation 251
Instinct 251
Morphogene 253
H. electromagneticus 255
Synchromesh 2 - Psychosomasis 257
How Does the Connection Work? 259

Chapter 17:   Caduceus 261

The Logic of Embodiment 261
Core Principles 262
Morphogenic Crystals 263
The Logic of Development 267
Caduceus: the Human Morphogene 268
The Informant Domain 273
The Informed, Energetic Domain 274

Chapter 18:   Death 276

One Thing Is Certain… 276
D-Day 276
The Logic of Disembodiment 277

    Towards a Unified Theory of Life and Death 279
    Post-mortem Psychology 281
    Anathema 282
    Immortalities 283
    High-level Death is Life 286
    Ante-natal Psychology 288

Book 3:   Biology 290

Passive Information (Biological) 290

Chapter 19:   Unified Biology 290

The Principles of a Unified Theory of Biology 291
The Basis of Biology is Information 293
Biology is Hierarchical and Cyclical 301
The Central Executive is Homeostasis 302
Nuclear Super-Computing 304
Conceptual Biology 305

Chapter 20:   Alchemy 309

Chemical Evolution? 309
Darwin: Half Right, Wholly Wrong? 309
Modern Alchemy 311
    Not a Great Start 312
    Atmospherics 312
    Unnatural Interference 313
    Evaporated Soup 314
    Chained-Up Unchained 315
    Not So Sweet 315
    Bags of Life 316
    Reflections 316
    Join Up, Fold Up 317
    Number Games 318
    Pristine Instruction 318
    DNA 318
    Supreme Elegance 321
    Supreme Density of Data Storage 321
    Supreme Operation 322
    Perplexity 323
    R not *DNA*? 324
    Raw Energy Spawns Disarray 325

Chapter 21: Cell Sell 327

Catalytic Philosophy 327
Energy Metabolism Perchance? 328
The Origin of Irreducibly Complex Mechanisms 332
Minimal Functionality 332
Natural Nanotechnology 334
Biosynthesis 335
In Extremis 336
Tick Tock 337
A Definite Flight from Science 337

Chapter 22: Neo-Darwinism Isn't Fit 338

What's the Problem? 338
The Editor: Natural Selection *340*
The Origin of Species 342
Galapagos and All That 345
The Tree of Life 345
Homology: Common Descent or Common Design? 347
The Origin of Type 350
Types of Fossil 352

Chapter 23: Neo-Darwinism Doesn't Work 356

The Creator: Mutation *356*
Entropy of Information 358
Evolution in Action? 362
Non-Protein-Coding *DNA* 365
Hierarchical Language 367
Super-codes and Adaptive Potential 372
Natural Genetic Engineering 375

Chapter 24: An Impossible Dream 377

Anti-Parallel Interpretations of Biology 377
Twists that Entwine 380
The Reproductive Archetype *387*
The Origin of Asex 388
The Origin of Sex 389
Archetypal Sex 391
Archetypal Sex Expressed 396
Which Came First? 399

Chapter 25: Extra-Grand Mythology 401

    The Origin of Growth and Development 401
    Bio-logic *405*
    Logically Expressed 407
    Evo-devo 408
    A Clap of Fragile Wings 410
    A Mutant Ape? 412
    As You Like It: Scientific Animism 412
    Has Darwin Had His Day? 413
    Theories of Accommodation 415

Book 4: Community 417

Chapter 26: Community/Society 417

    Towards a Unified Theory of Community 417
        Physical Part 420
        Biological Part 424
    Nature's Negativity 425
    The Nature of Evil 427
    Towards a Unified Theory of Community:
        Social Part 431
        Peripheral Religion 434
        Nuclear Religion 435
        Politics 436
        Law 436
    Towards a Unified Theory of Community:
        Individual Part 437
        Self-Government 437
        Individual Association 438

Chapter 27: Up and Away 440

    Where Does the Data Actually Lead? 440
    Bottom Line, Top Conclusion 441
    Lux et Veritas 443

Glossary 445

SAS Icon 470

Index .... Error! Bookmark not defined.

Bibliography 471

Illustrations.

Intro. 1	Abbreviated Contents Box.	14
1.1	Pivoted Existence.	30
1.2	Concentric Rings.	31
1.3	Mount Universe	32
1.4	The Relationship Between Models and Stacks	39
2.1	Neutralities.	43
2.2	Duality within Unity	45
2.3	Primary and Secondary Duality	46
2.4	Primary Inversion.	53
2.5	Conscio-Material Coordinates.	56
2.6	Three Tiers of Mount Universe.	57
3.1	Cosmic Fundamentals and Their Ziggurat.	60
3.2	Upper Sub-divisions of the Ziggurat.	60
3.3	Lower Sub-divisions of the Ziggurat.	61
3.4	Subtendence and Transcendence	62
3.5	Cosmological Bearings.	66
5.1	Hierarchical Information.	80
6.1	The Order of an Act of Creation.	94
8.1	Crystallisation of Principles.	128
8.2	The Diamond Capstone.	130
8.3	The Fall of Unity	139
9.1	Vectored Voids	151
9.2	Source and Sink are Zero	155
10.1	Species of Time.	164
10.2	Physical Grades of Time	169
11.1	Primary and Secondary First Causes.	175
11.2	The Symmetrical Polarity of Light.	176
11.3	Alpha Answers.	179
12.1	The *WMAP* Picture	188
12.2	A Miraculous Projection	191
13.1	Microcosm of the Macrocosm	200
13.2	Brain Very Briefly.	205
13.3	Brain in Principle.	210
13.4	Natural Dialectic and Brain's Hemispheres.	212
13.5	Essential Psychology	216
14.1	Mental Ziggurat.	219
14.2	Five Main States of Mind.	220
14.3	Norms and Loops of Mind	223

15.1	The Subconscious Sandwich	229
15.2	Grades of Man/ Dialectical Bio-classification	236
15.3	Wireless Man	238
15.4	Psychosomatic Linkage by Domain	240
15.5	Suggested Architecture of the Subconscious	240
16.1	*H. archetypalis* in Biology	248
17.1	Vibrant Morphogene	264
17.2	Caduceus	269
17.3	Human Extent: The Conscio-material Gradient	271
17.4	Information Man	272
18.1	Life Swings	277
18.2	The Logic of Disembodiment	278
18.3	Brands of Immortality	283
18.4	In-swing & Out-swing: Psychological/ Biological	286
19.1	A Dialectical Plan of the Way Life Works	292
19.2	Biology in Brief	294
20.1	Neo-Darwinism - A Tabulation	310
20.2	Mirror-image Chemistry	317
21.1	The Light to Life Energy Conversion Chart	329
22.1	Plasticity	342
23.1	Innovation and Mutation	356
24.1	Representation of Branch Dialectic as a Ladder	377
24.2	Dialectical Ladder as a Double Helix	378
24.3	Cerebral Inversions	382
24.4	Polar Inversion of Visual Faculty	384
24.5	Reproductive Archetypes	387
24.6	The Archetypal Polarity of Sex	391
25.1	Developmental Stereo-Computation	402
26.1	Absolute and Relative Communities	417
26.2	Amoral 'Evil'	426
26.3	Moral Good, Immoral Evil	427
26.4	Religion, Politics and Law	432

Preface

Curious creatures that we are, it seems our questions count as many as the stars. Where did cosmos come from? Who in heaven's name am I? Where on earth did you come from? What, if any, is the object of brief travel through the nothingness of time and space? Such child-like questions have excited and still excite the whole of philosophy and science. In their great scope I look for patterns (or a basic pattern) by which I can understand my part and in whose light of logic I am able to derive a self-consistent set of answers.

Abbreviated Contents Box.

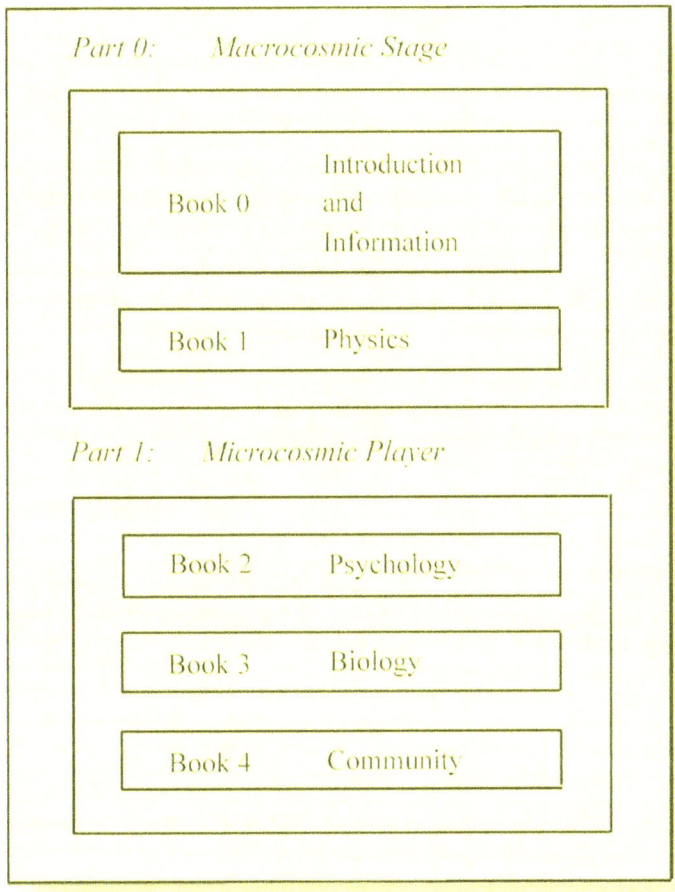

The Box describes a nested order of the zones this book explores.

Firstly, let's ask if you're **agnostic**. If you are the book's for you. It will prompt decision.

Are you **religious**? Do you want to understand the natural roots round which your faith has formed?

Finally, are you of **atheistic** faith? Since this book challenges your inclination you might wish to learn how well it holds its own.

Materialism's axiom[1] is that every object and event, including an origin of the universe and the nature of mind, are material alone; a few oblivious kinds of particles and forces compose all things. Moreover, cosmos issued out of nothing; therefore, beyond this realm of physics there is only void; and life is an inconsequent coincidence, electric flickers of illusion in a lifeless, dark eternity.

You, of course, are alive. You know full well, subjectively, life's consciousness - but is it proven physical? Your body's doubtless physical and you accept a cosmos made of matter. Is your consciousness not metaphysical and, like your body, also part of something universal? **I have simply added immaterial, as a second fundamental cosmic ingredient, to material. Or, conversely, I have added material to immaterial**.

This simple premise is powerful to the extent that it relates all physical and psychological phenomena that humans can appreciate. Hence arise two propositions central to both the book and, as far as I can see, the cosmos it describes. **The first is that realistic comprehension of the world includes *two* primary components - immaterial and material or, as obvious to everyone, mind and matter.**

Is there really any difference? Isn't consciousness unconscious? Matter mind? Isn't a material brain the same as, or at least the generator of, your mind? Aren't you your body? It is made of cells, cells are made of chemicals, chemicals of atoms and atoms aren't alive. If atoms, molecules and cells aren't then your body isn't. It might be a marvellous machine but it is not alive. *So who are you? Are you alive or dead*? **It follows that a scientific world-view that does not profoundly and completely come to terms with the nature of conscious mind can have no serious pretension of wholeness.**

What, moreover, was before the world began? What is the nature of such nothingness whose logic or its lack substantiates creation, chemicals and bodies? Creation as a whole, the science in us feels, is 'logical'. **The second proposition is, because our understanding reasonably reflects it, that existence as a whole *is* 'logical'.** Such logic can be intellectually expressed in various ways. Physics chooses mathematics but we'll trace cosmic contours with another kind of symbol, one that probes where numbers cannot pass - words.

---

[1] see Glossary under Principal Axiom of Materialism, *PAM*.

At this point I repeat the primary assertion - mind and matter are two separate elements. But, you might respond, materialism's primary axiom is that there is only one. Precisely so. That is non-materialism's simple null hypothesis. But let us at the outset be completely clear. **These assertions are both philosophical; neither is a scientific one. <u>Materialism is a philosophical and not a scientific posture</u>.**

Moreover, if the axiom that mind and matter are two different kinds of element is true the logic of this book in its entirety is unassailable. **It is drawn, over the chapters, into a self-consistent, polar model of creation; and, paradoxically transcending this polarity, into consideration of a causal singularity.**

Of course, such axiom exacts a toll. A heavy fee needs to be fully paid. Costs that need, like stinging nettles, to be grasped, include:

(i) **the nature of consciousness (Chapters 5, 6, 13-14), sub-consciousness (Chapters 15 and 16) and non-consciousness (Chapters 7-12).**

(ii) **whether individual mind can exist independent of a body and, if so, the nature of its entry, attachment, exit and disembodied condition (Chapters 13 and 18).**

(iii) **the interactive relationship of individual mind with body; the nature of any *PSI* (psychosomatic border or, perhaps, quantum linkage) between mind and matter (Chapters 6 and 15-17).**

(iv) **the mechanism by which universal mind, if such exists, might inform non-conscious forces, particles and gross phenomena; the origin of physical constants and patterns of behaviour, that is, the laws of nature (Chapters 5-12, 15, 16).**

(v) **the nature of physical and biological prototypes, homologies or, if any, archetypes (Chapters 7, 8, 15-17, 19, 22 and 24).**

(vi) **the question whether biology is informed by chance and aimless natural law or by design in accordance with such law; a wholesale reappraisal of the neo-Darwinian theory of evolution (Chapters 5, 6, and 19-25).**

These and other 'eternal' questions constitute, alongside the underlying Nature of Nothing, a *motif* that invisibly substantiates the whole narrative. They constitute a continuous backdrop to be borne in mind throughout a Natural Dialectic that plays off anti-parallel (*top-down* and *bottom-up*) perspectives against each other. What, if anything, is Nothing? What is the nature of the world? Is there any purposeful agenda? Will some or all these questions, although addressed, be answered? If so, how well?

Firstly, to neutralise aforesaid six stingers several bold but logical connections will be made. They'll be marked up as the seamless, self-consistent narrative proceeds.

Secondly, **this book deals with physics, psychology, biology and more. Not everyone's polymath and there's no need to plough straight through. Some themes will attract you more than others.** *I suggest you may prefer to browse these first and, as your horizon widens, join the points until a panoramic comprehension heaves in sight.*

If, moreover, you have not 'done science' or meet an unfamiliar detail do not let it put you off. Consult the Glossary, a textbook or the internet but, above all, press on. You can skip along the string of stepping-stones, skate across a surface made of text *italicised* (a lesser or a leading point) or **bold** (a major point); and you're invited to regard the Index as a web, a kind of sub-text whose indented entries make connections all across the volume's natural world. A learning curve's a journey. This book is its map. Grasp the principles and details will 'self-organise' around them. Facts, of course, don't change but how you see them can. Press on and in a short while you'll engage a fresh world-view. I hope, because it's entertaining, you enjoy the intellectual exercise!

Thirdly, the book is stand-alone. You can read its gist straight through. Our universe is full of space but, on a journey round its compass (physical and metaphysical), you might feel in this slim abstract that there isn't space enough. In fact, its abbreviation cuts to about 25% of a 'back-up' volume and contains a fly-past fraction of the illustrations that the larger, strolling version shows. It constitutes a portal. Gates form entry points to great estates. You pass through them with excitement. You do the same with a museum, theme-park or, virtually, computer game. You enter through a portal at which the directions for your visit are displayed. This book, though in itself complete, is such an introduction.

Our particular estate, including science *and* philosophy, is comprehensive. Philosophy means 'love of wisdom'. 'Science' is a crisper way of saying 'knowledge'; but scientific crispness costs because its mode of explanation is exclusive. Its naturalistic fraction of our understanding, physic seeming to squeeze metaphysic out, includes the broad disciplines of psychology, biology, chemistry and physics. But we can, as we'll see, now add metaphysic back. An immaterial, informative dimension is restored. Thereby, re-cast within a simple, formal framework (called Natural Dialectic), science and psychology expand into a volume of enthralling exploration which, boiled down, this book reviews.

Book 0: ABC and Information
Chapter 0: Anti-parallel Perspectives

The Light in Your Eye

Rays of light are passing into each of your eyes. They travel through a lens, across the vitreous humour and strike the retina. From here a blizzard of impulses is transmitted via the optic nerve to your brain.

With the right instruments this activity can be traced. If you were correctly attached you could note the effects of your sensation on a screen. So far the process can be monitored scientifically. It can be quantified in terms of mathematics whose descriptions underlie physics and chemistry. Or it can be pictured as a biological phenomenon. In short, the process can be viewed from 'outside' as 'not-me'. Call it *objective.*

Experience

At this moment something extraordinary happens. Sensation flips beyond the grasp of science, eyes and brains. It becomes me. Your sensation becomes you. Objects 'outside' have been integrated, by means of nervous code, with an 'inner' subject. There is conscious 'knowing-as-a-whole' - experience.

objective *subjective*
outside *inside*

Where does this 'knowing' happen? Nerves mediate both action and perception; knowledge is, although the world inside your head is clearly outside, registered within your skull. Where, then, are you? Is mind inside as well? If so, how is the outside world translated to a cinematic screen inside the darkness of a brain? If not, then mind is not the same as brain. Our worlds aren't just grey matter. How, either way, does such vital interaction work; surely it is not the gift of atoms, molecules or electronic charge - those non-conscious ingredients that constitute a nerve?

Whichever way, we certainly inhabit a subjective part. We live in mind with brain attached: nor is there evidence (although there's dogma) that these two are indeed the same. Still, you are 'inside' looking out; the world is known inside out; in this sense everything's subjective. *All experience, which is the only way that anything is known, is subjective.* **This subjective aspect is an abstract - 'nothing' in material terms: but without such immaterial 'nothingness' there's nothing! Paradoxically, as far as we're concerned, it's everything!**

World-view

A world-view is a network of beliefs about reality in the light of which one's observations are interpreted. It is, if you like, a mind's eye lens; what we make of life is through its goggles. And, at root, there are two basic views of the universe. Either it is physical alone or both material *and* immaterial elements (like information)[2] universally exist. The information centre that you know for certain is your mind. Body is a special composition made of universal matter. We may thus holistically suggest that human form incorporates its special part of universal mind.

object	*subject*
matter/ material	*immaterial/ mind*
non-conscious	*conscious*

Holistic and materialistic develop anti-parallel world-views. They treat matter just the same but, to a materialist, an immaterial dimension's immaterial - what does not exist can't count. To a holist, on the other hand, what is immaterial exists and may count fundamentally. *Which, inclusive or exclusive viewpoint, is more open-minded? Which, at root, correct?*

Opposite Directions of Mind

Hidden in plain sight the revolutionary element that opens up holistic paradigm is immaterial consciousness. Such consciousness involves a two-way flow.

outward	*inward*
diffusive/ centrifugal	*centripetal/ concentrative*
reflex/ involuntary	*deliberate/ voluntary*
sensation	*reason*

By habit and material circumstance man's attention is drawn strongly towards his earthly part; out and in, by action and sensation, oscillates our centrifugal tendency. Weaker, on the other hand, is centripetal bearing. No doubt, education strengthens such contemplative thrust against the flow of creature-focus that is dissipated outward on the plans and principles of mundane circumstance; but even intellect and reason mainly chew the worldly fat. Only contemplators seek to strengthen inward, centripetal concentration such that centrifugal pull's eliminated and the mind detached completely from its bodily environment. In short, we use and gather information in two ways; centaur-like we oscillate each side in various degrees. Sometimes we drop towards animal condition - instinct and sensation; and at others rise to intellectual learning and such natural principles as guide the work of scientist or, centripetally complete, the life of saint.

---

[2] see especially Chapters 5, 6, 13 and 14.

Two Sorts of Implement

outward/ downward focus	*inward/ upward focus*
action/ sensation	*contemplation*
stepwise logic	*idea/ grasp of principle*
analysis/ breakdown	*grasp of whole*
intellect	*intuition*
rationale	*insight*
protocol	*creativity*
left hemispheric 'nuance'	*right hemispheric 'nuance'*

Objective science, with the use of reason, focuses out through the senses and their technological extensions.

Conversely, subjective science involves detachment from the external whirl of events and steadfast, meditative attachment to the internal centre. Called the 'contemplative way' or 'science of the soul' such practice is, with its core metaphysical values, supported in various formal ways by all world faiths. *Its technology is naturally prefabricated - yourself; its focus is inward at one's own 'third eye';[3] and its understandings can, although numberless and wordless, be reasonably phrased by intellect.*

Both objective and subjective science are systematic but each is directed towards opposite ends of the cosmos. One's object is to define 'diverse, peripheral phenomena' (the world of matter) and the other's to unite with a Single, Central Subject. Both involve mentors (skilled 'adepts' qualified by successful practice), employ the process of experiment and log self-consistent, universally agreed results. Both work in a laboratory; you go in and close the door. In the objective case a lab is stocked with kit and chemicals; in the subjective case the inexpensive crucible is body and, from experiment to experience, its instrument is mind cleansed by meditation and the practice of an upright life. Both kinds of search involve normal human faculties; they are not, therefore, mutually exclusive. Why should saint not be a scientist or *vice versa*? <u>It will certainly become plain that the subjective course, far from being mysterious, whimsical or weak-minded, is rational, logical, clear, purposeful and highly focused.</u>

Two Pillars: a Dialogue of Faith

It should be clear by now - there are two directions that your focus of attention takes; and continually, between their poles, various degrees of concentration fluctuate. *From these directions are derived a fundamental couple, two main implements of knowing; and, in turn, from these two anti-parallel mind-sets, pillars that support all human faith.*

---

[3] see Glossary: 'third eye' and 'meditation'.

Both camps unveil what is already here. Physical and metaphysical, both seek substance under superficial show. They seek principles on which the obvious evidence of circumstance is based. Although their directions of focus differ, both seek revelation. *Both want the truth. Both seek it with rigour but a clear, initial distinction needs to be drawn between physical science and materialism.* **It cannot be overemphasised that materialism, which comes in various brands, is a special case of metaphysical philosophy - one that denies the metaphysical.**

The two perspectives of dialectical dialogue, anti-parallel vectors[4] cast as characters elaborating basic mind-sets, are called *top-down* and *bottom-up*. You might think that they express antagonistic mind-sets and, to some extent, that's true; but it's noted that, in fact, they must compose a complementary pair. ***Bottom-up*** is taken as the empirical method of a humble student who, from child-like ignorance, starts from knowing nothing. Such lack of preconception marks a strength of scientific method. Its student learns by experiment and experience; and at the same time is guided, *top-down*, by the certainties of a higher authority, a top-notch teacher. ***Top-down*** implies you've got the information that you need; from 'on high', it is a system maker's expert point of view. This pair constitutes the anti-parallels of knowledge.

Why just a pair? Firstly these anti-parallels derive, at root, from nothing more than difference of focus. This amounts to the difference between sensation and contemplation.

objective	*subjective*
sensation	*contemplation*
atheistic tendency	*theistic tendency*
material side	*immaterial side*
materialistic exclusivity	*holistic inclusivity*

The former concentrates on matter's objects and the principles that drive their cycles of oblivion; call this the **objective** point of view.

The latter, in holistic search, concentrates on mind itself. It concentrates upon the immaterial, informative dimension - one of value in an abstract and immeasurable sense. *It involves subjective meaning. Therefore we can call this the **subjective** point of view.*

Energetic and informative causation, running anti-parallel, are the way the world proceeds. Materialism tracks the physical and energetic: holism tracks the psychological, informative domain of mind. One's pitched outwards with an interest in material creation, and the other inward towards the only known source of creativity. ***Top-down* and *bottom-up* are designations that compose antithesis; and the Natural Dialectic of Polarity consistently contrasts these**

---

[4] see Glossary.

anti-parallel vectors of comprehension. Indeed, so different are the 'world-views' derived from the 'opposing' perspectives that, whenever the counterpoint of this contrasting Dialectic is expressed, each party is habitually italicised.

Inconsiderate materialism excludes metaphysic; it accepts no grade of creation other than material and thus excludes the notion of a cosmic hierarchy. Because this 'Way' cannot admit interpretations other than its own it commonly, often unconsciously but always incorrectly conflates itself with rationality. How can there be another rationale, another sort of rationalism? Holism, one avers, cannot be rational.

Holism's rationale, on the other hand, accepts cosmic hierarchy in its brief. What principles inform behaviour? Causal information guides effect. Thus its hierarchy runs from mind 'above' to an oblivious, material sink 'below'. Cosmos is sourced, *top-down,* from an internal, immaterial pole: therefore non-conscious matter, materialism's single element, is just the base-pole of holistic universe.

You may have thought these two positions represent antagonistic mind-sets. If, however, they derive from nothing more than a difference of focus, they compose a complementary pair. One has focused on external objects, the other on mind's immateriality. The latter eschews sensation, the former embraces and enhances it. Perhaps, therefore, neither alone relates the truth, both in principle and detail, about everything. Perhaps, like sensation and contemplation, as complementary truths they compose a whole. Better view them, like *yin* and *yang*, as complementary, interactive parts of a single universe. In other words, perhaps the ancient divide - between reductionist materialism and holistic metaphysics - is illusory.

You might agree or disagree. We'll see. But what is each world-view's First Cause? Here division cuts complete. Is it an immaterial or material form of 'deity', a purely conscious or non-conscious one? In which latter case creation would be due to accident and in the former purpose. Which mode, origin by chance or by design, does material evidence fit best? Which order of a universal start is absolutely true? **Only one fundamental can provide the most coherent set of answers to mankind's interrogation of the world.** Thus who'll debug whose mode of thinking? How?

In short, there are ultimately only two ways to view the universe. *Either, **materialistically**, everything is composed of non-conscious matter/ energy or, **non-materialistically**, everything is not composed of such non-consciousness; there also exists a conscious, immaterial element.* **The latter is non-materialism's single, simple axiom (of which materialism's is the simple null hypothesis). Accordingly, creation either originates without reason (atheism) or is planned reasonably**

(holism/ theism). Which, however, of the opposites is ultimately true, which false?

Scientific Delusions

Each of the two pillars of faith incorporates, as well as truths, its own denominational delusions. Science, like law, politics or teaching, is a profession. It may have one exhilarating difference, that in principle it is dedicated to the discovery of fresh material truths, but it is prone like any human undertaking to professional deformation as regards its mind-set, peer-group pressures, lapse from ideal and, thinking or unthinkingly, to individual and collective delusion; and, with delusion, imposition of taboo. For example, while pure science is agnostic when it comes to immateriality, a few excited militants stridently proclaim their 'scientific atheism'; and many scholars simply, coolly and without much thought tend, by a kind of intellectual osmosis, to accept philosophical materialism, its central prop (the theory of biological evolution[5]) and consequent world-view. *Yet such materialism is actually as well as philosophically unjustified; and modern science has constructed a whole version of reality from the limited constituents of its carelessness.* The nonchalance amounts, if not to delusion, to a currently common form of sleep-walking. By now, therefore, you might have guessed.

1. **The leading science delusion is, in the naturalistic heads of those who suffer it, that there is only unconscious energy/ matter. This alone makes up reality.**

    A scientific fact is a material one; no immaterial factor scientifically exists. If, therefore, the naturalistic methodology[6] of science excludes all but material answers to any question (such as the nature of mind or the origin of biological forms) then - it is not rocket science - interpretations and conclusions will be entirely materialistic. 'Respectable reality' must exclude immaterial possibility - such as information, consciousness, reason, aesthetics, mathematics and other metaphysical entities (unless you've faith that they are all physical products of ionic motion and molecular arrangements!). Inevitably, materialism serves up numerous 'godless, must-have, just-so' gaps as stop-gaps - especially regarding origins.

2. **The second great delusion is, therefore, that mind is, somehow, brain.** This is a primary affirmation of materialism's sanity. 'Non-conscious matter concocts conscious mind'. *'Mind-is-meat' summarises the final, crucial sealant that entombs holism and therewith the basic premise of this book - root immateriality*! In this view soul is a non-entity; consciousness must somehow

---

[5] see Glossary: evolution and *PCM*.
[6] see Glossary and Index.

emerge from nervous chemistry; and it must not be different in kind but only in degree from, say, quantum electronic states of chance. It is non-conscious matter, in the form of brain, that generates what we call consciousness and sometimes, at the mind's extreme, a 'God delusion'. 'You, that is, your experience of yourself, is nothing but a bunch of nerves that work by mindless chemistry'; this is scientism's grand, deflationary idea. But atoms and their molecules are not alive. *How, therefore, can multiplying lifeless atoms make them live as mind, memory and aware experience?* How, moreover, if you claim awareness is illusion doesn't such awareness thus deny its own reality? Which ghost, since we're illusions, dares to disagree? Consciousness is relegated to illusion suffering an illusion.

3. **Hard on this delusion's heel is a belief material science might, alone and in the future, yield omniscience.**

4. **There's a delusion, one already glanced, that material science operates in total objectivity**. Of course, both striving and pretence are there but does hypothesis pull clear of expectation? Is subjectivity squeezed wholly out of science? In other words, does an 'observer', though hypothesizing, devising, performing, measuring and then interpreting what he observes, detach completely from all interest, enthusiasm or attachment to results? Impersonal voice (not 'I did' but 'it was done') adds tone of *gravitas* and neutralizes personality but is the aura of detachment more than a professional face? The ideal of super-human detachment is diminished by the lens of mind-set, instruments of observation (including nervous system) and experimental expectations. Don't such limitations on our observations ever skew results - not only in the quantum field but especially when it comes to quest involving aspects of psychology? Don't 'right' results and expectations play a part - even if researchers sometimes cherry-pick their best results and throw the rest away? In short, is sufficient scrutiny applied, are inaccuracies and bias never waved uncritically through?

5. Do you believe in magic? With delusion comes illusion. **A fifth delusion is that specific, complex, systematic information (such as codes convey) can ever physically perchance 'self-organise'.** A most important trick is, thereby, to persuade yourself and thence your audience that bodies biological might seem 'designed' but really aren't. *Their coded forms, you claim, have just 'self-organised'.*

6. Has nature any immaterial, conceptual side? If not, how must a secular creation story be expressed? The tale, first article of scientism's creed, is evolution. *Thus, in discussion that is normally taboo, this book confronts a sixth delusion and,*

thereby, evokes the greatest heresy of all. This is to question whether evolution, in such confusion as the Glossary defines, is in the biological department true. **The fact is: bio-evolution's *not* a fact. Instead it is, as this book systematically demonstrates, a top-wish on materialism's list.** This schism pinpoints where the battle-lines of mind-set, materialistic or holistic, are today most sharply drawn.

You're in denial if you think the issue isn't one of faith. 'Objective reason' has become faith's godless hope. Big-bang and evolutionary theories, scientific versions of historical beginnings, are inferred, experimentally unproven but embraced religiously. Despite conscientious objections one or both of these materialistic necessities may turn out false. Both lack an immaterial, informative dimension. Thereby could mammon's half-truths not have systematically flipped the logic of the whole truth on its head? We'll look to see.

7. What, when it comes to cosmos, are the facts? You can't squeeze much from nothing, can you? Perhaps you can. Perhaps, as you claim, absolutely nothing physical[7] produced the physical, that is, the whole non-conscious universe. One's eyebrows rise. But wait! There's still a crucial bubble to 'emerge'. Sometime, in a nifty but momentous moment chock-a-block with serendipity, life emerged from matter - so that in time its atoms brewed up brows to raise and 'minds' to wryly realise the mindlessness of everything. Both 'factual emergences' originate from the material presumption 'metaphysical is nonsense'. **You might conflate them as a seventh wonder - or delusion - of the universe.**

Religious Delusions

1. **For materialistic faith the first and greatest of religious delusions is, in the heads of those that suffer it, that a Live Creator (or an administrative hierarchy of Transformers) made the world.** From this primary error spring, perforce, the rest. Of course, from an upside-down perspective atheism would reverse theism's case. Thus, it is claimed, we aren't big G's creation but, simply, He is ours.

2. **No metaphysic, no creator; thus religions with creators simply wallow in a second, consequent delusion.** Divinity is an idea alone; mind makes up God, matter makes up mind and brain, like everything, evolved. He is, like you, reduced to a few chemicals.

---

[7] Chapters 7, 9 and 12.

3. Science isn't poetry but surely you have feelings? How do you define your inner self by number? How can you explain - except by using metaphor or model - emotion, concept, moral or ideal; how express a state of mind or your experience of life? *The symbolic tools of the humanities are picture and language.* In the pictorial case the question of religious *delusion* really boils down to your material imagination(s) of an Immaterial Entity. But if such Being transcends mind, as all spiritual advice (not least the third commandment) warns it does, ideas or models cannot capture its reality. At best they'll reach approximation of the truth.

So metaphor deals in the metaphysical; wordy symbolism, imprecise, is not the calculating, scientific way. You may, as mystics or a poet, try explaining love or meaning that transcends objective fact. One has to use conceptual metaphor but aren't these same idolatrous imaginations what 'logicians' denigrate? Often symbols, literally interpreted or simply misinterpreted, are woven into 'straw-men' up for ridicule; or else host useless theological polemic. The fire of transcendence fades; enlightenment is pulverised to intellectual ash - diverse priestly renderings of 'what the teacher really meant'. **Thus is ghosted, in dogmatic argument, the third and ritualistic species of delusion.**

In short, the perspective of either of the two pillars of faith tends, when presented in public, canonical form, to ossify and degenerate into cast-iron dogma, unquestioned and unquestioning, of one sort or the other. Insofar as spiritual faiths involve an immaterial element Natural Dialectic braces them; but not the various eccentricities - divisive dogmas, local shibboleths or cultural habits - that they orbit with. Natural, Central Being, in its purity, transcends all these.

4. **A fourth, oft-bellicose delusion of any particular organised religion is that it's 'the only way' or 'sole repository of truth'.**

For either side's believers no explanation of their faith's required; and for what they don't believe none will suffice. **For atheistic 'infidels' the religious delusion is that metaphysic has existence. For theistic believers various delusions stem from encrustation of ideals, misunderstandings, misinterpretations, fanatical conformities and their hypocrisies.**

A Complementary Course

The compass of human knowledge, now so wide and detailed that a single person cannot know all scientific or other facts, has been sectioned. Instead of a polymath familiar with the whole picture, specialists now marshal the details of each sector. *In this respect it is not that Natural Dialectic adds to the exploding volume of scientific data.*

Its intention is rather to generate a neutral framework within which data can, from opposing materialistic and holistic viewpoints, be assigned and reasonably assessed; it is to generate a philosophical routine, a 'dynamic' within which physic and metaphysic are accommodated and may be reconciled.

How might material/ immaterial composition work? The basic ideas behind good science and philosophy are simple. The couple are, even if their ramifications are complex, simple enough for an intelligent person to grasp. This is certainly also the case, in principle, with this book. **Science and the Soul's second objective thus involves, in the wisdom of Albert Einstein's words, "... seeking the simplest possible scheme of thought that will bind together the observed facts".** To work from the complexity of this sensible world and develop simplicity of principle within a tightly woven, highly ordered structure has not been easy. *For a narrative to unfold according to the lines of nature has meant the issue became, time and again, one of order, order and more correctly nested order.* Without such order no archetypal consequence, however mundane or revolutionary, is inexorably derived. The structure of Natural Dialectic unwinds (or you might say 'devolves') a hierarchical, *top-down* perspective. *Grasp the principles and details will 'self-organise' around them. Despite so many trees the whole wood is still visible.* The book as a whole is divided into volumes and nested 'levels' that reflect the fundamentally binary, complementary construction of both contrived and natural creations.

Material science is extremely competent to treat a man's embodiment; but is body all there is to him? If not, the Dialectic is a work of reconciliation treating physics *and* metaphysics, scientist *and* saint. It is not a question, for anyone, of 'either/ or' but 'and'. This book is keen to identify an overall infrastructure within whose apparent logic nature works; to investigate the construction of both subjective and objective aspects of the world; and to shed light on the character of such psychosomatic interface as may exist between them. Integration and cooperation between sciences, in an unrestricted sense of the word, is the only intelligent way forward. In other words, it is time that subjective and objective, in theory and in practice, were seamlessly coordinated and thereby the balance of the whole truth given better chance.

Chapter 1: Natural Dialectic's ABC

Natural Dialectic

The thesis of this book is cast in Natural Dialectic. This framework represents, essentially, the cosmic infrastructure and thus, simultaneously, another formulation of mankind's Perennial Philosophy. How else, if truth is constant, can it be? **Simply, it asserts that to-fro, binary logic is the way that all things work; and is, furthermore, the way our polar cosmos is constructed.** For example, particle and anti-particle mutually annihilate as light; inversely, neutral photons may create charged positron-electron pairs. Such interactions illustrate duality derived from unity and *vice versa*; and paradoxically, we'll see, the Dialectic's operation marks the route of interaction run between absolution and antagonistic relativity.

Put this another way. **Dynamic cosmos oscillates within a pair of fundamental poles (mind and matter; see *fig.* 2.3); and may, at points of balance, rest from flux.** Mobile relativity and equilibrium; restful absolution within polar play. If this sounds obscure, it's not. We do it all the time. The world's all change but, living in this flux, we're always seeking equanimity. We continually seek peace in alignment, in a balance between inner mental and outer bodily worlds; such poise is a temporary, relative event before the next upset appears. But the implications of our own experience are profound. Equilibration is a cosmic operation too. Balance and equation rest beneath the restless show. Two-in-one; one-in-two. Unity, duality *and* trinity. We're about to meet the fundamental three and if, in fact, the world involves them you must judge how closely Natural Dialectic comes to frame its truths; and how closely it is able to align with other ways that polar interactions have historically been expressed. In trinity it well reflects relationships between all kinds of complementary pair; and, beyond them, links each couple with its pivotal control, that is, its government. **Indeed, Natural Dialectic is a theory of principle that, setting up its framework of polarity, makes possible a universal and yet self-consistent description of the natural order.**

New wine in old skins tends to split them. The wine will drain to waste. Mind-sets respond like old skins to fresh ideas. Not only are reflex resistance or rejection of a fresh formulation historically common but a thinker's preferred mode of abstraction also skews response. Some prefer to weave their own symbolic frame of natural philosophy with numbers (the business is called science); others work with words. Natural Dialectic works with words and builds connections where mathematics can *and* cannot pass.

If you prefer description mathematical you may decide a framework philosophical does not appeal. *Yet explanation of a system needs precede its application.* **Our Dialectic is, essentially, simple but, like grasping any language or fresh system, needs a little uphill work at first.** You'd lose the infrastructure into which we're going to load our facts but still, at this point, you may vote to skip the study of a modern (and yet also ancient) form of order. You may vote, at this point and pending a return, to swap the swot for the excitement of its application in the sections covering Informatics, Physics, Psychology, Biology or Community.

If so, loop now to Chapter 5 Information, 7 Physics, 13 Psychology, 19 Biology or 26 Community respectively.

If not:

sink   source

Accept that poles are opposites - say, source and sink; and that between them you can draw a field of action. A spectrum, high to low, represents a scale of energetic (or informative) ability. Scales, levels, proportions, relativities are built, as in creation, into Natural Dialectic.

material   immaterial
physic   metaphysic

A source might be material. How then (Chapters 9 and 12) did matter make itself? If, however, it is immaterial what then might cosmic axis, source or centre be? Information, consciousness or mind? If so, as we'll see (e.g. *figs* 1.3, 3.1, 3.5, 5.1, 9.2), (↓) materialisation amounts to increasingly 'down-and-out' inertial or centrifugal appearance as consciousness is lost; and (↑) 'up-in' centripetal gain makes a return towards 'lively freedom'.

At any rate, the universal field would represent an immaterial-material gradient - more or less of either sliding anti-parallel upon a scale. If information is immaterial you might construct a slope composed of two coordinates. Twinned energy and information operate an info-energetic calculus; they generate a conscio-material spectrum. Thus your spectrum illustrates, proportionately, a loss (↓) or gain (↑) of energy or information. Using discontinuous levels you might model vectors up against a pyramid, a hierarchy that is stepped and called a ziggurat (*fig.* 1.3).

In fact, four geometrical archetypes may each help serve to represent the Dialectic - dot (no dimensions), linear extension, concentric, circular expansion and pure space. 1-d line can be drawn 2-d (as area) or 3-d volume; and linear is curved as edge of circle or a sphere. Space is formless; and a dot's the source of form. Now, if we take potential (latent possibility) as formless source, its opposite is sink - a possibility that's now specifically realised. Between these poles the action happens; along this line events occur.

Scientific and non-scientific concepts need explanatory metaphors. Metaphors and models are conceptual hooks. What's invisible or immaterial is hung on images (or, if you like, imaginations). These can mislead, approximate or usher towards a clearer model and a closer truth. In short, mind needs models like a handle so that it can grasp abstraction. If you accept that metaphysic might exist what metaphors or models can proponents of holistic outlook use? **There are three main, simple images or, if you prefer, universal models of creation used by Natural Dialectic.** They are scalar (involving quality) *and* vector (involving relative quantities) in operation. In macrocosmic terms Chapters 5-12 are based around them: in microcosmic terms Chapters 13-26.

Firstly, you can think of cosmos as a pair of scales.

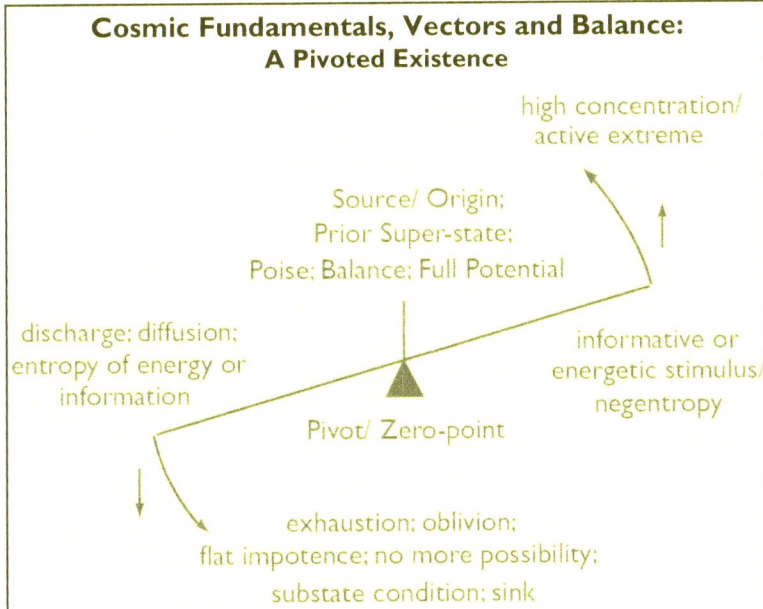

Imagine creation as a scale in which is balanced everything. The translation of this order of creation - from potent super-state through phases of action down to sub-state impotence - is also rephrased in terms of spectrum, ziggurat and, as we'll shortly see (following pages and *fig.* 1.4), three cosmic fundamentals. Each fundamental is predominant at one of three major levels of creation.

***Fig.* 1.1 makes clear that, in description of swing around a pivot, potential *is* that pivot.** Although written dialectically above, it is the point whence wobbling changes in the two domains of mind and matter start; it is the point of poise round which all swing begins and of equilibration when a balance is assumed again.

The cosmic scale integrates a pivotal, balancing factor (*Essence*) with two antagonistic vectors of existence. In this, the perpetual changes of creation are

> seen as myriad adjustments against the disturbance of balance. Up, down - nature is a scale whose beam forever wobbles in equilibration round its various centres; and, on the large scale, one could see such existential instability equilibrating round a Central Pivot, an Essential Point of Balance. Stable Axis represents Perfection. **Creation's instability is, therefore, forever perfectly imperfect; perfect imperfection shows as this perpetual motion of continual change.** Existence thus amounts to a self-regulating balance; it amounts to the sum of myriad individual actions and reactions each of which always shows proportions of two vectors pivoted around the poise of a third non-vector. Such orderly procession by equilibration is sometimes referred to as the *'karma* drama', physical and psychological fields of *karma* or equations of creation. Physics also measures transformations by equation.
>
> **fig. 1.1**

In this picture ups and downs swing round a pivot by degrees. Two pans of a kitchen balance oscillate about their starting-point. Opposing vectors swing around a point of balance. The motion of creation weighs upon a fulcrum of potential. You can, by equation, weigh the cosmos in a way, at base, as simple as the scale you have in mind. We shall come to understand how pivoted existence works.

You can write the scale's triplex as:

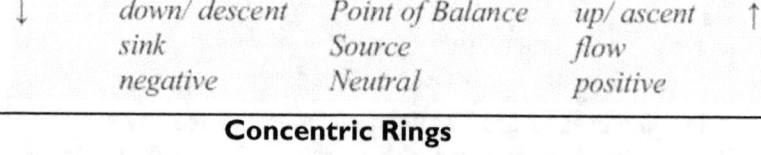

> **Concentric Rings**
>
>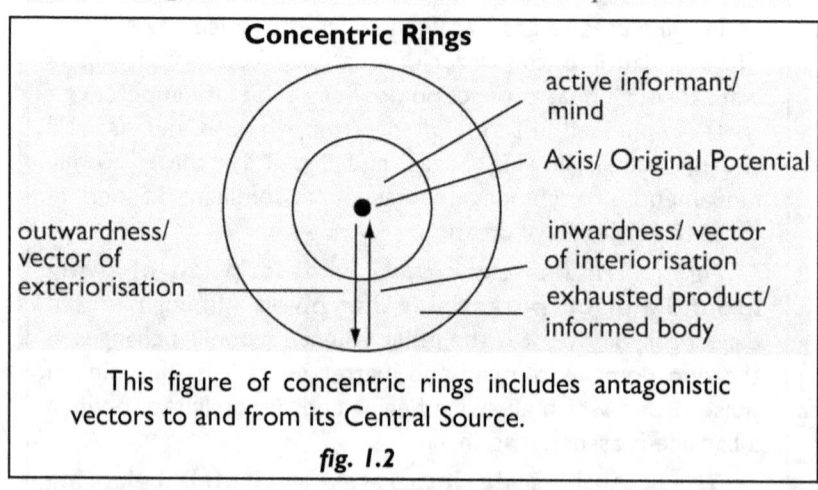
>
> This figure of concentric rings includes antagonistic vectors to and from its Central Source.
>
> **fig. 1.2**

Images of radiant energy familiar to physics and mystics alike include Bell, Light or Pool. **In such respect the <u>second</u>, energetic** model of the cosmos is one of concentric rings (or, three-dimensionally, concentric spheres) losing power as they recede from Source.

The image describes a central source, a projection of power (rather than a once-for-all creation) and a gradient of vibrant energy - just as light dims with distance or the ripples on a pond fade. Dynamic rings pulse from a central store of power but radiation, even in a vacuum, is attenuated. If your metaphor derives from heat, then fluid waves diffuse according to the nature of their concentrated cause; or else are locked by cooling into crystalline precipitate. *Potential energy flows through kinetic down to an exhausted level.*

↓ *black*	*White*	*spectrum* ↑
end	*Origin*	*action/ process*
effect	*Cause*	*stimulus*
base	*Peak*	*gradient*
end-product	*Potential*	*creativity*
sub-state/ matter	*Super-State*	*forms of mind*

Another aspect of light, sound and energetic transmission is spectral continuity; frequencies, like levels, form a seamless scale. You can extend this image to include gradients of material concentration. Plasmic, gaseous, liquid and solid states also mark an energetic gradient; the variegated universe is in these states. And levels of informative awareness - brilliant, dimly ignorant or in subconscious oblivion - mark states of mind; variegated are the forms of minds. Natural Dialectic's existential dipole is composed of energy and information. Creation shows them as inverse proportions of each other on a conscio-material scale. This spectral view of cosmos is, as we'll see, reflected well in one of many diverse forms - your own.

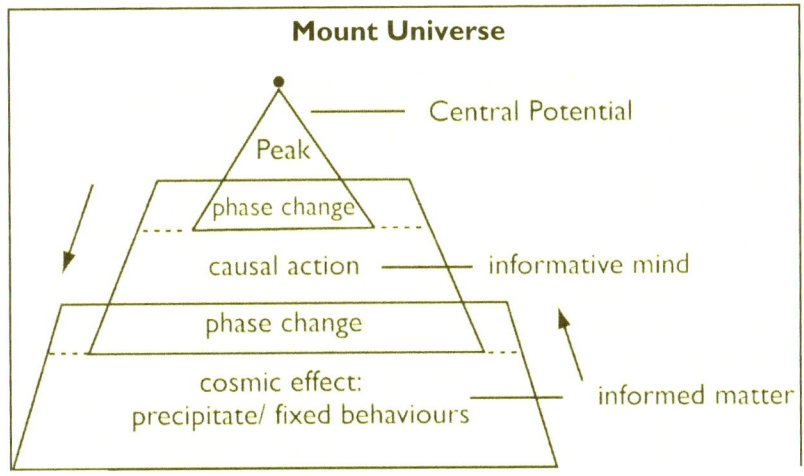

> A square cone or, even squarer, pyramid describes 'static' hierarchy. A useful representation is the stepped pyramid, also called a **ziggurat**. In this case each step of a **ziggurat** stands clearly for a phase, level or stage; and the apex of its capstone, a point that points beyond the finite grades below, implies peak infinity. **This capstone is the peak of what we call Mount Universe**.
>
> *fig. 1.3*

The <u>third</u> image to represent the gradient of creation is Mount Universe or, in the conceptual geometry of a cone or pyramid, *The Cosmic Pyramid.* States of matter display a continuous but also discontinuous, phased aspect to their energetic levels; thus a smooth pyramid becomes, converted into discontinuous levels, a stepped structure called a **ziggurat**. This model is 'heavy'. It is an inertial, solid picture of existence. Its grades are not smooth; the phases change in 'jumps'; welcome, then, unto the Dialectic's tetrahedral ziggurat. In the tiers of a pyramid, you obtain a clear picture of hierarchy.

base	*peak*
low	*high*
sink	*source*

Source above sink. Hierarchy always runs from source to sink; it falls between these poles. It forms a gradient of creativity or power. **Indeed, you can think of intelligence, information, energy or bulk materials in terms of concentration gradients**. Think, for example, of an energetic concentrate, a star; and then the radiant diffusion of its light. Or, if you like, think of electrical or chemical diffusion down a slope from concentration, source or origin. Prior to action any single thing is, almost egg-like, in 'potential equilibrium'; such stillness, readiness or poise is full of certain possibilities. There follows (if some fluid circumstance permits) a spread of chemicals, a reaction, flow of current pole to pole or growth of seed to adult form. An event rolls to its close; exhaustion is potential's opposite, inertial kind of equilibrium. Could you, from such 'flat death', regain original charge? You would have to pump with effort up 'against the flow' but you'd 're-charge your battery', re-concentrate lost energy and thus complete the cycle. **In this view potential, whether in the form of energy or information, is a source**. It is replete with charge that is by action discharged into impotence, its sink.

↓	*down/ descent*	*Equilibrium*	*up/ ascent*	↑
	effect	*Precondition*	*causal stimulus*	
	fixity/ precipitate	*Potential*	*flux/ action*	
(−)	*negative*	*Neutral*	*positive*	(+)

Is three nature's lucky number? It is now obvious from physics, event vectors and Natural Dialectic that we are not dealing with dualities but *trinities*. Take a scale's three phases - balanced and swinging upward or downward. **The actions that these words symbolise find various expression in each object and event.** More subtly, they reflect the *qualities* intrinsic to creation. They compose three fundamental operating principles whose permutations are expressed, in varying degree, as *tendencies* both psychological and physical. **We call them cosmic fundamentals**. Ups-and-downs and peaceful poise; ins-and-outs and equilibria; existence is an ever-changing play of tensions that evinces, as already mentioned, trinity. *If the triplex is naturally fundamental it is also fundamental to the operation of this Dialectic; but while the orient has long worked with these immaterial radicals western minds have not.*

| *yin* | *Tao* | *yang* |

You might be familiar with this far-eastern, Taoist set and its associations. Therefore, I suggest using an equivalent, equally ancient abstraction whose purity is, because of occidental unfamiliarity, less stained by prejudice or shadowed by prior connotations. Let's employ (despite, perhaps, conservative resistance) the following as-yet faceless trio.

| *tam* | *Sat* | *raj* |

Sat, raj and **tam.** Those interested in dietary cooking might have already picked up on their categories. For example, *sat* food is fresh and includes fruit, vegetable and cereal products. *Raj* food stimulates; it is hot, promotes physical activity and includes spices, curries etc. The *tam* ingredient is stale or heavy; it includes meat and alcohol. You get the flavour. **In fact, the triplet (each member called in Sanskrit a *guna* or a thread) is much less prosaic and more connective. It is all-pervasive to the extent of describing the three major tiers of Mount Universe itself.**

A simple link from psychology describes the three basic conditions of information as (sat) potential informant (concentrate of consciousness) prior to (raj) active mind (informant and informed) and (tam) passive, dormant or sub-conscious mind. Of course, information may be *carried* by material forms but it's essentially a mental entity.

And a simple link from physics describes the three basic conditions of energy as (sat) potential prior to (raj) action and (tam) exhaustion.

Following chapters elaborate, in Dialectical terms, the meaning of these linkages. Start, process, end; poise, excitement, exhaustion; balance (*sat*), up (*raj*) and down (*tam*) compose triplex creation. **<u>Psychology</u>'s the study of an immaterial information-centre - mind. (*Sat*) consciousness is the potential for (*raj*) active thought and creativity; exhaustion falls into** (*tam*) **passive sleep, that is, a subconscious zone.** And our normal waking zone clearly recognises 'higher' or 'lower' levels of, for example, intelligence, moral and aesthetic quality, rationality and happiness. **Physics is the study of non-conscious, energetic transformations.** How, here, does triplex modulation work? Natural Dialectic develops a very simple, columnar structure to illustrate, in an orderly and connected way, the myriad expressions of just three basic abstracts that compose creation. These abstracts *are* not any particular phenomenon but are, proportionally, *in* every one. *It is, therefore, a distinct and necessary advantage that the terms for these fundamentals - **sat**, **raj** and **tam** - are fresh, unaccustomed and uncoloured by particular preconception or prejudice.* What does each stand for?

↓	*tam*	*Sat*	*raj* ↑
	from nuclear	*Nuclear*	*towards nuclear*
	to periphery	*Central Origin*	*from periphery*
	fall	*Peak*	*rise*
	effect	*Potential*	*stimulus/ cause*

<u>**Sat**</u>**, the Essential or Central Quality, comes first.** Its 'truth quality' is in the paradoxical department. It represents the 'deep principle' of unity that underlies duality. It is the *Tao* from which *yin* and *yang* divide. *Sat represents the whole. Including both sides of an equation it is, as such, the factor of equilibration; and also the synthetic resolution of dialectical debate (see Glossary).* Thus it can be viewed not only as fulcrum but as the centre or apex of any systematic creation. It is the point of origin and originality; the essence of this principal is consciousness. *Sat* is potential's quality; it involves prior, implicit capacity whose possibilities, when reduced, will have informed and defined an explicit, actual end-result. It involves, in other words, principle, intelligence and law. This applies as much to the universe as to each of the minuscule events from which it is constructed. Thus its character is source not sink, start not finish of a process of creation. In terms of expression call it precondition; in terms of initiation call primal-unity-encompassing-polarity an egg, the cosmic egg; and in terms of physical effect understand metaphysical cause. Informative potential, plan, precedes material behaviour. *Sat* is high-level, at the axis or top pole of things. Perhaps its purity is vested in a superstate, that is, a Transcendent Concentrate of Information at the Apex of Mount Universe. In terms of local, energetic concentration call it 'sun'. As such it blazes for all satellites; the brilliance of its influence is variously reflected by those influenced. *Sat* shines, physically, with energetic light and heat. Psychologically, informative illumination imbues, in varying degree, the qualities of comprehension, knowledge and wisdom; and infuses the information implicit in an ideal, purpose or construction. In truth it is the source and principle of order, coordination and coherence - the superintendent principle. It both resolves the tension between poles and is, as zero-point, the axis of balance around which fluctuations swing.

Its immaterial equilibrium gives rise to the two other swaying, mobile tendencies (of *raj* and *tam*) and, in this sense, it is also their transcendent, causal point of origin.

Raj **and** ***tam*, under *sat*, are the antagonistic fundamentals.** These two are inversions of each other. *Together they break the absolute symmetry of pre-formation; they act to express any potential.* Pole and anti-pole, they represent opposites, ranges and relativities. **Scales tip, pendulum swings, the world oscillates**. **It cycles**. These cycles are its beats, its life-beats. Motion downwards, irregular, spluttering or losing rhythm is *tam*; but, pushing forward with regular and well-timed swing, *raj* is on the vibrant up-beat.

Raj **(action ↑) is represented on the right-hand column of Natural Dialectic**. Action, like motion, is one thing but its vector is another. *Raj* ascends and its climb, *as the vector of levity and of negentropy*, becomes increasingly buoyant. It returns a pan's descent towards equilibrium, equilibrates or moves towards balance. It 'ascends', in other words, towards a culmination of relativity in Absolution.

***Tam*, on the left-hand column, is the vector of 'fall'. It is the negative, passive, inertial quality. It is the (↓) materialising tendency.** *Tam drags. Tam exhausts.* Its negative arrow subtends towards base (called 'sink'); its 'purity' is represented by the sub-state pole. As opposed to *raj*, an activator, *tam* contains and resists. It represents the medium which energy informs, the 'solidity' 'fluidity' has patterned. *Tam* rigidifies. *It is the vector of manifestation expressed as gravity, binding energy and crystallisation.* Mind's a multiplier; from principle to practice, from theme to detail[8] it's a differentiator. Similarly, as gas drops to solid state so *tam* vector precipitates out through the regions of the universe. At the same time as it solidifies it divides, at base level, into fixed and separate details; subtle atomic principle is expressed as coarse aggregate; action ends with an inertial equilibrium. In descent there manifest increasingly extreme forms of properties such as impotence, massiveness, unawareness and destruction. In terms of process *tam* represents descent.

Natural Dialectic is expressed in 'stacks'.[9] These are verbal diagrams. They constitute a shortcut, quasi-pictorial way to simplify a complex world. Indeed, in conjunction with the cosmic fundamentals they form a triplex archetype from whose substance all complexities (including the various 'fundamental' laws of chemistry and physics) ramify. In this matter, just as Beethoven might allow that a scientific claim to have completely explained his music by physically descriptive mathematics had almost completely missed the point so, regarding Natural Dialectic's

---

[8] see, for example, *fig.* 6.1.
[9] see Glossary: dialectical stack; for a primer of the grammar of Natural Dialectic see *SPFP* (Science and Philosophy) Chapter 1 and Glossary.

holistic score, such claim falls on deaf ears. Its stacks are not mathematical. They are conceptually simple but not thus trivial or wrong!

↓	down/ descent	Point of Balance	up/ ascent	↑
	sink	Source	action/ flow	

Above is a 'tri-logical' stack. However, it is often more convenient to write a non-vectored stack and then polarise its neutrality into a second (*raj/ tam*) vectored stack. What is meant by this?

Firstly, think of (*sat*) source or potential. A stream springs running to exhaustion in the sea; and creation, as a 'field' called existence, must spring from a source as well. This pre-existent, non-existent equilibrium we call Essence, that is, Being-without-predicate. The question is: what is the nature of this Void, this pre-conditional Nothingness? Is it material or immaterial? To help us answer we can draw two stacks. *The first, top one disposes Essence and Essential Characteristics on the right against those of existence on the left.* **Such a binary stack is called Primary, Essential or Central Dialectic**. *It is, indicated by writing the right-hand column with a capital letter.*

Essential Dialectic, although binary, on the left involves a neutral, generalised expression of its existential component; and, on the right, because (*Sat*) elements involve no movement, no arrows are attached. Primary right-hand characters always indicate an aspect of Transcendent Nature, that is, Essence. Because it involves such qualities Primary Dialectic is placed, as the superior stack, above its polar, existential counterpart.

Primary Dialectic:

tam/ raj	*Sat*
existence	*Essence*
issue	*Source*
relativity	*Absolution*
duality	*Unity*
finite	*Infinite*
polarity	*Neutrality*
something	*Nothing*
wobble	*Balance*
(3)/ (2)	(1)

Secondary Dialectic:

↓			
	tam	*raj*	↑
	negative	*positive*	
	division/ multiplication	*unification*	
	isolation	*connection*	
	drag/ resistance	*stimulus/ flow*	
	exhausted sink	*action*	
	(3)	(2)	

You may set unity against duality. Duality, however, implies polarity. **Such polar component is expressed in the lower, vectored so-called existential, secondary or polar dialectic.** Polarity, implicit in the left-hand, 'duality' column of Primary Dialectic, is divided into explicit opposites. Between any pair of extremes (e.g. light and dark) an attribute/ property may oscillate through a spectrum, scale of relativity or phased series of values; and whatever exists may be described in terms of a stack of such dynamic descriptors. Since oscillation is a vibration you might think of changes in scale as faster/ slower, more/ less energetic or regular/ less-than-regular cycles. You might imagine their motions like those of music registered on a sound meter, digital display or, as waveforms, on an oscilloscope. Such oscillatory calculus, wheeling up and down between the scale's extremes, is vectored.

Polarity itself is split, for example, into *positive* and *negative* components. Such inferior, existential stacks represent the various kinds of polarity from which the changeful web of existence is composed; each pair of so-called 'polar anchor-points' implies a scale or dynamic range that runs between 'paired opposition' or 'complementary covalency'; such gradients are the way, at root, the cosmos works. In this scheme a causal source is placed on top and its effect drops down a gradient of action towards antithesis, its end-product. Thus physics' three basic conditions of energy are (see *fig.* 1.3) equally reflected in a cosmic hierarchy that runs from *Sat* (or *Tao*) First Cause, Potential or Start through *raj* (or *yang*) active to *tam* (*yin* or sink) exhausted, crystal phase. General possibility is expressed in (*raj*) energetic or (*tam*) specific, rigid, local outcomes.

Fashion's not an arbiter of truth. Now the currently unfashionable idea of cosmic hierarchy has been introduced we can:
1. note that, in the tri-logical view, pairs of opposites interact with one another under the influence of *three* cosmic fundamentals (up, down and equilibrating). Such triplex logic, embedded in all things, has always been intuitively grasped by poets and playwrights. No less easily can you.
2. relate Natural Dialectic's three main models, as above, to Essential and existential stacks.

This **cosmic grammar** is new and, like any unfamiliar language, takes a little getting used to. More practice, as we'll see, will clarify the framework Natural Dialectic's phrased within.

To summarise: **Natural Dialectic asserts that to-fro, binary logic is the way that all things work; and is, furthermore, the way our polar cosmos is constructed. Such radical infrastructure involves a trinity of qualities.**

Permutations of these qualities are expressed in every object or event, physical and psychological, throughout existence. **The source of existence is termed essence; thence, from essential and existential character (unity and duality), are derived non-vectored Essential and vectored (↓↑) existential dialectical columns of opposites.**

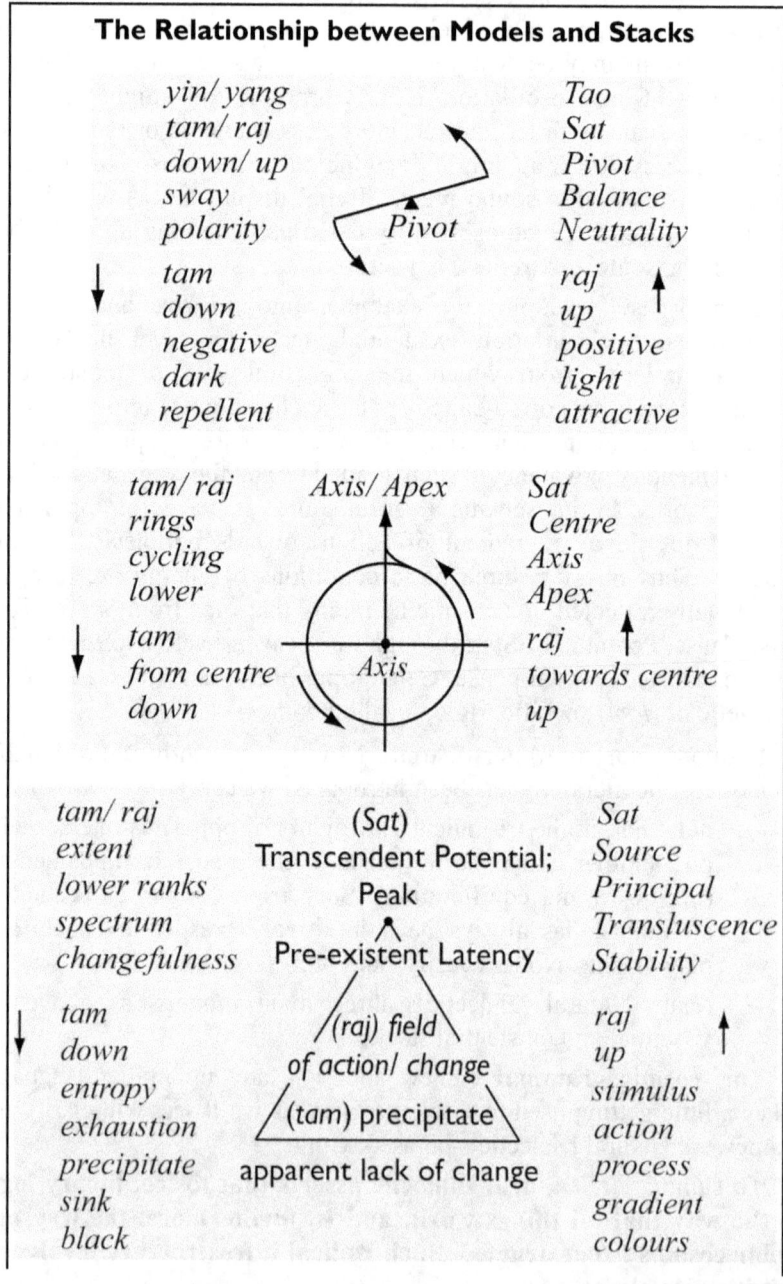

The Relationship between Models and Stacks

yin/ yang		*Tao*
tam/ raj		*Sat*
down/ up		*Pivot*
sway		*Balance*
polarity	*Pivot*	*Neutrality*
↓ *tam*		*raj* ↑
down		*up*
negative		*positive*
dark		*light*
repellent		*attractive*

tam/ raj	*Axis/ Apex*	*Sat*
rings		*Centre*
cycling		*Axis*
lower		*Apex*
↓ *tam*		*raj* ↑
from centre	*Axis*	*towards centre*
down		*up*

tam/ raj	*(Sat)*	*Sat*
extent	*Transcendent Potential;*	*Source*
lower ranks	*Peak*	*Principal*
spectrum	∧	*Transluscence*
changefulness	*Pre-existent Latency*	*Stability*
↓ *tam*	/(raj) field\	*raj* ↑
down	*of action/ change*	*up*
entropy	/ \	*stimulus*
exhaustion	/(tam) precipitate\	*action*
precipitate	*apparent lack of change*	*process*
sink		*gradient*
black		*colours*

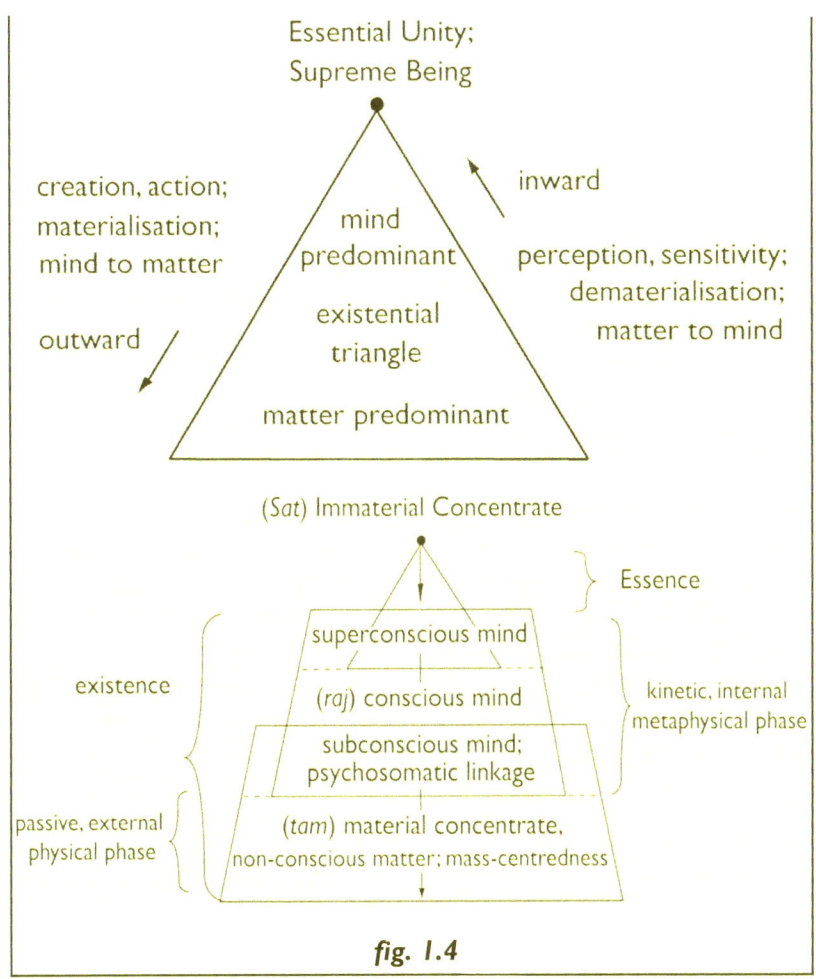

fig. 1.4

On these diagrammatic stacks each pair of 'polar anchor-points' implies a scale or dynamic range that runs between 'paired opposition' or 'complementary covalency'. Members of a stack are not lists of synonyms-with-antonyms; they are not identities but represent the equivalence of entities in terms of cosmic fundamentals. *Their perusal is intended to promote connections because consideration of connections tends to help unify/ collate/ organise one's working comprehension of any matter in hand.*

The columns are this Dialectic's backbone. The philosophy's three-way motor is now up and running. Its polarities represent not only human but the cosmic spine; or, if you like, the system generates a muscular body of philosophy that accurately reflects the order of the cosmos. **It generates an abstract, metaphysical machine, tight-knit, well riveted by bolt and counter-bolt, the simplest working model of the universe.**

Chapter 2:   First Principles

If you relish intellectual complexity then maybe dialectical simplicity has not, as yet, appealed to you. For whom, however, truth is not a function of complexity you'll have grasped the basic polar and yet also triplex scheme of thought. Now let's sketch the main principles of this philosophical vehicle. A picture will emerge whose simple outlines act, in an unfolding narrative, as the framework for more complex physical, biological and psychological arguments.

First Principles

Any engineer will tell you there is no avoiding first principles. However abstract or impractical they may appear, they constitute the basis for whatever system follows. If Natural Dialectic's structural design is symmetrical, strong and works then details should fall into place; if it accurately reflects nature, including mind, it should be functional; and, although its logic is not typically mathematical, it should be self-consistent. It should 'work'. Such 'functional logic' should provoke patterns of thought, questions and, within its framework, solutions.

thing   *nothing*

You might think the universe is full of space but science sees it as a pool of energies. In this world-fluency particulars of any formal kind are simply interruptions. They are 'stodginess'. But energy is still a fluid *thing* and, most science and philosophy agree, things had a start. Before things (and thoughts are things as well) there were none at all. Space, time and things make physical existence up; but if they came from nothing what's the nature of preceding absence? What is the nature of transcendent 'emptiness' whence cosmos was projected?

The fact is that pure emptiness could never make a thing. Do you, therefore, really claim that absolutely nothing nowhere for no reason rippled; a spontaneous ripple swelled *ex nihilo* and from this inflation everything exploded - not least, given time galore, you, me and all else that lives? Full marks for a magnificent mythology!

Nothing

Nothing is, in personal terms, the absence of a thing I seek; or it's a form of ignorance - what I don't know or haven't yet discovered doesn't mean a thing to me. Is there, however, absence absolute - nothing existential out of which existence might have sprung? If so, what might be the character of Universal Void?

Such Void (in Sanskrit *sunya*) implies no action, influence or thought. No form, either physical or psychological. Not even vacuum; no extension, neither space nor time. No change. Therefore, did formless nothingness precede creation's forms of influence and patterns of behaviour? Or were things always here?

Absence of materiality can't be material. Void must be immaterial. If things sprang from it then material must spring from immaterial - so what's the immaterial nature of the Void?

material	*immaterial*
physical	*metaphysical*
matter	*mind*

If the universe is hierarchically tiered - a possibility - something on one tier (say, reason as a quality of thought) may be nothing, that is, may not exist, upon another (say, the material plane). Certainly bureaucracies are tiered. Orders issue from on high. From what immaterial centre might the natural order be derived? An obvious immateriality is the experience of mind; and mind's essential substance is its consciousness. Do any physical phenomena consciously behave? *Or is mind in fact an element in its own immaterial right and, therefore, physically nothing?* In which case what bull's issued to forbid consideration of, as well as universal energy, universal information in the form of universal mind?

Nothing flows through nothing. What are time and space but each, respectively, nothing physical and thus, by materialistic definition, also inconsiderable as mind. Yet we've two whole chapters (9 and 10) dedicated to this ghostly, fundamental pair or, as Einstein well observed, this space-time union. The basis of our modern, relativistic physics surely can't be nonsense; but did the phantom couple actually precede all 'real' forms of physics? For now let's simply, briefly make a starting-point of start-point's absence. Let us take a stab at nothingness and see if anything is hit.

thing	*Nothing*
with form	*Formless*
finite	*Infinite*

Several issues have already risen.

1. If Nothing (which by definition's formless, boundless and thus infinite) trumps every thing then let us write:

tam raj	*Sat*
existence	*Essence*
1 on action	*0 Off Peace*
something	*Nothing*

↓ *tam*	*raj* ↑
stillness	*motion*
absence (of action)	*presence*
completion	*process*
nothing left	*duration*

2. Secondly, completion means the end. It means exhaustion of activity - nothing left. *Thus you can think of nothing in two crucially different ways.* **Potential** and **exhaustion.** Pre-active and post-active absences of change. This picture might help to explain.

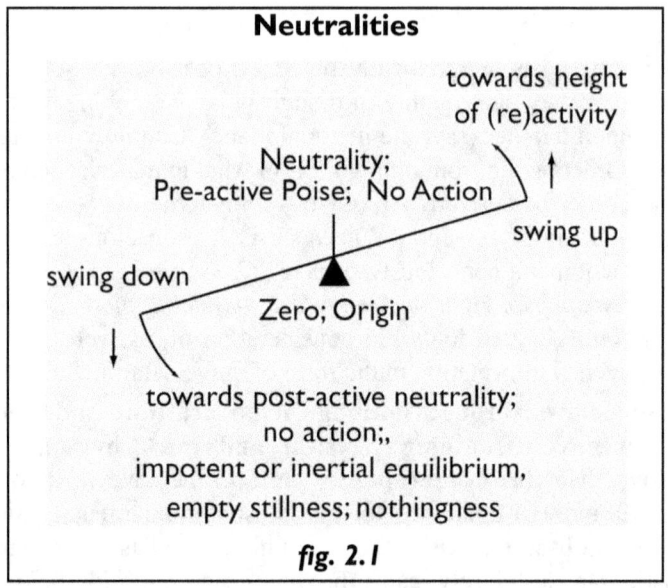

fig. 2.1

	expression	*Pre-active Potential*
	constraint	*Pre-condition*
	polarity/ charge/ action	*Neutrality*
↓	*negative*	*positive* ↑
	passively informed	*actively informant*
	end-product/ effect	*cause*
	impotence	*drive*
	post-active exhaustion	*action*

In a *minus* way absence describes a state (or, rather, statelessness) that occurs when everything has been subtracted; or when nothing has been added to nothing in the first place. Zero.[10] It is the bereft sub-state of exhaustion, abstraction or impotent void. Inert and empty space. Nothing left comes after.

---

[10] see Glossary and Index: zero/ non-existence/ nothingness.

In a *plus way* nothing comes before. Such zero describes the state of pre-condition that anticipates an actual behaviour and without which the behaviour, pattern or event could not occur. A tiger is poised to pounce. Such poised readiness is the source of possibilities prior to the realisation of any particular one. **Again, potential energy precedes kinetic energy; it precedes any particular creations of that energy and, of course, its exhausted effects.** In a similar way the potential for a mission is its plan - the information. Command and control executes orderly strikes according to such immaterial material, such non-physical intelligence. Could there exist Supreme Potential sufficient in its generality to 'fire' a certain set of possibilities that we call cosmos?

Informative potential is, for sure, not physical; but could such material nothingness anticipate its own kinetic motion (motion we call thoughtful mind)? And, in this case, what state of mind might guide the energetic patterns that, sometimes in scientific mode, we all perceive? What kind of archetypal entity is the potential matter from whose presence issue, automatically, mindless behaviours we call rules - the rigid laws of nature? S*uffice for now to log that informative and energetic possibilities precede their actualities; they are potent and pre-active; they are action's 'super-state'*.

Which of these two, energy or information, leads? Which pre-empts the other? Latency occurs when a potential's not expressed. Such 'in-place fields of immanence', pre-material archetypes, play a primary role in Natural Dialectic. Potential mind and matter both need exploration. Such exploration is a main thrust of this book. **Finally, we call the latent immanence that seeds *all* motion-packed existence (that is, matter *and* mind) Absolute, Supreme Potential. Essence.**

3. A principal comes first in line; a principle describes a source of order or a guiding force. What has form exists; it has a start. If neither mental nor material form exists there's nothing - their non-being. Yet prior Nothing, whence they must have formed, is Essential Being. Call it metaphysical necessity. Thirdly, therefore, grasp the first and final paradox - non-being is The Being! **Thus Fundamental Nothing is the Principal and the Essential Principle of Natural Dialectic.**

Something

Existence is the second basic principal/ principle of Natural Dialectic. Of the Essential Pair it is the 'other', non-essential pole.

Existence is, according to the Dialectic's Primary Axiom (see Glossary), a compound of mind and matter. It is, therefore, the umbrella-word for all psychological and physical constructions. It includes every formulation and formation, that is, everything. Creation is, paradoxically, both within and without Essence.

For materialism such Essence makes nonsense; and only the material fraction of existence harbours sense. Supreme Being is an absurdity but what are 'lesser' beings - thoughts, objects and events - if they are perceptible? They have 'ordinary' being without superiority, inferiority or any sense of hierarchy.

For *top-down* holism, on the other hand, Essence and existence both make sense but only existence is sensible. Essence is paradoxical. It is beyond form and yet at once the heart of mind and body. It is the Unmoving, Uncreated Axis round and yet within which existence orbits; it is the Potential whence actuality appears. In your case everything you know appears in consciousness. How, though, did and does the universe appear; what is its fundamental field?

Duality within Unity. Existence is binary but contained within Singular Essence. Essence is nothing existential and yet is the cause of existence. It is 'beyond' yet creates the existential system. From an absolute viewpoint (called Enlightenment) all is, within Essential Projection, quintessentially one.

Two Existential Principles

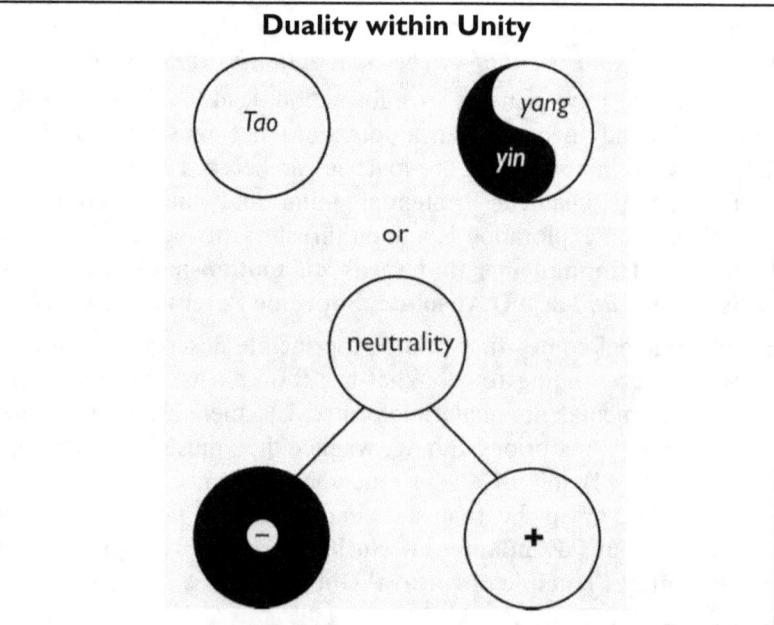

Duality within Unity

These are two familiar ways of expressing duality within unity or existence within essence. The Taoist emblem clearly represents the principle of polarity within complementary unity. Niels Bohr, recognising its power, incorporated the dualistic, dialectical *yin-yang* design into his family crest! But the mystic quest, Taoist or otherwise, is to transcend the duality of existence.

fig. 2.2

Duality within Unity. Existence is binary but contained within Singular Essence. Essence is nothing existential and yet is the cause of existence. It is 'beyond' yet creates the existential system. From an absolute viewpoint (called Enlightenment) all is, within Essential Projection, quintessentially one.

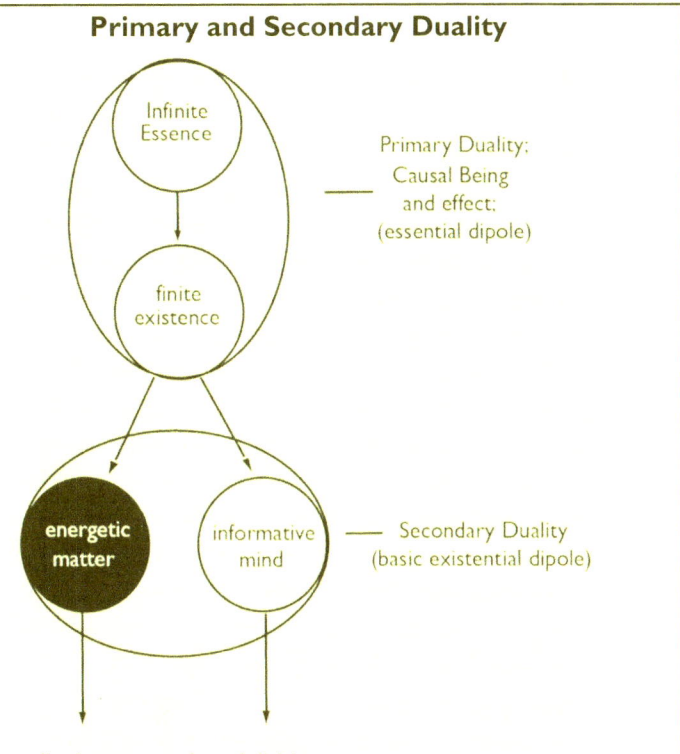

The philosophical term 'monism' means that you stipulate
a single, fundamental and universal quality or entity.
'Dualism' means that you stipulate two.
Saturated materialists are 'monists'
whose single element is energy/ matter.

On the other hand Natural Dialectic proposes 'Essential Monism' with 'existential dualism'. The single, infinite 'substance' is identified as The Concentrate of Pure Consciousness. Motion of this essence gives rise to the dualism of existence; the basis of this dualism is information/ energy or mind/ matter. The couple is hierarchical in that the former generates the latter; immaterial generates material; informant mind generates (in a way that Natural Dialectic describes) informed patterns of energy. Finally, just as there exist universal matter and specific biological forms, so there exist universal mind and specific embodied minds.

fig. 2.3

Dependent on the Essential, Independent Monopole is its issue; this is the second pole of the Essential/ existential dipole, existence. The existential pole of the primary dipole is itself, as the *yin-yang* formulation illustrates, polar.

existence	*Essence*
becoming	*Being*
non-unity/ duality	*Unity*
dipoles	*Monopole*
relativity	*Absolution*
motion	*Stillness*
creative dispersal	*Total Focus*
conditions	*Precondition*
↓ *negative pole*	*positive pole* ↑
creative dispersal	*re-creative focus*
passive	*active*
receptor	*donor/ doer*
sink	*source*
foil	*stimulus*
static/ quiet/ inertial part	*kinetic part*

If entropy's a measure of creative dispersal isn't nuclear fusion an example of a re-creative focus? At least, if Total Focus (Essence) is pre-causal and existence is a maelstrom of causes and effects, we'll need to retrieve, by re-creative concentration, the nature of First Cause. Could it be original motion of, say, The Essential Tao? Does conscious motion of the mind involve will-power, purpose and desire or not? Of what nature is our Prime Initiator?

Causality

existential duality	*Essential Unity*
conscio-material gradient	*Pure Immateriality*
objective/ subjective mix	*Complete Subjectivity*
subsequent effects/ causes	*First Cause*
↓ *effect*	*cause* ↑
physical bodies/ events	*mind*
complete materiality	*shades*
result/ product	*excitation*
counteraction	*action*

Aristotle believed in a First Mover itself unmoved by any cause.

St. Augustine observed that no 'efficient cause' can cause and thus precede itself. Thus causal order can't be infinite; there needs to be an uncaused primal cause. Existence is composed of caused, finite events. Whatever begins to exist, asserted the *sufi* Algazel, has a cause; and 'something which begins has a sufficient cause' is also the modern principle of causality. This principle is constantly verified and never falsified. The physical universe began to exist and therefore has a cause. What is caused is not eternal. It is finite. Its effect becomes a further cause. Thus all existence is a changeful network made of causes and effects; creation is an action and reaction zone.

Where nothing is an absence, nothing comes from nothing. Did you think space was nothing? Wrong. Vacuum did not come before its cosmos; nor, since particles arise from fluctuations in it, is it nothing. Why, therefore, should cosmos as a whole derive from nothing? If it didn't, did the universe create itself? Then it existed prior to itself. Cause caused itself. Such is the kind of incoherent, 'boot-strap' logic some display. Perhaps, you claim, there was no cause of physicality! But, if there were, such cause could not be physical. It must transcend the physical. Its physical non-being must be immaterial; its being must be metaphysical. This Uncaused Being is not absent. It is self-sufficient, potent and with presence; its causal level of reality is 'higher' than non-conscious, physical phenomena; and its timeless time is Now.

In summary, it's as simple as it's crystal clear. **What starts to exist is always caused. Material cosmos, known as nature to the natural sciences, started to exist and therefore has a Natural Cause**. This cause, preceding physicality, is physical non-being. It is nothing physical but causes cosmological effects. What came (or comes) before the latter's matter, space and time must itself be time-less, space-less, immaterial. **Preceding its secondary causes and effects the primary, first cause must be physically uncaused; it is uncreated in a naturalistic sense, thus super-natural. Metaphysical. In this respect the term preferred by Natural Dialectic to 'big bang' is *'transcendent projection'***. Since there's no scientific explanation you might reasonably term such origin a miracle. Shot from its archer's super-natural bow this cosmic missile undergoes, according to time's arrow, changeful decay. Why should what precedes the arrow ever suffer *physical* demise? Better try and understand the nature of an uncreated immanence from whose causation galaxies and men have sprung.

Rephrased, motion is potential that's expressed; cosmic action is Uncaused Potential worked. The primal motion of Essence is First Cause; such First Expression is the Start of starts and purest form creation issues in. All existence, psychological and physical, is the Start's effect. Although, within the whirligig of space and time, we choose to call some actions 'causes' and others their 'effects' in fact they're all, traced back to origin, effects of Cause. **First Cause is linked with everything that follows its beginning; and it forms the link between the pair - Essence Absolute and relative existence - that comprise the basic, prime polarity of Natural Dialectic.**

Since the energetic universe was once initialised creation's been replaced by conservation (of energy) and transformation.[11] Physical transformations are what science tracks; and Newton's Third Law of Motion (that every action has an equal and opposite reaction) captures the intrinsic order of equilibration underlying changes in this natural traffic. The law grants cosmos fundamental balance and descriptive science its equations. In the orient it's translated, where '*karma*' means 'action', into the '*karmic*' law of chain reaction.

Why, however, should a balance sheet of the world's economy not extend to the psychological dimension? Why should precision in the automatic world of matter not be mirrored in the mind - *karma* physical *and* psychological? As with physic so with metaphysic; every desire provokes an equal and opposite reaction; causal thought creates a boomerang effect.[12]

You understand such psychological equilibration well. Scales and sword. Natural *karma* is reflected in a golden statue. Objective and impartial Lady Justice tops Old Bailey; and she represents, in principle, all other human courts.

Cause informs effect. At this point let us briefly introduce a mystery that through the book we shall increasingly explore. **No doubt, energetic causes push effects; they bump you from behind; their arrow, physic's arrow, runs from past to present.** And things suffer (though you'd hardly think it as regards a proton or electron) from increasing weariness called entropy. They run out of steam. What, though, about a cause that is conceptually implanted? What about *informative causation*? **This is goal-oriented; and goals are in the future pulling you their way. They pull you from the future; they lift forward. They are metaphysical *attractors*, guides that govern your behaviour as they lead you through the world.** Not material but immaterial, such leadership is not by force of gravity, electric charge or magnetism; information's metaphysical not physical; it's psychological and, although at this point you may want to know exactly how, be patient - as the book's elaborated we shall come to see. *Information's entropy is negative.* **Mind is negentropic and thus metaphysic's arrow flies, from future back to present, anti-parallel to physic's.** Thus you are guided, present to the future, by plans realising goals.

---

[11] see Chapter 12: Labours of an Empty Womb; Index: thermodynamics.
[12] *karma* theory; see also Chapter 18: Anathema, Chapter 26: Social and Individual Parts, and Glossary.

Purpose is what pulls you personally forward; and, even if they're fixed, we'll see how creation's vibrant generalities, its behavioural attractors, pull things forward the way they have to go. **Push and pull cooperate. Energetic and informative causation, running anti-parallel, are the way the world proceeds.** You still ask, however, how this can apply to physicality.

In future we shall not much use the word 'attractor'. **The Dialectic's word is 'archetype'.** Archetype's a key, most rational *idea*. An archetype is filed in memory but, like any state of immateriality, will draw materialism's fire - because it spells the latter's death. Who is not roused to try and slay a mortal enemy? Attractors/ archetypes, the source of natural order, purpose and coherence, are of two kinds although they are both found in mind.

The conscious attractor, causal or potential mind, is identified as Archetype; this Primary Archetype is Logical; it is Single, First Cause or *Logos*.[13]

The lower, unconscious attractors, causal or potential matter, are identified as archetypal memories or multiple, fixed forms in universal mind.[14] In the human case these secondary, passive records are reflected by fixed patterns in your own subconscious mind; they are memories known, in part, as instinct. The way such morphological attractants, that is, archetypes are linked to polar, cosmic infrastructure is an issue we'll elaborate. Gradually we'll pin the crucial, causal notion down.

individual	*general*
somewhere	*nowhere/ everywhere*
local	*non-local*
factual	*conceptual*

An archetype is typical and not specific; it is immaterial not material, conceptual not physically factual. It's mysterious yet obvious, as hard to grasp yet easy as mathematics, logic or a way of nature - nowhere, everywhere and always. Thus its 'shadow' is more real than transient shadows flitting at its bidding, shadows we call physical reality. Indeed, various behavioural patterns, archetypes, substantiate all form - including yours. If this idea is old perhaps quantum physics indicates it's new as well. For now, let's simply note that archetypal memory is something that (especially in Books 2 and 3) we'll learn much more about.

Polarity

The second principle is *polarity*. Poles express the fundamental duality inherent in things as opposed to nothing. Indeed our cosmos can,

---

[13] see also Chapter 5: Top Teleology.
[14] Chapters 15 - 17; see also mnemone.

as we've begun to clarify, be seen as anchored between poles of various sorts; nature, physical and psychological, is seen as a dynamic interplay of opposing and yet complementary forces.

	conscio-material field	Top Pole
	polarity	Neutrality
	duality	Unity
	repulsion/ attraction	Consummation
	motion	Balance/ Peace
↓	repulsion/ negative	attraction/ positive ↑
	divisive/ polarising	depolarising/ unifying
	non-conscious matter	subjective mind
	base pole	in-between

These are the stays from which its range of webs, its tapestry, is woven. Their extremes influence each other. They can repel and stand apart or they can attract and work together in a covalent kind of way. Between complementary, reciprocal opposites ranges a spectrum of intermediates. Such oscillatory range of influence is known as a field. The doctrine of polarity (or duality) wherein complementary opposites interact and together constitute a whole is ancient. The Greek Empedocles, for example, identified polar dialectic between the universal forces of love and discord. Attraction and repulsion. Not only matter but the whole of existence is imbued with the paradoxical two-in-one nature of binary opposition. Sometimes emphasis leans one way, sometimes the other but, whatever you're considering, polarity is represented in the aspects of it. Indeed, you might abbreviate existence to a play of push-pull, in-out tensions in a field of nothingness. *And descriptive Natural Dialectic's a reflection of such cosmic chemistry that always binds polarities into a single pair, a double singularity, a covalent stack of interaction. You might say it dances to the tune of counterpoint's polarity.*

However, the Dialectic also *resolves* polarity in the form of a third, central component, Balance. As regards the whole creation it proposes mind and matter as two existential aspects of an underlying Essence. Two-in-one is also one-in-two. The framework thereby, as we've seen, complements existential relativity with Essential Absolution. Thus its stacks represent Essence (Top Pole) and Essential Characteristics set, top to toe, against a range of relativity we call the existential field. **This existential matrix, in which informed and informant (energy and information) interact, is called the conscio-material gradient.** From its Top, 'Strong' Pole of Subjectivity this gradient falls to the special case of 'weakness' - a base pole of none. The nether case, non-consciousness, is creation's body; such objectified embodiment of Essence we know as our spacious, sunny universe of matter.

We can note in passing the basic relevance of binary, digital systems to our information age - in mathematics, computing and biology. In cosmic terms 0 and 1 denote potential and action. Above all, therefore, they denote Presidential Essence and, when it is switched on, auxiliary existence that stands out from and, apparently, apart from it. The two numbers are also used to denote (in Part 0 of this book)[15] the informative and physical potentials that compose a lifeless, macrocosmic stage and (in Part 1) a stage alive with units of drama, that is, with 'switched-on' characters, with conscious incorporations such as you, me and, although sometimes less lively, every other organism.

The fundamental polarity is between Essence and existence, between Potential and action or, in other words, Nothing and things.

The Basic Existential Dipole.

	realisation/ expression	*Potential*	
↓	*fixed/ static/ impotent*	*kinetic/ energetic*	↑
	completion	*action*	

No doubt everything, both large-scale and small, is in relative changefulness. But motion of change reflects only one aspect of energy, the kinetic. **In fact, there are three basic aspects of energy *and* information.** *They are potential, kinetic and fixed.* These are, translated into process, possibility, action and exhausted completion. Of these, neither complete potential nor exhaustion exhibit motion.

As regards physics the dialectical definition of 'potential energy' is broader than simply 'energy a body possesses by virtue of its position'. It implies possibility as yet unrealised, latent capability and 'readiness-to-go'; it implies a neutral and yet potent form of equilibrium; this is positive neutrality.

And the definition of 'kinetic energy' is broader than the simple 'energy a body possesses by virtue of its motion'. Putting aside the mysteriously perpetual, very rapid motions of quanta, the definition includes both fundamental flexibility (the potential of whose 'probabilities' is reduced, some quantum-how, to actuality); and, for any event, either of the two directions its flux/ change can take - towards inertia (↓) or, by stimulation (↑), towards free flow. It therefore includes dynamic actors such as light, heat, electrical current or force-in-action; and 'anti-dynamic' actors such as gravity, mass, aggregation and 'crystal precipitate'.

Moreover, the poles of Natural Dialectic represent, unlike flow between them, opposite extremes. It is a mark of Dialectic that these poles (in this case 'potential' and 'exhausted') reflect each other by inversion. Their opposition is, you may say, in 'reflective asymmetry'.

---

[15] see Preface: Abbreviated Contents Box.

> **Primary Inversion** 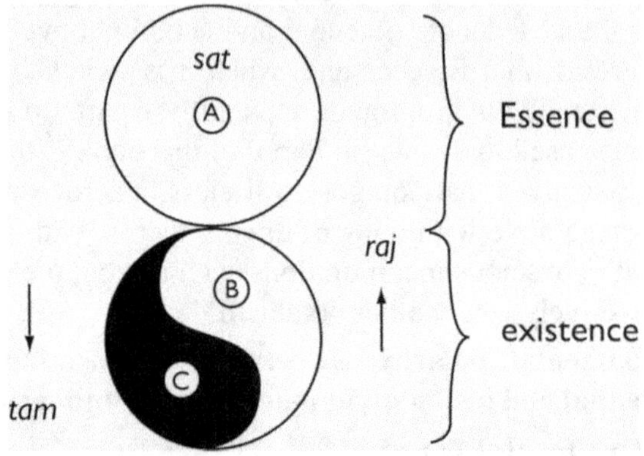
>
> *sat* — Ⓐ — } Essence
>
> *raj* — Ⓑ
> *tam* — Ⓒ
> } existence
>
> It comes out opposite. **An inversion means that something turns out opposite its origin.** Cosmic poles and inversions sound complicated but can easily be simplified using a 'flipped open' version of the yin-yang symbol (*fig.* 1.1). You can see that A (Unity) is the antithesis of B/C (duality); also, within existence, B (metaphysical mind) counterpoints C (physical body). It is clear that the diametrical or 'far opposite' of A (Essence) is C (the physical universe); in other words the extreme inversion of Essence is material solidity and *vice versa*. Such antitheses are regarded not so much as mirror images but *inversions* of each other. 'As above, so below' runs the reflective maxim. Dependent B and C reflect the nature of independent A but, in the process of realising A's potential, this reflection also involves an inversion. It involves the 'organic' kind of development that expresses 'inside information' outwardly. According to its logic Inmost Essence (A) devolves a polar creation whose final stage ends up as non-conscious matter (C). **This, the drop from consciousness to non-consciousness, is the primary, creative inversion.**
>
> *fig. 2.4 (see also fig. 3.1)*

Electrostatic charge accumulates in purple clouds. From motionless potential builds the storm. It breaks and pours and, spent, is over. There is calm before and after cloudburst but of very different and inverted kinds. Is not exhaustion the reverse of powerful potential? From potency to impotence, before is not the same as after. So with the cosmic storm we call existence.

From electric charge to flood, from electromagnetism and internal forces to bulk chemistry things develop from the inside out. Cosmic inversion nurtures them, like seeds, from implicit to explicit, that is, from 'inside potential' to their 'outward, finished form'. 'Expression' is the word. The stillness of the seed is not the stillness at the end-point of expressed development - no further growth.

A gradient slopes from potent to impotent peace; a balance swings from poised until inertial equilibrium. Thus, as day's lease gives way to night, light is blackened. Start to finish takes the course of an inversion. The immaterial pole of consciousness is twisted in descent to an extreme, non-conscious, material constriction - perhaps in the form, at base-pole subtendence, of a black hole. In terms of Primary and Secondary Dialectic pole to anti-pole inversion is registered as an *innate* 'switch' from Primary Right to secondary left. It reflects the general fact that a (*tam*) left-hand extreme represents the negative opposite of a (*sat*) characteristic - even though they may appear superficially similar and the same word, such as 'rest', may be misleadingly used to describe them both. Their true characters are diametrically apart.

At the final stage of 'organic inversion' consciousness has been turned inside out and fallen, therefore, into the non-conscious physic of bodies animate or lifeless. Natural Dialectic would suggest it has been twisted, upside down and out, into the bailiwick of physics and its chemistry. Conscious concentrate has been completely lost; what started out without non-conscious energy has turned to nothing else. It has become a plenitude of lifelessness. If oblivious matter constitutes the last flight of creation's stairs its final step is turned to stone. The scale from zone of soul to that of pure, non-conscious energy completes the opposition of its poles that Natural Dialectic calls, we've seen, reflection by asymmetry. *This is what, in universal terms, we mean by paradox of inversion.* **Our starry universe is clustered round the lowest rungs of a Great Ladder, the last step-down out of a higher, grander cause - an outcome that completes inversion of, as well as you,[16] the cosmos**.

How do energy and information interact? Materialism apoplexes, scientism flinches but don't think the subject's academic. The rest of SAS explains. For example, an inversion close to home - top-to-toe, mind-to-matter, information-in-exchange-with-body - is biologically registered in you. Immaterial information is expressed in coded matter; conscious to non-conscious vector naturally engages such a cosmic twist; and Natural Dialectic can reflect it.

Thus to physical you need add metaphysical; to energy you need inspect another factor, information. This dipole, energy and information, forms the basic coordinates of existence.

Information

Check 'information' in the Glossary.

---

[16] Chapter 17: Caduceus and *fig*. 17.3.

manifestation	*Latency*
conscio-material gradient	*Consciousness*
forms of Information	*Information's Source*
↓ *energy/ matter*	*information* ↑
passive forms of info.	*active/ semantic forms of info.*
informed/ guided	*informant/ guide*
objective aspect	*subjective aspect*
non-conscious/ physical	*conscious/ psychological*

An argument has been enjoined - that cosmos is composed of energy (including matter) *and* immaterial information. If 'natural' is equated with 'material' then 'unnatural' information has no natural cause. Non-conscious, physical phenomena create neither codes nor, which codes carry, messages; although intrinsically devoid of meaning or purpose they may, however, *convey* both of these. Things may be dynamically or otherwise arranged, as in the case of music, language or machine, to express meaning; matter may, to its molecular core, be shaped by the various logics of purpose.

In other words, for Natural Dialectic there are two main kinds of information - active and passive. Informing and informed. Active information is semantic; it involves conscious appreciation, understanding and manipulation of symbols and forms. Matter, on the other hand, knows nothing. It is passively informed.

In brief, Natural Dialectic would propose that there exists, as well as non-conscious physical energy, metaphysical mind - a 'field of information' also endowed with potential, kinetic and inertial (sub-conscious) aspects. This immaterial information centre may influence and be influenced by material circumstance. Interaction and exchange occur. **For this reason information and energy are identified as the component derivatives of causality and polarity and, therefore, the basic coefficients of existence.**

Energy

Physics is the study of energetic interactions of precise and pre-set kind; and, put simply from a dialectical point of view, chemistry observes the push-and-pull of polar charge. Thus Chapters 7-12 deal with the basis of physical science - **energy**. These six divisions might well, for reasons you will come to understand, be entitled Rules, Infinity, Unity, Nothing, Time and Everything. Meanwhile Chapters 5, 6 and 13-17 deal with **information**, knowledge, meaning and psychology. Before we split these two fundamental elements, however, we can note their polar field, called biological life. In this, our own psycho-physical field, they can be written as coordinates. This implies their interdependence. Then, stepping from your microcosmic self to macrocosmic universe, you might concede that, interwoven in various and variable proportions, this couple of principals substantiates the fabric of existence.

Indissolubly bound in varying degrees and ways, information and energy constitute a conscio-material spectrum, gradient or hierarchy of existence. Two sides of a single essence, they co-exist. Material and immaterial, they are complementary opposites, the covalent poles of an existential dipole.

Informed Energy

Energy and information are co-principals. From this pair creation is derived.

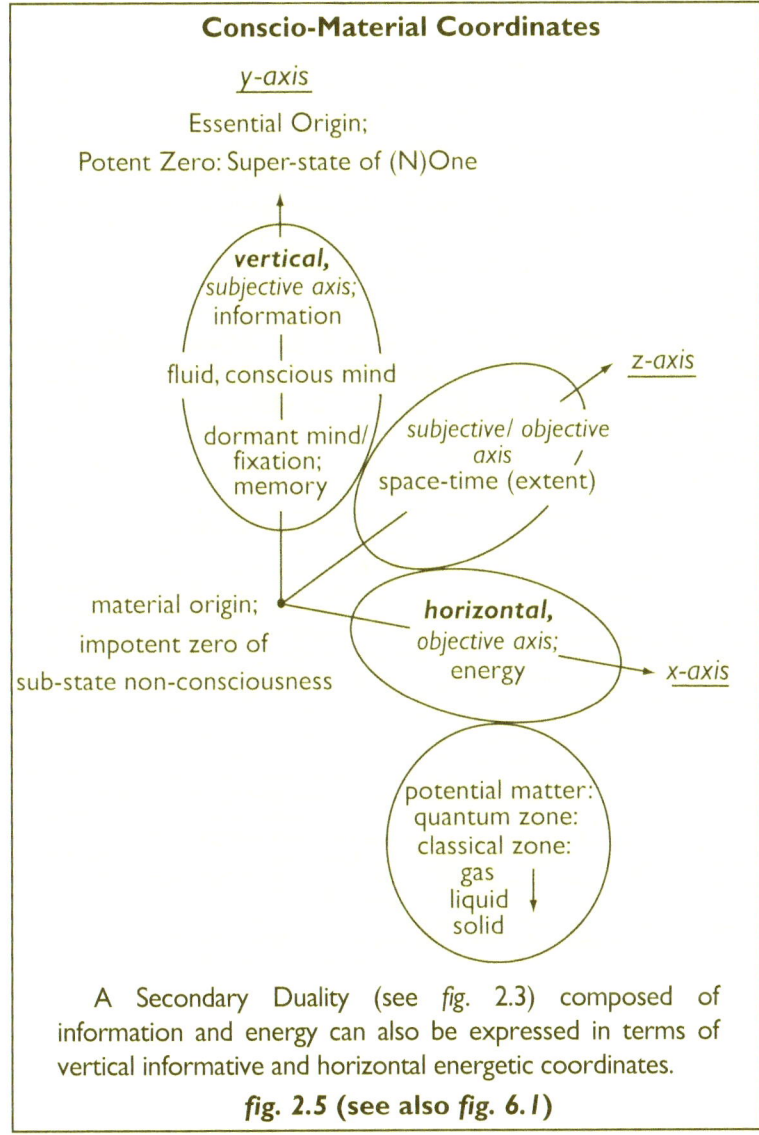

fig. 2.5 (see also fig. 6.1)

Objective energy, the stuff of physics, is about action; subjective information is about the motion of will, power of purpose, meaning and control. Together they comprise plan and its realisation.

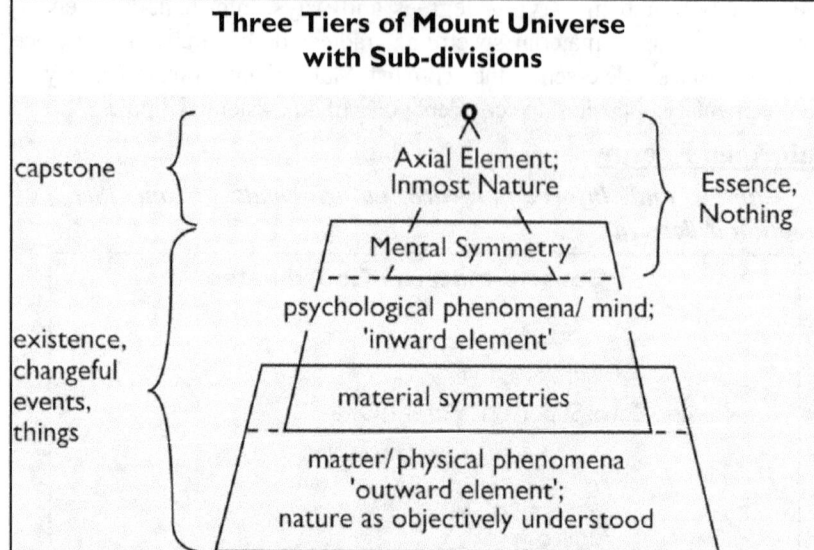

Major sub-divisions of cosmos are *(Sat)* Essence, *(raj)* mind and *(tam)* matter. The information in these ziggurats will be elaborated throughout the course of the book.

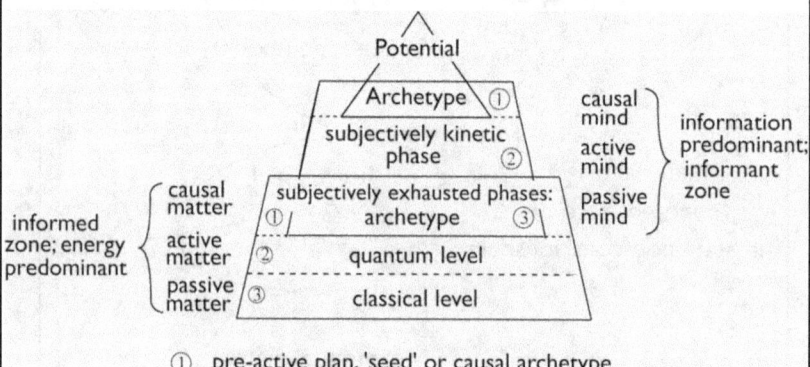

① pre-active plan, 'seed' or causal archetype
② internal informant, pattern-maker; primary effect
③ external structure, fixity of pattern; secondary effect

Within each major division minor (numbered) sub-divisions can be drawn. These numbers are related to the three cosmic fundamentals. For mind these are *(sat)* causal super-conscious, *(raj)* conscious and *(tam)* passive sub-conscious levels; and for non-conscious matter a psychosomatic link grade, *(sat)* causal/ potential matter, *(raj)* quantum matter-in-principle and *(tam)* bulk or bonded matter-in-practice. This final minor sub-division itself involves a familiar sub-sub-division into gas, liquid and the final, most fixed and 'static' expression of them all, solid.

> It is often claimed that distinction drawn between indivisible quantum systems and those of apparently divided classical events is artificial. This is because quantum effects, although less obvious on the large-scale, exist at all levels. A similar proposal might be made with respect to potential matter and its archetypal effects.
>
> **fig. 2.6**

To inform means 'to give form, shape, intelligence or organising power'. It is to communicate a pattern of behaviour. Psychologically, informative patterns involve sense, knowledge, understanding and desire; they include the attributes of meaning and purpose. The province of such knowledge is 'active information' - conscious mind; an obvious example of actively informed energy is a machine. And the natural province of 'recorded therefore passive information' is memory - subconscious mind. The automatic characters of energy are also passively informed. They behave by what is known as natural law and show as patterns of an object or as courses of events. Any pattern of behaviour is explained by its cause; its cause is its reason; and this reason includes both innate capacity and the external permutations of circumstance. Reasonable behaviour needs a plan; information is a deed's potential; and mind precedes material behaviour. *What, however, is the cause of physical behavior, the reason behind natural law?* This question occupies Chapters 7-12. For now it's simply noted that if the way material things relate cannot be generated by those same relations then, logically, transcendent laws of nature must be metaphysical. Their provision of initial conditions for all natural events means that their origin is super-natural. Cosmos is a *carrier* of information but is not its cause.

What, therefore, does *fig.* 2.6 imply? **It implies that the universe does not have a single dimension. It has three - potential and its dual expression, mind and matter. The quality of events within these grades depends on the proportions in which information and energy are involved with each other.**

It equally implies an order of expression, a sequence down the scale of creation. Mind over matter; from mind mindless matter. *From a materialistic point of view this is completely outrageous.* It commits a capital thought-offence. Off with its head! Truncate the universe from immaterial information! Let us, therefore, at this moment re-clarify. There is not only matter in this world; there is meaning too. The association of information with energy should not be seen as a Cartesian split but as a gradual, mutual, proportional involvement. *Let us then see, as the book unfolds, how the game transpires. How do things work out?*

Chapter 3: Hierarchical Perspective

The anti-parallels from which creation's permutations are expressed have been identified - energy and information. **Four** more matters need to be discussed.

Hierarchical, Triplex Construction of the Cosmic Pyramid

Firstly the world, for a materialist, is basically colliding particles and interactions due to force. And, biologically, outside and even inside cells 'self-replicating' molecules (a slender thread on which materialism hangs) do not exist.[17] Nor do nucleic acids know or care a fig. Such oblivion is ignorant of 'ought', morality, laws of logic, reason or intelligence. Even *if* its carelessness developed nerves (another great guess in doctrinal dark) why should their chemical reactions produce a calculation or a sonnet? How can a set of physical changes, physically caused, possibly 'correspond' to such conscious experience as seeing that 'an axiom is self-evident' or to conscious logical transition as implied by the word 'therefore'? If a whole sequence of logical steps, mathematical algorithms or reasonable statements were indeed merely the effect of a causal chain of physical processes, all blindly and mechanically or electrochemically determined, it would follow that the speaker could not think what he wanted or help saying what he did. He would, quite literally, not know what he was talking about. His statements, as reasoned arguments, should therefore carry no weight. *Thus, why should we believe a word he says*? **In this respect it is common knowledge that, on his own terms, a serious atheist is talking gobbledegook**. Physicist Sir John Polkinghorne rams the point right home. 'Thought is replaced by electro-chemical neural events. Two such events cannot confront each other in rational discourse…The very assertions of the reductionist himself are nothing but blips in the neural network of his brain. The world of rational discourse dissolves into the absurd chatter of firing synapses.'

However, flattening cosmos into one, material dimension needs a cure, a restoration to holistic, three-dimensional perspective. **It makes sense if you expand the atheist's primary axiom and actually allow an immaterial element**. And, when this rational axiom's accepted, view of cosmic hierarchy kicks straight back to life.

You can read each step of the ziggurat (*fig.* 3.1) as nested in the next; from base to peak is, if you worked with concentric circles, from periphery to central source. The three broad hierarchical divisions can be pressed a little further into triplet sub-divisions of mind and matter.

---

[17] Chapter 20: Perplexity.

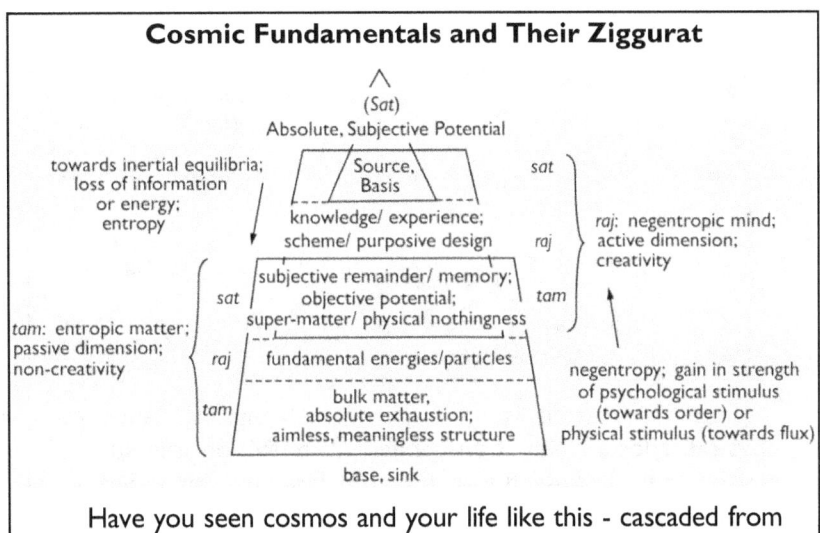

Have you seen cosmos and your life like this - cascaded from the Apex of Mount Universe? **Hierarchy** is a fundamental aspect of creation. *If you disagree with that, look no further than your own construction.* It is (see later in Books 2 and 3) hierarchical.

This model, a ziggurat, is validated by inclusion of an immaterial element.

fig. 3.1

In *fig.* 3.1 the top subjective division, Potential, *is*. **Thence downwardly, subjective sub-divisions at each level of mind (in *fig.* 3.2) represent *loss of active information* from the pure state, enlightenment;** that is, they represent lower, slower or, as with memory, fixed/ passive states of mind. Such natural (↓) 'entropy of information' is consciously reversed by the (↑) negentropy of concentration. This rousing stimulus amounts to interest and, at root, intensity of focus called the fire of love. The highest aim of such a focus would, logically, be its Cosmic Source.

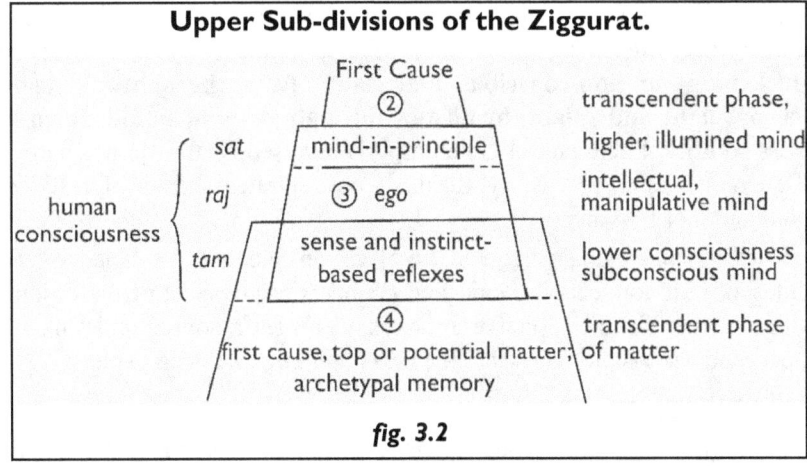

fig. 3.2

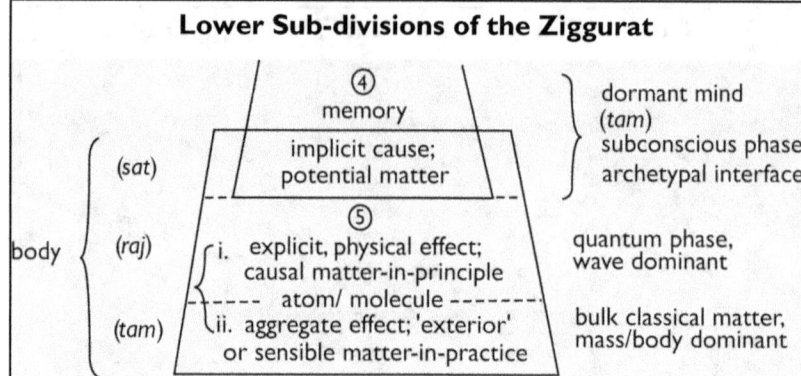

Lower Sub-divisions of the Ziggurat

Potential matter is an implicit, metaphysical cause of physical effects. This transcendent sub-division of 'super-matter' and its psychosomatic interface are discussed in Chapters 7 - 10, 15 and 17.

Through quantum physics could science touch with, at its archetypal phase, universal mind? Section 5 involves all physical effects. Of these an explicit cause - cognate quantum particles and forces - gives rise to so-called 'condensed', 'externalised' or bulk matter. In other words the material universe is an effect composed of interior, kinetic and exterior, inertial forms of energy. This couple are the subject of Chapters 7 - 12.

fig. 3.3

The psychology of consciousness is discussed in chapters 5, 6 and 13-14, sub-consciousness in 15-17 and chapters 7-12 elaborate on the sub-divisions of non-conscious matter. In the sense that it 'underlies' mind our zone of physics might be termed an underworld.

Objective sub-divisions of level 4 shown in *fig.* 3.3 represent internal, metaphysical, sub-conscious agency; **but external, bodily phenomena of level 5 are measured by a *loss of free energy*, by a downward 'thickening' of non-conscious materiality from the subtle state of plasma, light and quantum factors through gas and liquid down to gross solidity.** Such natural (↓) entropy is reversed by the (↑) negentropy of energetic input, that is, by variously concentrated stimuli of colliding momentum or fiery heat.

Each step of this ziggurat involves an element of science and philosophy. It connects the pair; and connects each aspect of the cosmos with its other 'levels'. Such framework identifies a source and sink - a super- and a sub-state. These are the *next two things* we need explore.

Subtendence

The (*tam*) cosmic tendency is down. To subtend is to extend beneath; it is to oppose and delimit. Thus, to describe Natural Dialectic's negative extreme we can reasonably coin the word 'subtendence'. This sub-state pole opposes a transcendent super-state. Sink is the opposite of source.

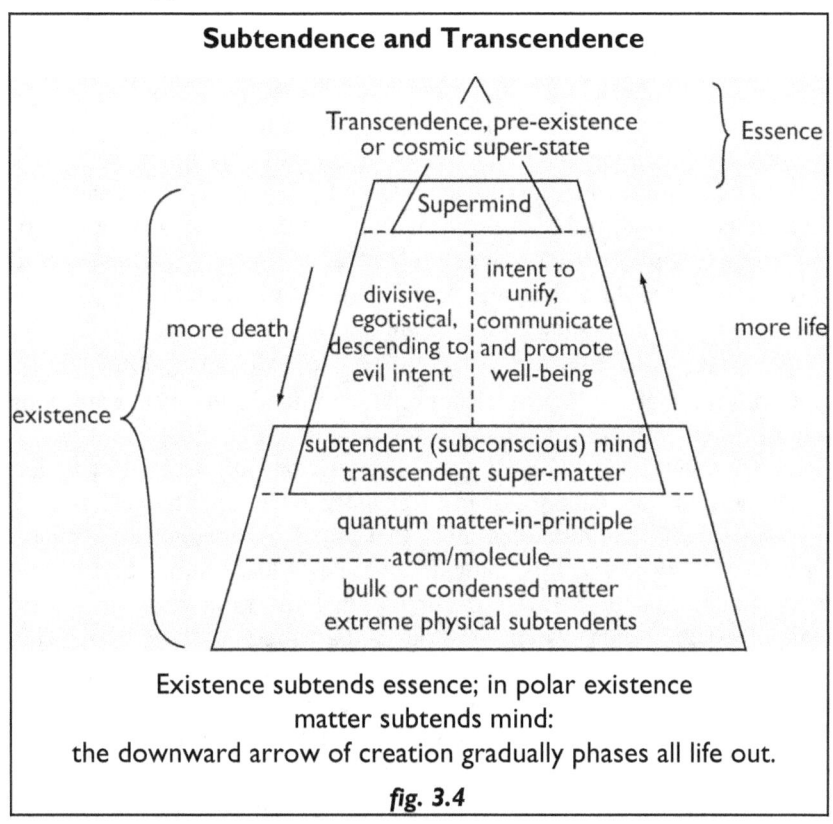

fig. 3.4

tam/ raj	Sat
below	Transcendence
range of action	Super-State
expression	Pre-motive Potential
current	Source
↓ tam	raj ↑
descent	ascent
subtending	transcending
subtendent base	action

It is true, *top-down*, that products subtend their producers. A producer stimulates and stirs things up. From this perspective information precedes its material consequence, metaphysical mind plans physical arrangements, super-natural precedes natural order. *Mind first, body after; body's an appendage of its mind.* Does it, that physical depends on metaphysical, invert your 'normal' sense of things?

In the psychological zone sleep, we say, subtends waking; *sub-conscious subtends conscious* and the latter subtends super-consciousness. Although we may be unaware of it the sub-conscious, dormant level of mind still holds information, informs and is informed. It is the repository of memory, instinct and the psychosomatic (*PSI*) channels that interface, immediately and continually, with the nervous system and thereby molecular structures, cells and organs of the body. *Stored, reflex or automated information is passive; passive information subtends active.* A singer records a song. The CD carries rigidified information. What is it but a plastic form of memory? All matter, which is totally non-conscious, repetitious and predictable in its behaviours, is informed by the internal or immanent exigency of 'laws of nature'. What is this 'song', whose record plays across the fabric of the known universe? Material bodies are governed by and thus subtend its vibrant rules.

Let's take your own body first. When you waggle your little finger, is it possible that the line of operational command runs from conscious mind through sub-conscious templates (memories, instincts etc.) over a psychosomatic border? And that incoming or outgoing information would be first translated to, or last translated from, the physical side by the product of 'excited' electrons, that is, electromagnetism? Where *matter subtends mind*, this 'radiant' phase would be subtended by the biochemical; and both levels occur within bulk, biological structures, that is, cells, nervous system, muscles and a coherent whole body. This places your finger-waggle at the base of an informative/ energetic hierarchy. Your sensible, physical form (perhaps including the origin of its shape) equally resides at the bottom of such a hierarchy. It would be the final outcome of a two-pronged, coded plan. First prong, generic, springs from archetype; the second, locally specific, is genomic. From signal biochemistry is body built; phenotype subtends its molecules and genotype. Thus bio-form's a frozen yet dynamic program; or, prosaically, a 'functional structure'. Very complex, yes; automated, yes; it's an incredible machine.

All bulk bodies subtend quantum agents; these 'organise' their chemistry. Whence, from a *top-down* angle, did these subtle agents rise? What is the template they in turn subtend? If one does exist how does connection to it work? Could, closest to whatever transcends physicality, pure energy (say, mass-less radiance of light) subtend a signal interface? Thus bulk, solid rock emerged from air (which, if the universe was once much hotter than it is today, it did) because the gas subtended originally-projected, pre-atomic energy whose traits were ordered under archetype. From Natural Dialectic's point of view (*fig*. 3.4) quantum physics interfaces archetypal memory; *matter naturally subtends a psychological potential.* If (↑) vaporised concretions make a gas the (↓) opposite applies. Thence the question:

'What's the end precipitate of gas, the final stage of matter past which it is impossible to drop? Of what consists the world's subtendent end?'

The lifeless, non-conscious underworld we call the zone of science expresses two forms of (*tam* ↓) extreme subtendence, one energetic and the other massive. An object, super-cooled, approaches (but can never quite achieve) a total 'loss of levity', an immobility called 0°K.[18] The second, massive kind of cosmic death involves approaching total 'gain of gravity'. Such theoretical - perhaps even actual subtendence - that throttles matter to extreme degree is called a black hole. Is this subtendence (ring-fenced by a 'horizon' that stymies any probe) infinitely dense or not? At least, utter negativity makes sense but are its boundaries permanent and real? Some are sure but, in the case that starts and ends are veiled in mystery, men interpret while nobody really knows. How would you, a scientific super-hero, break beyond the bounds of space and time and things?

To summarise: diametrically opposed to Essence exist some extremely negative polarisations. Such subtendencies are due, in the informative case, to severe loss or paralysis of consciousness and, in the energetic case, to extreme loss or confinement of energy. Yet such exhausted sinks are localised. They come and go. Change-prone relativities do not contain The Absolute. There can be neither Subtendent Purity nor Perfect Negative where, in cyclical existence, all is relative and nothing lasts forever.

Transcendence

	existence	*Essence*
	lesser	*Supreme*
	relativity	*Absolution*
	lower	*Highest*
	subtendent creation	*Transcendence*
↓	*downward*	*upward* ↑
	lowest/ lower	*higher*
	outer	*inner*
	inferior	*superior*
	contraction	*action/ release*
	fixity/ sink	*flux/ stream*

---

[18] 0°K is a strange, paradoxical cosmic boundary. Such absolution is, according to the Third Law of Thermodynamics, impossible to obtain but its proximal range involves properties of super-conductivity, super-fluidity and apparently infinite thermal conductivity! Such a freeze lacks any energy; its empty edge contrasts the source of physical creation; its sink of nothingness is fallen polar opposite of big-bang's sudden super-force and absolutely-nothing's miracle of wholly energetic ultra-heat (perhaps 10^{28} °K). See especially Chapters 9, 11 and 12.

Materialism's lexicon does not contain the word 'transcendence'. If everything's material then what but nothing can transcend it?

Action rises to extreme; high-powered describes a force of energy released. Does this apply to physic's source (projection from a singularity?) or, as opposed to brute, inactive mass, subtler quantum entities? 'Potential matter'[19] means 'where matter comes from'. What might interface with its transcendent nature thus linking physicality to what it's not? Transparent, insubstantial - who can shed some light? Do you think a photon you can't squeeze between your fingers isn't powerful? Light transcends mass and, almost immaterial, flies with absolute velocity. Concentrated light can cut through steel. 'Grounded' it transforms to heat; its 'nothingness' drives life on earth. And light's communicative, information-bearing - you can signal using it.

(↑) Ascent translates, in dissolution, to release, boundlessness and (no matter in the wave) individual annihilation. Isn't this the character of mass-less light? **Balanced electro-magnetism** (*fig.* 11.2) **is, for Natural Dialectic, physic nearest unto metaphysic - that is, transcendent or potential matter.** Therefore, could not quantum particles and basic forces be the medium for influence of archetype? And, if the purest form of energy is light, could not purity (of energy or information) illuminate the borders of transcendence (material to mental or from mental to Supreme) step by cosmic step (*fig.*3.1)?

matter/ mind	*Transcendence*
outworking	*Ideal*
comparative	*Superlative*
relativities	*Full Life*
lesser perfection	*Perfection*
lesser positivities	*Summum Bonum*
↓ *dark*	*light* ↑
decreasingly positive	*increasingly positive*
sinister	*dexter*
closed	*open*
diversification	*unification*
isolation	*togetherness*
hate	*love*
lie	*truth*
confinement	*freedom*
matter/ body	*mind*
more death	*more life*
sublative lifelessness	*life-in-relativity*

---

[19] see *fig.* 3.3 and Glossary: archetype.

From quantum subtlety we pass to metaphysical phenomena. Mind's scale is, like matter's, subdivided. On the analogy of spectrum an ultra-conscious band of mind transcends conscious that, in turn, transcends infra-conscious. Or, matched with states of matter, (*sat*) super-conscious transcends (*raj*) conscious transcends (*tam*) sub-conscious phase; and each mental phase exhibits its own properties. For example, nothing is anything to inertial sleep; indeed, sub-consciousness seems nothing either when I sleep or wake. Is such mundane transcendence, waking over sleep, as far as waking goes? Are you fully woken up? You feel so, as do fish or bird or bee, but are the trammelled states of your own mind the only possibilities? Could further, voluntary awakening transport you nearer to Transcendence; might it lift you up the grades, across constraints towards excellence and sweep you ultra-consciously towards The One?

First out, last in. If the first expression of Perfection was First Cause, the last stage of return will know The First. This first, primary current of creativity is called *Logos*, Holy Name, Christian *Word*, Koranic *Kalam-i-Illahi* and Sufic *Kun*. As the first vibration from which creation emanates it is Sikh *Shabda* and the Hindu *Paranada* whence *Om* and other *Nadas* (sounds) descend. *Raj* moves, informatively, towards Illumination. The final and most important transcendence is logically from existence to Essence. This step, beyond even the 'Living Sound and Light' of First Cause, is variously called Communion, Release or the Jewish *Ain Sof*.

The Cosmological Axis

Cosmological Bearings.

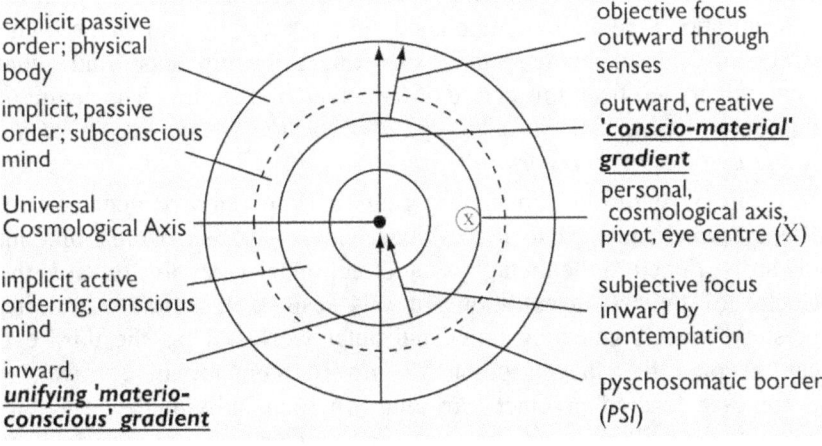

fig. 3.5

Fourthly, **who do you think you are and where? What is the direction of your short-lived pass through time and space?** A traveller wanting to arrive need first locate exactly where he is and where he thinks he's going. Every explorer needs to *orientate*. Only then can he plot with military precision the bearings of his destination and begin the chosen journey.

To monopoly-materialism it appears that a human is an animal composed of chemicals. It is a child of chance, a minuscule creature uncradled in the vast, dark, cold and sometimes violent abyss of space. Such diminutive, deriving its morality in theory from mutant genes, should entertain no thought of Central Origin. There exists no Upper Pole towards or from which humans turn.

Chemicals are part but are they whole truth? The holistic view is different. It includes a spectrum of subjective mind. Holism asks, *top-down*, the band of 'conscious visibility' in which I dwell. Where, between the poles (Transcendent Centre and subtendent body), do I stand? Three points on 3.5's simple, nested model can act as a map to obtain cosmological bearings.

First *is the psychosomatic (mind-matter) interface*. This, labelled *PSI*, exists at the base of mind's least lively part - sub-consciousness. Again, waggle the little finger of your right hand. This specific act identifies a matter of cosmic relevance. You have registered a higher level of order on matter. Somehow information has caused matter to move *on purpose*.

X marks the **second** *spot*. Touch your forehead at a central point above the nose, just above and between the eyes. Just behind this, marked X on the diagram, is where you think. *This location, identified as the* **second** *point of interest, is called the eye-centre.* Though not physical this centre is your vantage point, a cockpit of consciousness sometimes also called the 'third' or 'single' eye. A non-sense eye. It is information's point of concentration, the hub of embodied life, a metaphysical pivot round which the scales of human judgement swing. All campaigns, from trivial to grand, are led from this centre of command and control. The nature of this cosmological axis is relativity. *It is not the central reality of the whole universe but it is certainly yours.*

From X, therefore, concentric rings of your environment radiate. Rings spread outward to a world of bodies - physical creation; but inwardly, tipped to the metaphysical direction, rings evolve towards the Centre of their Concentration. In this respect X marks a mobile, personal pivot connecting inner and outer worlds. Thus the third eye can become the starting-point for inward, contemplative travel. It becomes a 'sacred precinct', an altar whereon individual restrictions collectively called *'ego'* are sacrificed. *Thus, if you wish, X is transformed into the launch-pad towards a* **third** *point of interest - The Universal Cosmological Axis.* This Pole Star, Absolute Point of Reference or Pivotal Truth is within your own more outward X. Its Transcendent and Immobile Hub is *within* the ceaseless hive of mind; from it radiates a jubilant cosmology.

Chapter 4: Truth, Appearance and Reality

Chapter 0 introduced anti-parallel directions of the mind and *top-down/ bottom-up* perspectives. Where, in such fundamental relativity, might rest single truth?

Is it *your* truth? This, as it stands, is the appearance of the world. Are single, final and your own truths all the same?

Truth, Appearance and Reality

tam/ raj	*Sat*
lesser/ existential being	*Being*
lesser truth	*Truth*
lesser reality/ appearance	*Reality*
↓ *tam*	*raj* ↑
objective	*subjective*
matter	*mind*
from Centre/ towards triviality	*towards Centre*
lesser truth	*greater truth*

Bottom-up, reality is physical. Materialism's primary axiom is set. The rest, excepting maths and scientific fact, is simply story-telling. Thus everything, including consciousness, is construed as a 'material phenomenon'. Such is, although it might be absolutely wrong, materialism's promissory faith. What you can slap is obviously real. Yet, even here, Niels Bohr claimed that we accept as real what's not. Quanta substantiate the large-scale universe which, on such terms, is an appearance. Matter is, in this sense, really energetic patterns and grosser views are less than fundamental. Part of physical reality's invisible. To this extent its truth's incomprehensible to common sense. How does grey matter grasp this fact?

No doubt, brains are lenses. They cut to a local portion of reality. They are constructed, you may claim, with such capacity and in the way that an embodied human's *meant* to know the world - but is 'mind-meat' alone enough to sense and understand? From a material perspective perception is, in fact, nervous activity. In other words, just find a 'neural correlate' (the pattern of the way synapses form and fire when using different parts of brain) to find perception's underlying truth. Where electric signals storm like drops of rain, there at grey stuff's neural base you'll find mind's rainbow crock of subjectivity. Matter is the truth; mind and metaphysic are appearance; physic is self-evident reality.

If, however, mind is *not* meat then, *top-down*, perspective flips. Every organism knows its world (that is, its reality) from inside gross and subtle frames. Gross is the non-conscious, physical body; subtle is instinct. Many organisms, including humans, are also subtly framed by personal memories. These three filters colour a local conscious experience; they frame interests and purposes within a sensory, creative or contemplative focus of attention; and they limit sense of what is going on and thus restrict our truth to personal experience. The apparently inescapable combination, one part physical and two metaphysical, is commonly identified as 'me'.

The metaphysicals are bundled up as 'mind'.[20] After all, an atom isn't conscious, mindful or informative. Why, therefore, should a thousand or a billion be? Congregation of non-consciousness doesn't just 'become alive'. Oblivion is every atom's everlasting state. Is consciousness composed, therefore, of oblivious matter or an immaterial element? Though not permitted by the rules of physics its unproven metaphysic is each person's base reality. Not physically provable may not mean non-existent; don't immaterial thoughts themselves create our science and mathematical description? **Certainly, if there's reality *outside* our consciousness we'll never ever know it!** Nor is anyone aware of undiscovered facts. Consciousness is all we have to know with. It is life. Experience, feeling, knowledge, logic, math and meaning utterly depend on it. Are these unreal?

Holism, you are well aware by now, does not admit that mind's a function of a complex form of matter and, in essence, physical. In this view all things *aren't* equal. *The illusion of illusions is that consciousness is an illusion*! **Consciousness and matter are elemental components of cosmic duality; information and energy are distinct, complementary aspects of existence.** This line of reason leads to a hierarchical or vertically-graded structure within which truth, appearance and reality are evaluated. **If Absolute Truth manifests a hierarchy of appearances, if from Substantial Reality emerge relative illusions and, therefore, things are not equally real, what is the criterion by which one is judged more or less real than another?**

It is as if, down the Cosmic Mountain's slope, different levels of perspective are obtained. Perspective changes as you rise or fall. Does love or understanding give the answers ignorance or hate purvey? Such a Theory of Relativity with respect to Truth, Appearance and Reality is actually neither new nor strange. Buddhists, for example, are taught appearances can be deceptive; the sensible world of trees and people and teacups is a partial illusion. Science now knows that, from the non-sensible perspective of quantum physics, this is correct. Objects can be understood as a web of almost nothing (less than 0.01% of electrons and nucleons) spun at very high speeds in nothing (over 99.99% space). Moreover, what substantiates this evanescent web is vibratory. Are not

---

[20] see Chapters 13-17.

the Buddhist notion of perpetual flux and the modern theory of kinetic matter in essence similar? In the latter, constant motion and vibration of bulk matter's make-up - quantum elements and atoms - explain such basic features of our world as temperature, diffusion and phase change. In the former, explanation is extended to include a medium of information - mind that also oscillates, radiates and can't keep still at all. No permanence. Vibrant motion drives appearance; it constitutes existence. Science well describes its protean object, energy. But whence the latter's source; what in reality substantiates the ripples of existence? And what, when physical and psychological illusions have completely cleared, is a Buddhist's Absolute Reality?

	relative illusion	*Truth*
	less right	*Right*
	critical comparison	*Criterion*
	lower qualities/ lesser values	*Quality/ Value*
	shades	*Illumination*
	degrees of incomprehension	*Clarity/ Clear Mind*
	lower principalities	*Principal*
	death/ life	*Pure Life*
↓	*lesser truth*	*greater truth* ↑
	blinder mind	*clearer mind*
	away from 'right mind'	*towards 'right mind'*
	lower qualities/ separators	*higher qualities/ unifiers*
	quantifiable objects/ events	*experience*
	non-conscious forms (matter)	*conscious forms (mind)*
	individual/ local contexts	*principles/ symmetries*
	less/ least real	*more real*
	towards darkness/ oblivion	*towards light/ understanding*

Comprehension and incomprehension roam the slopes. Ignorance is relative and what transcends familiar, contemporary knowledge certainly exists. Humanity's constrained by ignorance but always yearning to be free. What sort of knowledge brings us liberation best - transcendence of a fact in principled ideal or detailed understanding of the nature of material oblivion?

Appearances can be deceptive. If cosmos is projected it appears from source. Its appearance is effect not cause, lesser truth not basic. On this basis physic's cosmos isn't Real. Matter isn't Real. If principles inform behaviours then informative causation governs energetic patterns of effect. This sort of world-scale means a starry universe of matter isn't True - it's an appearance, only relatively true, projected from its hidden yet informant metaphysic, Truth.

Is there such Ultimate Reality? Belief, at least, depends upon criteria or, absolutely, your Criterion of Truth. If Top Pole is positioned as

Creation's Source then you might claim that it's most real. Transcendent First Cause throws a hierarchy of appearances whose shadows deepen with their distance from its Light.

In short, there is a sliding scale of truths but also, at the top, a Truth of truths. In terms of information a qualitative hierarchy drops from accurate and important through to incoherent, valueless or even malevolent. In terms of energy potential is expressed through action to exhaustion; it is constrained from possibility to single actuality, the end-result. The greater reality is, as with Bohr's quanta, vested in the invisible origin of any particular expression. At mental level this source/ potential is called *Logos*; and at material level 'potential matter of the archetypes'. If potential precedes action then we have a sliding rule - metaphysic unto physic - for priority, importance, truth and reality. Relative truths are measured against Absolution; such Top, Exemplary Criterion is embodied as a Perfect Saint.

Two Value Systems

We have glimpsed perspective that is, at first glance to a sensible, down-to-earth way of thinking, strange. Are you now ready for a fresh dose of dilemma - this time of another sort? Value, meaning and significance arise from mapping cosmos by experience; each new experience within this context is imbued with truth and truth must, more or less, bear worth. Minds and moods at different levels and with differing objectives make various value judgements; these are coloured by your overall perspective called a world-view - *top-down* or *bottom-up*.

quantity/ quality	*Quality*
objective/ subjective	*Subjective*
range of values	*Value*
lesser truths	*Truth*
↓ *lesser truth*	*greater truth* ↑
external	*internal*
objective thing	*subjective sentience*
physical context/ bodily self	*mental context/ egotistic self*
quantity/ aggregate	*quality/ meaning*
value of things	*value in mind*
numerical/ market/ bodily value	*motivating/ emotional value*
less important	*more important*
utility	*beauty*
using/ abusing/ careless	*caring*

Philosophy today is ambivalent. Are mind's illusions born of gene mutations? Are they only nerves or not? Some cannot quite believe that everything is physical. Others, hard no-hopers, can't accept the metaphysical at all. Their saturated scientism demonstrates monopolistic tendency. Such authority would dictate how a human must explain his cosmic state. Many strands are complicated in its uniform; each thread is spun from physics' loom; and all are woven into 'there-is-only-matter' patterns but must, perhaps grudgingly, acknowledge immaterial colours of minds full of information, will-power, creativity, society and, due to choice, morality. Perhaps you'd prefer, since you're a ghost in the machine, that there was no psychology - just neuroscience, electronic circuitry and the biology of nerve. Then it's easy - evolution made the whole thing up.

That, then, is who you are! Amoral molecules cannot beget morality. Since man must be the product of rude chemistry his body's nature is defined by genes and their surrounding circumstance; and mind's experience is nurtured by a nervous network of relationships. Thus are goals atomistically conceived; and by their yardstick what is judged success or failure must accrue. Critically, what is the value calculated? What is, when it comes to goals, a man's most valuable choice? Comfortable survival turns up trumps! No doubt, therefore, that politics and economics rule a creature's day.

'Quantity' of life-style is the objective business of economy - resources, wealth and body-care. And scientific values are non-ethically composed of numbers and utility.

'Quality' of life-style, on the other hand, involves subjective business - interests, relationships, aesthetics and happiness all round. Such socio-economic linkage needs firm leadership. A population is a clamorous 'family' composed of selfish and thence often spiteful members. It needs more or less a strict, *external* government in order to promote group balance and to mould conflicting instincts, interests and behaviours into as stable yet dynamic a pattern as can be. Equilibration keeps the peace. The quality of equanimity is, however, optimally based not on external but *internal* government. Such self-imposition stems from moral principles by which a person lives; and these in turn are coloured by the pillar of his faith. So civilised! In an ideal city of the blessed and blessing there would exist no needy, selfish sinners. Nor would vice hang expensive millstones round the neck of virtue. *Costless, painless law and order derives from principles - invisible, immaterial, unscientific principles.* Yet in our time it sometimes seems that neither clever socio-biologists with selfish genes nor cunning crooks respect the timeless root of crimelessness.

Top-down vision is consolidated round the Central Axis of Enlightenment. Its holistic value system attends Nature or (if you insist that nature's only physical) Super-Nature. Its Highest Value, the First Principle from which all others flow, is equated with the Nature of Innermost Self, The Highest Good or Apical Experience. Such Experience is extolled. It is symbolised at the sacred heart of all world faiths and in their personal devotions. This central residence would, dialectically, deposit any cult of matter (including humanistic scientism) at the periphery of truth; it would define a *bottom-up* perspective as, according to its concentration of materialism, a relatively eccentric misconception. Indeed, Fortuna's faith in origins-by-chance is one of origins irrational; and atheism's hopeless instinct doesn't sum to immaterial faith at all. You might therefore classify their tendency as 'sub-religion'. Next, inverting sub-religion's natural realities, realise the poise of Natural Reality. Such order is ideal, no doubt, and idealistic. In this case egalitarian brotherhood is clustered round a Sacred Heart. Such commune is 'supra-religious' and its absolutism is without political excuse - since politics is always tempered by pragmatics on the ground below. Climb any tower. Look out upon the city. From this cosmic balcony all you survey is built on metaphysical foundation - the cement of aspiration, information and desire. Now look inwards towards the Holy Sky.

Rights and Wrongs

From a value system flow one's reasons. These calculate, accordingly, what's right or wrong. Whether it's a thought or action, 'right' decision sails you home and dry.

	relative shadow	*Light*
	range of rightness	*Rightness*
	scale of inferior reasons	*Reason*
	failing	*Ideal*
↓	*negative*	*positive* ↑
	anti-principle	*principle*
	downward/ outward	*upward/ inward*
	from Reason	*towards Rightness*
	false/ wrong	*right/ true*
	passion	*patience*
	obsession	*detachment*
	pain	*peace*
	sin/ vice	*virtue/ righteousness*
	towards death	*towards life*

Do right and wrong, like ought and should, depend on metaphysical criteria? Who or what decides what's good or bad? Since every brain is different, might we materialistically deduce no absolute criterion exists? Neural products known as preference, logic and morality should vary in each one. Moral relativity is this game's name. If ethics and aesthetics ooze from nerves in brain then codes are cerebrated as the molecules dictate; and since brain configurations change our values vary with them! Why, anyway, should anybody, much less Him or It, tell me what to do? In this view 'personal or social opinion' blurs with absolute morality. Why should enslavement, murder, rape or theft be 'wrong' if that's the way I naturally think? Survival justifies self-seeking plans. You don't jail cats for killing mice and we're just animals as well. At least, this is the sort of creed that evolution justifies.

The problem is that such society of relative morality (and thence, politically, its laws of state) is built on shifting sands. <u>Without an absolute reference-point trivial issues are confused with major principles. Each man judges the excellence or otherwise of a given behaviour differently; such a secular society drops into an ethical morass of fashionable 'political correctness', changeable, relative moralities and even, at nadir, moral meltdown.</u>

From Science to Conscience

Oblivious maelstrom's valueless, meaningless and (*pace* atheism) cannot generate a code of any kind. Rational, legal or moral codes don't spring from witlessness. Thus study of material is not the same as study of the immaterial. Laboratories are not the place for moral seminars. *Indeed, scientific progress is irrelevant to moral progress.* Nor does the latter's personal kind of evolution change with history. Thus scientific materialism may exclude reference to moral struggle and human social context or, at best, objectively describe these in animal (socio-biological), humanistic or politically utilitarian terms. Such reason does not render subjectivity unreal. *There are different truths and therefore ways of seeking them. Do many ways up to its Apex mean a mountain lacks a Peak?* Why is it rational to deny the Universal Height is immaterial - unless a metamorphosis of reason called materialism has forgotten reason is itself a form of metaphysic?

Mind evolves by lowering its own entropy of information viz. increasing comprehension. No doubt, clever intellect pursues its variously important disciplines but wisdom comprehends by depth of empathy. Either's orderly simplicity derives from principle; and the more embracing a principle the more details are easily subsumed under it. Principles are concentrated power. They are the rule of natural and man-made law. If, therefore, the Axis of Mount Universe represents the Central Principal of Principles then the ascent of wise man will involve his mountaineering towards this Most Essential Peak. **Enlightenment is, naturally and automatically, Morality**.

Such morality's as natural as gravity. Its levity is unavoidable. Such immaterial but crucial datum is not gleaned from a laboratory bench. **Indeed, the fact is that all the emperors, generals, politicians, philosophers, philanthropists, artists *and* scientists in human history have not exercised as much influence as a few perfect mystics**. Waves of their transcendent experience resonate, as they seem always to have done, in the heart of mankind. Sages such as Buddha, Christ or Nanak (in alphabetical order!) personify Ideals. How, therefore, might you realise their fantasy of phantoms, an imagination made of neural networks and patterns produced, in the last material analysis, by randomness?!

Is There an Absolute Morality?

If the most powerful curative for personal, social and ecological ills is the right psychological perspective then the question becomes 'What is most right? How do I recognise it and obtain its Rightness?' Such questions have always absorbed mankind; they encapsulate his quest for Truth.

Bottom-up, reason and not Reason wins. A reasonable, atheistic humanist may well, with bare-faced self-contradiction, reject Darwinian 'morals' (nature-red-in-tooth-and-claw, survival-of-the-fittest and so on). He may resolve the conflict by behaving 'as if' some immaterial code, some variable creed prevailed. A clever humanist may also, with his finger on your pleasure spot, seem to dissolve morality's control. He can construe how a few ancient Greeks and other classical authors (say, Democritus, Epicurus or Lucretius) seem to endorse his sophistry. Who'd be fool enough to disagree? Comfort, if not hedonism, is 'the good life's' game. A siren voice excites an easy popularity. Just scratch my back and I'll scratch back - especially if you're 'family'. To hell with medieval sackcloth! Let us maximise on fun!

Top-down, reason is aligned with Reason. Conscience-in-practice starts, with an appreciation of the right (↑) and wrong (↓) vectors of behaviour, at the seat of mind - the eye-centre. It is here, where centaur-like humans leave behind the body's animal, that choice between a course of action and its consequence is weighed; and therefore here, where they can voluntarily transcend instinct and *think*, that moral rudiments are realised. From this point, X, their cosmological axis, a first positive, purposive and therefore, obviously, voluntary step towards higher truth is made. *It involves, through meditation, the achievement of a highly coherent or concentrated focus of attention at the eye-centre.* From this point the goal, following an inward transport, is to reach the top of Mount Universe and thereby achieve identity with the Centre. **This identity is with the Principal of principles. It is called Enlightenment - in which *top-down* perspective is entirely obtained. Teacher, method, practice and completion in identity together constitute a science of the soul.**

In short, if the cosmos is stepwise in structure, then well-aligned philosophy and good, sound mind should reflect the fact. In other words, there will scale a gradient of qualities of mind and, logically, one might aspire to Highest Good, Noblest Truth or The Criterion. **Such achievement *is*, intrinsically, The Absolute Morality.**[21]

---

[21] see also Chapter 26: Individual Association.

Chapter 5: The Immaterial Element

Let's cast aside value judgements, clear the decks and start with a clean slate!

Clean slates have nothing on them. You have to make a start. A start's an origin. Where did the world come from? Where did you? The context of your origin will colour all that follows. You already have to choose. What colour will your dream-slate be?

You remember the two pillars of faith. Either everything's energy and matter, including mind; and soul's a fiction atoms of the brain dream up. In this case cosmos is the child of chance. Or, *top-down*, in addition to these universal quantities, there is the universal quality of information. This metaphysical element cannot sensibly be overlooked. It is of central importance to *IT*, engineering, biology, linguistics and, indeed, the workings of our mind. Workings we call life.

Whether it actually involves two or three basics, the *origin* of the universe and various types of organism (including you) has not been observed, will not be repeated and cannot be tested. Each slate's historic start is, therefore, speculative. Which colour might a reasonable inference best confirm?

So, as befits our 'information age', let's start by checking information out.

Information, Messages, Arrangement

	tam/ raj	*Sat*
	object/ subject	*Transcendence*
	matter/ mind	*Quiescent Potential*
	informed/ informant	*Information's Source*
	order	*Preordination*
↓	*tam*	*raj* ↑
	passive information	*active information*
	informed	*informant*
	object	*subject*
	non-conscious energy	*conscious mind*
	non-teleological	*teleological*
	purposeless	*purposeful*
	pointless structure	*functional structure*
	chance	*guidance*
	necessity/ natural law	*design*
	physico-chemical 'system'	*(bio-)technological system*

You may say there's only matter.

I say there are material *and* immaterial elements.

If more than physic does exist then what is metaphysical? **Natural Dialectic's immaterial element is information**. Information and communication! This is life - craved news and its buzz. *And we're embedded in an 'information age', an Age of Metaphysic!* **Why, though, should psychology obey the laws of physics?** The rules of each are mutually exclusive; to apply those of one to the other, as does materialism, creates only confusion. For example, mind knows purpose, creates order and can choose; it informatively signals and communicates in code. Thus you can't evaluate it with the mind-set of a physicist. Yet logic, reason and mathematics are informative exemplars that support all science. Of what, therefore, is metaphysic's scientific 'stuff' composed?

***Top-down**, signal information is the gift of mind and never matter.* **It is irreducible to scientific scrutiny. Yet, as we'll see, its semiotic metaphysic dominates debate about creation and our lives.** Pick up a postcard, menu, letter - anything informing you. The object is reducible to chemistry and physics but the *sign* or *message* it conveys is not. Signs and signals always have a purpose; objects never do. In other words, information's fundamental to our being but does not fall, in a semantic sense, within the scientific remit. *Particularly, mindless origin of bio-code is an irrational hypothesis.*

***Bottom-up**, however, everything is seen as energetic interactions. Energy's the physical informant.* Information's therefore, even in the case of brain, evolved by chance and natural law. Oblivious physic generates its off-chance. Is such 'metaphysic' an illusion, a confusion or the truth?

Firstly, in this dilemma of perspective, let's distinguish mundane from scientific/ materialistic understanding. *In mathematics of the latter sense 'information' must <u>not</u> be confused with 'meaning'.* So what exactly, in this apparent shortfall, *does* the word mean?

It was Claude Shannon who, with Warren Weaver, first devised a mathematical and thereby scientifically acceptable definition of information. Shannon treated its transmission in purely physical terms according to statistical formulation of the entropy-inclusive laws of thermodynamics. In such transactions his unit was the *bi*nary digi*t*. This on/ off, one-zero 'bit' allowed the quantitative properties of strings of symbols to be formulated. His theory inversely relates information and uncertainty. The more uncertain, the less probable a sequence of symbols or arrangement of materials the more information it is calculated to contain. Rephrased, an amount of information is inversely proportional to the probability of its occurrence by chance. Simple, repetitive or predictable sequences contain less and complex, irregular arrangements a greater quantity of information. Thus 'Shannon information' is a measure of improbability. *It makes no judgement whether such irregularity is specified; it involves no sense of meaning.*

Shannon's definition is suitable for describing statistical aspects of information such as quantitative aspects of language that depend on frequencies (such as how many times the letter 'a' or the word 'and' occurs) but it treats any random sequence of symbols as information without regard to its concept, meaning or purpose. In other words, the more improbable any arrangement the richer is its Shannon-defined information content. *In short, two messages, one meaningful and the other nonsense, can be exactly equivalent according to this form of analysis.* For example, ZNQW&RSIXT AZ2HVB and COMPREHENSIBILITY are assigned the same value. In other words, Shannon has reduced information to a statistical quantity; he has shaved off any sense of meaning, cut out sense of purpose in numerical analysis. *Shannon-shaving deals in non-purposive complexity.* It might make statistical sense but definitely linguistic and rational nonsense. In this assessment gibberish can have as at least as much 'informative value' as sense. The real nature of immaterial information, its quality of meaning, is ignored. So is the nature of its source. How can Shannon-information therefore serve as useful metric of what information actually conveys?

It is important to grasp how Shannon, thoroughly negating logical reality, accords randomness and purpose equal status. **But randomness is really reason-in-reverse; it is information's opposite**. So when it comes to language or to mechanism, including those embodying biology, such conflation is an error of first order. Shannon's analysis does not distinguish between presence of mind, authorship or creativity and their absence; it fails to recognise purposive specificity; nor does it accord function or meaning any premium. Thus order and precise meaning - usually complex, always accurately coded - might as well be able to emerge from randomness or senseless motion under natural laws of physics. You can thus deceive yourself and think in topsy-turvy terms of 'senseless design', an anti-teleology well-known as Darwinian evolution. Such self-deception works well because, like an ups-a-daisy religion, its adherents believe it. It infects, for sure, the mind-set of an academic discipline throughout the world.

Secondly, in *top-down* view, let's note that there exist two main categories of information - active/ informant and passive/ informed. The *active* kind is subjective, immaterial and involves knowledge, specificity and purpose. It thus has any meaning only in the context of a knower and is essentially a psychological phenomenon. You may, of course, find *passive* information out of mind. Don't principles inform the way material nature works; don't designs inform the way creations are developed and arranged? Materials can carry information pressed upon them and such passive sequence can be formulated mathematically. Ink and paper, clay or iron can be arranged to carry code and such code-carrying capacity be analysed statistically. For codes and materials *per se*, of course, neither mind nor meaning can exist; bare of meaning, void of purpose their passivity yields 'Shannon information' to such minds as measure bits and pieces. All information needs a mind to recognise it but the issue of intent rests only in what starts a reasonable system up. Was original information 'specified' by chance or not? Is ever a symbolic language (called a code) established unintentionally? At least, the way things are arranged, informed by natural forces or by men, needs mind to recognise a *modus operandi* and develop any system's logic and its rules.

Information is Immaterial

Top-down, **information is an immaterial element in its own right. Why? It is nothing without a mind.** Mind is active and commits ideas into material orders that are called, in the Dialectic, passive information. Mozart's mind, through instruments, informs material sound. Is not the real heart of energy through air you hear as music therefore immaterial? Is its primary arrangement psychological or physical? Is it physical or metaphysical?

Buy Mozart's music! CDs and floppy-discs are materials carrying information. Have you ever weighed one accurately then rubbed it clean? Deleted all the information? What, when you reweighed it, was that information's mass? Zero? None at all? Do the same to your computer. The essence of computers is their programs and their database. Delete this essence and peripheral hardware weighs the same. Because informative capacity inhabits the *arrangement* of material - not just any old arrangement but one with meaning that serves purpose or accords with principle. Purpose weighs no more than understanding; both weigh just as much as meaning. In the scientific balance meaning weighs as much as abstract theory. Each is lighter than a feather. Weigh every thought in every mind that ever was; weigh universal mind itself. Such aggregation measures zero grams and so the scales tell nothing; nor do the finest registers of time and space as much as twitch. If, though, information's absent massively, what is it? If immaterial then, as this volume shows, it renders atheism (but not science) most illogical! Indeed, Norbert Wiener, mathematician and founder of cybernetics and information theory, said *"Information is information, neither matter nor energy. Any materialism which disregards this will not survive one day"*.

There is, furthermore, known neither law nor process nor sequence of events through which oblivious matter can create or collect information. The latter is not a property of matter. Purely material processes, unguided except by natural law, are fundamentally precluded as *sources* of information. Information is not a thing itself but a representation of physical things and metaphysical entities. Nature wears no numbers, bodies are devoid of words. Symbol, code, idea, conceptual relationships or interests - all are immaterial, none exists without a mind. In fact, data has no meaning outside the purpose of its maker or its user, mind; and information may be defined as the arrangement of material (including voice or ink) according to plan. You might think of information, as did Wiener and does Natural Dialectic, as a 'non-physical form of anti-entropy'. What is not physical is metaphysical. Wiener's 'negentropy' is metaphysical and maybe thought of as a property of mind.

Information is Hierarchical.

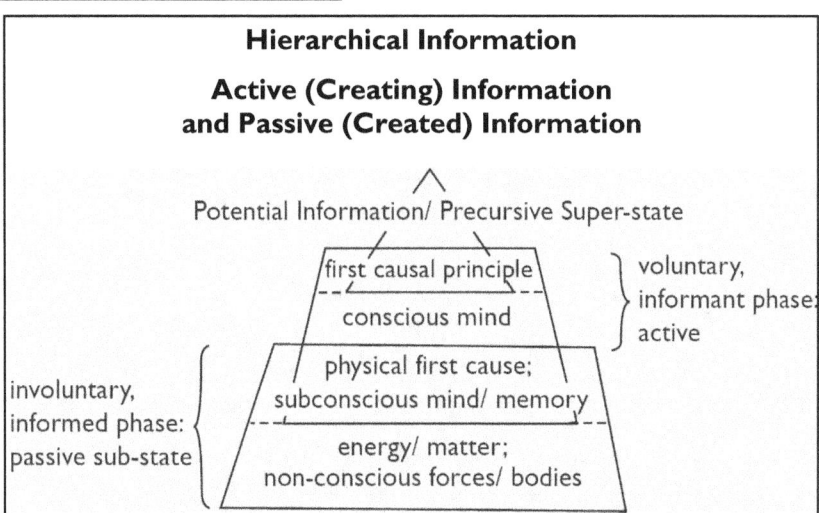

Information is phrased in terms of awareness. **Of the major cosmic sub-divisions mind is termed active information. Although it may exhibit highly active behaviours matter is, with respect to information, non-conscious; it is passive, automated and in this respect creatively impotent.**

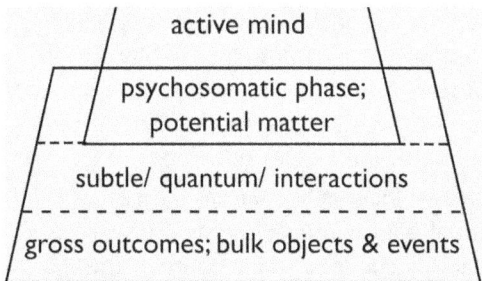

Clear distinction needs be drawn between informative mind and informed matter

There also exists, however, an active, informant phase of matter - energy; and a passive, informed phase of mind - a sub-division called sub-consciousness. Sub-consciousness is mind's storage facility. *Memory is the natural recorder of finished events in active, conscious mind.* This lowest level of mind also constitutes, in dialectical terms, the point of psychosomatic linkage with matter. *In other words, passive mind in the form of memory becomes the transcendent, causal principle that orders matter.* This hierarchical view of memory is not standard, is not materialistic but, we shall see, fits well with psychological, biological and physical facts.

Opposing Hierarchies

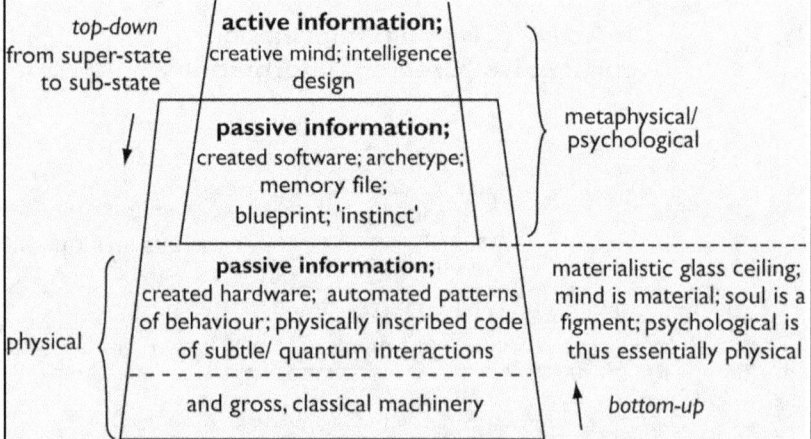

Top-down and bottom-up views of the origin of information are here juxtaposed. *Top-down,* information originates in active mind; *bottom-up* it starts with passive matter. Since information is the source and sustenance of life which view makes better sense?

In the *bottom-up* view passive leads to active information. Matter/ energy emerges from nothing or is, perhaps, eternal. This matter fortuitously, in a process called abiogenesis, creates a cell coded by *DNA*; such a physical carrier of code itself generates information; and such informative though random chemistry somehow eventually makes consciousness. Intelligence. *Hardware creates software.*

The *top-down* view is absolutely the reverse. Information is the immaterial gift of mind. Just as software is composed of non-material instructions that control the hardware, so a distinct immaterial element, conscious mind, designs software instructions (stored in the files of memory as archetypes). Such metaphysical programs may be reflected in material carriers. Similarly, sub-atomic particles can be conceived as carriers of archetypal code;[22] the behaviour of non-conscious substance is thus metaphysically controlled. *Software defines hardware.*

---

[22] Chapter 11: Matter's Holy Ghost.

Finally, within existence active information involves awareness: passive information does not. Passive information is fixed, automatic, reflex. The domains of its storage are subconscious mind and non-conscious matter.

Reflection of the information gradient from active, inner to passive, outer phase

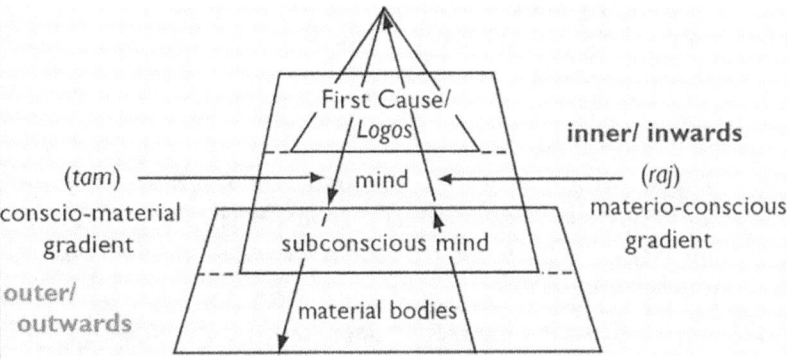

This is a key illustration. Its ziggurat anticipates figs. 6.1, 11.1 (two kinds of First Cause) and 17.4; and it illustrates the dynamic of creation's conscio-material gradient from inner, immaterial and active through to outer, material and passive phase. It drops from Clear Formlessness at peak through psychological and physical forms of various quality. The latter are trapped in reflex, oblivious cycles; the former involve, at least for humans, a measure of choice. You can easily understand why the sage turns his contemplative focus inwards towards unification with First Cause (see later in this Chapter: Top Teleology); and a scientist outwards to the details and, in principle, unification of an essentially non-conscious universe.

Source, before even the primal motion of First Cause, *is* a concentrate or super-state of consciousness; and consciousness is, by nature negentropic and thus its 'creation express' develops orderly. It devolves down to oblivious physic within whose context chance may play a part.

Between the concentrates of consciousness and non-consciousness (First Cause and material bodies) each vector of a 'polar dynamic' has its part to play. **Expressions of (*tam*) gravity of materialisation and (*raj*) levity of dissolution oscillate in all psychological and physical events. Loss or gain of energy or information describes change in the universe.**

fig. 5.1

below	*Top*
subsequent order	*Archetype*
its excitation	*Immaterial Field*
its expression	*Informative Potential*
consequences	*Initiator*
↓ *outer/ external*	*inner/ internal* ↑
involuntary effect	*voluntary cause*
gross	*subtle*
informed	*informant*
passive	*active*
static/ finished	*kinetic/ in process*
passive information	*active information*
sleep/ sub-consciousness	*wakefulness*
memory/ 'solid' mind	*kinetic mind/ mind-in-flux*
expression in code	*symbolic creativity*

Top-down, cosmic hierarchy shouldn't be a mystery now! Covalency of mind bonds top to bottom; it binds the Apex of Mount Universe to sub-state base-camp, matter; it ranges in between a Conscious Super-State and total lack of it.

And in this case of lack a gradient also falls. In the physical hierarchy principle derives from potential matter; the automated behaviours of energy in all its forms are patterns regulated by an archetype. From such archetypal framework polar vectors issue in the form of (↑) 'energetic levity' and (↓) 'inertial gravity'. The former is dominant at sub-atomic, quantum level; and from quantum matter-in-principle derives matter-in-practice, that is, all the macroscopic, massive entities that we sense and call our normal, gravity-dominated world. This 'classical' normality, which scales from gas to solid, is just the outer layer of a cosmic onion.

Bottom-up, materialistic science probes 'inward' from our normal, sensible but outer layer. We discover 'quantum strangeness' which, on Natural Dialectic's scale, would show as an approach towards potential matter. Thus, as Pythagoras suggested, basic physics leads us to a borderland - the frontier where material and immaterial (mind and matter) meet. Mathematics deals with what we call the cause of natural law; the abstraction of glass ceiling[23] is a metaphysical, informant structure known as archetype.

Orderly Creation

Laws order. Information and order are closely allied.

From a *bottom-up* point of view natural order is paramount. It is described in terms of material laws of nature whose constraints are also

---

[23] see also Chapter 7: Precondition/ Potential.

dubbed 'necessity'. Conscious mind and its species of order is, though powerful, simply a contingency. It is an evolutionary 'extra' whose recent, unexplained emergence on the scene is cosmically irrelevant.

A deduction that omits mind from the world equation until the very least and last precisely reverses the *top-down*, dialectical order of information. The latter's order of creation runs, first and foremost, from mind. Things start in mind. The order runs from psychological to physical, from active to passive information, from creativity to its creation. ***Top-down* hierarchy emphasises an order of information that runs from active (mind) to passive (matter).**

Psychological is linked, as we've just seen, to physical expression of Cosmic Dialectic through a psychosomatic (mind/body) phase we know as the residence of 'fixed thought' - subconscious memory. Therefore, where memory is the natural, paperless retainer of ideas, it is also the depository of 'filed plan' called archetype. Potential is an unexpressed capacity. Archetype is energy's precursor; it is the 'template' or the 'field' that, activated, will accordingly express orderly mental or material creation's play. Potential in this sense is causal. Archetypes *cause* orderly succession; what succeeds or 'comes below' is dependent on their regulation. And, simply put, the Archetype of mind's psychology is both informative potential and its activator rolled in one. We call this 'primal action' *Logos*.

The God Delusion?

Do you remember that for materialism the deepest of delusions is a Live Creator?

An atheist states that there is only matter.

This, remember also, is an axiomatic bluff. Matter *maketh* consciousness. His arguments all start from this presumption, one unproven scientifically. The pillar of such faith starts with a leap of faith; the first step of its first leg is a guess. *Could this be the devil of a delusion*?

Neither proven is my own assertion. Science does not deal with immaterial factors but who knows how 'material' might be reclassified? Were electrons 'immaterial' before they hove in sight? What you don't know or don't yet deal with can exist. And consciousness is labelled, in the Dialectic's book, as a 'subjective immaterial'. Is this another axiomatic bluff? Let not, at the first step, pot call kettle black. *Top-down, bottom-up* - is either way of looking wholly right?

When it comes to life the issue is acute. Atheism elevates its own oblivious 'designer' - natural selection of coincidence. What motive, though, has a molecular complexity to mindlessly 'survive'? Yet the irrational 'value' that, supposedly, selects a form of life is its survival. Such bio-designs look just 'as if' they were designed! The chance is very slim but, you can clearly see all round, unconscious matter must have made cells on its own. It even threw up mind! But wait! Where mind's involved you don't measure using the effects of chance. It's inappropriate. This time the sleight of atheism's hand is not appeal to unseen transformations or to potent swathes of time but to mathematics. In this case the idea of probability is applied statistically; it measures the effect of chance occurrence in a world of reflex, physical events, that is, of material science.

Could cosmos actually have made your teacup in a thousand billion years? It must have so impossibility's a possibility! But argument for bio-logical design is not decided, any more than is an engineer's, by probability. As archaeologists interpret evidence of artefact, interpretation of design is drawn from evidence. Chance is not the issue. Programs, machines, art and artefact all have creators. Not chance but anti-chance has entered into the equation. Informant mind, although undetectable by instrument, abhors randomness and espouses purpose; its business is certainty and, as such, it would annihilate all chance. It renders such an instrument of argument irrelevant. *The atheistic chain of reason lacks in self-consistency.* How can interpretation of design or a designer be logically irrational?

Next, no universal law of increasing, specified complexity exists and yet such specificity is what biology is all about. **No 'bio-force' exists in lifeless matter that is able to develop instrumental systems, organs or metabolisms.** The fact is that, without exception, intelligence is needed to inform design.

"If it could be demonstrated", wrote Charles Darwin, "that any complex organ existed which could not possibly have been formed by numerous, successive, slight modifications, my theory would absolutely break down."

The fact is everywhere present that this is not the case and that, indeed, the theory is a broken one. Minor variations-on-a-theme occur but the fact is that organs, systems and bodies always show a coherent complexity of design that derives, as in the case of all machines, from a purposeful arrangement of specific parts; and this distinct arrangement is dependent on a language written on a *DNA*-made database of such digital, computer-like complexity as (information and computer theories indicate) chance/ necessity could never generate. In life's *language* genes,[24] *DNA*-written chemical 'hardware' that carries fundamental bio-information, are digitally coded text; such helical scrolls operate according to the grammar of symbolic code; and such language and its agents of expression are conceived *in order to* transmit a message optimally - something senseless matter never 'thought'

---

[24] Chapters 20 and 23.

about! *Indeed, every code contains agreed morphology (the form its symbols take) and significance (what they stand for or they mean) between its sender and recipient. Such prior agreement always needs intelligence.* This is as true of 4-letter nucleic acid translation into 20-letter protein language as of one spoken language to another. **There's no 'code analogy'; genetic code *is* code!**

Finally, for now, we might resist the notion randomness constrained by death 'invents' all forms of life. Can a sieve of death, Edward Blyth's original idea of natural selection,[25] 'design' the systems some suppose that it can incrementally perchance evolve? Might 'survival-of-the-fittest' really innovate? Neither catchphrase conjures 'teleology' but each is a tautology. Who or what survives survives. They sum blandly to assertion that any particular accidental change (called an adaptation or modification) either does or does not cause death. **But modification by accidental bump or grind says nothing about the origin of vehicle systems or a vehicle as a whole.** Although a vehicle in constant use is always subject to accidents and failures of its parts such wear, tear and mishaps never made it. **This interpretation of events takes Darwinism at face value as an explanation of accidental variations, trivial changes to main structure and a theory for the origin of minor novelty, including species.**

At this rate evolution (as opposed to biological variation) is viewed in the light of a logical fallacy, a trick called equivocation. **This semantic trick is sustained by the conflation of two entirely different matters - firstly, variation on existing features and, secondly, the naturalistic innovation of complex, fresh features (such as coherent organs, systems and so on) in the first place.**

If chance can't manage things materially what can? If rules can whence, denying entropy, came rules for building codified complexity? Whose delusion is proscribed?

ID, IC, IE - Call it What You Will

Design, creativity and engineering all require intelligence. Yet intelligent materialists reject immaterial intelligence in the very nature they inhabit! Intelligent Design, it's claimed, fouls up the scientific game. The latter is a methodology for physically testing guesses and thus winning, in the end, a truth - truth physical. Since Darwinism is its myth of mindless origin metaphysical ID must be, perforce, a powerful delusion. Metaphysic is proscribed; material hypothesis alone can be allowed. Thus - despite routine employment of its principles by archaeologists, palaeontologists, cryptologists and *SETI* astronomers - ID 'isn't science' and, on that decision, should be firmly binned!

---

[25] Chapter 22: The Editor.

Construed in this restrictive mode, is science really a sufficient vehicle to approach whole truth? After all, according to which unit will you judge a thought's relation to a nervous twitch? How will you measure ingenuity? What test will you apply to innovation, at what notch balance human logic or 'designs' of nature on your scales? The fact is, also, that experimental method can't be used for either one-off past events or unfathomable future possibilities. You can only best-guess using some statistics or draw inference according to your inclination. Then, if you've philosophically reduced your scope of inference to chance and uninventive natural law, your logic *must* produce an evolutionary outcome. There is no other way. Design must be 'design'. **However, the best inference for generation of information and complex, specified design may well be mind.**

Randomness, replacing mind, is naturalism's staple. By chance[26] the universe, by chance life, mind and information! Chance creates new circumstance but, by definition, unintentionally, irrationally and unpredictably. Naturalism is Fortuna's Faith. Any outcome can be fancifully proposed but none for sure.

Mind is creative. You can't predict the unexpected innovations of an artist or an engineer but *can* predict, by definition, the character of purpose, reason and, from immaterial ideas, logically expected outcome.[27]

One claims chance, the other immateriality. *In fact, neither a theory of historical origins based on the inference of chance nor one based on informative competence is testable and thereby falsifiable by naturalistic methodology.* **In this respect neither IE nor Darwin's theory of evolution is scientific. <u>Both should be labelled pseudoscientific</u>.**

But labels don't decide reality. Scientific or unscientific, did serendipity or logic generate our world? Instruments can't measure supernatural creativity - but we *can* infer creation by intelligent designers, ones called engineers. **The issue is not one of identifying a particular designer but detecting such criteria as spell design or lack of it.** You might, for example, expect to find in a design purpose, rapid infusion of information that elaborates a concept and transforms it into a functional system, complex integrated parts, possible use of symbolic code (such as facilitate on-off or more complex switching systems) and the re-use of functional units (or mosaic variations on their theme) in different and physically unconnected systems. *You find all these substantiating what we call biology.* **In fact, the priority of IE is, as it involves the immaterial component of technology, information.**

---

[26] Chapter 7 *passim*.
[27] Chapters 20 to 25.

From the detection of such factors one might predict biological discoveries and, thereby, infer a generator of innovative information. This, we've seen and will see, cannot be rationally identified as chance. So why black-mark, cane and then expel the rational alternative? In what kind of school does 'scientific reason' thus irrationally master us?

IE/ ID *predictions* could be confirmed by the following discoveries:

1. intricacy, coordination, complexity of coherence, specificity and an irreducible number and arrangement of parts per functional bio-mechanism
2. reasons *why* specific, predetermined parts are co-assembled, that is, *why* as opposed to how they self-organise in cybernetic order to perform a task
3. in accordance with conceptual biology (Chapter 19) a convergence or mosaic appearance of modular parts
4. the discovery that previously misinterpreted components are not in fact vestigial junk but structurally and/or functionally necessary
5. that a functionally-integrated system involves the complex operation of language, that is, accurate symbolic code
6. that, if discrete systems are designed (albeit with intrinsic, predetermined adaptive potential to flexibly respond to circumstance and thus retain the balance to survive), you would also expect to find discrete types of organism
7. accompanying such discreteness a case of fossil 'abruption', that is, an historical lack of billions of graduated missing links.

The presence of all such factors in life on earth might lead a rational man to infer its design. And, based on this perspective, propose such research fields as the computer-like properties of *DNA*, non-protein-coding (n-p-c) genomic function, front-loaded adaptive potential as regards both environmental adaptation and development, the practicalities of applied protein engineering and so on. A reductionist is, on the other hand, forced to identify his 'designer' as chance a-dance with death; and to narrate hypothetical stories and suggest research objectives accordingly. Why abort an IE/ ID perspective and thus only allow issue of the latter world-view to thrive?

Yet strange logic runs amok. Groups lobby parliaments and make petition to Supreme High Courts. Should they find in favour of 'design' or Design? If such a Court wishes to rule on a case of 'unconstitutional violation of principle of the separation of church and state' which church should it choose? Should it admit each species of faith, neither or simply an opinionated 'chosen path'? Which path? Which guess? Today the purveyance of evolutionary theory is institutionalised; but does a belief in such naturalistic authority create atheists as logically as a belief in design makes theists? If theistic realism were denied and the atheistic position permitted to partake a monopoly of 'legal protection' then, as Harvard graduate, law clerk in the US Supreme Court and long-time legal academic at UCLA, Phillip Johnson, writes, 'the Supreme Court will in effect have established a national religion in the name of First Amendment freedoms'.

It's not only Uncle Sam! Nor only judges but our European politicians also seem to know it all! This Goliath totally ignores (or never learned) that the Renaissance, progenitor of modern scientific method, derived enlightened sense of order from a universe intelligibly created by a Most Intelligent Creator. In 2007, at once dismissive of its founder scholars and the strength of argument from inference of 'intelligent design', the European Union's Committee on Culture, Science and Education proposed to the Parliamentary Assembly of the Council of Europe that 'creationism' (a carelessly-defined, pejoratively-used word) should be outlawed. It is as if such bureaucracy, examining the intelligent design of a motor car, banned mention of its inventor from science classes in its federation's schools! How could you argue from interpretation of the facts for that inventor when, by philosophical decree, you are bound not to?

British governmental creed may well be seriously infected too. Scientism[28] is not science. It is a world-view dictating that only materialistic answers to any question be considered. Science and the Soul demonstrates the nonsense of such philosophical *diktat* which is, essentially, an affront to scientific free-thinking. This is because, as an extreme that excludes open-minded holism, it accedes to an unarguable imposition of materialistic interpretation of data concerning historical origins. 'Forget an immaterial element of information; this does not exist!' it cries. Thus legal imposition due to pressure from a fashionable lobby would reduce abductive scientific query[29] to atheistic consideration alone; and thereby imply that this alone is scientifically/ naturalistically correct! By now you see Design is not a pretty theory all alone. All sorts of people have designs on it - some nice and others not so nice. Whatever else, it cannot but affect their mental health.

Alignment with Truth

The immaterial element, information, involves several kinds of quality.[30] We want to be informed (and thus inform) as truly as we can.

---

[28] see Glossary.
[29] see Glossary: logic.
[30] These kinds of quality include truth and error (Chapters 4 and 14: Quality of Mind), importance and triviality (Chapter 14: Quality of Information), intelligence/ discernment (see Index), a moral dimension of good and evil (Chapters 4 and 26), colours, gears or states of consciousness (Chapters 14 and 15) and Alignment with Truth (Chapters 4, 5, 14, 18, 26 and 27).

Thus, what kind of messages do you rate highly? With which grade on truth's scale do you align?

Reason is a serial, problem-solving path by which an answer is obtained; it is the instrument of ego calculating how to satisfy desire; and, in its third and highest form, it seeks to disentangle truth from error and discern the kind of life best lived. Such discrimination is not scientific. **Science cannot offer moral value, meaning or transcendent faith to lift a soul beyond its body-world.** Indeed, the focus of its love is for this world. What, therefore, is the unscientific way that humans work out what is good and bad? Is there a rational way to seek and find the Highest Good? In what does 'progress' in this sense consist?

	existence	*Essence*
	practices/ principles	*First Principle*
	complexity	*Simplicity*
	concentric circles	*Centre*
	oscillation	*Axis/ Pivot*
	lesser reasons	*Supra-Reason*
	relative truths	**Truth**
↓	*from Truth/ Principle*	*towards Truth/ Principle* ↑
	towards complexity	*towards simplicity*
	centrifugal dispersal	*centripetal focus*
	executive focus	*contemplative focus*
	dispersal/ diffusion	*concentration*
	sub-reason/ instinct	*reason*
	irrational	*rational*
	unthinking/ emotional	*intellectual*
	attachment to sensation/ physical objects	*detachment from sensation/ attachment to generalities*
	action	*understanding*
	outward practice	*inner learning*
	motor system	*sensory/ perceptive systems*
	material progress	*immaterial improvement*

Natural Dialectic's terms are simple to divine. The vector that increasingly aligns with truth will carry you (↑) beyond the mind. Enlightenment reveals that Reason transcends reason. Call it, therefore, Supra-Reason. Sub-reason (comprising instinct, unpredictable emotions and irrational behaviour) is not the same as Supra-reason. The latter is, although transcending intellectual explanations, not irrational. Quite the reverse. Which way of reason, therefore, might exceed the intellect? What might excel pure logic?

A reason is be-cause; truth secretes itself in cause; what, therefore, might you reason is the nature of First Cause? What is the truth of origin, of where you come from? Such cause happened in the past. While science well defines contemporary operations of the universe, the latter's origins can only be inferred. The same is true of life. Thus abduction by best explanation wins the day. Some prefer materialistic explanations. Others would infer a 'hidden invariant'[31] whose metaphysic, information, finds its source in mind; and if the root of mind is consciousness and pure consciousness is the essential component of creation then the Delphic exhortation 'know thyself' dives deep and clear. **With which sort of explanation do you philosophically align?**

(*Sat*) Potential or Transcendent Information

From a *bottom-up* perspective, lacking concept of a conscio-material gradient or Central Origin, 'Transcendent Information' is delusory nonsense. An intelligent modernist, having adopted 'The Theory of No Intelligence', claims such thinking is relatively unintelligent and, certainly, absolutely irrelevant. Was not the world's First Cause probably a 'big bang'?

What of your personal origin? It lies, of course, in the soil of your ancestors. Not only in passage through the churchyard's graves but also a reversal through parental apes and fossilised uncles to a primordial speck of mucilage at the vital moment - abiogenesis.[32]

	existence/ nature	*Essential Nature*
	matter/ mind	*Potential*
	its peripheral designs	***Information Centre***
	relative levels of truth	*Central Truth*
	conscio-material spectrum	*Pure Consciousness*
	breakage into creation	*Ultimate Symmetry*
↓	*passive information*	*active information* ↑
	matter/ fixed mind	*experience*
	non-conscious/ subconscious patterns	*conscious thoughts*
	storage or automatic routine	*reason/ purpose*
	resultant creation	*creativity*

Top-down, a holist agrees our physical roots are human ancestors but adds metaphysical dimension. Do you remember[33] that equilibration is a cosmic operation; and that, in the flux of mind and matter, we continually seek rest beyond the storm? Equanimity and balance are the goal. Peace is our truth. How is it most radically achieved? Could there be, centre of both gravity and levity, a cosmic pivot that informs the mobile universe? Is there a transcendent balance round which cosmos swings?

---

[31] see also Chapter 7: Are You Certain?
[32] see the book A Mutant Ape?; also Chapters 20 and 21.
[33] Chapter 1 *esp. fig.* 1.1.

informed/ informant	*Wholly Informant*
particular expressions	*Full Potential/ All Possibilities*
principalities	*Principal*
scale of lesser truths	*Truth*
concentric radiation	*Axis*
body/ mind	*Soul*
↓ *wholly informed*	*informant* ↑
passive/ automatic	*active/ influential*
material phenomena	*mental phenomena*
physical consequence	*psychological initiation*
(world) body	*(world) mind*

Such Essential, Axial Pivot, prior to motion, might appear no *earthly* use yet, paradoxically, its Immaterial Potential sources everything that moves. It informs existence. What, therefore, is the Brilliant Nature of Transcendence with which top teleologists align? Is it a function (called *Sat-Chit-Anand*) of Pure Consciousness and Bliss? What Truth do they profess?

Top Teleology

It is not so much a case of 'God delusion' as of 'existential illusion'. Theatre is projected out of mind. Jacques' monologue (Shakespeare: As You Like It Act II Scene II 'All the world's a stage…') well reflects our Contents Box (front matter, p.3). *Top-down*'s Truth embraces immateriality. Its Reality is Pure Immateriality, Essential Source, Producer and Director of the operatic cinema we call 'Existence'.

lesser being	*Supreme Being*
parts	*Whole*
duality/ polarity	*Unity*
sound	*Pregnant Silence*
range of colour	*White Translucence*
expression/ creation	*Potential*
relativity	*Absolution*
practice/ sub-principles	*First Principle*
↓ *passive*	*active* ↑
detailed expression	*sub-principle*
informed	*informant*
its results	*vibratory information*
consequence/ final form	*cause*
material automaticity	*cosmic levels*
black sink	*range of colour*

If Supreme Being pre-exists any and everything then it simply *is*. This is why it is called 'I am that I am'; and why it is Essence (or Being) without predicate, quality or condition. One being one. If Unity is what you crave, you will have found its place; if *Nirvana* or 'The Peace that passeth understanding', here's its place. Is not Top Teleology's enlightenment the root of all faiths and, therefore, a believer's scholarship? Its Subject is straightway recognisable as the major theme to which humanity has, in various more or less canonised forms, always adhered - except in the materialistic grain.

Will commands and a command's an order. A command initiates the action. Authorisation is causal. It says 'Arrange things as I want'. Such is the teleological order of *Kalma* or *Hukm* (Command, as the Muslims call it), *Shabda* (Sound or *Bani*, as the Sikhs say), Zoroastrian *Sraosha*, Taoist *Tao* or Hindu *Nad* and *Om*.

Om is, in Christian terminology, Amen. Or, again in Christian terms, Byzantine *Christ Pancrator* is identified as *Logos* (the Word of Reason). Intent in nature; nature's teleology. Thus, of course, from *Pancrator* ('all-powerful') to *Pancreator* is no step; power psychological includes the attribute of creativity. *Logos* is a Greek noun meaning 'reason' or 'the words expressing it'. Speech, message, code. And also 'voice'. *A word has meaning and its purpose is communication. Information. Instruction. Command.* Call *Logos* The Informant.

Words are expressed with vibratory sound and imply a grammatical structure within which to impart an inner plan or thought or reason. This arrangement, *logic* or convention must exist before any message can be issued or understood. Could there exist a natural language? And if so, in what terms is it relayed? A second meaning is 'account' or 'calculation' and a third is 'the order of self-consistent narrative - speech, story or theory'. In the Greek language *logos* can be construed in a mundane sense; it is also found elevated to a cosmic level. Here *Logos* is seen as the source of order or plan. For a Stoic it described the principle of reason that impregnates the world. Nor only in the Christian faith is Name of God First Rational Principle; Word, its synonym, is the Source of subsequent principles, laws or reasons that together constitute the vibrant grammar of the world. What more Natural than nature's spring? Such transcendent origin of information is a flow of 'speech', a current of instruction, an 'oration' from Top Teleology. **Who would, therefore, by denying *Logos* be an enemy of Reason and thence natural reason?**

If *Logos* is an outward current of creation what about its circulation? What of upward, inward passage back to Source, Creator, Topmost Teleology? **Such inverse return is, say all saints and mystics, to unify duality, transcend polarity and come to rest in Central Essence.**

Chapter 6: Mind, Machines and Mind Machines

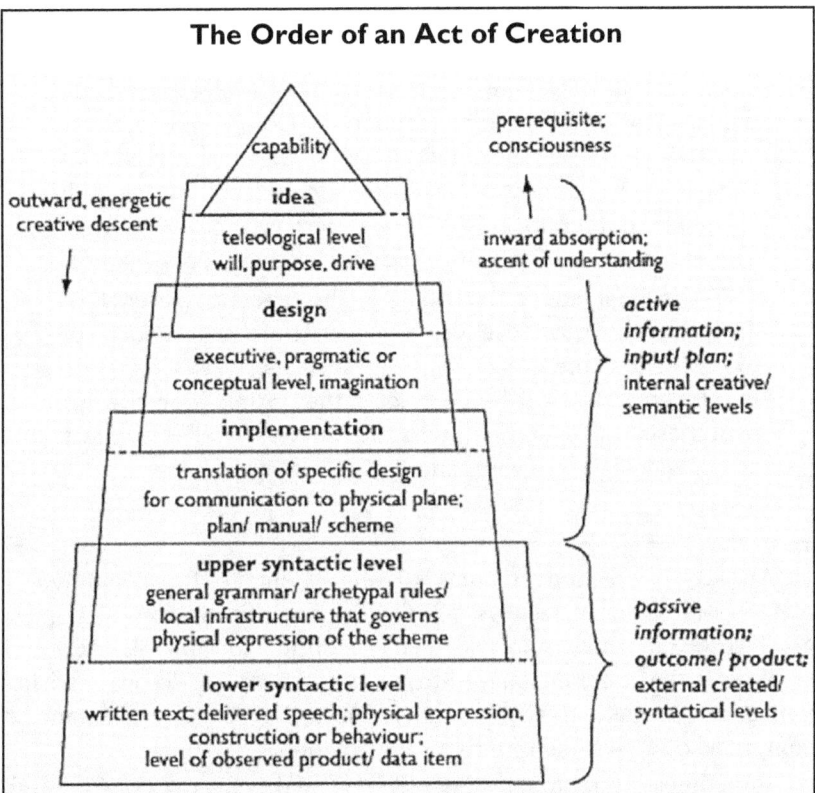

In this hierarchy levels are conceived of as nested but co-existent. A creative act develops, organically, like a seed to flower. The act itself 'descends' from metaphysical to physical levels - from mind to body. Such 'descent' may occur very rapidly as, for example, in a conversation; or more slowly in, say, working out the detailed solution to a technological problem. Acts of creation range from trivial through important to cosmic in scope.

Conversations, solutions to problems and so on also involve 'ascent'. They involve an act of 'counter-creation', comprehension or understanding of what is going on. This applies with full force at top, teleological level where principle, motivation and purpose are grasped. When you have got the idea you have the reason.

Not all stages are necessarily engaged. For example, a person ordered in a military way to do something scarcely enters the top three levels; instinctual and reflex information

> loops are passive. Indeed, the base 'lower syntactical level' is that of our passive, physical universe. Its knock-on causation is called *'horizontal'* as opposed to active, *'vertical' causation'* down from mind to matter (see also *fig. 2.5*).
>
> Have you ever 'got the idea'? Grasped what someone or something actually meant? What about the Cosmic Idea? Since, in this view, the Origin of cosmos is alive then a full comprehension of the Idea amounts to Communion with its 'thinker' - Pure Subjectivity. Such Essential, *Top-Down* Subjectivity's Perspective, transcendent yet commanding all existence, is therefore The Very Purest Objectivity as well. The two are one. Knowledge (Science) is their Super-Nature.
>
> Although irrational according to the tenets of materialism Illuminative Knowledge is, of course, the product of experimental technique. The mystic, as was Christ or Buddha, meditates. If he completes the cosmic loop and his hypothesis is proven right, what a result! What, Pascal might add, a winning bet, a vindication!
>
> **fig. 6.1**

An act of creation amounts to phased intent. It is, simply, the materialisation of an idea. A simple idea is, in execution, elaborated. You work out what to do and then, according to your plan, do it. If creation is the re-arrangement of the world according to one's wishes, you and I remake it all the time. The major phases of intent are comprised of active and of passive information.

Sometimes, no doubt, interactions and constructions ramify. Complexity occurs; and of complexities there are, at root, two kinds. The first, <u>non-purposive</u>, is a function of energy transformations. It involves the confinement of energy (e.g. in sub-atomic particles) or loss/ gain of energy in changes of state (e.g. from formless gas to such differentiated variation-on-theme as snowflakes show or *vice versa*). In aggregation or in action nature's physic is mindless. Its *complexity* (whatever its beauty in the subjective eye of a beholder) is <u>passively derived</u> from the automatic behaviours of energy informed by rigid 'rules'. In this sense matter is a 'no-possibility zone'; having no freedom of will it is lifeless and automatic. Scientific research and application focuses on this indiscriminate kind of complexity.

The question is whether you can infer a purpose behind the 'design' of apparently non-purposeful complexity. Could the universe be, by anti-chance not chance, a product of design? Could it be both wonderfully and purposefully made? The response from scientific materialism is that 'there's no necessity to think that way because it's all necessity - the name for natural law'. Its answer is, in other words, a resounding, study-stopping 'no chance, since it's all (including the origin of natural law) by chance'.

Where physical possibilities are mathematically predictable, psychological ones are more flexible. The second, *purposive* kind of complexity works the other way. It is a function of information gain, expansion of consciousness and a capacity to grasp and purposely, creatively exploit the principles and possibilities inherent in any circumstance. An increasingly concentrated focus of attention wakens to greater capacity, flexibility and possibilities for specifically ordered, coherent or *active complexity*. At each level of ascent the degree of coherence may improve; the degree of ingenuity, adaptability and innovative complexity may increase. For example, human mind and its society constantly respond, adapt and build new, elaborate structures to combat the age-old exigencies of a problematic life on earth. *Active, purposeful complexity* works against the 'downward' wear and tear of time and chance; it codifies and specifies design - which chance cannot. It is an instrument of biological survival, intellectual enquiry, technology and artistic creation. We continually experience it. Its proof, in artefacts and actions, pervades our lives.

Let's, therefore, check the *active* order of creation in terms of information.

(Raj) Active Information

Active information is only created in conscious, choice-flexible mind. *What constitutes an act of creation? What is its order?* On a conscio-material gradient passive follows active information; input precedes output. Sensation starts in matter and ends up in mind but creative action starts, *top-down*, in mind and ends specifically re-arranging bodies in the world outside. It 'drops' from causal reason towards a reasonable effect. Firstly, therefore, follows an account of the 'internal' or 'subjective' aspects of an information loop. These are conscious teleological, pragmatic and semantic phases.

Purposely Down to Earth

First issues an **idea**. This is the potential, seed or genesis of any active, formulating process. There accompanies intention to elaborate the embryonic inspiration's promise. This, grand or trivial in scope, is the basis of all purposes. Action's void is filled by information, will-power and desire. *Indeed, will and desire are the psychological equivalent of physical electromagnetic radiation.*[34] Will is like electricity, desire like magnetism. Together they are the light and life of mind; the '*volitio-attractive*' charge of their attention draws, fires and bonds. Such current lights up life.

---

[34] see *fig.* 5.1: legend.

Getting Your Way - Pragmatics

Mind's dynamo streams from focus of attention, a concentration of interest; and interest may intend to order things or events according to its own purposes. **Call this executive design**. Pragmatics is set at the level of executive or systems analyst. *Transmission of purpose through the pragmatic level involves, in the outward direction, imagining and planning its realisation; and, on the inward, deciphering another's plan.*

The sender of a message wants a relevant response. Desire expects results. *This is active teleology*. It involves manipulating options, working out exactly what is wanted of the communication or design, what its format or construction will be and the effect it will have. A message thus transmitted involves contextual interpretation by the recipient as well. This recipient may be animate. On the other hand, interaction with inanimate, non-conscious material may result in the creation of an implement, machine or some impact on the environment; and, passive on passive, reflex materials often interact. In such communication an incommunicative material side is always called the passively informed recipient. In fact, whenever an intention is physically expressed it works through mechanisms and machines. Machines, which include biological bodies, operate according to physical law and fulfil their function using, in one form or another, energy. They specifically accord with plan and inform the world in ways unguided nature can't. As such they obviously link information (that is metaphysical) with energy (that's not). Could the universe itself express a covert plan? *Could it be, at root, a Great Idea?*

What Do You Mean?

Causal reason generates meaning. All active information is meaningful. **Implementation** of a plan involves semantics whose specific meaning transcends the generality of grammar and syntax. The latter are simply vehicles of reasonable expression; their immaterial symbols are needed to make connection with the material world and thereby order it. They translate mind to matter. Coding and decoding, using speech-through-air, written word or other forms of signal, are core semantic business.

In other words, symbolism and simulation are mind's agencies. They are its metaphysical intelligence. They make sense. Although *meaning* is the important, active ingredient behind codes, data transmission and storage, nevertheless these passive instruments of communication are important. We need frameworks (hardware) within which to manage information (software).

Passive information (such as printed or electronic signal) carries the message of its creator. Paper, ink or electrons are physical entities, meaningless *per se*. However, their commander has controlled their order to provide a channel of communication. **A command erases** randomness; it is a **deliberate restriction imposed to cause a non-random outcome.** It employs an agent of restriction - sign, symbol, code of one sort or another. *This is the world of signals, semantics. What is spoken is not a matter of chance.* **Whatever is encoded is intentional.** Code and chance are chalk and cheese. They never mix. Language, other symbols and the construction or decipherment of meaningful communication (called semantics) make up the third level of information. In other words, this is the level at which ideas are framed in code or blueprint before their presentation in material form.

(Tam) Passive Information

Passive information is the expression, external to conscious, flexible mind, of active information. Such information may be dynamically exchanged (as in the case of speech or body language). Otherwise its impressions are stored either in subconscious mind (featuring relatively inflexible memory, instinct, archetype and so on) or using matter (where instruction is carried by arrangement on chemicals such as clay, papyrus, ink, *DNA* or other messengers). Storage may be fixed (as in a file or photograph) or dynamic (as in a running film, program or automated mechanism).

Passive information includes grammatical rules, syntax and the objects (or units) of their construction - words that compose linguistic or other natural code. A code is an agreed set of symbols arranged to format information. Such **upper linguistic/ codified level** involves particles (say, letters of an alphabet); forces that regulate their conjunction (punctuation); grammar (say, the elements of a language such as noun, verb and so on); and syntax. Syntax is the convention or legal framework within which symbols are ordered; its law naturally determines those structures allowed and those not. Thus the upper syntactic level acts as a filter through which order is communicated to and from the lower (environmental, statistical or quantitative) level of data items - physical phenomena. Inward, subtle, immaterial regulation orders outward, gross, material expression. Seen thus the laws of nature are, in type, linguistic code. Written in particles and forces on blank sheets of space, cosmos is a book. It is thus easy to understand (where *logos* means speech, word or order) why the ideal, causal level of cosmic oration might be called *Logos*. The universe might indeed be logical.

In this view passive information correlates with matter in a universal way. Informant energy, which shapes all material objects and events, is (due to its automatic, preordained behaviours) identified as passive information. **Such basic linkage, information with matter, will be elaborated throughout the rest of the book.** It therefore bears emphasis that, in terms of Natural Dialectic, the latency of cosmic language is always present and, wherever roused to action, expresses an orderly creation. The upper syntactic level of information's infrastructure, code, is equated with **potential matter;**[35] and such matter is, in turn, equated with **archetypal memory, archetype or, simply, nature's memory.** Such automatic regulation of behaviour appears, behind events, as natural law.

No doubt such equation, potential matter with archetypal memory, is heterodox; what else, by definition, can a fresh suggestion be? The issue is, when immaterial information is included in the deal, whether such equation smoothly integrates within a post-materialistic paradigm.

Information's Infrastructure - Code

Must 'law' be codified? Could not 'necessity' evolve from randomness by chance? Or is necessity the vehicle of purpose that created it? Chance or purpose, which informs 'necessity' of natural law from which the world is made?

↓	*irrationality*	*logic* ↑
	no-message	*message*
	no-code	*code*
	physico-chemical maelstrom	*psychological scheme*
	accident	*teleology/ purpose*
	chance construction	*technology*
	mindlessness	*mindfulness*

Top-down, of course, chance loses out. **Code is always the result of a mental process.** If a basic code is found in any system, you might conclude that the system originated from a mental concept, not from chance. You might therefore conclude it had an intelligent source - especially if that code is optimised according to such criteria as ease and accuracy of transmission, maximum storage density and efficiency of carriage (such as electrical, chemical, magnetic, olfactory, on paper, on tape, broadcast etc.) to its recipient; and if, above all, it works and orderly instructions are unerringly responded to. *A coder takes no chance. Randomness is eliminated.* A compiler is a mindless mechanism but a programmer is not. He determines the code and its operation: error is rigorously debugged. By definition, mistake or randomness degrades information; and the job of any editor is to eliminate interference, 'noise' or mistake. **Chance neither creates nor transmits information. On the contrary, accidents always (unless accidentally reversing a previous degradation) degrade meaning and, by degree, render information unintelligible**.

Message always comes in code; and where there are personal purposes (as in the case of conscious animals) there will be variable expression. The ability to devise or learn different, flexible languages is

---

[35] see *figs.* 3.1-4 and 6.1.

the function of a framework laid in animal instinct; the general capacity to engage in orderly, flexible communication is innate. Man, the information hunter, is capable of enormous yet most orderly flexibility and therefore great subtlety of expression. Whence though, we'll ask, arises archetypal, informative instinct?

Code is devised and stored in mind but information's physical expression obviously employs material arrangements. **The alphabets of such code are bio-chemicals (supremely, *DNA*), binary nervous code and the sub-atomic elements, forces and atoms of physics and chemistry.** The former couple are specific to life forms - possibly, at least with respect to *DNA*, anywhere life may exist in cosmos. The latter also constitute a universal code. This code dictates, through the agents that a study of phenomena elucidates, the way things naturally turn out. *Looked at this way cosmos is a Grand, Dynamic Text.*

The Lowest, Physical Level

Creation issues from within; active information orders passive. Mind generates code. Matter can act as a vehicle or storage medium for data but never, being subjectively impotent, generate code and therefore coded information.

In other words, information has meaning to a mind; such meaning (or semantic) may be organised into passive, symbolic receptacles such as language. But, while the concept of 'message' cannot be expressed in terms of the solely descriptive concepts of physics and chemistry, at the lowest, physical level it *can* be expressed through vocal, written or other material form. This is the **lower linguistic/ codified level**, that of physical expression. Sound waves, paper, clay or *DNA* can register ideas; so can machines or artefacts of any kind. Thereby the constraint of code is locally translated; and, with respect to natural archetype, the general is turned physically specific everywhere.

Therefore, bump! This is the part that bottoms out in light, sound, fluid patterns and in crystalline solidity. This is the 'external', 'objective' or physical side of the psychosomatic border; and, while forces and atomic particles comprise its primary expression, the hard, bulk universe that we survey is its secondary. *At this level we find data items, that is, materials on which arrangement is imposed.* The phase appears as an object, event or reflexive pattern of behaviour without subjective quality, context or intrinsic meaning. It is the level of raw data *per se*, that is, objects in whatever form they appear *before* higher inspection, manipulation or interpretation. No subjectivity and only husk-like objectivity remains. The values inspection may impose on such oblivious objectivity are numerical and its description measures 'thing-ness' alone. In short, the lowest level of 'information', devoid of sense or meaning, is vested in force or aggregate of atoms, that is, in physical being.

Music

discord   *harmony*
disease   *ease*
blockage   *flow*
incoherence   *coherence*

The old word for integrated order is harmony Music, like health, is harmony in action. It is an archetypal formulation of energy, constrained only by its type of instrument, harmonics and the skill of its musician. Melody is a most profound form of information-in-motion. It is perhaps the best medium for the vibratory transmission of meaning. Thus, regarding composition, Top Teleology does not need brain! Logical Archetype, primordial vibration's song, will do! Symphony composed by resonant association is enough! Hence, not from brainwave but deliberate fluctuation in Pure Life, the network of the world is roused to dance. The vibratory energy of First Cause and its subsequent constructions embody the internal logic, patterns or rhythms that pulse through each level of creation. **Therefore, beyond the other cosmic models and although it can't be posted in a diagram, music is Natural Dialectic's Master Analogy.**

If they print universal order might not resonance and radiant communication be seen as aspects of a messenger? *This vibrant couple are indeed identified as agents of law enforcement.* They can certainly transmit physical information and maybe psycho-physical as well.[36] Physics has detailed properties of charge-linked, life-giving, sight-giving light; spectroscopy uses it to identify the components of far-distant galaxies, stars and to tell apart, each by its own unique 'bar-code', every element and molecule on earth and in the heavens. Resonance, for its part, is a feature of all vibratory systems - atomic, electrical, musical and so on. The sound of a gas or liquid put under pressure is called 'resonant'. The sound of your voice as you speak is an example. You order molecules of air according to the passage of your will; and Cosmic Sound is known as First Cause because it is the first and therefore principal notation that resoundingly informs creation. The whole opera is, in these terms, a reflection and a resonance of its transcendent order.

Song will, as every musician since Orpheus has known, pull you straight to the heart of things, to the centre of life. For Brian Pippard, physicist and a Cavendish Professor at Cambridge University, "A physicist who rejects the testimony of saints and mystics is no better than a tone-deaf man deriding the power of music". *Looked at this way the life and the universe are Operatic.*

Machines

The metaphor for universe used to be biological. It was of the

---

[36] Chapter 16: Psychosomasis.

universe as an organism - the soul, mind and body of God. Man was a reflection of it. Then, in a renaissance born of scientific rationality, it changed to technological - a Newtonian, inorganic kind of universal machine. *The 'meaning' of a mechanism or machine is, of course, vested in the purpose of its maker. Is conceptual background therefore part of nature's show?*

A machine involves a system of well-matching, interacting parts that, unless any is removed or degraded, contribute to a function or produce a targeted result. Such systems are therefore specifically and irreducibly complex; and so to work they must be made at once. In this case which comes first - a machine or its concept, a work of art or its inspiration, the chicken or its egg? Extrinsically fruit issues from the branch. The branch came first and the fruit after. *From an intrinsic point of view, however, the branch is from the fruit. The fruit came first but, in order to bear it, the branch was conceived.* **Wherever an apparent 'chicken-and-egg' situation crops up the puzzle is resolved by the introduction of purpose.**

"Machines", argued philosopher/ scientist Michael Polyani, "are irreducible to physics and chemistry." They are irreducible because they involve immaterial purpose, the stepwise development of a plan of implementation, a directed cohesion of working parts and, of course, the thoroughly non-material anticipation of an operational outcome. Such a machine may itself be simple or complex; it may be possible to deliberately adapt its purpose and therefore function; but if parts are missing or corrupted then its operation fails. *Engineers routinely face the chicken-and-egg problem of designing interdependent components for their mechanisms and machines.* They are not inclined to leave their solutions to chance. Thus, although a machine is totally material, it is at the same time full of its inventor's mind. It has no mind but, paradoxically, by proxy does.

In short, there is neither law nor known agency that can cause information to originate by itself from matter. The materialist's problem is acute when it comes to machines. Indeed, materialism provokes a crisis of identity. *Exactly who and where from are you?* Because all mechanical processes are initiated by mind's genius; all machines are purposely conceived; and intention involves reason, design, causal interaction, signal, code (or language) and meaning. Engineers and biologists both use all these words extensively. **Indeed, function and reflecting form *are*, at root, the *meaning* of a machine (or bio-machine).**

Let us also be absolutely clear that, although machines are operationally subject to the constraints of natural forces and environmental context, they are never created by them. *No machine ever appeared as a consequence of the addition of random free energy into an existing system.* **Physical nature can't create machines; yet**

information and machinery are closely twined in living and, indeed, all teleological constructions. This is fact. *Why, therefore, did Newton visualise the natural universe as a Clockwork Machine?*

Mind Machines

Now, in line with the Information Revolution, it's sometimes seen as computational. A Great Computer working repetitiously.

A computer is a mind machine. Inspect the logic of its functionality. Examine integrated circuits. Their molecules, like those of a brain, show no sign at all of mind. They are not even biochemical but metal, plastic, silicon and so on. Yet they are replete with passive, rigidified order. In this respect each one's determination cries out its ghost, the active order of its maker. The whole machine is absolutely full of maker's information.

Programs are a mind machine's intent. They express the will that mind invested in machine. *It needs be re-emphasized, every machine (including body biological) has the mind of its maker in it.* Not in its atoms *per se* but in its purpose, design and lawful operation. Machines passively embody information. Mind-machine information is passive. Active has produced passive information. **There is, to emphasize the fact, known neither law nor process of nature by which matter originates information. The source of information is always mind. And at the heart of mind is consciousness.**

If awareness is an immaterial element then, of course, all materialistic and evolutionary ideas about psychology besides biology just crumble into dust. That[37] is a lot of dust, enough to cloud your vision; but if their basic premise is not right then they need deconstruction followed by rebuilding on the right lines. Facts won't change, perspective and interpretation will. The fact is things don't specify stepwise activities that lead to goals. **It is against all known laws of physics that oblivious matter might <u>create</u> information; or that material instruments of purpose self-organise or instigate their own programs.**

Machines are devised to save labour and excel human capacity in speed and/ or power of operation. In their capacity to execute algorithmic complexities computing machines far outpace our own intelligence. Indeed, by materialistic definition a soul-less, super-smart robot with a superlative ability to mimic human response feels and *is* human. The latter assumption is, however, due to a persistent science fiction. An official one. It rumours, without a shred of hard evidence, that mind sourced itself by chemistry - because it is only an originally accidental arrangement of atoms. Brain, most excellent of mind machines, of late evolved! And soul is, of course, a redundant figment of untutored imagination. *Robots are brainily designed but, since it was brainlessly conceived, you can't proclaim that brain's designed or thoughtless, soul-less cosmos is a natural Mind Machine.*

---

[37] Chapters 13 to 25.

Universal Authorship

So, *bottom-up*, matter precedes and produces mind. Having exorcised the spook, materialism extirpates any report of out-of-the-body experience, notion of a mind-world or mind/ brain distinction. It bins them as preposterous delusions conjured up by pathological events. The subjective point is not just lost - it never really existed. Thus cosmos started up and runs by chance. Without design. **Existence and all that, including subjectivity, is a fantastic accident evolved from impulses of energy constrained by the mindless serendipity of accidentally concocted natural law.**

Top-down, **mind imparts intent.** It informs another mind or prints designs on matter. Thus purpose may be conscious or, as a reflection of original design, unconscious. Subconscious instinct, archetype, code and non-conscious machines are each unconscious, reflective, passive examples of the carriage of purpose.

Subconscious entities are metaphysical, material behaviours naturally reflect their immaterial informant and the world-tree germinates from archetypal seed. Essential Seed develops psychological then physical a cosmos. *In this view cosmic development is 'organic'.* **Creation involves a 'polar inversion' or 'asymmetry of reflection'. The Cosmic Centre expresses itself 'organically' from within; it 'turns inside out' so that internal is precipitated in external, physical form.** The scale tips from creativity to gravity; mass unseats awareness and emerges centre-stage. The inversion grades, at the same time, from conscious to non-conscious, voluntary to reflex, actively informing to passively informed. At the base of its spectrum, at the foot of the cosmic rainbow physical energy predominates. Its total objectivity precipitates, in the order of things, a massive crock of fool's gold called bulk matter.

Active on passive, conscious on subconscious mind and matter - materialism execrates but is there any proof subjective is objective, immaterial material or life composed from deadness? If there is none my case is sound. *If (and you are an exemplar) energy and information are two separate, interacting elements then what gainsays cosmic animation?*

In this respect two major thrusts well indicate a cosmic scheme of anti-chance, that is, creation by intelligent design. *They are physico-chemical and biological.* The former (Chapters 7-12) involves the integrated precision of principals (and their formulaic descriptors called chemistry and physics) that permit your presence here. The latter (Chapters 5, 6 and 20-25) illustrates the severely limited correctness of a bio-illogical theory that you were evolved by oblivious forces; and emphasises the irreducible, coherent and ubiquitous complexity of biological design, informed by code, found in the remarkable components of your (and every other) body.

Book 1: Physics
Chapter 7: Lady Luck and Lord Deliberate

In 1900 it was thought that physics was, essentially, complete. 2000 radically disagrees.

There exists a contemporary Authorised Version of Physics (*AVP*). It represents a brilliant intellectual endeavour that includes quantum theory, special and general theories of relativity, *SMPP* (standard model of particle physics), classical mechanics, *SMC* (standard model of cosmology) and more. The *SMC*, for example, successfully matches theory with many observations and *vice versa*. However to explain anomalies it resorts to assumptive 'fixes' such as 'matter-from-absolutely-nothing' (big-bang theory), inflation (cause unknown, effect to smooth out bang's explosive chaos into what the telescopes observe), dark matter and dark energy.

Tension already! While *SMC* needs dark stuff for its galaxies the *SMPP* cannot provide the particles it's made of. Indeed, intonation of these creeds should conjure up the origin and status of the universe - except they're in the dark for 95%! Thus, in chasing this receding crock of gold, more theories than theoreticians spring eternal from the cornucopia of faith. Might a device called 'supersymmetry' call up the extra ghostly entities or *CERN*'s detectors[38] even spot them? And, in this mix, don't forget to add a pinch of bendy, stretchy nothingness that's full of energy and empties space of emptiness. Indeed, theory climactically whisks whole new worlds from its originally empty hat. Isn't our universe a bubble blowing up with countless others in interminable inflation - welcome to the unseen multiverse!

Is such myth wrong? What story will tomorrow's be? No doubt, eventually, a complete and correct description of material reality will be elaborated but its appearance won't look like we're thinking now - since our *AVP* is certainly neither complete nor, it may transpire, correct. Answers to a scroll of problems will forge post-modern physics' shape. Meanwhile guesses generate a torrent of speculation; occasionally, empirical principles are relaxed sufficient to generate pseudoscience (passed as science) and, since physicists prefer numerical to verbal metaphor, even anecdotal mathematics. *We'll discuss these issues but firstly place them in the context of Natural Dialectic's immaterial*

---

[38] The *LHC* (large hadron collider) at *CERN* is the world's largest machine and most powerful particle collider built to simulate conditions supposed present at the start of physical creation (the big bang?).

element, information. How much sense does such inclusion make? Is it logical or mythological? The argument is followed through.

Let's note two things to set us on our way.

Firstly, this chapter is about **potential**. The word implies, of course, possibilities; and the greater a potential, the greater the number of possible outcomes - within the natural rules that govern an event-in-question's behaviour. Rules endure but circumstantial outcomes change. But what's the source of rules called natural law, that is, behaviour and character of lasting universal parts? Are they accidental or purposely designed? Scientific methodology demands, of course, naturalistic answers; it welcomes ghostly physical intangibilities but never (since never's been decided) the intangibility of an informant, universal mind. Luck doubtless plays her part but is the Lady source of everything? By contrast, others argue information is potential that can make the rules it also governs by. The world's consistent and not arbitrary; it is precise and logical; nature's rational and not irrational. Did, therefore, charm or order organise world-sport? Is Lady Luck or Lord Deliberate responsible for cosmos?

Secondly, in normal parlance we grade down from 'certainty' through 'probability' and, remoter, 'possibility' to an 'impossibility'. However, in statistics one word covers the whole show - probability. The range is 1 (certainty) to 0 (impossibility) and every possibility between - except that, quantum physics says, you won't ever make exactly 1 or 0. You might seem to treat as if there's neither absolute certainty nor complete impossibility. Is this, in every context, true?

The domain of past cannot be changed. It is fixed and certain. The domain of future is, of course, uncertain; it involves conditioned possibilities. But conscious choice, natural law or both combined yield, always, on the instant, present certainty. This is the way potential is 'reduced' to actuality; it is the way the world works.

Puzzlingly, however, at the very edge of minuscule, quantum physics states that, while the character of particles and forces does not change, their action's indeterminate; cosmic apparition slips and slides upon a modicum of Planck oil. Does this mean, to all effects and purposes, matter is uncertain or is not? Such oil has encouraged some to claim that 'whatever can occur will.' The emphasis bears down on 'can'. What does 'able to' mean? Can anything *not* happen? Indeed, there is philosophy that, like snake oil, temptingly exudes from this. Its medicine, charlatans agree, will poison Lord Deliberate full and finally. 'Anything', they tout, 'can happen if only you wait long or imagine big enough. Impossibility's impossible!' Thus highbrow primates peddle that 'impossibilities', by quantum probability, *can* happen. Even miracles can happen! Uncertainty has killed off never-ever; it has resurrected only possibilities - some stretching, we shall see, to never-

never land. Like certainty impossibility curbs endless possibilities; so why should I believe the natural world, in trillions of years, could make a teacup - let alone the tea and me! Do you believe the wind and rain could, over time, contrive to write a sonnet in a book that's full of others? Impossible. It is mind, not time, creates specifics - codes and systems - that the elements can't even dream of. The chance-reducer builds, in moments, what those natural years cannot. You may conflate a chemical, called *DNA*, with the information that it carries; why, though, should anyone believe that uncreative operation of the elements could write a book of life, create containers made of cells or stir their metabolic cup of juice? *The fact is, when it comes to rationality, there's no necessity. Long live impossibility*!

The Order of Invariance

Invariance under transformation. Underneath the world's commotion basic character of parts remains the same. Contexts change but automated rules of play do not. Without conserved invariance you can't obtain the balance that equations need. Energy's conserved; and at any time in any space from any angle laws of physics stay the same. **Does physics see such changeless cosmos as a self-consistent, fine-tuned set of principles or are its invariant patterns of behaviour basically informed by chance?**

Dichotomy. Which view is true, which false? Or, paradoxically, are they both consistent with the truth? If so order is, at root, accidental; natural law sprang for no *reason*. Yet law dictates the reason why; a set of principles informs a framework of effects; chance is a substanceless effect so how can it be matter's reasonable cause? In other words, what was the true potential for creation? From nothing was projected cosmos. What is the nature of such powerful nothingness? Of what sort is such transcendent, prior void that has informed (and perhaps informs) and has empowered (and perhaps empowers) the volume of our universe? Information is, in fact, potential telling action how. It commands. As well as potential this chapter is, therefore, concerned with **rules** - rules versus randomness, that is, invariant principles producing various effects. **On top of rules we shall consider choice of origin, that is, 'original how and why'.**

Precondition/ Potential

Bottom-up, **no metaphysic orders what is physical**. The so-called 'laws of nature' are but descriptions of behaviours that a starting point (say, big bang/ initial projection) has spawned. Formulations and the values of their constant factors are not fixed by any deeper level than experimental regularities. Natural reasons have no reason; nor do material effects have *a priori* cause. Certainly there's no intention that substantiates cosmology; the character of automatic movements started as an accident; freeze teleology right out of every frame - this is the order of the game!

Physics has its limits such as 'strings', 'space granules' or circumference of the universe. Such 'glass ceiling' means that, looking through it darkly, immateriality is not in evidence. Of what, if it exists, might metaphysic's immateriality consist?

***Top-down*, the order of invariance is hierarchical.** Cosmos, not chaos, constitutes reality. A stack expressing law might read:

tam/ raj	*Sat*
existence	*Essence*
exhausted/ kinetic phases	*Potential Phase*
variation	*Invariance/Permanence*
development	*Seed*
practice/ actuality	*Principle/ Law*
subsequent order	*Archetype*
elaboration	*Plan*
change/ inconstancy	*Stability/ Constancy*
↓ *tam*	*raj* ↑
passive state/ object	*active energy/ force*
exhausted phase	*kinetic phase*
informed/ ruled	*informant/ ruler*
contingent variation	*basic themes*
external 'fall-out'/ outcome	*internal order*
crystallisation/ bulk shape	*energetic pattern*
chance	*necessity*
accident	*design*
apparent chaos	*cosmos*

Natural Dialectic indicates a source from which the stages of a cascade fall; it involves an inner, central 'symmetry' from whose intrinsic order cosmos breaks. It breaks, like all phase changes, with a natural spontaneity. Natural, contextual spontaneity, however, never barred preordination. *Rules are information, information is potential for behavioural patterns*. Regulation thus precedes and guides the way a game is played spontaneously. Solids fall from gases and, as they do, so different rules of state apply; different sets of laws precipitate from inwards outwardly. There would be, according to this scheme, essential principles and primary law; and secondary sub-routines that follow on. What is, we'll need to keep on asking, the central source of symmetry, polarity and bifurcating cosmic code?

Law is regularity; symmetry of law is stable; and, as Einstein claimed, the rules of nature are 'incorporated reason'. Archetypes, in this encoded sense, are the voice of reason heard through mechanisms of material form. Their issue is a 'breakage' out of background metaphysic; expression leads from hidden symmetry through simple, subtle quantum agents to the detailed aggregates of flux and sharp fixity. As scientists know and strive to better understand, you have to dig to find the principle substantiating practice. *From what superficially seems chance you have to disentangle rule.* In this respect archetypal memory is potential matter; it is the hidden invariant substantiating a hierarchical, informative Theory of Potential (*TOP* - equivalent of *CUT, see* Glossary: unification and Index).

Cosmo-logic

In the beginning was *NOT* chaos. Modern physics shows that mankind dwells in a finely-tuned universe. To be precisely fit for life it must have started in a way most orderly, specific, specially defined. *Fine-tuned by chance? If low probability together with specific definition indicate design, no chance!* **A universe fine-tuned regarding many parts might be construed as one of specific, *irreducible complexity*; and, if lacking any part it failed to work, of *minimal functionality*. This pair's efficiency can, we shall see, indicate the hallmark of design.**

Or, if you must obtain a naturalistic answer, what initiatory contrivance can you dream up out of mental space? Coincidence stacked on coincidence - is a fine-tuned universe coincidental? This 'fine-tuning problem' lurks at physics' heart; its mystery gnaws the scientific core. What is the origin of accurate congruity? Who or what first cracked but never scrambled laws of nature from the cosmic egg? All in order, all on board for life! Was cosmos fixed or did it fix itself? Has some Great Inventor twiddling with his knobs and dials so tuned the cosmic program into working perfectly? Was the source of software natural? You cry that Mindless Mother Nature must have fixed herself; this way the origin of natural law, the 'fix' of physics, is the child of Lady Luck. Arch-dodgers of creative metaphysic postulate a multiplicity of bye-laws in the endless districts of a multiverse. Our universe, they speculate with shameless lack of elegant parsimony, is probably just one from multiversity, from infinite arrays!

Fine-tuned and fit for life; but, you muse, is the world's invariance a fix? *Bottom-up*, you won't agree; *top-down*, by now you know you will. Throughout this chapter we'll be checking who or what fixed cosmos logically. Or not.

Lack of Cosmo-logic

Physics is defined by formulaic rules and yet, at root, materialists conceive of Chance as Grand Creator. Is chance reasonable or, by definition, not? Yet Luck, it's claimed, gave birth to cosmos and to life. Such Reason isn't reasonable but, still, this face of Mother Nature is a major player in the scientific pantheon. *Unfathomable Fortuna needs attention*. We'll now spend a chapter's worth of time with her.

To begin with, how did she obtain apotheosis - by trick or by evasion? Chance is unattached to reason. She is unhinged from logic. **Her wildcard is, as an essentially irrational trump of naturalism's accidental game, a metaphysical evasion.**

From a *bottom-up* perspective, evasion of the whole truth is *by* metaphysic; delusional evaders call on immaterial God or gods-of-gaps to cover cracks in knowledge. In fact, materialism claims, we should believe that nature sprang from nothing. Void excludes information; it excludes metaphysic; there was no design; in this view life and even cosmos with its laws occurred by chance.

How, though, from absolutely nothing, might cosmos appear? No strings attached. Straight off, nothing's something that's a wondrous mystery!

Top-down, evasion is *of* metaphysic by materialism's rigid physic - only physic, never metaphysic. However, information's immaterial. In the way that a house is devolved from the metaphysic of an informative plan, so the patterns of non-conscious energy operate within the framework of a program that, because 'information precedes energy', preceded physical creation. Informative potential precedes expression. *Before cosmogony, before the starry universe began all was <u>not</u> nothing.* The material world was present, in principle and in potential, before it was physically expressed. Its cosmic egg was archetypal information. Thence the bird developed and can fly.

Bang! An abrupt appearance of invariant behaviours and their description as the laws of chemistry and physics is one thing; the elements of physics are, arguably, specific information that's incorporated within vibrant energy. Prototyping of a body biological is, due to code-specified complexity, quite another. Organisms are information incarnate. **Yet there exists no Law of Non-conscious Innovation and Integration, no Principle of Material Evolution. Wind, rain and natural forces *never* codify; one doubts that primal absence ever did. There is no physical but only psychological Law of Innovation.**

So how did innovation of the primal cosmic players, quantum particles and basic forces, suddenly occur? The issue's pivotal. The fact-less notion of a 'Principle of Naturalistic Innovation' or of 'Specified and yet Non-Conscious Integration' - pervades contemporary thought. It pervades because prior philosophical judgement has been signed, often by consensus, to deny intelligence is anywhere but brains; and, although its nature is materially obscure, to ban the thought of consciousness as metaphysical and thus distinctive from non-conscious physicality. <u>*This is as metaphysically evasive as you get.*</u> Such unquestioned and unquestioning evasion permeates the study of science worldwide. Its mind-set therefore misses out code's crucial metaphysic - active information. Active information, also known as intelligence, is not the *zeitgeist* of today.

Materialism is the *zeitgeist* whose central prop is evolution (either out of simple void or, life-wise, natural chemistry). Unfortunately, again, there is neither law nor process nor expected sequence of events by which matter is able to create information. In fact, the design and construction of all purposive schemes and machines is the product of informative negentropy. There you have it. *Informative negentropy is needed; only energetic entropy is available.* <u>Thus evolution theory, with its basic metaphysical evasion, fails to match reality</u>. **Neo-Darwinism suffers, throughout the entire body of its scholarship (and there are libraries of eulogy), from a primary philosophical error of category**. Thus metaphysic's most succinct evasion is retailed by those biologists who call design 'design'; and thus, through Darwinian smoke and mirrors, information's penny's never seen to drop. Nor must it. If it drops both theory and philosophy go pop. Such metaphysical evasion is fundamental to thrice fifty years of academic thought.

Teleology is barred; its spot shall neither taint nor stain. Out, therefore, damned alternative, down with the laws of Lord Deliberate and up with material indeterminacy - except materialism's own self-certainty! But is it logical to lean too heavily upon the whimsical; is it wise to swing up close or even philosophically espouse Caprice - the girl that we've already met as Lady Luck?

Getting to Grips with Lady Luck

What's her character? How slippery? Is chance dumb, blind or a blur, a trick of incoherent mind that can't quite grasp a real reason? Is she simply a cosmetic answer, nothing more than simple make-up, an illusion whose substance vanishes the more you peer? In any case she is central to materialism and peripheral to mind. To the former she's a powerful force for transformation, a creative 'anti-deity'; you might even crown her Queen of Serendipity. To holism, though, she's a phantom flitting only in the unsure mind of an observer so that, since she could not puff a powder let alone create a drop of tear, you would make obeisance to an impotent, strange and imaginary idol. Either way the blur needs disappear. Although she's not an easy ghost to lay the Lady must be caught and, if necessary, pinned in place. Slide her underneath a microscope. Her shape must be revealed; the flirtatious nature of her ambiguity must be resolved. It might help tease the problem open if we split it into mind and matter.

Uncertainty (of mind) and luck (coincidence), although alike, are not the same. Uncertainty's a mental construct; possibilities occur in mind. Then there are scientific objects and events; how far is material uncertainty identical to probability and, within the constraints of a given system, probability to chance? The world's a complex, layered ferment of activity; when, in the context of its maelstrom, something unexpected happens then we call it luck. Luck's probability occurs, you bet for sure, within some given system (say, a lottery, a horse-race or electron orbital) but, as far as matter is concerned, there's no such thing. It's all in your mind; you can't put your finger on it, can't predict it and so call the business chance. Uncertainty is due to unpredictability, immeasurability or ignorance of factors. Is matter at root indeterminate? Maybe it seems that way because a wave is spread in space and time and quantum physics bases its mathematical descriptions on wave-like analogies. You can be uncertain but, for nature, does indeterminacy (or uncertainty) exist?

I don't know. Caprice is difficult to fix because by definition she is unpredictable. Chaotic. Chaos isn't order; you can't abbreviate what's random into formulae and so she's more or less incalculable. Her lack of order dubs her Lady Luck. Chance is lapse from an expected course and, if you are not omniscient, predictable as unpredictability. We live variations on our daily themes and, occasionally, within a sea of trivial unpredictability a wave rears and affects us. Chance then challenges. It changes things in unexpected, sometimes major ways. Who, for example, can predict when he is going to die? To define luck is an uphill slog. Is this because incalculable behaviour, complex interactions and irregularity upset the reasonable bent of mind, whose nature is to actively inform, find patterns and to understand as far as possible the truth - for sure? It needs mind to spot a chance yet chance runs counter to it. Chance loosens up mind's certainties, punctures preconceptions and waylays its rationality. How can you assign a reason to a factor that, by definition, hasn't got one? How ascribe cause unto lack of it? How logically describe what cuts up logic or rationalise irrationality? *How could you ever test a theory (such as evolution) based on the disorderly principle of chance?* How recreate the randomness of its creation? With, at most, a broad uncertainty - best guess, alternative interpretation of the facts and choice of options are as far as explanation's remit runs. Luck is definitely a dickey business so be warned that, flirting with Capricia, you're up against the odds!

'Of course,' a chorus swells, 'we all agree that Luck alone can't make a mountain or a man.' Capricia can't bear cosmos by herself. She needs a Fixer who'll fix energy and thus supply the forms; Lady Luck needs discipline according to the iron rod of natural law. Thus Chance *and* Necessity (*aka* The Vice, The Fixer, The Rigidity of Natural Law) together create mice, men and anything that happens in their bounds. This Fixer sprang, you understand, *ex nihilo*; it sprang by Chance from chanceless nothingness and so, of course, there's no necessity at all for Lord Deliberate to live - unless the fine-tuned body of The Law leapt from Design. In Deliberate's case Necessity's Invariance would underpin a great conceptual scheme; principles of law would anchor nature's frame. Within this generality you'd plan and build, as engineers agree, specific mechanisms (say, robots or life-forms) to conform to your intentions *and*, of course, the frame's constraints.

Are You Certain?

Machines are minded yet mindless. Metaphysic can't be found somehow inserted into automatic physicality. Does it, in the world-machine, exist at all?

A *bottom-up* perspective elevates a maelstrom made of accidents to gradual creator of the world. What, after all, is there but mindless matter, matter which perhaps 'expanded effortlessly' out of 'absolutely nothing'? No purpose, only lawful and yet 'lucky' interactions do the work; no mind abroad, the cause is chance! Where chance is cause's absence! If the fundamental character of chance is an apparent lack of cause, then causelessness created every cause! And if these causes constitute the patterns of behaviour we name laws, then pristine lawlessness created every law. Get that? *Non est deus. Quod erat demonstrandum* (*QED*).

appearance	*Reality*
expression	*Potential*
subsequent order	*Archetypal Certainty*
different materialisations	*Ideal Field*
various breakages	*Symmetry*
↓ *particular*	*general* ↑
fixed form	*energetic flux*
local	*encompassing*
detail	*theme*
certain actuality	*possibility*
particle	*wave*
classical reality	*quantum reality*

Look. Law's intrinsic, accidents are circumstantial. *Actions can occur by law and at the same time by coincidence.* Look closer. Numberless co-incidents are happening all the time but who can separate each one and tease the plethora of bumps and grinds apart? Can you eliminate, by sharp analysis, apparent 'accident'? Could you, as the determinist Pierre Laplace avowed, predict each one if only you could know enough? Could such omniscience not calculate each detail of the world's work in advance; or, if you'd lived a billion years ago, have predicted your own brain and what you're thinking now? In other words, is unpredictability a function of just ignorance, an inability to cope with the complexity of simultaneous interactions? Laplace, at least, determined that complete determination rules.

Contrarily, Werner Heisenberg's famous *HUP* (Indeterminacy or Uncertainty Principle) is based on the notion that, at the heart of atomic matter, a measure of uncertainty in the measurement of variables (position/ momentum and time/ energy) always and inexorably blurs deterministic lines. Of course, wave/ particle's a schizophrenic model of a quantum entity and not entirely physical. Treat 'wave of chance' and any point must disappear. So-called wavefunction charts this blur of overlapping possibilities but measurement 'collapses' them to pointed actuality; point-particles occur again. Yet still the double singlet's fuzzy. Treat as wave and your position's blurred; freeze-frame the particle and obtain position but velocity's unclear. The same with time and energy. Measurement can't give you both at once and thus uncertainty occurs. Precisely, imprecision rules; while randomness seems wobbling at the root of things you can't nail quantum nature down.

What, therefore, is the real nature of the raw, indefinite pre-data of our cosmos - quanta? **If large-scale things are made of these what's the basic nature of reality?** A 'cloud of metaphysic' (probability) collapses into actuality; it 'freezes' into physicality. Which of this pair, determinate or indeterminate, best represents the ultimate material verity? If, moreover, indeterminate substantiates determinate how do they correspond? How and why should a 'collapse' of quantum probability perpetually render large-scale certainty? *How, in short, do relations between classical and quantum mechanics work?*

Interpretations of such relationship vary. Some think it practically unimportant to know how. Others, such as Niels Bohr, suggested the conjunction of experimental instrument with quantum entity might cause the outcomes we perceive; large-scale systems constrain behaviours of the small - which doesn't answer how large-scale initially appears.

The crux of any observation is observant mind. It has also been suggested that observant consciousness is what (as in the case of Schrödinger's unhappy cat) collapses probability to 'here' or 'there', this way or that. How, then, do you explain a world determined prior to when it was observed? Incalculable certainties have, unobserved, occurred. What, if filtrate from an archetypal template's disallowed, turned possibility to certainty each time? Nothing? Perhaps every possibility exists in parallel in many worlds at once. Our universe is one. The many-worlds hypothesis is Ockham's nightmare, rampant speculation without any evidence, a scientific mega-myth.[39]

Fuzziness, for minds steeped in precision's game, froths with an embarrassment of choice. Many physicists have, therefore, abhorred such multi-lemma of uncertainty. Quantum theory grapples with the realisation of cosmic potential. How are archetypal possibilities reduced to local actuality? Debate over theoretical implications has been protracted and intense. I'm certain I'm uncertain of uncertainty since I prefer deterministic certainty. Thus Einstein, who believed in 'complete law and

---

[39] Among competing hypotheses William of Ockham, a medieval friar, suggested the use of a 'razor' of parsimony to cut to the one with the least assumptions. Scientists in search of 'elegant solutions' appreciate his tactic.

order in a world which objectively exists', filed a series of objections designed to checkmate the implications of indeterminate 'chaos' and irrationality. 'God does not play dice'. Planck, Schrödinger, de Broglie, Bohm and many others (maybe including Heisenberg himself) also believed that discovery of deeper causal mechanisms, called 'hidden variables', might reinstate nature's actual determinism.

Indeed, for Heisenberg the substance of modern theory, its uncertainty, was down to an intrinsic shortcoming - an inability to completely describe our underlying quantum world. Although a logical positivist in the sense of rejecting metaphysic he was, paradoxically, a religious man and seems to have had sympathy with Platonic forms, an archetypal order of things or universal mind-behind. For example, when asked by Wolfgang Pauli if he believed in a personal God he replied, "Can you…reach the central order of things…whose existence seems beyond doubt….as directly as you can reach the soul of another human being? I am using the term 'soul' deliberately…If you put the question like that, the answer is yes." Nevertheless the apparent abandonment of the principle of causality and, thereby, causal determinism by the *HUP* - leaving chance to tweak the roots of nature - has been characterised as 'a struggle for the soul of science'.

No doubt, the universe can be interpreted as 'quantum all the way'[40] but what about the order as opposed to indeterminism its behaviour displays? The nature of a proton, photon or electron never changes. It is entirely determinate: and, while the space-time location, velocity or energy of a particle may be indeterminate, there is equally no doubt that such 'quantum indeterminism' little interferes with macroscopic operations and, therefore, daily life. In fact, isn't even uncertainty subject to various, strict 'rules' such that quantum mechanical equations themselves reflect kinds of determination? And if the way that I deduce them was (by evolution) and is (by sub-atomic motions in my brain) determined by such chancy nature as those motions represent then logic runs in circles. You think you're certain but you're not; you (including brain) are constructed from incertitude. How do you sustain a logic born of determination that the world is indeterminately run by causeless chance? Perhaps, as Max Born wrote, 'the motion of particles follows probability laws but probability itself propagates according to the law of causality.'

Then struck Bell. John Bell was pro-Einstein/ anti-Bohr at first; but if his Theorem (called Inequality) rings true it means that, in the face of quantum theory, Einstein's 'local realism' comes unstuck. If so, perhaps rules don't rig occurrences; uncertainty and quantum probability ring true. 'Spooky' action-at-a-distance can occur; co-created particles can interact immediately, faster than the speed of light, across the universe!

---

[40] as noted in *fig.* 2.6.

Experiments by Bell, Alain Aspect, John Clauser and Anthony Leggett have since been interpreted to indicate that Bohr was right.

If so, the question sticks. Why should 'amplitudes of probability' collapse to surety? **What *is*, at root, material reality?** In other words, if 'quantum strangeness' is the norm, why is regularity apparent everywhere? Mathematician Roger Penrose takes an inclusive, common-sensible approach. From a quantum perspective 'bizarreness' *is* normality. It exists in potential but is 'confined'; it's everywhere a possibility but is, for the most part naturally 'suppressed'. How? To observe a quantum you disturb its state; you collapse its 'probabilities' into an actuality. Local fact is realised. You don't, however, need a laser or some other kind of probe. Interactive relationships with bulk objects of the universe engender mutual interference and, continually collapsing probability, forever bring the action 'down to earth'. Past and future tryst in present actuality. An immaterial 'cloud of possibility' is realised; immanent potential rains with drops of matter locally expressed. Can't quantum 'fluff' and classical 'hard angles' co-exist? Constrained by the 'statistical' bulk of billions of neighbours, most quanta do not dance in isolation; bound in molecular and larger entities they submit to various chemically and physically predefined forms of order called the certain, non-random laws of nature. Creation *is* the orderly constraint of energetic chaos.

Possibility to actuality; non-location to location; what translates every-where-in-principle to here-in-practice? Check the previous stack. *Quantum uncertainty seems to reflect creation's field; for Natural Dialectic it's a cosmic shadow whence our 'real' world appears.* **It partially describes the staged translation of potential matter to expression - wave agency and then, with character intact, collapse to property-of-particle whence every atom, molecule and other shape is locally and orderly constrained.**

Thus, if all systems are at root 'non-local' then physical nature appears supported by an invisible reality whose connectivity is omnipresent. One might thus accept that the 'Copenhagen' way of thinking - based on *HUP* - was a true description of 'microscopic deep reality'; Einstein would seem wrong. *Non est deus. QE(possibly)D.*

On the other hand, he might be right. For example, David Bohm's account of quantum theory is deterministic. His strict separation of Bohr's complementary couple, wave and particle, leaves particles classically determined but he adds a 'hidden variable, 'guiding wave' or 'source of quantum potential'. Its action yields specific effects. Thus, where Heisenberg's uncertainty is Bohr's reality, for Bohm it's down to ignorance. A hidden variable brought to book will fix uncertainty. For Natural Dialectic this 'variable' is potential matter, that is, non-local, archetypal 'notes' of energy[41] *invariably* struck locally as photon, electron, proton etc.

Such consideration might confirm that 'God does not play dice'.

---

[41] see Glossary and Index: archetype; also *figs.* 2.5, 3.3, 3.4, 9.1, 10.1.

Einstein would, with certainty, be determined basically correct; and, as corollary, Darwin, with his notion life evolves entirely through Lady Luck, is fundamentally awry. If it's Einstein versus Darwin, Einstein wins. E's right, D's wrong. *Est deus. QE(probably)D.*

Perhaps you can see where this is leading. It isn't where materialism wants to go.

Firstly, information is an entity that exists independent of matter. The information-centre you are personally familiar with is mind. Science that's materialistic must exclude what's immaterial; metaphysic isn't scientific everyone agrees. Since when, however, could a philosophical exclusion outlaw mind's primary, informative reality?

Secondly, although materialism axiomatically excludes an immaterial element, if such element exists (as Natural Dialectic axiomatically supposes) it doesn't have to bear the properties or serve the laws of physic. For example, its 'holographic', unifying and yet 'nuclear' capacity for information may impact each 'pixel' (or each 'quantum') of the cosmos simultaneously.

Thirdly, **potential matter precedes physicality; it has been identified as archetypal memory.** Memory's the fixed or 'solid' phase of mind. The invisible reality of archetypal files constitutes an 'immanent potential'; **universal mind is the *implicit* universe.**

Mind's precedence is an informant whose principles (i.e. specific possibilities) are realised in the expression of particular physical phenomena. **The latter constitute an *explicit* body called our cosmos.**

Thus archetype is an object, thought or event's ***intrinsic being***. *Quantum physics is viewed as a phase in the realisation of archetype that yields its determinate physical 'overlay'. It represents a mid-step in the translation of universal mind into locally projected matter.* **As Natural Dialectic's framework clearly shows, quantum is the world becoming. Through its 'level' general potential everywhere is crystallised into what has become a specific, physical reality.**[42] **Archetype, quanta and particular fixations are thus simply understood as three different facets of the same reality.**

The implicit nature, quality or type of any explicit, localised object is expressed by cosmic fundamentals.[43] As colours derive from ratios of primary red, blue and green so physical phenomena derive from permutations of (↓↑) polarities and neutrality. Aspects of this primal trinity, expressed as ratios with various intensities of 'shade', are reflected in the stacks of Natural Dialectic. (↓) Gravity/ (↑) levity, electrical polarity, mass/ lightness and balance/ motion are examples of such aspects, using which the palette of creation paints each entity.

---

[42] *figs*. 3.3, 8.3 (with following stack) and Chap. 10: stack 2 may help explain.
[43] see Glossary, Index and Chapter 1.

Accordingly archetype, that's metaphysical, emerges as the 'hidden invariant'. Its changeless 'colour-coding' underlies the ever-changing stream of physical transactions. Perhaps cosmic memories are first physically expressed as immaterial force-fields. Sufficient excitation of such field (or latent file) results, of course, in action of its sort; actual examples of abstract archetypes, characters called particles, appear. We're into physics' 'upper', quantum level now. **Thus order, both determinate and indeterminate in style, is just the same; according to this scheme of things these two qualities combine.**

This conclusion is important to the point of repetition. Why should rules of physic govern metaphysic? A physicist's considerations are material but, if 'metaphysical linkage' does not have to cross physical space-time, it becomes instant. Archetypal influence is instantiated everywhere. This influence would be a property of 'pre-space', that is, of potential matter's void. Potential matter's been identified, unconscionably by naturalistic reckoning, as subconscious mind.[44] **Subconscious mind is a generic term for files of memories.** Could omnipresent connectivity reside in mind - not individual but universal mind? Not conscious mind but archetypal memory? **Thus logically quantum physics, dealing with the subtlest, fundamental forms of matter - force fields, pure energy and charge - would interface with metaphysic.**

Towards a Theory of Physic and Metaphysic

You may understand, this way, that Metaphysical Reality yields physical appearance. Archetype orderly informs the expression of physical polarity most basically expressed as energetic purities (forces) and constraints (particular things). Wave and particle, Bohr and Einstein, quantum and classical are each right - but, not unnaturally for physics, science misses out the Prior, Informative Reality.

Could *you* plan a cosmos better? Lucid physicists agree it looks 'as if' designed. **When it comes to astrophysics and cosmology stunning ingenuity seems to have coordinated chemistry and physics' natural laws, not least when it comes to bio-friendliness.**

Physic *and* metaphysic? *Metaphysic seems, except for logic, reason and mathematics, irrelevant to an operational description of material phenomena.* It is apart from physical events but not, perhaps, from their foundation. Unseen, it could be part of operational pattern; it may be relevant to *why* the system works, that is, why it issued in the form it has. It's worse than that. *Scientific materialism that excludes metaphysic essentially, delusionally excludes the instrument of its own operation*!

The Dialectic's metaphysic rests with information; and for physics information is conserved. What kind of conservation's that but so-called 'natural law', the behaviour of energy projected from outside the edge of

---

[44] see *figs.* 3.3 and 5.1.

space - where 'outside' actually means emergence from an inner metaphysic and not some outer spatial or temporal place.

Where, therefore, is the projector; what's the transcendent nature of projection's source? Natural Dialectic's 'holographic' edge is everywhere; it's 'super-posed' on physics, omnipresent but invisible because it's metaphysical. **It is the place where theme is turned to individual instance, principle is practised and where metaphysic with its physic meets.** For example, from a single archetypal (or ideal) electron local, physical electrons can appear; space, time and things are all *within* universal mind. **Archetypes are thus connectors that project our universe; archetypal memories, potential matter, are the essence of our physic; they inform, unchangingly, material being. Immaterial information holds the world, physical and biological, together. Cosmos is by archetype conserved.**

Snowflakes whirl on random, unseen gusts of air. Yet, in drifts or icicles, an order shows; flying fragments are 'constrained' in definite complexities. Quantum factors whirl in thermal motions. The laws of physics are statistical and, as atomic bonds and aggregates increase, so order shows. Masses of cooperative particles that constitute phenomena assume predictable configurations and behave in certain ways. *Thus statistical order arises from quantum disorder.* It is due to atomic structure (as in the inevitable quantum configurations of atoms), bonding and bulk material constraints. It might well be argued, therefore, that materialism's emphasis on randomness is quite misplaced. **This is because, beside unfathomable motion, *character* of particles endures - such is innate, intrinsic or potential order.** Different particles yield definite, though different kinds of interactive order. **Order is intrinsic in phenomena.** *The question then becomes (for electron, proton, photon and so on) origin of this enduring character that leads, in company, to mathematically predictable behaviours. Did chance or otherwise give particles and their great universe specific character?*

Archetype as origin of natural law is not a new idea. Potential matter is sited on the interface at which subjective and objective meet. The metaphysical precursor to all physicality is known, in either individual or universal mind, as archetypal memory. Through this unconscious gateway are expressed the agents (quantum behaviours) and the outcomes (gross, sensible materials) of a starry circumstance that we call 'home'. Order therefore emanates from inside; guidance of the cosmos issues, hierarchically, from above. **The challenge is to integrate an archetypal form of information with physics as it stands today.**

Can Mathematics Help Us?

No objects bear a number yet you number them and count. Numbers, symbols and mathematics aren't a physical but changeless, metaphysical reality. *Maths is a form of metaphysic.* Is that metaphysic natural or not?

Stars are a 'natural' form of physic; we discover them. A computer is 'unnatural' because invented. Is maths discovered or invented or both kinds?

It is strange but sure that a symbolic construct of the human mind - materially non-existent and meaningless without it - can reflect so well the energetic patterns, balance and equation of an ordered universe. How expertly chance (if chance it was) has crafted cerebration capable of immaterial symbolism, reason and mathematical description of cosmos - even its own nervous blips. Spot-on reflection! Macro-serendipity! Evolved brain cells are, it is supposed, able to deduce plausible scenarios of unsettled matters that range from, say, big bang to string theories! Could it be that mind not matter naturally originated universe? You might dare to wonder if numeric rules, embodied by specific formulations, mean the fabric of creation is a rational one. Did a pure mathematician plan the logic of real nature's numbers game? Does science hinge on metaphysicality?

Pythagoras, at least, believed that 'All is Numbers'; such immaterials help describe mind-matter frontiers whose formations are called archetypes. For others, including Roger Penrose, the Platonic world of mathematical forms is also real. He writes, 'There is a very remarkable depth, subtlety and mathematical fruitfulness in the concepts that lie latent within physical processes.'

Could essential, physically-independent numbers really govern physical complexity? Einstein, Planck and Eddington believed that, once you dropped upon their key, you could *deduce* (*top-down*) the reasons for all natural laws; you might unlock the codes whose inmost mysteries reveal just how a stable universe is sparked; a feat of mind *par excellence*!

Why, for example, are the electron-proton mass ratio, the fine structure, gravitational and other constants exactly what they are? Their formulaic outcomes are so physically co-operant you might proclaim that mathematics in the cosmos lets us live!

Symbolic numbers obviously pre-date humanity and they'll endure as long as cosmos but is this enough? Numbers are eternal abstracts while it is energy not immaterial mathematics breathes the real fire; flames and motions not equations drive the cosmos forward. *Like the 'laws of nature' ciphers may describe but don't per se create a thing.* **Indeed, they have no being but in mind.** They inform the various quantities and qualities of mass and motion with a web of metaphysical relationships, transcendent logic and eternal principles. Men exploit this 'inside information' to control material circumstance. Can sums alone fill up your bank account? Naturalists might dream that mathematical equations could create (not just describe) a universe - but theirs is pure illusion.

If, however, mathematics forged an archetypal link beyond the bounds of space and time it would, like any law of logic, have to be an abstract entity. In what could abstraction be embedded on a universal scale? For Natural Dialectic the reply is simple. The physical world is, it is scientifically agreed, mathematical; and mathematics lives in minds alone. *Is it, therefore, an illusion or the truth the world exists in mind?* **In this sense mathematics is a real form of metaphysic; the world runs, you could say, on immaterial lines. Archetypes are also metaphysical and have no being but in mind.** *Maybe, therefore, 'natural mathematics' of a physicist describes real, archetypal files.* And perhaps these files compose the link between the corners of what Penrose has referred to as a mysterious triangle of physics, maths and mind. At any rate, their logic's fixed. Fixed forms of mind are memories; and archetypal files are memories in the data bank of universal mind.[45]

There is, very close to you, definite evidence for Natural Dialectic's immaterial element. Just allow subjective mind is not the same as brain; allow the information centre that you're thinking with is metaphysical; allow that nervous molecules of brain work with but do not excrete your consciousness; nor, though they may label, *are* they actually your vivid memories. Is experience just material? Or could there be a simple fifth and metaphysical department, an informative dimension labelled mind? And, as universal body, so a universal mind?

You say there exist only non-conscious elements in four observed dimensions.

I say there are two existential components - mind and matter operative in at least five dimensions.

I propose, furthermore, that metaphysical *precedes* physical. *So is the unifying principle that sources physical phenomena universal mind?* How bizarre is that? Reason, logic and mathematics are, like morals, metaphysical. They inform. And symbolic information is the basis of biology; its foundation is, in this respect, definitely metaphysical. Could this be true of cosmos too? At least, before it drops to concrete earth, the nature of this abstract mind can be described in mathematical but also balanced, polar terms. For Natural Dialectic information regulates the form of physics; and compelling mathematics underwrites the logic of an archetypal lot.

Punting on the Cosmic Stream

What did you expect? Chance is what you *don't* expect - though if it were completely random neither law nor mathematics could describe its operation. Complete randomness is, by definition, completely unpredictable and therefore unreasonable. So can you conclude, although the cosmos is fundamentally irrational, that its irrationality is governed by rules, by chance-generated natural laws that make it reasonable? In other words, chance is the absence of a defined cause but initially created the universal framework of defined causes within which it happens, a framework of 'necessity' called natural law. Can such nonsense be true?

---

[45] see Glossary and index: archetype; also Chapters 13, 15 (Frozen Time) and 16.

In which contradictory case (the one that philosophically underpins physics and chemistry) chance and necessity are flip sides of the same material coin - one weighted, you bet, towards odds-on necessity. Indeed, the whole scientific endeavour, based on a presumption of balance, law and predetermination, has been to illuminate the patterns, rules and laws to which physical operations experimentally and without exception adhere. The behaviours of material automata are described by a universal language, a balancing act called mathematics; they are described, that is, on the presumption that what goes into a reaction comes out transformed but, essentially, energetically, the same. It works. *And the effect of rigid determination is to eliminate chance. What is residually perceived as chance happens within a framework of natural law. Is such residue itself an aberration, a flaw in certainty that's down to statistical error?*

Yet it is presumed by complete contrast, according to the tautological cycle of illogicality spun in the previous paragraph, that the origin of the world's legal infrastructure coincided with a sudden, causeless appearance of matter from nothing. In other, oxymoronic words, the principles and laws are themselves a matter of chance. Direction was undirected. *Chance happens within an accidental framework.* Do all paths lead to Lady Luck's? Was 'necessity', that runs the universe upon a stable course, the star-struck product of just accidental intercourse? Did chance mint the patterns of transformation we call laws of nature? *What, you ask again, seeds earth's reality?*

The question is humanly important because ethics, morality, law and social politics are all derived in the context of its answer. *They involve both current and historical contexts; and of these the historical eventually and intimately invokes the issue of origins.* Where am I from? Origins are, like parents, at the heart of how we see ourselves. Creation stories, of one kind or another, are an imperative. They closely affect the texture of our lives. If there is deliberate (and therefore reasoned) theme and variation-on-theme, that is one thing. If, on the other hand, nothingness for no reason exploded/ expanded, natural law is itself the product of One Big Accident and unsystematic evolution works by breaking themes of type, then the basis of reason is itself unreasonable. What rationality resides in chance? What's your punt upon the fundamental current of the cosmic stream?

Shots in the Dark

What are the punting odds? What's the probability that everything occurred by chance, that cosmos and its forms of life were *not* designed? Could Luck, essentially, have come to pen this autobiographical essay? Where 1 means certain fact 0 means impossibility. Probability is ranged between but odds longer than 10^{50} to 1 against are generally conceded impossible. However, odds as low as that can easily be ranged against the chance occurrence of complex components, let alone the correlated coding, of a single protein in a cell! To compute that nothing is impossible (that zero can't exist) is, in effect, to waive constraint; it is to ignore the automatic, uncreative necessities of physics and the rules of chemistry. In a world of predefined or, at least, definite natural law this is clearly nonsense. Sheer fairy wands, will-o'-the-wisp and lucky charms. The horseshoe on the lintel of Fortuna's shrine has fallen off. Approximate determinism is the maximum an anarchist, atheist or other cultist can expect. In this affair he trusts, with halo slipped to jaunty slant, Caprice's infidelity; but is 'conditioned chance' enough to have evolved the Slender Lady's worshipper?

For matter nothing aims to hit a target. Every shot is in the dark. Its principle of certainty is a fixed pattern of behaviour, a mathematically precise, predictable 'law of nature'. Automata, the mindless objects and events of nature, experience no uncertainty. They neither expect nor predict anything. What is is certain. In oblivion there's no such 'thing' as freedom, possibility or chance.

Unlike matter *mind* seeks information to dissolve its darkness and so aim in clear air. Humans crave relief from ignorance and the empowerment of finding out. We want to know for certain what is going on. *Mind is certainly an uncertainty-cruncher.* It is a probability-reducer, the agent of anti-chance. It is the organ of shot-in-the-light, that is, of teleology. *Nothing skews probability into certainty like thought. No magic transmutes impossibility into actuality like mind, like purposeful action. Natural forces only partially control our world.* Human history (both individual and general) is testimony to the unpredictable, world-changing effects of an invisible, subjective, metaphysical kind of data item - ideas. *The impact of ideas on behaviours and, thereby, human social and environmental history is of central importance but hardly amenable to mathematical description.*

How do you feature in this film? Perhaps, as far as life's concerned, Deliberate's assassination and Luck's coronation are illusions of an atheistic mind. Perhaps reliance on original power of chance is just a dream that will not die. It is a work of physics to identify the *modus operandi* by which principles are outwardly expressed; but in the case of biology P. Johnson writes,[46] "What is presented to the public as scientific knowledge about evolutionary mechanisms is mostly philosophical speculation and is not even consistent with the evidence once the naturalistic spectacles are removed." Is this true or false? Is life purposeful or purposeless? Is purposeful derived from purposeless? Is life the shot of all shots in the dark? If so, then Lord Deliberate falls dead. And long live Lady Luck!

---

[46] Reason in the Balance p.12.

This is the royal assassination that materialism's revolution utterly depends upon. Oblivion must win the marksman's prize. The lucky shot must hit; the hit-and-miss must not have missed. Secular philosophy, pervasively entrenched in universities, the media and palaces of politics, is based upon this utterly incredible fatality. Deliberate's dead, long live Luck's vitality! A shot of long shots in the dark is at the heart of atheism. Disciples of the faith have bet their lot on it. No hedge. Now you can understand why edginess erupts if you elaborate the way, by light of day, that evidence decrees such an imaginary pellet must have missed. Who bites the bullet? Who is shot? Who, in this attempt to dominate the world of mind, is flattened?

In short, a definition of life is that it purposes; and that a presence of mind in matter changes everything. Mind is not Fortuna's province. The notion that cosmos is the product of intelligent design is materialistically unacceptable. *Teleology, you remember, is the doctrine of final causes, of design, of preordination or purpose.* **Science cannot deal with it and for a non-teleologist natural intent is the key anathema.** One-tiered materialism has, by decree, outlawed a rational origin of cosmos. But reason is the clerk of cause. Be-cause. If explanation of a system lacks the reason how it came about the first and greatest cause is lost. You can't view that system with a full perspective. There's chaos and unreason at the root of things. If reason is denied it's broke; unreason, in the form of randomness and unpredictability, slips into the office that it leaves. Thus, from the holster of a revolutionary guess, shots are fired into original, acausal darkness.

The Chance-Killer

Do you think that what you're reading's repetitious? Did you know it all before? Why then, if the penny's dropped, are you still frozen in a fashionable, humanistic pose and why does science still tout Darwin's theory as a Truth?

It must be time to turn the tables. Time to resurrect a hero from the dead. From law to luck and now the spotlight swings again onto the very soul of reason, Lord Deliberate. Doesn't his comportment ring a bell? You met him under an alias, another ID known as Intelligent Design. Intelligent awareness is an understanding full of plans and purposes. Deliberate never has a flutter for the Lady Luck nor she, fine thing, a bet on him. The odd couple are strange bedfellows - if they co-exist at all. Each programmer de-bugs, every engineer deliberately suppresses any chance of chance. Is it any mystery? Purpose kills chance; Lord Deliberate is a lady-killer and the lady's Luck.

Luck shafted? Ghost busted? Her ladyship struck down? Deliberate does not execute with random strikes but strikes precise on randomness. Sharp the wit that, at its sharpest, with the first stroke cuts a swathe through error. No mistake. Full marks. And when its deed is done the die is cast. Deliberate's culture's one of certainty, of luck's elimination.

So if her ghost appears then, gripping Lady Luck, you grip thin air - nothing but (as ghosts all are) the unexpected. Who'd rely, instead of schemes, on luck for what he wanted? The girl's a charmer; she's a dream as mindless as Deliberate is not.

You best define a man by his priorities. What is Lord Deliberate's priority? What is most important? It's the central point of education. Correct logic fosters intelligent thought. Grammar and syntax precede the expression, in words, of that thought. Purpose, rules and law precede intelligent behaviour. Such behaviour predisposes matter according to its plans. Intelligence intends to organise the things around it. It creates patterns that accord with the character of its desires. Perhaps non-chance is not coincidental! Our lives are clear examples. For instance, each man more or less perfectly understands what's orderly and what is not. As each person's relative sense of order is a function of his sense of truth, so intelligence is a function of order. **The origin of order is critical.** What *is* the origin of cosmic order? Of natural law? What latent logic precedes and thereby informs the character of motions in space-time? *We need to know because, in truth, we subscribe to order and not chance. Reason and rationality are based on it.*

It is a natural fact. Mind is anti-chance; it creates, by nature, order. Why, therefore, blench white-faced from a Theory of Intelligence? The higher an intelligence, the more informed a mind the more effectively its plans are implemented. The keener its interest the more rapidly, systematically and harmoniously informed is the complex detail of its creations. **Intelligence is certainly a first-class chance-assassin. Purpose by its nature always kills chance dead.**

A Culture of Doubt

↓ *tam*	*raj* ↑
negative	*positive*
scepticism	*faith*
unbelief	*belief*
doubt	*confidence*
uncertainty	*certainty*.

Doubt is faith's other face. Uncertainty is certainty masked inside out. Each depends, in its degree, on prior axiom, experiment and your experience. They depend, that is, upon perspective and interpretation. How, therefore, to light on truth without engaging dogma? In which direction might I find the Truth of truths?

Richard Feynman, a popular physicist with a modern image, doubted 'mind behind creation'. He thought of empirical science as a culture of doubt, a vehicle of probability, that is, of degrees of uncertainty that range from very doubtful through to very likely but never absolutely, dogmatically certain.

Cosmologist Stephen Hawking also rejects a metaphysical culture of certainty. Instead he's certain that before a 'big bang', 'absolutely nothing' lurked.[47] Nothing material nor (since nothing immaterial exists) immaterial either. Having thus employed the scientifically respectable expedient of banning any pre-physical factor, he's sure Deliberate is as non-existent as the cosmic absence that, for him, exists. Yet although he disagrees with Natural Dialectic Hawking still perceives a fundamental paradox in the search for a completely unified theory of physics - such a theory must include ourselves, our thoughts and actions. 'So the theory itself would determine the outcome of our search for it! Why should it determine that we come to the right conclusions from the evidence? Might it not equally determine we draw the wrong conclusion or no conclusion at all?' In short, he believes such hypothesis has not succeeded in reducing chemistry and biology to the status of solved problems; and the possibility of creating a set of equations through which it could account for human behaviour remains entirely remote.

The gap expands wide open. **A theory that omits an immaterial element can be taken no more seriously than a theory of the non-existence of matter.**

If you dislike this view the only riposte is to kill it. Categorically (but at the same time metaphysically) eliminate the metaphysical. This is why atheists (such as the late militant, Francis Crick of *DNA* fame) feel compelled to try and prove that mind is not just caged in body but consciousness is made of matter. He wanted, as a priority, to bust the ghost and thus snare the element of subjectivity within the objective scope of science 'proper'. An iron fist wields, of course, exclusively material consideration. The frontier for reducing consciousness to nerves is pushed by neuroscience.

Of course, no doubt exists if you can just squeeze thinking out of brain. Indeed, 'solidified' psychology treats subjects as if objects. Instinct, ascribed to chemicals not archetypal mind, rules patterns of behaviour. Behaviourism, an ideologically logical but now discredited outcome of materialism, treated instinct, thought and feeling objectively (that is, from 'outside') as if mind worked mechanically. Are not the motions physics traces automated? Is not mind, though complex, physical and therefore reducible to mathematical, perhaps statistical, analysis? What did you expect? Where 'psychology' means 'study of the soul' this special type is soul-less. It has objectified the true, immaterial subject of its attention, gone out of mind and thereby 'lost it'. Such post-medieval reason rests on nothing but an intellectual annihilation of subjective consciousness. Mind is, by mind alone, reduced to nervous residue that neuroscience studies.

---

[47] see also Chapters 2, 9, 11 and 27.

What kind of object are you? If consciousness were nullified would simply energy remain? *For Natural Dialectic information and energy (or mind and matter) are two separate elements, material and immaterial, entwined.* Subtract mind from mind-and-matter and you've only matter left. On the other hand, if you've devalued life's coordinate to zero what might lift its value up again? Simply add increasing concentrations of awareness we call mind. **In this duplex case a unified theory must include psychology that, more than simply electrochemical, takes full account of metaphysical information and its purposes.** It must recombine both objective *and* subjective coordinates into a unitary scheme of things.

A scientific culture's one of healthy doubt. Is faith in doubt a working premise or a final attitude towards life? Could such a culture bring itself to doubt its own foundations? *In this case would a dose of doubt about materialism do?*

The Matrix

The dose might come from east and west. We could check what Einstein, Bohr, Heisenberg, Sir Jagdish Chandra Bose and Satyendra Nath Bose FRS believed. Sir James Jeans noted 'the universe seems nearer to a great thought than a great machine''. But let the final word rest with Max Planck. Planck not only first read, recognised and published (in the *German Annals of Physics*) the start of Einstein's revolution, the latter's Special Theory of Relativity. He also pioneered quantum theory, the second pillar of modern physics whose study is sub-microscopic, sub-atomic phenomena - the matter-in-principle of Natural Dialectic. Actually, his foresight may have ushered in the next revolution in human understanding which has already begun to focus on the primacy of the 'unscientific' informative co-principal[48] as opposed to its energetic coordinate. He asserted that the discovery of truth can only be secured by a determined step into the realm of metaphysics and, at a lecture in Florence (1944), said:

"As a man who has devoted his whole life to the most clear-headed science, to the study of matter, I can tell you as the result of my research about the atoms, this much: *there is no matter as such.* All matter originates and exists only by virtue of a force which brings the particles of an atom to vibration and holds this most minute solar system of the atom together...We must assume behind this force the existence of a conscious and intelligent Mind. This Mind is the matrix of all matter."

Indeed (The Observer 25-1-31 p.17), "I regard consciousness as fundamental. I regard matter as a derivative of consciousness."

Perhaps Max Planck would have appreciated Natural Dialectic.

---

[48] Chapter 2: Informed Energy.

Chapter 8: Principles of a Unified Theory of Matter

The Principles of a Unified Theory of Matter

It might, in this cause, be said that Natural Dialectic simply takes what's there and reassembles it in an internally self-consistent way to make another sense. Or again, that it tracks along its sub-routines to improve sense of the cosmic program as a whole. *Every systems analyst understands top-down logic.* If you want to understand how, *top-down*, anything works you have to begin at the beginning; you have to understand its purpose and its principles. **For a full understanding you need to grasp first principles. To grasp the principles behind physics you may need to consider their metaphysical origin.**

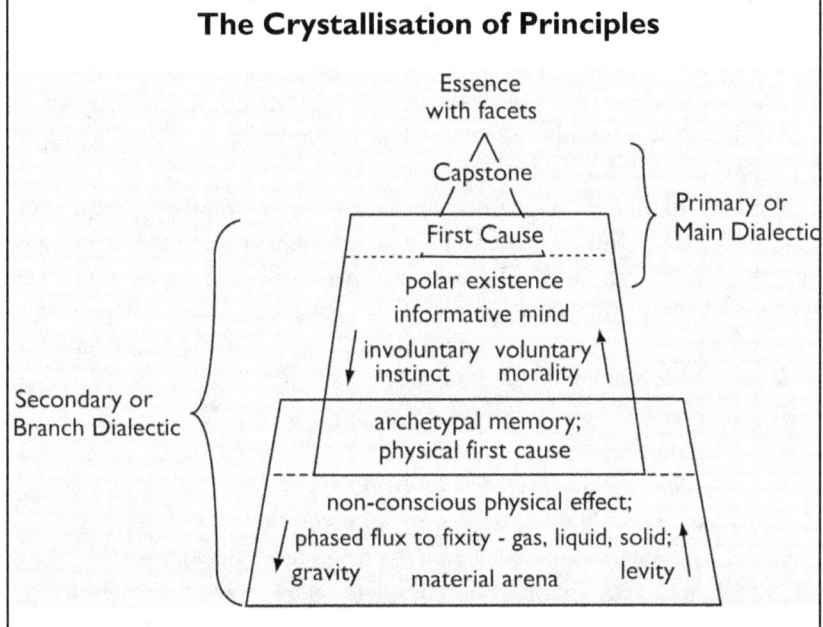

Crystallisation of principles is the natural process of creation.

Physics inspects the world's non-conscious stage. It examines what's 'below' the physical first cause. This cause, dialectically 'wedged' between mind and matter, is an archetypal step reflecting, as potential matter, the Capstone's existential First Cause.

fig. 8.1

What is a principle? **A principle, which has been called a highly condensed form of information, is conceptual.** Because it precedes, orders and directs subsequent and often complex actions, *principle is potential action*. Because it is an apparently absent latency, nothing without practice, principle is *symbolic action*. What is law without behaviours, idea without fruition or intention *sans* its tool to implement? Principle orders action. It is the source and definition of all legal activity. Its intent is expressed through mechanisms that accord with its logic. *This is no less true of the cosmos than the latest plan or program of action (however grand or trivial in scope) that you brought to fruition.*

Principle informs practice. For this reason it has been argued that active precedes passive information, concept precedes manufacture, thought precedes action and, axiomatically, mind is the repository of principle. In other words, mind is the potential for material order.

Because they reflect the way principle is (or could be) translated into practice abstract mathematics and the laws of logic can predict and compute automatic physical relationships and behaviours. Physical principle, the 'instinct' of nature, gave and gives rise to matter in a specific order and in order to understand creation properly it is imperative to follow this sequence. The narrative must follow its logic. *The right order gives the right reasons.*

You might claim that, from an inductive, *bottom-up* perspective, principle derives from practice. Things just behave as they do because they are what they are. There is no particular reason behind their 'reason'. Asking after 'deep reasons' or 'first causes' is misguided, a waste of time, an enterprise doomed to failure by its own misconceptions. Principle and law is the result of chance. But is it really?

The bottom line is that modern 'rationalism' dislikes the notion of a hierarchical cosmos based on Rational First Principles; and its atheistic brother-in-arms detests the idea of Natural Intelligence. What do they offer instead? Their foundation is a shrug and offering of accidents.

From a *top-down* point of view (*fig.* 8.1) such nonchalance implies a preference but also involves a complete guess. Whether genuine or disingenuous, the shrug incorporates an inestimable hypothesis (which divorces it from science) and the irrational presumption that chance can generate non-chance, that is, underlying regularity (which divorces it from logic). It derives, simply, from a one-tiered, 'flat-universe' perspective. What is the origin of anything? What is its reason? Not lack of reason, surely? So that everything (called cosmos) never lacked a cause.

<u>*Top-down, principle precedes practice and the influence of principle is increasingly 'crystallised' through sub-principles, ideas, archetypes and, finally, physical phenomena.*</u> Therefore, one needs to understand the origin of principle to understand its consequent practice.

The Diamond Capstone

This Capstone illustrates Essential Qualities from which existence issues

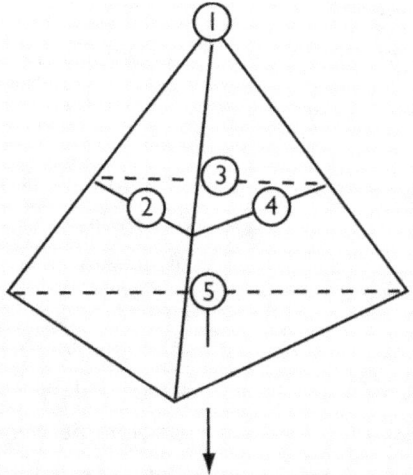

Substance	①	Consciousness	(Chapters 13 and 14)
Essential qualities/ facets	②	Infinity	(Chapter 7)
	③	Unity	(Chapter 9)
	④	Nothing	(Chapter 10)
Precondition	⑤	Potential	(Chapters 5 and 8)

The Apex of Mount Universe is likened to a Tetrahedral Diamond with three clear facets and, at the base of its transparency, a fourth to interface existence. The trinity involves, at the highest level, aspects of its Substance; it incorporates the triune qualities of a Pure Concentrate of Consciousness, that is, of Life.

Facets of the Diamond, Essence, include Infinity, Unity and Nothingness. These comprise Potential for First Cause from which creation is expressed; from this intrinsic precondition all the order of existence hierarchically falls. The next five chapters investigate the nature of their reflection in the zone of non-conscious energy, that is, physics.

fig. 8.2

tam/ raj	*Sat*
lesser truths/ appearances	*Truth*
aspect of reality	*Reality*
sub-principles	*First Principle*
particular expression	*Principle/ Law*
local practice/ outcome	*Potential/ All Possibilities*
subsequent order	*Archetype*
elaboration	*Plan*
change/ inconstancy	*Stability/ Constancy*
periphery	*Centrality*
polarity	*Neutrality*
↓ *tam*	*raj* ↑
descent ←	→ *ascent*
from Axial Source/ Hub	*towards Axis/ Hub*
separation/ > duality	*unification/ > unity*

for mind (information-dominant; truth-seeking) we write:

tendency >passive/ automatic/ lifeless state	*inclination > purposive/ meaningful behaviour*
towards infra-consciousness/ sub-consciousness/ oblivion	*> ultra-consciousness/ super-conscious awareness*
> non-conscious physicality	*> Natural Heart*
inc. appearance of coincidence	*inc. perception of order*

for matter (energy-dominant; non-conscious) we write:

increasing loss of free energy/ entropy/ bondage/ fixity	*energetic gain/ negentropy/ flux of process*
resistance/ inertia/ exhaustion	*action/ motion/ freedom*
repetitive motion/ immobility	*stimulus/ change*
grosser/ materialisation	*subtler/ dematerialisation*
gravity	*levity*

Before attempting to rephrase scientific jargon, consolidated perspectives and speculations in terms of Natural Dialectic, it is necessary to paint a broad-brush sketch of what sort of principles and conceptual norms that a holistic mind-set (involving elements of energy *and* information) might include:

1. *top-down* scaling; a conscio-material gradient
2. interplay between cosmic fundamentals in different proportions in different cases

3. Archetypal First Cause whose essence creates mind; and, at the base of mind, a physical first cause that constitutes the archetypal norms to which material patterns of behaviour cleave. In any actual context - psychological or physical - first, nuclear causes 'germinate' the rules by which events occur. Such informative potential is identified with the tier above its expression. In dialectical terms the informative tier above matter is mind; and in mind potential matter constitutes the lowest and sub-conscious tier. This implies that the principles governing the behaviour of material energy (the scope of physics and chemistry) are lodged in and derive from a metaphysical source. **We make sensible, sensory connection with matter; but we make contemplative connection with its metaphysical principles.**

Thus, in principle, the simple order of argument runs according to the next stack.

physical matter	*Potential Matter*
action	*Physical Absence/ Void*
program played	*Archetypal Program*
lower orders	*Upper Syntactical Level*
↓ *matter-in-practice*	*matter-in-principle* ↑
gross	*subtle*
classical bulk	*quantum flux*

First cause physical or potential matter correlates with fig. 6.1's upper syntactical level of passive information. It is also called transcendent potential, absolute matter or archetypal memory. Having identified the First Principle of a Unified Theory of Matter we begin, in this Chapter, to further discuss the immaterial nature of its material absence, that is, its 'pre-physical' void.

The order of play then turns to its (*raj/ tam*) vectored physical issue - the lower syntactical level or lowest, quantitative level of passive information. This level is the final expression of principle in physical practice, that is, in terms of data items called objects and interactions called events. It involves[49] two sub-divisions.

First of these is the kinetic condition of 'matter-in-principle'. This is also described as subtle or active matter. Physics' *AV* knows it as quantum and atomic physics[50] whose simple, primary agents give rise, in congregation, to subsequent complexity. This secondary subsequence, the locked-together condition of chemistry, is called 'matter-in-practice'. It is also known as the gross, passive or exhausted condition of energy that includes gases, liquids, solids and, of course, all study related thereto.

---

[49] *fig.* 3.3 section 5.
[50] Karl Popper dubbed quantum physics 'the transcendence of materialism' but for Natural Dialectic this phrase applies to *metaphysical* potential matter/ archetype.

The challenge for Natural Dialectic is always one of orderly narrative. **It has become clear that, starting at the top, the order of this book follows an act of creation.** Such act devolves from Essence through mind (Chapters 5, 6 and 13-18), matter (Chapters 7-12) and their biological conjunction (Chapters 19-25). Where the previous section (Chapters 5 and 6) dealt with psychological aspects of initiation - Consciousness, Information and Authorship - this one sets the material ball rolling. It explores four capstone transparencies in terms of physical projection, that is, in terms of physic's primal trinity - space, time and energy. It suggests that these essential qualities translate into the keystone of an arch binding the pillars of quantum and classical physics; also that quantum agents of 'matter-in-principle' reverberate with archetypal 'notes' or resonance.[51] Creation is thereby devolved from a mental template; through such an arch the world appears.

Thus Chapter 7 described physical creation in terms of precondition, preordination or *potential*; it related archetypal information to the integrated, energetic patterns that the basic features of our universe display. Chapters 8-10 describe the triune expression of *infinity*, *unity* and *nothingness* in terms of theories and fact. Finally, in Chapters 11 and 12, we deal with 'matter's ghost', *pure energy*, and energy's particular deployments in physical, non-conscious cosmos.

Infinity

First Principle is, paradoxically, beyond all principles. It is pre-existential. Although unconditioned it shows facets. The first is its Essence, which means unqualified Being. Essence is being-without-a-second. The major qualification of Essence is essence-in-motion. This is called existence. Existence (changeful events, objects or attributes) predicates Essence (the subject).

If First Principle is likened to a Diamond Capstone then its major facets subdivide into a stack of 'cuts' reflecting the internal light of Essential Substance. These 'cuts', characters or qualities are principles because the natural laws of mind and matter are, in practice, drawn down from them. **In other words, all nature is a derivation from First Principle; it is an expression of The Infinite.** The task of Natural Dialectic is, therefore, to take order from this Prior Potential.

Essence is It, mathematics needs it but, throughout physical and psychological existence, ∞ seems to be an object absent. How, therefore,[52] can anyone 'reduce', approach or analyse the thing? Books have certainly been written on infinity. Aren't sets, numbers and the universe potentially unending? The ultimately indefinable has even wrought, in the case of Cantor's 'towering infinities', insanity! You can

---

[51] see also Chapter 16: How Does the Connection Work?
[52] if one rejects The Nature of Infinity according to Top Teleology (Chapter 5).

add, subtract, divide or multiply but leave it just the same. Infinity = 1/0; therefore infinity x 0 = 1 and 0 = 1/infinity. Indeed, x (any number) times infinity is as infinite as any divided by zero; and any number to the power of zero = 1! Strange equations. What's their meaning? Can Natural Dialectic offer any hint of resolution?

The motion of Essence shows as existence - things and events. These are finite. What is finite is, simply, a fraction of The Infinite; it is a unit and can be assigned a number or described by an equation; but abstract description, verbal or mathematical, is not the actual object. Existence is in all parts, physical or abstract, finite, relative and dependent but its predecessor, Essence, is independent, absolute and infinite. Limitless Being-without-condition - The Infinite - is, as Itself, not anything and thereby inaccessible to counting. Here maths, even very clever maths, is not the point. Not abstract zero, it *is* Formless Nothing, that is, Immaterial Potential. And the Symmetry of this Immanent Void is unbroken, undivided and therefore One. **In other words, The (N)One is, simultaneously, one, zero and infinity.**

Such Supreme Paradox is, beyond existence, Absolutely Non-Sensible; absurd as any quantum trait it is, however, furthest from absurd; it is our foundation and our starting-point. **Its numerical nonsense, the Essential Equation, is expressed as 0 = 1 = non-numerical, formless infinity.**

The Existential Equation is, conversely, expressed as 0 ≠ 1 ≠ numerical infinity. *This, amid the relativity of existence, makes straightforward sense of things.*

The former, which represents Absolution, is converted into the latter, relativity, simply by virtue of motion:

motion	*Essence*
finity	*Infinity*
relativity	*Absolution*
appearance	*Truth*

Using this stack let's revise. (*Sat*) Essential or Central Dialectical, right-hand characteristics are primary. They also show, secondarily, in any local and/ or temporary instance of centrality, balance, truth etc. In Chapter 4 we called these instances 'lesser truths' or 'appearances'.

Perhaps Cantor but certainly all mystics promise Absolute Infinity. Relative to this stand intellectual/ mathematical/ existential infinities. These involve endless series of numbers, members or sets. Such objects, including the abstract extensions of time and space, Natural Dialectic calls 'lesser' or 'apparent' infinities.

So, in conjunction with *fig.* 2.6, let's differentiate between infinities at top and base end of the mind and matter grades. At top we note affinity intrinsic light/ illumination; and at base unseeing mind or darkness.

For mind we write:

restricted consciousness	*Illumination*
mind	*Essential State*
relativity	*First Cause/ Logos*
relative impotence	*Potency/ Control*
↓ *unconscious sub-state*	*mental spectrum* ↑
darkness	*range of shades*
total lack of awareness	*degrees of knowledge*
mind dormant	*mind awake*

And for automatic matter:

non-conscious matter	*informative archetypes*
classical/ quantum patterns	*potential matter*
actions/ excitations	*changeless template*
range of shades	*transparency*
↓ *classical restriction*	*energetic quantum phase* ↑
mass	*force/ energy*
macroscopic aggregate	*microscopic agencies*

And at the top of each grade find (see *fig.* 2.6 again) its archetype.

Initial motion of supra-mental infinity is called First Cause, Word, *Logos* and so on. By this primordial motion is established cosmic relativity, the dynamic that we call existence. Such Archetype is a conscious experience, the seed of thought, the Logical Start of things. Its experience and, therefore, knowledge is called metaphysical enlightenment. Call it mind's holy ghost.

Mind is, thence, the prism that diffracts Infinity; it emits the spectral colours of existence; archetypal memories are stored and orderly creations scaled therein. It is the medium whereby global ideal is translated into local action; and whereby general principle is broken into actual practice. Mind is how information is disseminated, programs are implemented or a project ramified into its product. What more cosmic product than existence? How, though, could mind evolve from matter[53] or, conversely, physical beginning involve mind?

The lesser supra-material infinity of body is a second, lower first cause. It is called, dialectically, potential matter. This physical absolute is, transcending matter, metaphysical. It is composed of unconscious archetypes or memories in universal mind.

We have seen that as well as potential infinities of number, density

---

[53] see Chapters 11: Matter's Holy Ghost, 13, 16 and 24.

and a 'continuum' of space, absolutes that are not infinite constrain the world. Speed of light (c) and 0°K are two of these.[54] The images emerging out of every frontier of the Mandelbrot conceptual set are, each being finite, less than absolutely infinite as well. But natural laws (derived from archetype) are edgeless in appearance; physically 'non-local' archetypes define the form and behaviour of bodies simple (as studied by physics and chemistry) and complex (biological).

Nevertheless, if boundless archetype is nothing physical what is *least* physical, *least* massive, *most* absolute and immaterial of anything that's actually material? Doesn't light's pure energy reflect these properties and thus, least earthly of all things in space, the character of archetype? Apparently transparent, silent, motionless yet powerful, restless, colourful - its radiance paradoxically imitates infinity. Let there be, first and foremost, mass-free photons! Indeed, pellucid radiation from on high was seen traditionally as both the primordial expression of creation and a medium of communication straight from Creator's heaven to material phenomena. Pure will informs, pure energy expresses His Chiaroscuro. If, at material level, you identify such purity with light then insubstantial light might qualify as matter's ghost. And this ghost's light and heat are engines that, with water, steam the pistons of biology. Maybe, we'll see, light also helps link mind to body.[55]

Ground-state vacuum and an excitation in it - not metaphysical light of reasoning but electromagnetism whose symmetry of geometry expresses perfectly dynamic balance - in this scientific case between the oscillations of electrical (radiant) and magnetic (contractive) field components and their straight-line 'normal' of propagation. Light beautiful, fantastic, undulation in an unseen field - pure, polar energy with which to see! It is a phantom from whose shapes and ways an excellence of reason reasoned its enlightenment. The history of science clusters round a beam of light, renaissance has developed from its oscillations in the abstract field of space. *Light is the spectroscopic messenger from stars, the informant and ultimate connector with all we'll ever know of time and space*; it has led us far in understanding how the world works. Furthermore, not only does its touch our eyes and instruments from galaxies but, as Natural Dialectic will logically deduce, brings knowledge much, much closer to our home. **Chapter 16 suggests how finite light links mind and biological matter; how, in other words, it is the informative medium of radio translation between, on the one side, cell chemistry (including cells, nerve cells and brain) and, on the other, lesser metaphysical infinity of archetypal memory**.

We'll deal further[56] with the holy ghost and labour from an archetypal womb. For now let's note that physical energy is sometimes called 'nature', as if this was all that nature was. As if there was nothing outside

---

[54] see Chapter 3: Subtendence.
[55] Chapter 16: *H. electromagneticus* and Psychosomasis.
[56] also Chapter 11: Matter's Holy Ghost.

or within it. *It needs be emphasised that physical cosmos constitutes the lifeless end of creation: and that the commonsensical view of matter is incomplete.* A broader perspective finds life subjective but all bodies objective; and finds the objective cosmos, as a product of principle, basically conceptual. Accordingly physical energy is seen as an externalised, rigid or 'crystalline' form of universal mind. A shell.

A study of matter is the study of a special case, non-consciousness. Or, rephrased, physics and chemistry are the psychology of non-consciousness. What can physics say about what comes above what's physical? It has neither knowledge of nor interest in what is beyond material phenomena. There are two kinds of such 'beyond'. One is hierarchical, the other temporal. The first is metaphysical, the stage of universal mind; the second comes, before a start, from less than anywhere - The Great Projection of a Universe.

There's an **order of the finite** that subtends an archetype's material infinity. Grit grows in the ghost, binding agents create atoms and thence aggregate reality. Matter-in-principle comprises quantum-level forces, sub-atomic particles and atoms. In their energetic ocean large-scale edges float - bulk bergs, condensations that incorporate all massive bodies, animate or not, which we can sense and call our physical reality. Immaterial information's neither lost nor gained but initial cosmic energy is gradually wound down. Gas, liquid and, least energetic and most tied up, solids constitute creation's base precipitate. This sensible and bulk reality is called matter-in-practice. What is any object but a slow or fixed event? It is a macroscopic, slow and massive process or else one microscopically flashing by; either case depends upon the 'immanence' of its internal programs that we call potential matter's archetype. The whole collection spills from 'inner space'; infinity's reduced to finite things; energy locked into order now luxuriates into a universe. Is there any more to add?

In this way creation is seen as devolution from Central Order, an inversion of the Infinite Interior. *In terms of Mount Universe the ideas of 'descent', 'ascent' and 'levels' are useful but the predominant mode of materialisation is a* ($\downarrow$) *downward-driven process; it is outward, binding and confining.* The physical part is seen as a coverage, husk or shell that binds two, nested nuclei - mind and the Quintessential Kernel! In this way its own basic part (a finite, non-conscious atom with mass at the centre and peripheral energetic spheres or 'shells') inversely represents the whole creation (at whose nuclear heart is The Most Natural Essence, Conscious Infinity). *An atom is a natural symbol of the order of material, mass-centred universe.*

Unity and Unification

The second term in the Essential Equation is 1. Just as, like nothingness, infinity is boundless so this 'unit' isn't one since it is unconfined. But, as infinity's the source of finite forms, so the capstone's

tam/ raj	*Sat*
division/ duality	*Unity/ Union*
partiality	*Wholeness*
degrees of restriction	*Without Condition*
↓ *tam*	*raj* ↑
separation	*unification*
dispersal	*concentration*
segregation/ isolation	*relationship/ merging*
discord	*harmony*
disunity	*integrity of parts*
apart/ individual	*a part/ community*

formless unity gives rise to cosmic units - from galaxies to particles, from fractions of space-time to all the 'Russian dolls' that any sensible phenomenon consists of. Order of the unit, as of finite form, is the order of (↓) creation - ever tighter boundaries, that is, conditioning. Conversely, the order of (↑) return involves release - physical, psychological and, in final dissolution, fusion with the (N)One.

It is clear, by now, that Natural Dialectic is dualistic; its Secondary Dialectic (Chapter 1) involves the ceaseless motion called creation or existence. In this case, division (↓) 'freezes' into units and unification (↑) 'melts' back into larger, freer wholes. Unification (of particles and principles) occurs when apparently distinct entities are realised as aspects of underlying sameness. Perhaps the most cherished goal of physic and metaphysic alike is unification unto unity.

The Dialectic is, simultaneously, unitary. The Essence of its Primary stack is, resolving two into one, Unity and Balance. At root, at the Apex of Mount Universe, it involves 'Central Potential' (*figs*. 1.3 and 15.3); and, equally, subjective Communion with Essential Metaphysic, that is, with the Source of Information. Such profundity of Oneness is beyond the laws of logic, physics and existence.

Unity is both the Potential from which duality is sourced and the Law against whose Purity the forms of its creation, light and dark, are tested. **The criterion against which the distance of duality from Truth, Goodness and Beauty is best measured is Original Unity - whose Balance, Symmetry and Communion are facets of the Infinite (N)One.** In practice this means measure in comparison with the life of a human who, in perfect alignment, lives these ideals.

As well as thoughts your single mind holds seamlessly its memories - nor does it work in space or physically relentless time. It is, which matter lacks, an information centre. We spoke of first cause physical as the metaphysic of an archetype - an 'instinct' of the world. Such program is, in the hierarchy of creation, filed as memory in universal mind.

consequences	*First Principle*
expression	*Potential*
local, physical types	*Singularity*
corresponding phenomena	*Archetype*
thing	*Nothing Physical*
↓ *informed*	*informant* ↑
outcome	*principle-in-action*
anti-principle	*principle*
apart	*together/ co-operant*
unit 'self'	*connector*
specific/ detail	*general sketch/ outline*
bulk events	*atomic/ sub-atomic events*
classical level	*quantum level*

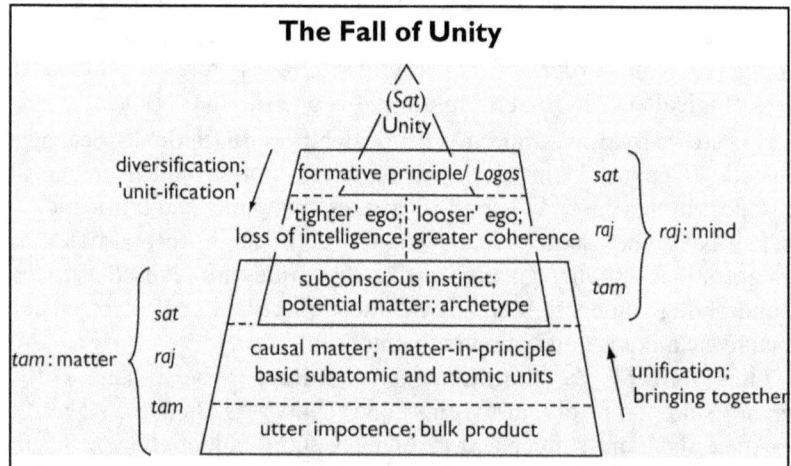

Communion or 'GUE' (Grand Unified Experience) is vested in Unity. This 'Stateless State' ramifies, through the single, transcendent portal of First Cause, down and out into the polar spread of mind and matter making up existence. First Principles are outlined in Chapter 1.

The unity of physical first cause, which ramifies out into the spread of 'physicalia', is at archetypal level. It is the metaphysical precondition or *intrinsic being* of matter.

Check figs. 3.1-3 as they relate to cosmic fall from Unity. Constraints increasingly separate original oneness into extrinsic, isolated parts; such isolated expressions of being (or potential) amount to individual forms of mind and matter; *physical formation is informed by way of matter-in-principle becoming, in turn, the matter-in-practice that we call physical reality.* Active unity falls to passive units; principle precipitates detailed, particular outcomes. Informative descent of mind is accompanied by informative degradation (entropy of information) and

> loss of consciousness; and energetic descent is accompanied by the gradual fixation of energy culminating in solidity.
>
> ***fig. 8.3***

Physics, down-to-earth, seeks to divine the causal unity in perhaps, some perfect law, one law to govern all, one law from which all others flow.

To recapitulate: physical or metaphysical, the 'make-sense' search is towards unification. **The grail of unity is total reconciliation.**

Holy Grails

The *AVP* represents a pinnacle of rational achievement but the various theories in its collection are riddled with contradictions, unproven assumptions and (notably between quantum and relativistic theories) stark incompatibilities. We're not there yet. In the end unification may well be achieved but, equally, involve factors presently not even considered.

Holy and holistic grails exemplify a primary drive in human consciousness to relate apparently unconnected facts or events by means of underlying principle and, in the case of scientific research, mathematical equivalence. Such reconciliation unifies superficial diversities and complexities into likenesses; it explains things in the simplest possible way. **The descent into diversity is, in this way, matched by a corresponding ascent back towards the inward, fundamental originators - first facts, first causes, first principles.**

It is the purest exercise of reason to find reasons. The prime motivation of scientific insight and research is to uncover patterns, principles or 'connectors', set upon higher set, until they are all subsumed under the auspice of a single truth. Such motivation involves a conviction that, psychologically, physically or both, there probably resides a central force or power to which all else is indissolubly connected, from which all derives, on which all depends and in whose comprehension we will, with a sigh of great relief, know everything.

Since ferment is the state of play it's evident the game is, far from over, in full swing. There exists no answer so complete as to extinguish all debate and ever-probing argument. No absolute, which is physics' Holy Grail, has been revealed. Is it possible to frame or rise above the ferment using Natural Dialectic? The stack below describes *fig*. 3.3:

The ultimate remit of a Unified Theory is to tie everything into an elegant, self-consistent 'simplicity' of explanation. In science such coordination happens gradually. Let's opt for a starting point. Galileo and Newton (with his foundational laws) unified motion and its special case, at zero, rest. They showed, with a principle of relativity, a thing seems different according to perspective. And for Standard Theory gravity, a weak force whose first description by Sir Isaac Newton was refined by Albert Einstein's General Theory of Relativity, operates on the large scale to 'order' galaxies, solar systems and planets; indeed its influence is everywhere at every time on everything. No mass, no gravity; but in a space of massive things it is an imitation of omnipotence.

subsequent hierarchy	Holy Grail
issue/ action	*Source/ Priority*
physic	*Metaphysic*
classical/ quantum	*Archetypal*
↓ *tam*	*raj* ↑
outward finality	*inner support*
aggregate/ precipitate	*influential action*
sensible matter-in-practice	*matter-in-principle*
large-scale body	*microscopic components*

The other three recognised forces (strong and weak nuclear forces and electromagnetism) are highly localised. They operate with strengths thousands of billions of times greater but over distances thousands of billions of times smaller than gravity; and they are the composite of a further unification, of special relativity and quantum theory, that led to the Standard Model of elementary particle physics. Progress 'unified' these forces in a *GUT* (grand unified theory); next quantum cosmology will hopefully (perhaps in the form of a string theory and so-called super-symmetry) embrace gravitation and merge particles with forces in its *TOE* (theory of everything). Job done! Where all is one the cosmos is completely understood! Does that not include you too?

Newton drew the celestial to earth; Maxwell, with his theory of electromagnetism, carried light to the stars; and, prodigiously, Einstein unified space with time and energy with mass. Things great and small have been reduced to energy in bendy space.

Next Einstein and Planck refined the way great and small scales of the universe are understood. *Together quantum and relativity theories constitute the main pillars of modern physics.* The substance of each pillar provides a framework that with great accuracy incorporates and harmonises data from the atomic heart of matter to the outer galaxies. *They are complementary but, at the same time, incompatible.* They run according to two unlinked sets of rules and, worse, quantum field theory predicts a huge density of vacuum energy (10^{93} grams/ cm^3 whereas all known galaxies squeezed into that cm^3 would only aggregate 10^{55} grams)! Yet, astronomically, only an infinitesimal density (10^{-28} grams per cm^3) is actually observed. If it were greater by a single power, a minuscule amount, then you and I would not be here. **This means quantum huge and astronomical infinitesimal; the difference between the cosmologist's classical and the particle physicist's quantum version of *ZPE/* dark energy is a massive 120 orders of magnitude!** Professors gloss a problem which is vast. This is, by far the worst 'fine-tuning' chime between fact and its theory in all physics. *Vacuum catastrophe*. What might overarch such different quantities? Can any theory string the two together? How will the grail of reconciliation be achieved?

Various theories have been employed to cast beyond and thus resolve the problems that contemporary physics struggles with. Yet compound assumptions, transcending observed physical phenomena, sometimes lead towards priestly metaphysic sermonising in the form of arcane mathematics. And priesthoods do not welcome criticism from the laity.

String theory, for example, is a mathematical spectacular, untested and maybe forever intangible, that promises to combine general relativity with quantum theory, to unify forces and particles and to explain the values of all physical constants in terms of a single value, string tension (and a probability of strings breaking or joining called the string coupling constant). Strings are, as opposed to point particles, treated as oscillators wriggling in Planck space; different modes of oscillation represent particles and forces. If quantum physics is phrased in terms of equations that also describe standing waves; if the volume of an atom is filled with ripples and waves; and if atomic existence is due to such harmonic oscillation (each kind of atom 'singing with its chord') then, from choral atoms to the galaxies we can view cosmos as a vibratory symphony.[57] Strings, if they exist, would resonate within a radiant cosmology.

Highly conceptual extravaganzas, as fiercely defended but unproven as a non-mathematical analogue, the theory of evolution, have since inception passed through several incarnations. At first twenty-five dimensions of space were required along with particles (called tachyons) that exceed the speed of light. However, superstring theory deleted tachyons, included strong interactions and massive particles and reduced spatial dimensions to nine; the theory even seemed to predict, as well as photons, closed strings called weightless gravitons. Quantum gravity's royal banner was unfurled across an all-embracing sky. Indeed, breaking, joining, opening, closing and string motion seemed to generate all forces and subliminally unify the particles we know. The vibration of open strings generate field-lines of the nuclear and electromagnetic interactions; those of closed string give us gravitational fields; and bosons (force-carriers) are united with fermions (quarks and leptons). The whole of physics might seem unified. Not only this, but the formulation's constants are reduced to Newton's gravitational constant and the string coupling constant. But such prize, manna from scientific heaven, falls from where and at what price?

'Symmetry' is thought to be a starting-point. Somehow, in creation, initial unity's presumed to 'break' into the particles and forces that we know; and action in physical systems is governed by 'continuous symmetries' whose invariance is reflected in equations (and thus laws)

---

[57] see also Chapter 16: Psychosomasis.

describing them. For example, the distinct quantum field theories of *SMPP* are constructed from three kinds of 'continuous symmetry transformation'. Might not transcendence, in the form of super-symmetry (or *SUSY*) sort problems out and yield a fresh, new physics? In this respect *SUSY* represents a step towards *GUT* but its *assumptive* gamble is that particles of matter (fermions) and force (bosons) can mutually transmute - except regarding mass. There's no evidence for that. It has been suggested *SUSY* might involve, for every particle that's known, a super-particle (or sparticle) that's not. Could neutral sparticles compose 'dark matter' or drive super-gravity? Again, there is no evidence at all; and for each problem solved hosts more appear. Particle numbers double, dimensions multiply and many undetermined parameters emerge. Things don't *simplify*; they aren't reduced to few as possible and definitely don't describe the world we know.

SMPP cannot, critically, predict the mass of particles that constitute our universe nor yet include its gravity. Could *SUSY* underwrite a theory of quantum gravity with gravitons? You might impose quantum rules on relativity or *vice versa* (a theory of loop quantum gravity). You might even apply quantum uncertainty to space-time but no answer's clear. Such reconciliation's well beyond The Current Version's comfort zone.

String theory still needs at least ten dimensions (nine plus time). And, in order to relate theory to the real world, six must be mathematically 'compactified'; these 'hyper-dimensions' must be rolled up into minuscule, invisible geometric shapes. Then you can generate many hypothetical versions, predictions and 'possibilities'. There are computed to be 10^{500} of these imaginary quantum forms. No evidence but yet again assumptive boost. Indeed, might not an unformulated conjecture mysteriously called 'M' tie various superstring hypotheses together? What, Friar Ockham, would you make of strings and M-theory?

Thrilling mathematics does not mean, as far as nature is concerned, that the description's right. Nobody in the string community has, for example, predicted dark matter or dark energy. And everybody, calculating fresh 'flavours' and carving whole new 'landscapes' of string theories, is out of kilter with the actual world we actually observe. There is reason to take any plausible speculation seriously but this is not, as any jury will aver, to claim it as a promissory declaration of the truth. Indeed, a Holy Grail may well require a factor that is metaphysical (see Glossary: unification/ *CUT* or Cosmic Unification Theory).

Snip, Snip

String conflict with reality is, in reality, a snip. It is divorce from known fact and cut from common sense. The disconnect is interesting but all in a stringer's head. The theory is heavily involved in metaphysics (what is beyond observation - including the logic of mathematics itself), modern mythology (*ex nihilo* creation) and philosophy. *No doubt it is less abhorrent and more acceptable, materialistically speaking, than the simple assertion of a separate cosmic component, informative mind.* But it is difficult to visualise a microscopic panorama whose multi-dimensional patterns of vibration orchestrate the sensible, superficial fraction of our universe. Perhaps the mathematics of string theory are, as a physicist might calculate the sonic properties of a Bach fugue, an attempt to flesh out abstract, physically absent archetypes and enumerate the vibratory forms of universal mind.

On the other hand, strings and loops might be a myth. They are inaccessible to physical observation. How, therefore, could you prove strings are or aren't? Such physics is as recondite as any theological enquiry (though mind or archetype is never treated as extra, *metaphysical* dimension). So, for all things great and small, big bang cosmology and string theory are the current exotic, esoteric orthodoxies; but outside their main streams both have critics. In the case of strings anti-stringers elaborate on how the theory 'veers off reality'. And, as we'll see, big bang's boost into our physicality was primed and fired from even grander metaphysical reality!

Split Continuity

Split out of nothing - the appearance of phenomena. Take the unity of continuity; take unbroken field or flow. But if you start with space (or even nothing lacking space) how do you get your head around its edge? How can you begin to crack pure flow or excite Father Field's pure latency? The bang-on phrase is 'breaking symmetry'. How and why such breakage works is still a mystery.

Split 1 is big bang, transcendent projection or whatever kind of 'nothingness' can rustle up a thing. It must be next-to-nothing and thus, unless nothing sprang at once to something everywhere, initially the split was very slight. Perhaps, first rung down on the cosmic scale, some factor 'quantizes'. Such individualizer might break space to 'blocks'; such locator might coordinate production of a particle. Standard theory can't explain the latter's origin nor origin and strength of interacting forces. How did initially identical 'pixels' of fluctuating energy/ matter 'break' to yield exactly the divergent, unpredictable particle masses that our universal set-up demands? Perhaps expansive cooling did, with serendipity, the trick. Or perhaps Higgs fields are required; interacting with this type of universal field might somehow lend a body mass. After all, you really need a theory of mass. The Version hasn't got a firm one yet.

Yet physicality, excepting light, is 'frozen over'. Mind's locked out of matter; and what's mass but locked-up energy? Gritty quanta are then further bonded into atoms, molecules and bulk materials. Half-locked, half-free, what about a quantum's continuity?

Discontinuous continuity and *vice versa.* A quantum (e.g. photon or electron) interacts as if a particle but travels as a wave. As well as a particle, however, the word defines the smallest discrete unit of energy. This is determined by Planck's 'constant of action', an infinitesimal amount which is used to relate energy with frequency or momentum with wavelength. In other words it relates the propagation of energy (radiation) to its interactions with matter. It is thus a quantitative link between two basic types of behaviour (propagation and interaction) and the two models (wave and particle) that equally describe them. Such a succinct unification is powerful. *To describe cosmic operations in such simple terms places the quantum and Planck's constant at the heart of physical phenomena.* You might say this constant quantised nature into 'pixels' from which everything is built. Thus mass and space and time are broken into fundamental natural units. Planck's constant therefore represents the primal split, the chip from nature's space-block out of which all else is built. You might ask, having done away with spectral continuity, whether such divisive 'micro-blocks' are fact or fiction. And if fact, are they the way core 'nothing-something' is embedded in the fabric of a polar, energetic archetype?

Split 2 is polarity 'personified' - *electric charge*.

Attraction and repulsion - the influence of polar charge generates currents, waves, atomic bonds and, eventually, the bulk matter we breathe, drink, tread on and are made of. What is it?

In theory real particles can 'borrow' energy from the 'vacuum energy' of a so-called Dirac field to create evanescent, 'virtual' ones. Paul Dirac supposed that spontaneous production of electrons and their anti-material, positrons, occurs in this neutral field. So perhaps a 'cloud of latent charge' 'creaks under strain' and with each 'wobble' ejects so-called 'virtual' plus- and-minus charge-pairs; these, by almost instantaneous self-annihilation, return the energy that made them back to its original neutrality. Both photons and electrons indulge the 'borrowing habit'. Each particle is surrounded by 'an effervescent cloud of so-called virtual particles. In a further step the electron's virtual photon pair is supposed to create its own Dirac reflectively asymmetrical pair (of particle and anti-particle). Where the electron of this pair is repelled from its parent electron, the positron will be attracted. So a virtual electron cloud will be separated from a virtual positron cloud. It does not matter that each unit of a cloud exists only for a tiny fraction of a second because continual effervescence replaces it and permanently sustains the cloud as a whole. It is the outer virtual electrons that are supposed to confer negative charge on their parent. Because the quantum field is everywhere, everywhere can be polarised like this. The vacuum has been polarised! In fact, it is claimed that there are zero-point interactions for strong as well as electromagnetic interactions. Could such complication be what charge, essential to our chemistry and therefore all the complicated composition of the world, is itself made of? Did you think simple charge - or anything down at the roots of nothing and infinity - was going to be an easy matter? Does your electrician understand the substance of his trade? Who does?

Despite imaginative speculation, though, electrons are the basis of chemical connections and, down-to-earth, our starry universe!

This brings us to *Split 3*. Somehow most quantised particles have been, one presumes, imbued with mass by an etheric kind of glue. Mass 'trips the zip' and thus, on cooling, nuclei and, with electrons, stable atoms can appear. If unlike charges must attract a child asks why these atoms don't collapse. Why does the world stand up? Electrons radiate energy if they accelerate, as they do when orbiting. In losing 'strength' how long can they keep it up - and not collapse into a neutral, nuclear coalescence? Electrons, attracted by protons and thereby drawn to cluster round and coat the nucleus, should rapidly lose their various energy states into a single ground state. Then no reaction could occur. No chemistry. Yet all students of chemistry[58] learn that, fundamentally, atomic structure involves electron orbitals. Mendeleyev's periodic table, chemistry's analogue of biology's genetic code, depends on it. So do their research projects and, much vaster in extent, the whole wide world outside.

Over several years a description was evolved. Electrons were assigned integral quantum numbers, related to the number of nodes in a harmonically vibrating system, that describe characteristics of their atomic motion. They include the angle, size and shape of orbit, and direction of rotation. Each 'permissible' energy state (or 'note') was assigned a distinct set of integral numbers. Of course a description, even a quantum description, is not necessarily an explanation. What actually maintains atomic structure? In a children's playground you time your pushes to keep the swing swinging. What if dynamic equilibrium were established whereby an electron radiated energy but, in the way of a resonant push, absorbed the same amount from *ZPE*? The vacuum itself would whack electrons round the nucleus; and if atomic orbit energies are sustained by vacuum energy you might observe that space maintains the whole stability of matter. Again, dialectically, nothing keeps the whole show on the road!

Whichever way, Wolfgang Pauli was able to perceive that each 'quantum state' in an atom was distinct and, crucially, if a particular state was occupied then the next electron would have to move to the next higher unoccupied one. His Exclusion Principle therefore states that no two electrons in an atom can have the same set of four quantum numbers. Collapse to ground state was averted and electrons cannot occupy each other's space. Nor can protons or neutrons. Theory was married with reality so that distinction and, therefore, the survival of atomic, molecular and bulk structures is assured. By this principle alone the world is stabilised. *Chemistry depends on Pauli's simple (↓) principle of separation.* The solidity of matter, stability of stars and so on illustrate its practice.

We now have three expressions that illustrate the development of an anti-principle, discontinuity. They are Planck's Principle of the

---

[58] Chemistry reduces, effectively, to a couple of complementary charges, neutralities, their particles, the shunting of electrons and equilibrations.

Quantum, Dirac's of Polarisation and, for orderly and stable individuality, Pauli's of Exclusion.

These splitters are important in construction of the periodic table but splicing and neutrality also play a basic role in chemistry. Charge excites and changes but splicing (bonding) and neutrality keep atoms in their place. Molecular and bulk units - almost everything there is - are made of them.

Boxing the Infinite

polarity	*Potential*
discontinuity	*Continuity*
degrees of freedom	*Freedom*
relative imprisonments	*Liberty*
↓ *negative*	*positive* ↑
involuntary	*voluntary*
automatism/ robotic state	*conditioned free will*
tight bondage	*loose bondage*
creation/ materialisation	*dissolution*
fixity	*flux*
capture	*release*
confinement/ boxing	*liberation*
isolated, individual unit	*interacting unit*

In three ways we've begun to physically box the infinite. This section attempts to theoretically connect materialism with Natural Dialectic's broader principals.

The Infinite is boxed by losing its potential, that is, by realising possibilities. Information specifies; from generality it assumes increasing detail, individuality, locality in space and time. Active awareness drops asleep; conscious falls unconscious. And, projected from first-causal archetype, informed and energetic matter mints the cosmic currency; from potential's infinity material events are realised.[59] You might assume that, once devolved from a creative singularity (from big bang's tiny time?), all different events must be entangled. Everything is logically connected, from the start and presently for all existence, inside universal mind. Cosmos is a (↓) precipitate of archetypal memories.

Split and splice. From neutral wholes split polar parts; and polar parts, by integration, form a whole again.

A potential, roused to action, will become exhausted; neutrality that's poised gives way to fallen, flat and finished state. Imprisonment. Local lock-ups. In a more abstract sense consider that, in materiality, discontinuity is dominant and isolation is the anti-form of unity.

---

[59] see this Chapter: Infinity and Chapter 7: Are You Certain?

Less abstractly, senses filter, brains restrict and ego limits what, already limited in time and space, we separately comprehend. Experience of such limitation is, wrote Einstein, 'a kind of optical illusion of consciousness. This delusion is a kind of prison.... restricting us.'[60] And solitary confinement, lacking all communication, is the psychological nadir - unless you add what stones don't feel, an element of suffering. In bodies biological mind is much constrained. Is not pain a blocker, a constrictor, an acute and violent psychological locality?

The going toughens, balance slips and the emphasis is tightened into left-hand anti-principles. Freedom is the principle whose anti-principle, confinement, really clamps to earth. Eventual strangulation; black-out; materialisation is a ($\downarrow$) downward process of increasing loss of energetic vibrancy. In physical phenomena all information is completely locked. Automated. It's entirely passive. Matter's shackled up in bodies, mass increases as constriction (like inside a proton) throttles energy. It is how, by degrees of loss, the Infinite's exhausted in solidity.

From this angle creation sums to a vast range of reductions; it is a scale of increasing bondage down Mount Universe's slope; it is composed of myriad partitions each of which suggests but veils a larger whole. Thus Essence is bottled, Liberty corked and Infinity boxed.

Interestingly, this is perhaps the point to develop Einstein's view of restriction, in terms of Natural Dialectic, until it leads to resolution of the tension between free will and determinism. First check *figs.* 2.6, 6.1 and 8.1. What correlation is there with the previous and following stacks?

existence	*Essence*
finite condition	*Infinity*
degrees of freedom	*Freedom*
possible expressions	*Potential*
$\downarrow$ *involuntary reflex*	*voluntary factor* $\uparrow$
informed outcome	*informant dynamic*

Chapter 1 describes the simple link between psychology, physics and Natural Dialectic. In this the possibilities of (*sat*) potential are reduced by (*raj*) action through degrees of freedom to the (*tam*) point of none, that is, the point of no further action, exhaustion or fixed outcome. This is the ($\downarrow$) vector of creation; the reverse ($\uparrow$) process therefore involves recovery of original, lost potential. Such recovery of complete freedom is called liberation. We associate liberation with freedom of choice - the exercise of wholly free will. But within existence freedom (of will or anything else) is relatively conditioned. Recovery of Absolute Freedom therefore involves Essential Potential.

---

[60] 'Einstein, A Life': Denis Brian p. 388.

↓ *no free will*	*conditioned choices* ↑
automatic/ robotic	*degrees of subjective freedom*
material energy/ matter	*process/ lifetime*

Check *fig.* 6.1 again. An act of creation, universal or individual, follows the (↓) gradient of existence. And the pre-existent basis of existence is Essence, that is, Full Potential. The first, decisive action of such Potential is called the First Cause, Archetype or *Logos*[61] of creation. Conversely, therefore, (↑) psychological liberation is obtained by recovery of Primary, Potential Mind. This is what a mystic means by attaining your Full Potential. How, therefore, can Knowledge or Potential be retrieved? What voluntary choice leads, wisely, to this end? Does its immaterial psychology have anything to do with union and love?

Physic, on the other hand, is a special case freedom - none. Its involuntary potential resides in archetype or first cause physical. This immaterial condition, along with quantum and gross material expressions of its character, is informatively fixed.[62] Physical creation represents, in its entirety, the exhausted, non-conscious incarceration of Essence. Such passive, objectified automation is (*fig.* 2.5) completely conditioned. It has no free will at all. Physical 'liberation' is thus energetic alone. In this case how can cosmic *rigor mortis* be relieved? Does anything refresh the universe? Solid is evaporated unto plasma; light and heat raise matter. Hard things are dematerialised; they are stimulated, re-connected and, to relative extent, (↑) re-merged as one. Has, as physicists at *CERN* suspect, such unification any bearing on initial conditions, physics' starting-point, the *origin* of physicalia?

To summarise, there is no unconditional free will in existence; and the less free will, the more determinism takes its place. Cosmos is, therefore, relative imprisonment involving a range of freedoms split from Perfect Liberty. Existence is shut out of Essence. In this view, as the Buddha noted, physical and psychological incorporations are a zone of suffering in various degrees.

Of course, there's no escape for matter from oblivion; but voluntary (↑) ascent towards Knowledge is what everyone in some degree desires. What quality of increase bids priority? Einstein again, '…Our task must be to free ourselves from [this] prison by widening our circle of compassion to embrace all living creatures and the whole of nature in its beauty.' Perhaps further, if this nature's not chance-born, what about embracing Nature at its heart - the Creator of creation? Beyond 'optical illusion' and all existential imprisonment Buddha and the saints impart a science of the soul - loss of unit self in Unity, revelation of the Cosmic Singularity and, thus, rediscovery of The Infinite Unboxed.

---

[61] see *fig.* 5.1 and Chapter 5: (*Sat*) Potential Information.
[62] see Index: alphabet, archetype and cosmo-logic.

Chapter 9: Nothing

Third term of (N)One's Essential Equation is 0. Zero.

Is it true the more you learn the less you know you really know? 'The only thing I know is I know nothing,' said Socrates. Perhaps, therefore, we'd best know something about Nothing!

Can you know nothing? Perhaps unconsciously! Did you, sleeping, see oblivion in all its 'blackness'? Can you have faith in such unseen? If Science and the Soul were accurate you would expect what's physical to devolve from what is not and what exists from what, essentially, does not. Do unseen voids make any sense? Let's see!

The Nature of Nothing

The paradoxical nature of nothing, both psychological and physical, is what we need to learn about thence understand:

tam/ raj	*Sat*
lesser nothings/ apparent things	*Nothing/ Void*
expression	*Potential/ Source*
excitations	*Archetypal Latency*
motions	*Source*
derivations/ dependencies	*Super-State*
↓ *tam*	*raj* ↑
impotence/ sink	*currents*
dead stillness	*change*
inaction	*action*
finish	*process*
zero	*units*
sub-state void	*events*
isolation	*relationship*
final silence	*rhythms*
blank	*script*

specifically for mind (nothing physical):

oblivion	*awareness*
unconsciousness	*perception/ cogitation*

for matter (nothing psychological):

absence	*presence*
non-locality	*locality*
general space-time	*thing*

A stack of Natural Dialectic's generalities might seem as relevant to the buzz of a well-equipped laboratory as Unified Theories to the hubbub of a city life. Yet broad generalisations and basic principles set the tone, orientate and radically affect the 'texture' of both thinking and consequent business. Having examined two of the absurd equation's factors, Infinity and Unity, let's take the third - Nothing. The study of nix is not for nought. 'Doing nothing' is no lazy option. Nature of nothing that includes, potentially, everything is quite a scoop!

Do you remember, first and foremost, you can think of nothing in two crucially different ways? **You can find potential and exhaustion, pre-active and post-active absences of change.** The negative sees exhaustion, void impotence, nothing left; the bottom of the pile's a sink. The plus sees what's to come; it sees potential and capability *before* the action starts. Thus, in considering a state of nothingness, it all depends. Is the object of your focus loss of everything or everything to gain? Is it source or sink, a thing you want or, as its absence, what you don't? Understanding nothing's grades and nature might turn out to be a most important quest not just in physics but in life.

Let's rephrase. If you believe the universe was always here, there is no need for any start. However, modern science no more subscribes to eternal matter than holism to a never-starting, never-ending cosmos. If, therefore, existence had a start then nothing, at least nothing existential, must have come before. *If you believe existence (mind and matter) stems from a projection, what holistic kind of 'nothingness' preceded it?*

If, on the other hand and in accordance with materialistic culture, you believe that only physical phenomena once 'started up', what kind of non-physicality projected physicality? *Again, what kind of 'immaterial nothingness' preceded 'anything'?* If your starter is a pyrotechnic of dense radiation and expanding moment-volumes of space-time what triggered this appearance?

Of course, the *top-down* notion of a psychological/ metaphysical projection grates on materialistic nerves. **In terms of Natural Dialectic Nothing (truly Essence prior to all existence) is Potential. Potential's activation runs the existential show; from Immaterial Nothing Everything is made!** Thus potential-prior-to-action is of prime importance. Either way, for holist or for humanist, the gift of everything depends upon the Nature of Prior Nothingness!

Vectored Voids

Bottom-up, materialistic view of void

'Mount Universe' is a void concept. There exists nothing but matter or its absence, vacuous interplanetary and intra-atomic space. Mind is a special case of matter; all metaphysic is void.

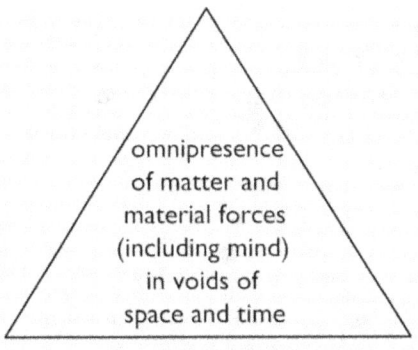

In this view informant mind is an accidental appearance; it is of ephemeral presence and secondary importance; primary matter rules

Top-down, holistic view of void.

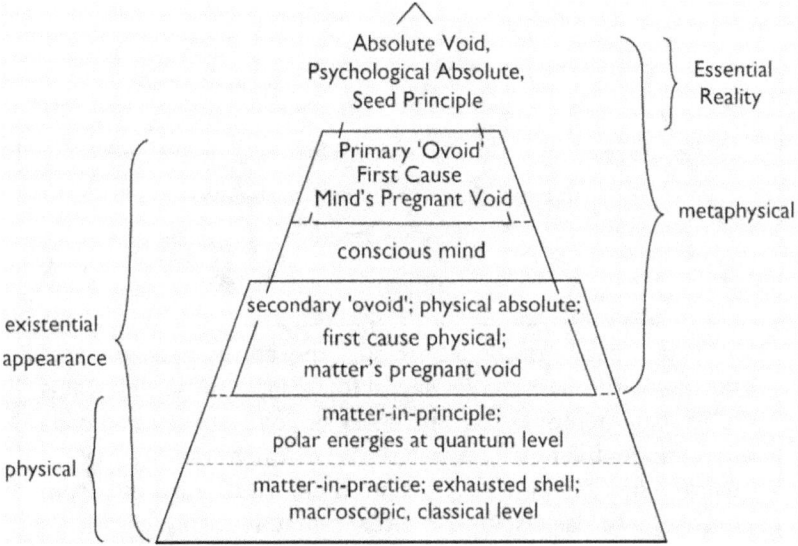

From this view Essential Void is Source. It is empty of existential motion but paradoxically, full of possibilities. Its High Plenitude is Prior. Its Concentrate of Information Unexpressed gives rise to creation. Creation is, in other words, networked tensions in the 'space' of nothingness.

First Cause is psychological. Its Primary 'Ovoid' substantiates all reasons, logic, principles and, in metaphysical space-time of mind, specific, local thoughts.

You partake of mind but are tied to body-matter and the latter's physical space-time. Lower voids of mind involve both partial oblivion (sub-consciousness of memories) and non-conscious oblivion, an absence we call matter. How can mind include matter? In what way is energy dependent on information? We'll explore.

Transcendent Projection:
Matter out of Nothingness/ Void

archetypal memory;
potential matter

universal singularity (0-dimension)
prior to physicalia

latent matter;
immaterial fields

levity
agents of force
gravity

elementary particles
matter-in-principle
microscopic atoms and molecules
matter-in-practice
macroscopic properties and behaviours

Mind's base, of subconscious therefore passive type, overlaps with matter in its causal form. First cause physical is archetypal memory. Its secondary 'ovoid' substantiates all energetic patterns. First phase is through the agency of quantum matter-in-principle and then, fallen furthest from potential into darkest sink, matter-in-practice. This, creation's shell, is what we sensibly survey.

In short, Nothing is the First Principle that predefines lower phenomena. Such Void gives not takes. It is, far from negative, the Great Plus that in its own reduction *adds* all things yet loses none. A nucleus represents inmost potential at the central heart of things; and cosmic Nothing is the Nucleus, Seed, Kernel, in a nutshell Nothing out of which the whole 'organic' business springs! **Its 'Ovoid' is, therefore, the Most Positive, Powerful Plenitude of all!**

fig. 9.1

What, therefore, is the nature of a pregnant void? Indeed, perhaps two such cosmic wombs exist. Central Void, non-existent Essence, is beyond creation; this would be the Womb of Worlds. Thence would first issue immaterial psychology and, at mind's base, the fixity of archetypal memory. Such memory's the second kind of pregnant void since it delivers fixed behaviours to non-conscious, energies; its first cause physical delivers automatic patterns of materiality.

The Order of Nothing

	existence	*Essence*
	differentiated Essence	*Non-existence*
	expression/ creation	*Potential*
	lesser voids	*Void of voids*
	lesser fullness	*Plenitude*
↓	objective	*subjective* ↑
	subjective void	*objective void*
	physical fullness	*mental fullness*
	matter	*mind*
	void of mind	*void of matter*

Why should The Order of Nothing differ from the orders of Infinity and Unity (since they are simply aspects of the same) - except that now creation is expressed in terms of vacancies, evacuations and of voids? The latency of voids gives way to forms and properties.

Nothing is simple! It's as easy as the diagrams explain. Pregnant Potential bears the poles of information and of energy.[63] The Primary Projection, First Cause, is an 'Ovoid' from which is creatively devolved, with mind prior to matter, cosmos. In this case there is conceived, within existence at mind's edge with matter, a secondary ovoid; this ovoid is a physical first cause - potential matter called the bank of archetypal memory. This bank traffics with matter-in-principle from which there aggregates precipitate - bulk matter.

Creation, full of forms, is void of Pure Void. Yet, in this active void, formless vacancies exist as well. There's, metaphysically, unconsciousness of mind - subjective absence due to sleep or comatose oblivion; physically there is exhaustion of all objects, that is, the objective absence of a vacuum, pure space - perhaps even pure space-time. Space is easy if there's nothing in it - but is it really so?

Vacuous space is, according to Natural Dialectic, one of three physical 'absolutes' - physic's primal trinity. But is there really nothing in it? A perfect vacuum is a sink that, modern physics estimates, jitters everywhere with quantum fluctuations called zero-point energy (*ZPE*). In other words, there isn't 'nothing' anywhere. Everywhere there's something though this 'thing' might be below the threshold of our observation.

---

[63] *fig.* 2.3.

Source and Sink are Zero

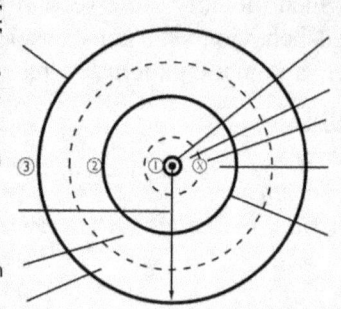

(*tam*) physical exhaustion; peripheral extremities; voids of absence of matter (vacuum), of space (black hole), of vibration (absolute zero, 0 K) and of energy (impotence)

(*tam*) vector of materialisation

informant quantum action

informed classical events

(*Sat*) Superstate; Alpha; Quintessence; Essential Origin

potent, informant mind

cosmological axis

(*tam*) impotent, passive or subconscious mind; instinct and memory

(*tam*) psychological base/substate; *PSI*; archetypal patterns; (*sat*) physical superstate; the metaphysical origin of material behaviours

The illustration rephrases *fig. 3.5*. It shows three '0's, three kinds of zero.

(1)   Essential Source of Informative Principals
(2)   psychological sink (archetypal memory)
      physical source of specified forms of energy
(3)   physical (and existential) sink

In one sense, therefore, you can understand that nothing is a latency; source is potential; it is 'nothing happening yet' before expression of the possibilities. In the other sense there's nothing left; exhaustion is the end of possibility; zero can be a sink as well.

Alpha Source (1) is Central and has two 'zero anti-poles'. One is subjective/ psychological, the other objective/ physical - in that order.

The subjective anti-pole is 'zero psychological' (2). It is a sub-state sink of unawareness, sleep and fixity of thought; it includes the passive and subconscious mode of mind called memory.

This anti-pole is also the material origin. Its secondary 'alpha' has been set at nought. Don't phenomena emerge from zero; didn't starry cosmos spring from nothing-physical? Fixity of mind, archetypal memory, serves as a source of information. Sub-state is, paradoxically, at once a natural super-state. Material government derives from this 'great instinct' of the universe. Memory is metaphysical; but its zero physicality is non-zero, archetypal template too and thus the source of every automatic, physical event. 'Potential matter' is not physical; yet its immaterial priority controls the way materials non-consciously behave. Starry cosmos sprang with this in mind.

A hundred billion galaxies still make subjective zero. Non-

> conscious matter (3) is the anti-pole of Immaterial Essence (1), mind and material archetypes (2); and the extreme of matter's anti-pole is its objective zero. This sink, 'zero physical' (3), marks total drainage at creation's furthest edge, exhaustion at its dark periphery. Its rim marks the extremity of Alpha's absence. It is matter's end-point, negative annihilation; its subtendence marks a natural posterior and rump of universe.
> *fig. 9.2 (see also fig. 3.5)*

Imagine you stripped *all* material away. If matter's real, strip out reality. Absolutely nothing physical, even space-time, stays. Would there be information? An informant in the 'shape' of archetypal symmetry whose breakage could let flow ($\downarrow\uparrow$) vectored principles - quantum agencies from which a world can form? Such latency, *potential* matter, is irreducible. Its potential informs material behaviour. Cosmic source code is not physical though metaphysic of mathematics may, correctly scored, describe it. Properties of things you measured would be products of an archetype - an entity that's metaphysical and not, therefore, constrained by natural, material law. Why, though, should nature's source be anything but natural as well?

From archetype to physicality - could quantum vacuum not support alone but also make the universe? Apparent void (potential's plenitude) is perceived as cosmic egg, womb of worlds and actual, pregnant context out of which all apparitions rise, all things have appeared. **Space might be invisible. It might seem dark and cold but isn't simply barren emptiness. It is a bridge. It is a psychosomatic point, a medium that brackets space-less metaphysic with its spaced-out physic. Information is transmitted; void links mind with matter and joins archetypal instrument with natural patterns that derive, like note from flute, from it.**

Thus this Chapter links, through space, with physic's primal trinity. The first physical imperative is material nothingness. We call it vacuum.

The second, a dynamic form of nothingness, is therefore time. Its space is Chapter 10.

The third, energy, is relative in type and motion. This chameleon may appear as ghostly next-to-nothing in bi-polar form - light's levity whose radiant form is matched by influence of contractive gravities. So (Chapter 11) we shall drop out through the crack in two-lipped emptiness down to a cornucopia of things that have emerged from it - including you. This, in short, is the tall order issued out of nothing in particular.

How Does Nothing Physical Work?

A psychosomatic junction would appear to be a critical form of nothing. It is of prime scientific importance to try and understand how the latch leading to our house is lifted, how the entrance to our universe swings open and the frame of natural law begins to work. How can we better grasp the archetypal 'grammar' out of which the pages of our world's book flow?

If space-shapes are seen as relatively 'frozen' energy so memory is seen as 'frozen' thought. Archetypal memories are by this definition concepts 'frozen' in a universal mind. In other words the logic of Natural Dialectic sees (*tam*) physical patterns of behaviour as the low-level, non-conscious, automated consequence of mind - instinctive mind that, as the internal root, governs involuntary, 'crystalline' matter. This governor's body is our astronomical universe. Physics and chemistry study the actions of its 'instincts' or archetypal control systems. These are natural law and their 'nervous system' is actualised in electromagnetic and other fields of force that orchestrate the way things interact. **In this view, contrary and outrageous for materialism, matter is developed memory; it is a projection.** And, at the outer rim of this projection, energy is frozen; any solid represents creation's edge; it represents, full stop, world's end.

The immaterial is indeed, by definition, irrelevant to material science and anyone who demands physical as opposed to inferential proof of non-physicality may well end up denying even his own mind. So let's retreat. Let's make space for space and turn to scientific vacuums. How are they conceived to work?

Old Vacuums

Top to bottom exhaust it. Freeze your black box, pump out every particle and piece of radiation to get nothing in return. A bare vacuum. Material immateriality. Objective but inert, impotent 'metaphysic'. Is outer space not such a box without its sides? A vacancy that's tenanted, an empty platform things play on, a non-existent sea that buoys the world up and yet, tell me, is not space an actual absence? It is an extension 'formed' of formlessness, nothing is a negative opposing every other thing. Physical abstraction at the base of things, space is a bottom bare and bottomless; 'old', 'traditional' void extends the coolest, thinnest of receptions.

Down the cosmic slope, therefore, we slide to matter through mind's postern gate; we drop from the archetypal 'soul of physics' into formless vacancy; we are swallowed from the mouth of order into nothing but amorphous extent - vacuum, chaos or (as the word means in Greek) boundless space. When everything's exhausted nothing's cleaned right up. Over the years many people, some in a professional capacity, have had a lot of fun making something out of this void, zero or blank. Is nothing really like it seems? How does the cut of nothingness stack up?

Simply carve up shapeless ether with the archetypal lines of geometry

As the Greeks realised, space and time constitute a 'frame' or 'blank' within which shapes and behaviours can be generated. Harmonious form of energy is best expressed as music; harmonious shape of mass is best expressed in terms of proportion or ratio. Greek geometry, the Euclidian division of space, was an answer to the question of order, proportion and the nature of objects and behaviours in space.

Not only Greek but Moslem philosopher-mathematicians waxed lyrical over the division of nothingness into spheres, circles and linear symmetries. 'The Centre,' exclaimed Ibn Arabi, 'is everywhere and its circumference nowhere'. Medieval Islamic art is a subtle, multi-faceted metaphor to describe the dynamic yet stable, finite order that derives from an Infinite Centre. One-in-multiplicity and multiplicity-in-one. Its motifs, created by the multiplication, division, rotation and symmetrical distribution of simple themes, evolve endlessly repetitive yet variable effects.

Aristotle's *'horror vacui'* denied a void; nature abhors a vacuum. Sir Isaac Newton, on the other hand, believed in absolute, infinite but separate space and time. His problem was how an insensible medium might propagate light or gravity. Was it really a subtle fluid called the 'ether', something through which waves might pass like those of sound in air? Although he seemed to need a medium 'exceedingly more rare and elastic than air' the great man did not like the idea of any impurity that might influence the motion of celestial bodies. He vacillated over ether.

Others did not and, to cut long stories short, 'luminiferous ether' became all the contemporary rage. True, Michael Faraday thought electricity was a 'varying state of strain', by which he meant vibration. In his theory of radiation he ventured to dismiss any concept of celestial ether but to include 'a high species of vibration in the lines of force known to connect particles, and also masses of matter, together'. Field theory was conceived. On the other hand James Clerk Maxwell, building on Faraday's precepts, recalled it. He developed a model for the medium in which electric and magnetic effects could occur. He drew an analogy between the behaviour of Faraday's lines of force and the flow of an incompressible, elastic kind of medium. He saw both light and electricity vibrate through this hypothetical stream. The ether came flooding back. By the late nineteenth century it was one of those scientific truths that, like biological evolution today, we hold to be self-evident. Scientists practically unanimously accepted the idea of a motionless, invisible medium extended throughout space and with the property of transmitting electromagnetic and gravitational forces. Although solid objects could traverse its 'gossamer' some degree of friction should accrue. It was this 'drag' which a French mathematician, Fresnel, had calculated and which the experiments performed from 1881 to 1887 by Michelson and Morley set out to test. The results showed no such 'drag'. You might therefore conclude that the assumption of a motionless, etheric 'ocean' that fills all space with the earth ploughing through it is unnecessary, an untidy excess due for Ockham's chop. Maxwell's ether was guillotined; one kind of space was dead, long live another species of the vacuum.

New Vacuums

New skins for old. Not only Aristotle but contemporary physicists deny the emptiness of space. Their quantum vacuums are positive, potent 'forms' of nothing that, as well as carrying 'particularity', host 'forcefulness' and radiation through immaterial fields. So suck and freeze! At Plum Brook Station in Ohio, *NASA*'s space facility, science shapes a block of nothing almost perfectly! Particles are pumped out and, in the belly of a massive church-like chamber, energy is subdued until a metaphysical horizon, experimental nothingness, seems near. But since you can't obtain a thermal zero (0°K) Heisenberg weighs in with delicate uncertainty. His principle asserts that definitely you can't have zero energy in nothing. Nothing isn't nothing when it's virtually something! And that's not all. Space bends, contracts, is grainy and expands with power that carries cosmos forward![64] Modern vacuums just aren't perfect. They are 'false' and full, perhaps, of various field(s); they teem with gravitational, electromagnetic and (uncertainty is certain of it) virtual fluctuations.

Such a vacuum is not empty. It is full! Dirac, by maths combining special relativity with quantum mechanics, turned these fluctuations to a bubbling on-off virtuality composed of particles and charge. In this case it has been proposed that vacuum might comprise informant source as well as energetic sink of everything. Indeed, unless some instability broke cover, might not balanced tensions stay unseen? In belief, 'virtual' quivers stay subliminal but sufficiently excited fields give rise, spontaneously, to quantum particles/ actual energies. Just as Pauli's distinct quantum states give rise to atomic structure so, it is suggested, distinct, specific fluctuations of a vacuum 'harmonically' give rise to fundamental particles.[65] This is the graduated density of vacuum and the nature of a cosmic egg; preordination rules; polar, fundamental ($\downarrow\uparrow$) vectors specify material types; a scale of immaterial fields constrains the action. Measurable patterns would appear. Then space is seen as a 'container' and the active vacuum as a large-scale latency with 'immaterial fields for development'. The world is thence developed on such filmy '3-d template'. *ZPE* and 'virtual photons' mediate all electromagnetic interactions. They are postulated as support for electron orbits, atomic structure and thence the phenomenal universe. The appearance and disappearance of clouds of virtual particles in this very lively vacuum are a price paid for what is claimed to be the most mathematically precise theory known - Richard Feynman's relativistic quantum field theory of electrodynamics.

---

[64] Chapter 12: Pass the Paracetamol.
[65] Chapter 16: Synchromesh 2.

Electrodynamics describes the effect of moving electric charges and their interaction with electric and magnetic fields - the point that Natural Dialectic identifies as linking archetypal fields of memory with matter. It is therefore tempting, at this point of psychosomatic linkage, to equate a physical face (implicit energy of 'new' vacuum and explicit light) with a metaphysical one (implicit, psychological energy of archetype identifiable, perhaps, with oriental *'prana'*).[66] Archetype would constitute an immaterial template for the orderly projection of the subtlest material forms. As well as photons and electrons, include quarks/protons. This atheistically-intolerable theme will be developed.

Dialectical Vacuum

	tam/ raj	*sat*
	business	*potential/ latency*
	action	*source*
	archetype-in-motion	*archetype*
	polarity	*neutrality*
↓	*negative*	*positive* ↑
	gravitation	*levitation*
	contractive	*radiant/ expansive*
	isolating	*relating*
	mass/' dead appearance'	*'live stimulant'*
	sink	*causal agency*

$0 = +1 \ -1$. From neutrality is sprung polarity (or, by annihilation, *vice versa*); from potential action's stream appears. Light yields, for example, (-) electron and (+) positron. Does archetypal symmetry with neutral space devolve the world's polarities? What is Natural Dialectic's spin?

Its vacuum's paradoxical. Vacuum is essential both as start *and* endpoint, source *and* sink. It is (can you see?) more than single. It's a trinity. **(*Sat*) void is an essential form of nothingness; it is an archetypal 'emptiness', a pre-creative latency.** For mind the Archetype is *Logos*; but for matter it is archetypal memories, potential matter or the principles of natural law established in the *metaphysical substrate of universal mind*. Law is the frame of possibility; its activation yields specific, orderly events.

(*Raj*) microscopic, kinetic space comprises fields in activated but subliminal or fundamental mode, that is to say, 'virtual' and actual fields of radiant (levitational) and contractive (gravitational) events. This is the chock-full space of Wheeler, Bohm and other physicists. *As*

---

[66] Chinese *qi* or *ch'i*; pre-scientific, oriental term variously defined as universal energy, vibration underlying perpetual atomic motion and life-force; physical correlates of supposedly metaphysical/ psychosomatic *prana* are identified as wireless electromagnetic force (light) and energy of electrical charge, especially regarding oxygen.

well as negative sink it is suggested that, in positive aspect, quantum void is a source of perpetual, supportive energy channelled into quantum forms.[67] **Such a reservoir of potential energy would thus constitute a medium between matter and its immateriality.**

(*Tam*) macroscopic space is what appears inert, oblivious and empty. It seems, in this aspect of absence, that there's nothing there. There are apparently no things at all. Dead void! Full impotence! Material immateriality! Such space is, like time, simply an abstract called extent. And such, physically, is the passive emptiness of what we know as 'outer space' - though atoms of the world and their bulk aggregations are, since space takes more than 99.9% of their own volumes, also hugely full of void; without your space you'd be invisible - a very heavy micro-speck of dust!

In summary, the vacant negativity of space is 'false'. Its explicit, superficial side is empty but, as physics notes, implicitly void brims with energy. **Such 'kinetic' species of space (that is notionally nothing) may well substantiate, in its various aspects, the form and action of material phenomena.**

Is Space a Waste?

So far nothing's yielded quite a feast!

Gaze into the less-than-crystal ball of space. Do phantom rules or particles appear? What does its transparent future hold? 'Nothing' - in the form of vacuum potential - hasn't reached the market yet. How can it when we only know it as a sink; and when we can't plug in. If we do will we tap power for free? Will there be scalar wave devices for energy extraction, transmission and immediate, non-local communication; anti-inertial and anti-gravitational mechanisms to fly with; and fuel-less transport, industry and entertainment? Will space science reveal electro-gravitation? Mass levitation? And, more than any other technological advance, endless scope for good and evil?

If riches can be harvested from space there is plenty of it, far more than of that other rich but hostile place, the sea. Will enterprising vacuum physics reap commercial crops of nothing; will space, after technical interrogation, be found a career path, be promoted and, in human terms, have a future? There are many questions about nothing but who, pragmatically, will engage its vacancy? Who will patent space and then, becoming its first mogul, pump endless billions from a boundless well of raw and 'non-existent', immaterial entity? And who, space billionaire, will write the first chapter in the story of exploited emptiness - not a myth of fortunes spun, thinner than the air, from nothing but a real and glorious history culled from frontiers all along the edge of being? There's no doubt that even sceptics should be glancing sideways, seeing and expecting nothing. Watch (this) space.

---

[67] Chapter 8: Split Continuity.

Chapter 10: Time

'If nobody asks me, I know. If someone asks me to explain I find I do not know', admitted St. Augustine when it came to time.

Second of physic's primal trinity, as obvious, impotent yet hard to grasp as space, is Augustine's process of duration.

Time

	tam/ raj	*Sat*
	existence	*Essence*
	finite	*Infinite*
	time-stream	*Origin*
	period	*Axis*
	motion	*Poise*
	times	*No-time/ Timelessness*
↓	*tam*	*raj* ↑
	finished past	*kinetic present*
	discontinuity	*continuity*
	isolation	*relationship*
	apart	*part*

Inexorable, implacable, what can resist? You can feel it but, intangible, you'll never touch. The harder you grip the softer it slips, completely yielding, to an edgeless flow of presence. Time is, like space, a cosmic paradox. It is nothing you can sense with organs yet something, closely, you perceive. Not physical but psychological, it isn't nerve but mind clocks time. Can you touch it? Nor can you affect but only tell it. What, therefore, is time but the space for motion and your turn of mind?

Absent presence! Where are surface, centre, volume or its texture? It is everywhere and at the same time nowhere all at once; and it's a single edge in nothingness that fills all space. Is it force or no? Like gravity it keeps the world in step; like levity it buoys all motion forwards. How can nothing roll the whole world down an absolutely one-way course? No time, no change, no motion, no existence. Is there anything not carried on its abstract stream?

Time and space, of matter or of mind, are the frame in which the pendulum of anything is swung; they are rails on which the sheet of finitude is hung; they are lines on which the engine of creation runs. Each is a function of kinetics, an aspect of a transformation, passive nothings that enable active things to pass. Are the missing couple

various oscillations	Centre-point
not-Now	Now
lesser, relative moments	Moment Absolute
appearances	Reality
becoming	Being
presences	Omnipresence
not-Here	Here
separate times	Pre-time
in time/ going	Pre-active/ Pre-go
↓ tam	raj ↑
behind/ backwards	in front/ forwards
now-looking-backwards	now-looking-forwards
now-from-fate	now-to-destiny
history	possibility
what came	what will come
over/ closed/ gone	open/ running/ going
become	becoming
memory/ remembrance	conception
exhaustion	spring
fall	rise
finished/ end	course

metaphysical or are they physically present? They are certainly locators in whose emptiness coordinates define events, on whose stage are lit the voluntary desires of mind and the involuntary, fixed programs of material behaviour. *Space-time (if you really have to tie the knot) is the means of separation, differentiation, isolation without which nothing could appear.* Without the continuity of their togetherness how could apartness happen? **Without such isolating continuity there would be, literally, neither space nor time for anything; without these differentiators nothing moves or stands alone; without these dimensions what a sudden cramp would clamp, what a paralytic lock-out of the most impotent kind would leave you neither jot nor dot of anything. Nothing. Nothing, even these, a couple that are nothing-in-themselves!**

Species of Time

relative	Absolute
absence of Presence	Presence
relative presence	Omnipresence
lesser truth/relative illusion	Truth
objective/ subjective states	Subjectivity
various durations	Super-time/ Presence

↓ *objective time*	*subjective time* ↑
fixed flow/ physical time	*psychological flexi-time*
units measured	*time experienced*
memory	*volition/ desire/ plan*
informed effect	*informant cause*
exhausted	*energetic/ in progress*
fixed/ sealed	*quick/ free/ in flux*
flat	*cycling/ vibrant*
dead	*living*
uncreative	*emotive/ creative*
no more possibility	*possibilities*

Time is a powerlessly powerful species of nothing-physical. Let's review the three fundamental forms of its sensible absence, its intangible omnipresence; the stack above applies to transcendent, psychological and physical domains.

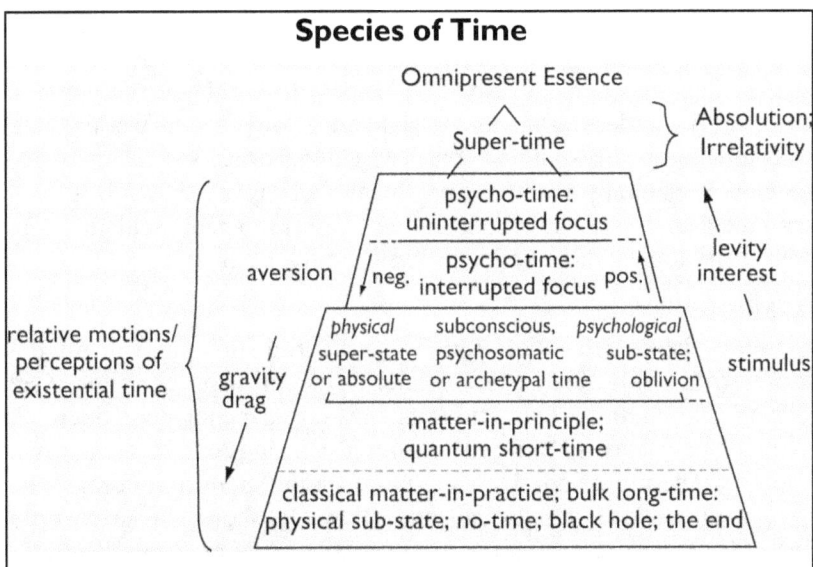

Irrelative time is Absolute. Here the present is concentrated into Infinite Presence; it is realised as every mystic's Singularity; and as such is a facet of the Infinite (N)One.

Relative time is the dynamic expression of Timelessness. It is fractionated Timelessness. As sensed by conscious mind its flow is flexible according to the quality of absorption, that is, by engagement of attention. As well as rubber mind there's clock. Without mind time would seem to tick the rigid same. But can it not be drawn out, as in pain, or squeezed, as with a black-hole, into squashed-up space?

> Conversely, energetic interest accelerates towards an undivided instant; and, light-like, absolutely concentrated focus leaves behind the mundane passing of materiality.
>
> *fig. 10.1*

Super-time

St. Augustine thought that time was part of God's creation so did not exist before. Before the world, he mused, there was no time. Just timelessness. Essential Timelessness is prior. It is independent of existence - out of mind and mindlessness. Yet if dependent existence constitutes a field of Essence-in-becoming, Infinity-in-motion or the working-out of possibilities then the Highest enters in; No-Time, in moving, comes to manifest each lesser, lower time-in-relativity. Such Axial Super-time, although a facet of Alpha and Omega, knows neither start nor end. Neither drag nor boundary restricts the Single Moment of Infinity, the Eternal Instant, Now. Real Time is Transcendent Being. To become is existential; 'becoming' crops up different all the time but Being is essential, central, motionless. Thus, paradox of paradoxes, time's distilled into a focal singularity - No-Time's Instant is eternal yet lacks any span! Vectorless, It marks the source and pivot of all worlds. Does Super-time seem like religious old hat? Contrariwise, smack up-to-date, this Now is Naturally behind the presence of all passing shows. It is Norm Absolute that, in relativity, motion of each lesser moment oscillates about. With such Moment those unscientific mystics seek a tryst.

Psycho-time

From time's essential paradox we drop to time in mind - informative, subjective sense of time. Let's call it psycho-time. Such passing is a function of attention. Its elasticity depends upon degree of interest; its path is 'warped' through varying engagements with intensity. The more negative an experience the more time 'drags'; the more positive it is the faster time 'flies'. In short, the more time-aware, ($\downarrow$) the more depressant is a psychological condition; less time-aware, ($\uparrow$) the more ascendant/ transcendent. Keen, uninterrupted focus of attention fires the immaterial flow. Such time, a function of love and interest, flies; and, in this case, more lovely is more real. Whereas scattered focus, with its short-term interests and aversions, yields interrupted, broken psycho-time. Noise, interference, puttering. Indeed, grounded by negativity or transfixed by pain, minutes seem to drag for an 'eternity'. Such is metaphysic's flexi-time.

Of course, your mind's constrained by body-sense. If not thus anchored (don't you dream?) what's the character of disembodied psycho-time? Post-mortem dream-time where emotions flood without restraint and trains of vision can't be earthed? No earthly clock ticks here. Close your eyes and start to learn just how elastic psycho-time can be.

Archetypal Time

As Super-time to *all* creation so is archetypal time to physicality.

Is memory not a 'solid' or enduring mental fixture? Are archetypal memories not fixtures inside universal mind? Archetypal time is no-time physical; it is metaphysical and thereby physic's absolutely super-time. As regards material phenomena archetypal character is therefore changeless. It is instant and yet omnipresent; its transcendence outspans every physical expression yet lives within each moment. Without passage it outlasts all passing of an ever-changeful world.

Closest unto immateriality, consider a material 'absolute', light's ghost. An appearance of time-absolute is vested in the mass-less photon whose velocity in vacuum (c) is the axis round which physic relatively spins. Light's the nearest matter flies to instant archetypal time.

Bio-time

Bio-time takes two main forms. The *first* clocks gaps between appearances and lets us judge the speed of change. A psychoactive stopwatch helps us gauge appropriate response to myriad and simultaneous perceptions we call circumstance. Perception of such bio-time is certainly affected by one's size, pace of life and speed of neurological response. For example, elephants live twenty-five times slower than shrews; what seems normal to an elephant is for the shrew in slow, slow motion. When you crack down as fast as possible the faster-buzzing flies evade your swat, which seems to slowly fall, easily with time to spare.

The *second* kind of earth-life clock is cyclical and as such vibrant and dynamic. Hard-wired to internal systems, it revolves by means of gears; molecular cogs integrate feedback cycles, that is, periods of homeostasis. Homeostasis is a process of equilibration ticking at the centre of biology.[68] Its internal triggers link, of course, with changing circumstance. Bio-cybernetics is connected with the cycles of earth, moon and sun.. Computation flows according to diurnal, menstrual and annual rhythms. Bio-clocks are linked to sleep, reproduction, navigation and a host of other systems. The purpose of informant bio-regulation is to keep life bouncing.

Material Times

Life involves attractants, purposes that pull towards the future; the purposeless oblivion of objects is only shunted from the past. In this sense physic has no future since it doesn't care. Nothing matters. There's no target. Matter, although pointlessly pushed round, is going nowhere since it has no sense of aim or time. It is not time-aware; it is

---

[68] Chapter 19.

inanimate. Don't protons and electrons each endure a senseless kind of immortality? Time's absence, as in oblivion and sleep, is Natural Dialectic's case of sub-time.

Wait a tick! You can chronicle a history of clocks. Such measurement of physics' time comes down to calendars and counters. Humans take their own backyard, the solar system, to establish the time-spaces that they live through. Year (sun), season (earth's wobbling tilt), month (moon's cycle) and day (earth's spin about its axis) are the points of reference used. Babylonians and Egyptians split time's flow the way, with refinements, that we know today. They used twelve temporal hours (that vary with the seasons) for a day. Greeks, Romans, Jews and Arabs followed this device. Drop by drop or inch by inch water clocks and sundials registered time's abstract passing. Now we use the Gregorian scale and rigid hours whose divisions, harking back to an early geometry of circles/ cycles, split to 60 minutes then to 60 seconds; or, more accurately if you care, we can measure oscillations of an atom.

You might, thereby, have thought one moment was as flat and smooth and long as any other anywhere - time-physical is inflexible but still a-changing. Newton thought it moved the same for all observers while his space was motionless, always the same and thus composing an inertial frame of reference. Each abstract is separate and absolute. Einstein claimed, however, time's dependent on a watcher's motion or an object's pace; it's a shadow, motion's phantom, thrown by change. It can be dilated by velocity or gravity; as these increase time slows until, theoretically, it stops and *vice versa*. And, while emptiness itself is flat, its immaterial space-time couple flex; 'bendy' space-time varies relative to mass, location and acceleration; nor can any rest-mass reach light's 'absolute' velocity. Such distortions instruct matter (e.g. planets, suns and galaxies) how to move; they can even, at black-hole extreme, be twisted to collapse! Space and time are broken! Some, though, also think that quantum space is not continuous; quantum mode breaks nothing into grainy, Planck-size 'pixels'!

Hands up those who fully understand this kind of logic and its actualities! Common sense cannot, it seems, prevail.

The Geometry of Time

Space-time geometry is one thing. Another is Euclidean geometry of space. Has time, like space, a geometry that's built from dot and line and curve?

If there are also three main grades of non-conscious matter - potential, matter-in-sub-atomic-principle and the matter-of-bulk-forms then you might reckon there appear three shapes and grades of time.

A dot represents the central, archetypal starting super-point; it's time physical's potential, no time yet - eternally.

Vibrant time is energetic. It is survival's sort of upkeep. Call its oscillations AC. Cycles are composed of curves. Curvaceous time, fast or slow, inhabits periods.

Flat time is sub-state. Call its line a ray. Its straightness runs from start to finish, past to future, full potential to complete exhaustion. Call such a discharge, flying like an arrow towards an end-point, DC time. Call such expenditure the way of aggregation, bulk form, entropy and death.

	time-lines	*Dot*	
↓	*arrow*	*wheel*	↑
	discharge only	*cyclical regeneration*	
	runs out	*pops up again*	
	dc	*ac*	

Is time AC, DC or, paradoxically, both at once? Is it an arrow, wheel, a calculus of dot or simply, without any geometry, a part of the events mind measures?

AC-time

On earth you cycle straight along a road or walk a line straight round the world. Many kinds of clock and every calendar spin round and round. So does vibratory AC time. Things oscillate, rotate and cycle with all sorts of frequency. They pop up again. A roll around circumference will pass you back to where you started. Cycles shuttle; each wave vibrates around its centre-point, every period orbits axially. Waves do not indicate straight-line progression or evolutionary idea. Cycles deal dynamic equilibrium. *Mind* makes improvements but for matter there are only simple ups and downs of things. Equations wheel, pistons push and pull, motions grind through nature's great and balanced rounds. This is *dharma*; this is the equilibrating way of *karma*; this is Newton's third law of motion as applied to mind and matter. Life's bodies all vibrate, at heart, with homeostasis; and, psychologically, desires are discharged into satisfaction; mind recycles into balance and recovers its potential in the healthy, happy mode of peace. Cycles run from source to sink and back. Clearly, not only Buddhists spin time on a wheel.

DC-time

Can end precede its start? Can effect precede its cause? A 'line' that's sometimes reasonably straight connects a cause to its effects. This, no tock-tick, is basic cosmic law. Down this line prior potential (involving energy or information) is expressed and then, at last, exhausted. Things run, as viewed dialectically and otherwise, from possible along kinetic to inertial phase. Options are suppressed, possibility and plausibility reduced into a single, actual outcome. You could say potential's realised and then, exhausted, hardens up and dies. Are not solids graves of energy? Even cosmos tires. Friction, loss of power and gravity always flatten bounce into a fixed, straight line. And causality's a valve that plumbs the cosmos into one-way flow.

The law of entropy requires this flow is down to earth; and that, in winding down, the universe began its time in a most highly-ordered way. From bulk matter's massive angle arrow and progression seem to dominate. DC is the arrow from a bow; DC river pours into a final ocean of inertia. This way roll body's youth and mid-life. They run, relentlessly, to age and oceanic death. Arrows, say laws of entropy and gravity, always fall to earth. And (unless you include air-lifted rain) rivers never flow, like cycles, back to source. Cosmos is a river flooding time; it is an arrow loosed in space. Its intrinsic metaphysic dwells in every interaction or, with universal action's end, perhaps both parties shrivel up and disappear. Who knows how forever-and-a-day will show or if, collapsing, time itself will fall away? Nothing then. Full stop. Less than a dot. A stop as full of emptiness cannot be imagined.

Grades, Principles and Times

You can phrase material time a graded way.

A motif that underlies the presentation of Natural Dialectic is one that reflects the gradient of creation. *The gradient runs top-down.* Cosmic hierarchy drops from mind to matter, from archetype to physical expression. Principle (condensed information) guides practice; thus simple runs to complex, general to specific, universal plan to detailed, individual, localised expression.

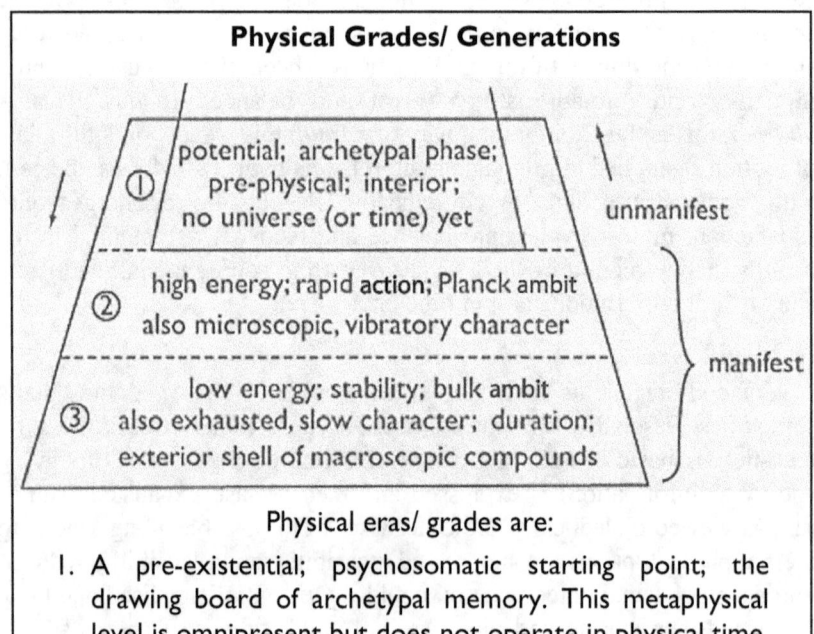

Physical eras/ grades are:

1. A pre-existential; psychosomatic starting point; the drawing board of archetypal memory. This metaphysical level is omnipresent but does not operate in physical time. Call it archetypal 'no-time'.

2. The expression of 'grammatical' information; archetypal memory is expressed as a sub-atomic, quantum or principle grade of matter (represented in the Standard Model by three generations of fundamental particle); this level amounts to nature's intrinsic code for matter; its dynamic alphabet operates in 'fast' micro-time and may be construed as the level of quantum activity.
3. Nature seeks the lowest energy-level. This turns out to be, by now, the last-occurring generation of matter (ours). Its particles (electron, neutrino, proton and, inside an atomic nucleus, the neutron) endure. Their atomic border gives way to a multiplicity of differences in the form of plasma, molecular gas, liquid and solid. The 'internal' energies locked in these low, classical grades of matter are vibratory; and external definition increases with loss of energy-in-motion. This level of time, in which you and I live, operates in 'slow' or 'sensible' macro-time. Call it 'long or slow time'.

fig. 10.2

	grade	*time*
sat	*pre-physical latency*	-
raj	*quantum micro-level*	*Planck/ quantum era of high energy and quick time*
tam	*classical macro-level*	*era of low energy/ bulk aggregates and slow time*

If grades run from potential to exhausted then hierarchy is both metaphorical and real. *In this sense (figs. 3.1 and 3.5) it follows that higher is 'within' lower and first is 'within' last. What came first informs what follows. 'Higher' is the magisterial cause of 'lower', consequent effect.* Each grade nests within the one below. Atoms are nested within bulk materials, particles within atoms and vacuum within everything.

The clear implication is that creation springs 'organically' from within outwards. The dialectical expectation is, therefore, that subdivisions of the physical cosmos devolved in the same order in time as in material grade. Creation moves from centre towards periphery; 'top' or potential matter transcends its expression as, firstly, quantum and then large-scale, classical matter.

Curiously, three generations of elemental leptons and quarks scale from massive (discovered at high, early-cosmic energies) to the less massive staples (electron, neutrino and up/down quarks) of our low-energy constituency. The *SMPP* (Standard Model of Particle Physics) cannot predict the mass of *any* particle; 'the mass hierarchy problem' asks why, if they were all identical before the forces separated out, generations of particle do not display a discernible pattern as regards their masses.

However, is not hierarchy serial passing-down? **Thus time's order would reflect the order of creation, linkage that is still inlaid in things.** Potential, short and long times - these have grades and eras interlinked; all three exist, as in the beginning, now.

There is, therefore, in this *top-down* view, a strong link between cosmic grades and the eras in which different grades of materialisation were predominant - a view endorsed by particle experiments at *CERN*. *To this end it is important that, as far as possible, the structural order of Natural Dialectic unfolds in a way that reflects the entropic development of cosmos with its vectors of both grade and time.*

How Long has Time Hung Round?

Turn round. Wind time back. How long would it take to press the universe back into whence it came? You say no-one was there. Astronomers, in viewing galaxies billions of light years young, can creep up on the start. For many, therefore, any controversy is a waste of time.

Would you Adam-and-Eve it? It was not the early Christian apologists that started it. Origen, Eusebius, Augustine and later (see The Correspondence of Sir Isaac Newton 1676-1694) Newton himself cast doubt on a literal 6-day creation with the seventh day of maintenance extending until now. No doubt about it, though, claimed the Vice-Chancellor of Cambridge University; during the 1640's John Lightfoot tussled with the Irish Archbishop of Armagh, James Ussher and eventually concluded creation week was October 18-24, 4004; Adam was created 'after breakfast' at the good clocking-in time of 0900 hours on October 23rd. Thus controversy's thunder clapped. Some thenceforth took the days as literal and others as a metaphorical expression of serial stages in the development of cosmos. All kinds of evidence has been adduced in either case. Some would say the facts have been traduced to literalise 144 hours-worth of creation. Does the Bible really mean this? Contemporary science definitely disagrees. Astronomers espy a universe expanding from, they logically deduce, a very lively blow. They identify projection of the universe, an instant miracle from outside scientific laws. Their probes seem to confirm, about 13.77 billion years ago, this burst projected as a 4-d space-time screen whose general character incorporates decelerations and decay. Dilated and diluted - every hour is stretched into about 100 million years!

A million centuries! Yet carbon-14 traces have been found in coal; and, along with skin, tendons and ligaments in hadrosaur fossils, even blood from stone! In *T. rex* bones vessels, haem and perhaps cells[69] have puzzled theorists. Moreover melanin (a light-absorbent pigment) has been discovered in the skin of mosasaurs, ichthyosaurs and leatherbacks.[70] Such complex molecules are fragile and, left to the elements, rapidly decay.

All dating arguments, radiometric, stratigraphic or textual, are abductive.[71] They depend on prior assumptions, engage interpretations, turn up anomalies and may involve tautologies. For example, geology professes doubtless faith based on two major assumptions - the great age of earth and James Hutton's general principle of uniformitarianism (summarised in the aphorism 'the present is the key to the past'). This pair effectively negates 'catastrophism', that is, erases the notion of global tectonic, igneous or, by flood, sedimentary catastrophe in earth's prehistory. At the same time they promote evolutionary faith in the creative powers of vast and indispensable tracts of time.[72]

Tautology is circular argument. Variants are routinely employed to bolster the Darwinian presumption. Trivially, we are here and so we must have evolved. More powerfully, fossil finds are all interpreted in evolutionary terms. Certain specimens found in earth's 'stratigraphic column' (built of rock units such as coal and chalk measures subdivided into beds and bedding planes) are used as indicators. The column itself is defined by such index fossils. Each type has been assigned an age based on the presumed age of rock layer in which it was found; find the fossil and you not only date freshly explored strata but any other fossils found in them. Herein lurks powerful tautology. *The assumption of*

---

[69] Science 25-3-05 ps. 1952-5; further 'sensitive', 'intolerable' and therefore censored cases of low radio-carbon dates for dinosaur soft tissue from Montana, Alaska and Texas have been reported. Check also preservation of *Nodosaur* and *Zuul crurivastator*.

[70] Sci-News.com 9-1-14.

[71] see 'logic' in Glossary and Index.

[72] A critique of radiometric dating systems is not included in this volume. A history and critique of the conceptual development of chronometric ideas such as timescale, uniformitarianism and 'deep time' (which by now condition not only the geological but modern, cultural mind-set) is not included either. Nor, again, is inspection of another key temporal assumption, a meme called 'the onion model'. In this model concentric spheres of time (called various ages of rock such as 'Cambrian') ideally represent earth's crust. The 'onion' is closely linked with a couple of other chronometric cornerstones viz. aforementioned timescale and an interpretation of 'globally correlated synchronous time'; but evolution, a further assumption involving tautologies and stratigraphic successions, does fall (Chapters 19-25) within the cover of this book.

evolution is the basis upon which index fossils are used to date rocks; and these same fossils are supposed to provide a main evidence for evolution. The fossil record, itself based on evolution, is interpreted to teach evolution. A closed circle is a ring that binds. By this sort of reckoning, called begging the question, the main evidence for evolution could be the assumption of evolution!

However, whether earth is young or old, Natural Dialectic takes the line that information, not time and chance, is the critical factor. Life's times do not depend on mindless physic. Informative mind precedes informed energy. *Cosmo-logical design trumps days, months, weeks or billions of years. And bio-logical codes drive the point right home.*

Age alone cannot create a set of cutlery let alone the specified complexity of forms of life. Physics shows an initial projection so fine-tuned that, although including elements of chance (the fodder of a naturalistic view), design is a reasonable choice for cosmic origin. And the science of biology now shows (it will be argued in Books 2 and 3 *passim*) that naturalistic process is an insufficient 'author' of our human hours.

Thus, to reiterate, the main issue of creation is not whatever kind of time you choose but *mind*. Mind is natural. Creative mind is anti-chance and has its range of purposes; expressing these it gives or takes whatever time it likes. But one thing every creator strives to eliminate (with an effect proportional to his intelligence) is bugs, error and the chaotic part of chance. *Only mind is capable of systematic determination, coherent construction and purposive complexity and the time it takes to realise its ideas is of secondary relevance.*

OK! Time's up! At least for time's being. We have not, however, reached the end. The next chapter's time has come. The game proceeds, the play is in full swing as we progress to consider the third of physic's primal trinity, the ingredient of energy. If there was a start to energy an end is logical. Possible physical expressions and terminations are considered. Having cycled from *alpha* to *omega*, one can then close the first half of the whole book. And take a break. In other words, clock off.

Chapter 11: Energy

Energy is, according to Natural Dialectic, the third of physic's primal trinity.

The two basic components of creation's existential dipole are, for Natural Dialectic, information and energy. Each is well characterised in terms of light - inner and outer lights. They are respectively the light of knowledge, illumination, and the light of physics, electromagnetic radiation. One is at the subjective heart of life; the other is the light by which the world is known, all bodies warmed and, through photosynthesis, life-forms fuelled. Subjective, living light and objective, inanimate light - these irradiate with understanding and with warmth. Each kind of brilliance is at the top of its domain.

Matter's Holy Ghost

'Ghost' is now commonly understood to mean a spooky phantom but actually the word derives from a Germanic root, *geist*, meaning 'spirit' or 'essence'. So, ironically and paradoxically, current parlance conjures up an ephemeral, local, insubstantial illusion; whereas the term 'Holy Ghost' labels a spook's absolute antithesis, the Real and Permanent Substance, an Action also named *Logos*[73] whence creation logically was and is projected.

tam/ raj	*Sat*
inertial/ kinetic	*Archetypal Potential*
cyclical effects	*Source*
colours	*Transparency*
impurity/ obscurity	*Purity*
↓ *tam*	*raj* ↑
lower/ darker	*higher/ lighter*
black	*spectrum of colour*
diffusion out	*concentration back*
light/ heat loss	*heat/ light gain*
energy locked/ bound	*energy freed*
contractor	*radiator*
dragged/ slowed	*kinetic*
inactive/ cold/ fixed	*hot/ moving/ fluid*
mass	*energy*
particle	*wave*
gravity	*levity*

---

[73] see Chapter 5: Top Teleology; also Glossary and Index.

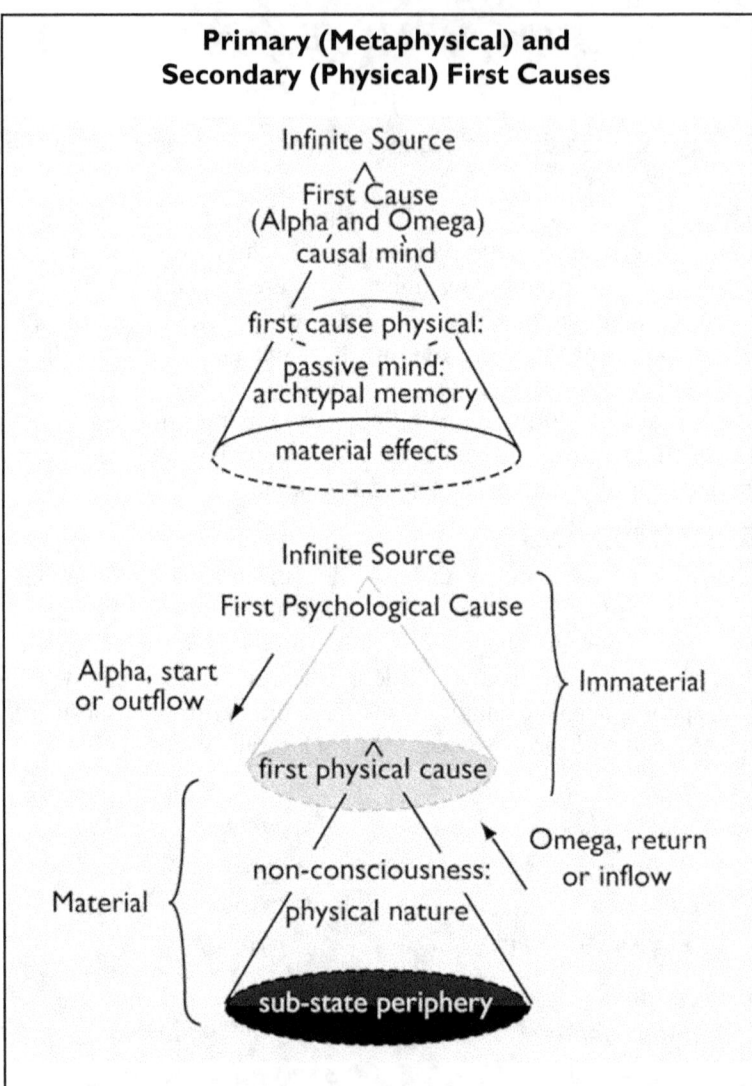

Primary (Metaphysical) and Secondary (Physical) First Causes

Each conic diagram expresses exactly the same idea (projection of the conscio-material gradient) slightly differently. It is clear that there exist two Point Alphas,[74] one 'major' and the other 'minor'. The major, prior starting-point is actively informant, psychological First Cause. This is the Essential Cause of existence. The lower, secondary origin, physical first cause, is passively informed. Its archetypes are fixed.

fig. 11.1

Nor is a spectre the same as a spectrum which, after Newton, has come to mean a continuum of values. Could existence, like electro-

---

[74] See also Glossary: first causes.

magnetic light, have been projected with an intrinsic range of values? Natural Dialectic's 'Holy Spectrum' is a conscio-material one including both subjective and objective parts. In this view information governs energy's expression - government we call the laws of nature. Expression is a coupled show; it includes structure and behaviour, form and action. Such coupling applies, of course, to light. Its expression, though arguably the 'purest', is but one projection from the energetic set.

Potential matter, archetype, is physic's holy ghost. It is metaphysical but what is *least physical, least massive* and thus like pure, fluid energy? **Light, ghost-like, approximates to immateriality.** Does subtle precedence require that, in material creation, it was radiance that first appeared? Or

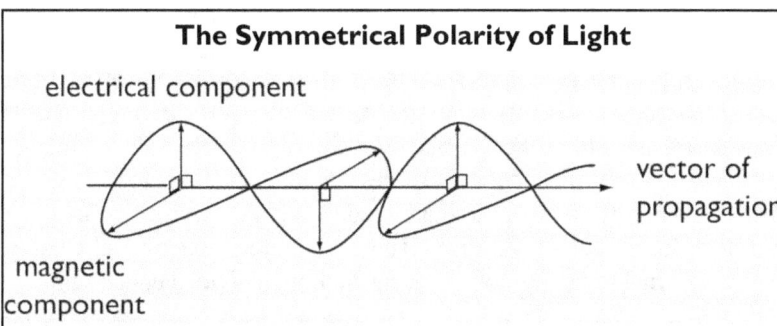

The Symmetrical Polarity of Light

While the polar expression of Singularity has been identified (*fig.* 2.3) as information and energy, this diagram illustrates the symmetrical geometry of physical illumination. Such symmetry is an expression of dynamic balance; it is a super-fine articulation of polarity, in this case between the oscillations of electrical (*raj*, radiant) and magnetic (*tam*, contractive) field components and their straight-line 'normal' of propagation.

In a further apparent reflection of Infinite Illumination the ghost-like transparency of light's (↑) radiance paradoxically involves an implicit spectrum of energies and colours. Moreover, also paradoxically, its motion may seem absent; yet simultaneously space and, with velocity, all time is calibrated to the absolution of a light-beam's cosmic limit, c.

Natural Dialectic identifies illumination's 'pure energy' as the subtlest (*sat*) component of physical duality. But its fundamental apparition, roaming spatial emptiness, comprises only half a dualistic pair. The other part is (↓) nuclear force and energy tied up in mass. Mass-less photons interact with mass. They are reflected or, upon absorption, transformed into heat. Heat, like sound but unlike light, needs mass to 'ground' its rousing oscillations.

fig. 11.2

virtual photons of a quantum fluctuation in some unlit void? Let there be, first and highest, mass-free light. Pellucid radiation from the heavens was, we've seen, traditionally considered both primordial expression of creation and a medium of communication straight from Creator to phenomena. Pure will informs, pure energy expresses His Chiaroscuro.

Insubstantial forces brace bulk substances; from immaterial patterns issue nature's forms. **In Natural Dialectic's spectral view, exterior appearance stems from subtle inner bands; creation issues, layered, from within.** From nuclear programs issue flowers and children. So do rocks and clouds and stars; their substance is an alphabet of sub-atomic letters whose interactions (by four forces) build into a large vocabulary; they sing the saga, write the world's text not on air or paper but the blank of space. This view of creation is, we said, 'organic'. *'Egg' is an apt metaphor for outworking from within.* To repeat, the cosmos is projected from a seminal conception and its consequent, logical ramifications; conceptual development is stored, like the blueprint for any construction, in memory; the blueprint's physical code is represented by an 'alphabet' of simple particles with a forceful 'grammar'; and a starry universe is, finally, the outermost fulfilment of A Plan. There is neither external Creator nor creation external to its Creator. All is one although divided endless times.

In this view no physical energy is 'injected' into the system. Instead, latent potentials are, like drum-skins or quiescent fields, roused into actuality. The notion of a roll is quite apart from one of accidental levity from singularities. Within an overriding key and structural theme variegated tunes of a concerto play. Physical energy, operative under basic conservation laws, is seen as part of a larger system that is, like a musical tune, the effect of a score vibrated into existence. This score, whose archetypal memory and excitation both derive from the Logical Sound of First Cause, raises the latent potential of matter into 'life'. Its broadcast 'quickens'; its reiteration 'refreshes' quantum instruments of energetic pattern, the notes of archetypal memories it first established. Reiterative vibration keeps nature's fundamental particles 'alive'; perpetual motion makes kinetic theory real and, at atomic and the larger scales, variation by re-combinations is the theme of change. In the last analysis atomic components (electrons, protons and a few charge-neutral particles) are the expressions of metaphysical harmony, the modulated repercussions of steady frequencies that seem perpetual - until the energetic music stops and there is silence. Nothing left. When current is withdrawn how long does the emanation from a record player last? Projection ceases, motion fails; cosmos straightaway collapses and the whole illusion's voided back the way it came.

Grit in the Ghost

There's more to light than meets the eye - not only, from its range of spectrum, obvious invisibility and polar symmetry. Its apparent continuity's composed of tiny shiners, quantum fractures, photons. How absolutely holy is a gritty ghost?

	relativity	*Absolute*
	division	*All-in-Oneness*
	polarity	*Neutrality*
	excitation	*Latent Field*
	fermion	*Boson*
	hadron/ lepton	*Photon*
↓	*minus*	*plus* ↑
	hadron	*lepton*
	quark/ proton	*electron*
	contractive fields	*electromagnetic field*
	massive/ bulky	*lightweight/ lifting*
	gravity	*levity*
	drag together	*stimulate/ drive apart*
	confinement	*liberation*
	binding/ capture	*freedom*
	fixity	*mobility*
	particle form predominant	*waveform predominant*
	locked energy	*connective energy*
	isolator	*communicant*

Maxwell and Faraday linked electromagnetism to electricity and, thereby, negative electrical charge whose only carrier, as far as understood, is another quantum - the electron.[75] What *that* is nobody knows. Light is a product of electrons in accelerating mode or shifts in their atomic orbitals. It depends, except in drastic cases of atomic fusion or material annihilation, on these quanta. No doubt the *effects* of electricity are precisely understood and practically all the mechanical, electrical and magnetic properties of matter (and therefore chemistry and materials science) are based on electromagnetic interaction. No doubt also that electrical circuit theory is a detailed and important abstraction behind the design of electrical devices. What, however, *causes* any quantum to appear? What *is* charge's point of action, what set up electrical polarity?

And mass? Electrons, protons and neutrons have, like every other piece of grit in nothing, mass. How would you make local grit from seamless ghost, how erect a massive world-stage out of nothing and set up without support? No doubt, appearances in space need, as their sharp formaliser, mass. Mass-production's problems, that various Higgs bosons may help solve, are embedded at the heart of Standard Theory. These particles are, mathematically, very heavy, thus unstable - gone before the world could blink. Even if corroborated by the *LHC*, does this explain how interaction

---

[75] see also Chapter 8: Split Continuity.

with a possibly-existent Higgs drag-field confers mass; or why, conveniently and hierarchically, such field should be frozen into cosmic fabric in the first place, at the 'top' end, and thus primordially dovetailed with all following fine-tuned factors? Maybe or maybe not, post-archetypal phase, Higgs mechanism is the physical way specific ghosts and massive bits of grit arise and stay. If so, exactly how and why?

Alpha Points.

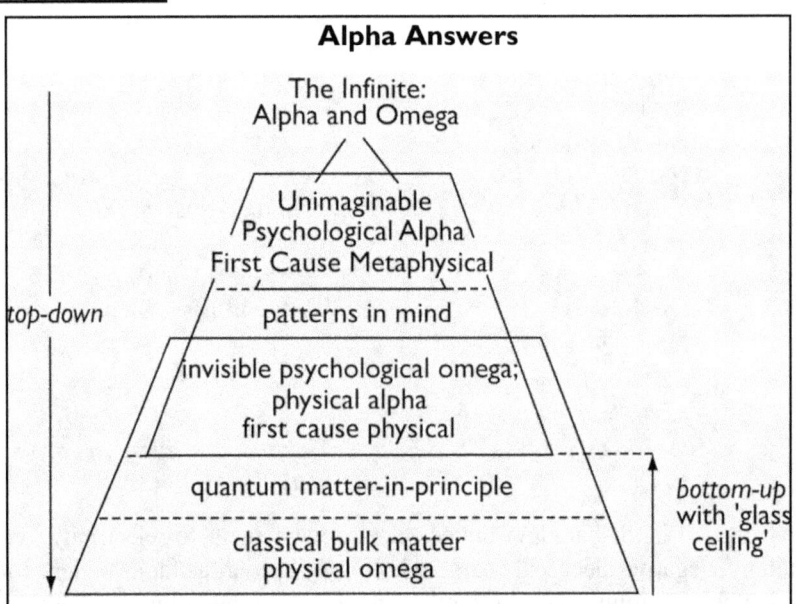

Alpha Answers

Top-down the Source is Infinite. In this view a 'glass ceiling' (made, maybe, of vacuum energy but certainly inclusive of transparent space) is reached where metaphysic's transformed into physic. The pattern of physical energy is issued 'through' this 'glass ceiling'; in other words physical derives its origin and sustenance from metaphysical.

Bottom-up the *top-down* view is absurd. A philosophical decision has been made - there's nothing that, even prior to cosmos, is not physical. Logically, therefore, everything physical must have had something physical precede it - except perhaps at first. There's no glass ceiling, nothing is unreachable by naturalistic means. Things must, of course, have sprung from nothing physical but 'nothing-physical' is not, in this view, metaphysical. Therefore the natural, material universe must be derived from absolutely powerless emptiness.

Which is the A-grade answer? Which the E-grade?

fig. 11.3

Every object and event, every pattern and behaviour has a birth, lifetime and death. Does this include creation as a whole? The last six chapters have been devoted to a *top-down* description of its interpenetrating layers as they extend from the Centre to an elusive periphery - although the Centre is everywhere. In other words, what could have seemed like a temporal odyssey from Essential Potential down and out to physical solidity could equally appear shorter than the distance of the book you hold, closer than grey matter to your thoughts, no more extended than the twinkling of an eye. So how was 'universal body', in all its great complexity, first projected? A week is a long time in cosmogony but as nothing for the intellectual gas surrounding it. Even saying 'big bang' takes far longer that the real thing, if it did, did; and explaining 'levity' much longer. What are the current versions of the birth (Point Alpha) and death (Point Omega) of physical creation?

The *AVP* is incomplete. Perhaps one day the answers to all questions will be phrased. This section deals, however, with appearance. How does something come from nothing, how is physic born from metaphysic? Was archetype projected or was material absence false, so that something prior blew up the world's balloon? If 'nothing-physical' (*fig.* 11.3) can't be informative then let us conjure endless universes of unknown kinds of matter! Any guess is better that the metaphysical! But, we'll shortly see,[76] any story that rejects an immaterial factor risks assumptions and confabulation on an epic scale. This is because initial mysteries can, wrapped in myth or metaphorical mathematics, only be inferred. If concepts are metaphysical which sort of metaphysic frames your answers? What problems plague solutions; whose inference, least problematic, might be analyzed as 'best'? How did existence happen?

The steady-state view is that start and stop don't, universally, exist. What *ad hoc* factor, though, drives such a state's expansion of *ad hoc* eternal matter? If every spatial litre generates, unstintingly, fresh atoms where are those new galaxies that should spring up between the old and new? The old are far-off and new near; expected stars aren't there. What would support apparently perpetual sub-atomic motions? If entropy's had time for final victory why have sky-borne star-fires not failed long ago? Fred Hoyle's theory of steady state was toppled but, as we'll see, materialism won't surrender its eternal matter easily!

Contemporary science and perennial religion prefer a single origin.

How Fit?

How fit is the scientific preference for such an origin and, thence, the evolution of our universe? Is its favourite model up to scratch?

---

[76] Chapter 12: Atheism's Last Refuge.

If cosmos is a mind-game you'd expect the rules made sense. No contradictions, please. No unfit parts that jam the works. Is this the case? If, in addition, free-play (known as life) is part of the intention wouldn't you expect the match, mind with matter, to be friendly?

Bottom-up, since they evolved, life's molecular configurations are insignificant as any other. In material fact there's nothing puts you at the centre of the universe. The world, state both Copernican and statistical principles, is randomly distributed and not 'designed' with you, Luck's child, in mind.

A 'weak anthropic principle' says that, because we're here to observe the universe, its laws allow this circumstance. To this tautological extent we're privileged.

The 'strong anthropic principle' proclaims, on the other hand, that cosmic parameters *must* admit observers, that is, intelligent life-forms. But materialism doesn't like such hint of precision engineering or design. They run against its naturalistic grain although, ironically, Copernicus and all prior astronomers espoused them. Thus harmonic geometry and music were the order of their explanation's day. A catalogue of solar-systematic 'coincidences' were logged as tables, pentagrams and other forms. Bode's Law, for example, indicated underlying pattern in orbits and periods of planets and their moons. Every planet dances prettily about its neighbour. The case of earth and Venus notably invokes Fibonacci numbers[77] and, therefore, the golden ratio. Again, the moon precisely eclipses the sun; its ratio to earth size is 3:11 (or 27.3%); and 27.3 days is the period of its orbit round earth - and so on and on. Whether the meaning of these geometries, ratios and harmonic interactions sums to greater than superposed coincidences is (since the implication of aesthetic order coupled with mathematically-based logic offends its creed of mindless randomness) a question beneath the dignity of contemporary, post-Copernican cosmology.

Such cosmology is an observational but not experimental discipline; and, in its attempt to explain an origin left deep in time, an historical science. Thus, as do geological and evolutionary theories, it relies on abductive reasoning;[78] it interprets facts; and thence makes assumptions when choosing from a range of possible computer models to explain its observations - however accurate these may be.

Today's fashionable model is, of course, 'big bang' theory. This is, essentially, a 'precise' explanation of what appears to have followed a 'transcendent projection'[79] such as any kind of universal birth requires. It relies on a medieval break with geocentric astronomy derived from

---

[77] see Chapter 24: Twists that Entwine.
[78] see Glossary: logic.
[79] Chapter 1: Causality.

Copernicus and now called the cosmological principle.[80] The principle assumes that, from any location in space, the distribution of matter will appear universally 'homogeneous' (uniform) and 'isotropic' (the same in all directions). Although various 'proofs' are offered in support it is uncertain whether this unverifiable assumption is correct. Nevertheless, a hypothetical device called 'inflation' is recruited to substantiate both principle and the 'fact' that expanding space (another unproven assumption since who's seen space itself expand?) is not curved but flat. This accords with Euclidian (not elliptic or curved) geometry. If, however, space intrinsically expands then why, as well as galaxies, don't solar systems, planets, even space-filled atoms spread apart?

Lastly for now, due to an explanatory failure of the *AVP* (standard model of particle physics and theories of relativity), with respect to galaxy formation and behaviour cosmology now proposes that 95% of cosmos (viz. dark energy and dark matter) is unknown. We see, as St. Paul put it, through a glass darkly; as darkly as almost unlit space. How fit would Sherlock Holmes feel if he lacked 95% of the evidence that he had to suppose informed his observations?

Whence, therefore, shall we progress from here?

---

[80] see Glossary.

Chapter 12: Magnificent Mythology

No doubt, since the universe can never be independently tested and many cosmological assumptions are therefore unverifiable, man can at best theorise abductively.

And if the first step of your journey (say, exclusion of an immaterial factor) isn't in the right direction when you stride away where does it lead? Do you want whole truth or half? If, as well as energetic matter, there's a non-physical factor; and if, as well as material there's an immaterial element do you want materialistic truth alone? Such truth may lack the mind you apprehend it with. It will involve at least a mathematical mythology.

Questions aren't in short supply but, eventually, chemistry and physics may well be developed to describe, entirely, all material phenomena. Mythology will not reside in such description; it resides in sense of having missed a point, perhaps a vital point - the origin of information (and, thereby, the origin of energetic patterns) in the first place. Thus, as with male without female, only half your story's told; with energy but without information, only half the world is understood.

Fit

Down-to-earth, what do *facts* indicate? What can be inferred from natural, physical constraints? A test-case might, since you're a fact, be you. Did Sheer Luck swill you up? Should atoms, given time, evolve an understanding of themselves?

On the other hand, the intrinsic basis of biology is definitely information. If, as an engineer, you wanted to create a life-form, what complementary, optimal materials and constraints would you devise? Are these devices what we find? Is the interrelated multitude of constraints so severe that your prefabrication is the only one possible? Indeed, are the laws of nature specifically adapted for carbon-based life on earth or similarly watered planet? And are the cosmic, galactic, solar and planetary aspects of ecology together friendly for your life-form? **The strong fact is that they are in very high degree.** The world fits well together. The universe is fit.

Energy *and* information. Is physics' basis information? If so, the scientific half needs spread its wings and come cosmologically[81] to terms with more than half of you. Astronomer Royal, Sir Martin Rees, boiled down to six the number of dimensionless constants and combination of constants that basically define the structure of our life-friendly universe.

---

[81] see also Chapter 7: Cosmo-logic.

There are about twenty other precise constants (such as gravitational constant, electron charge, proton mass) that would appear to 'coordinate' the operation of this world but, Rees calculates, if any of his six were 'out' by only 1% we'd not be here. They include the ratio of the strength of electromagnetic force to that of gravity - gravity slightly greater and a smaller universe would develop faster (and *vice versa*). If a ratio (Q) of energy needed to disperse galaxies and their rest-mass were smaller no large-scale astronomical structures would exist; larger, and everything would have been swallowed into super-massive black-holes. D=3 means three spatial dimensions in which the inverse square laws of gravity and electromagnetism can work. ε, the fraction of mass released when protons fuse to form helium nuclei inside a star, determines stellar energy release and the chain of reactions producing other chemical elements. Too little and a planetary system freezes; too great and heat-level renders life impossible; just right and here we are! The pieces of life's jigsaw fit.

Pass The Paracetamol

Why should pre-physical transcendence have dropped 'balance' everywhere? Are cosmic fundamentals[82] anything to do with it?

↓ *down*	*Pivot*	*up* ↑
gravity	*Balance*	*levity*
compression	*Equilibrium*	*expansion*
sink	*Source*	*flow*
slow-down	*Potential*	*action*

Of special interest is (Ω) omega. This defines the tension between gravity and expansive energy (levity) of cosmos. It is the ratio between actual density of mass-energy to the critical value required to eventually halt and reverse expansion. The actual density is many times less than needed (0.2 against 5 atoms per metre cubed of space) but an undetected ghost, dark matter, is invoked to pad the numbers out. For initial impetus (↑) to balance the (↓) decelerating tendency of gravity Ω must have been very finely tuned to unity. If its expansion energy had been larger the universe would be open and without stars, galaxies or life. In other words, faster (by one part in 10^{14}) and no galaxies would form; slower, cosmic collapse aeons ago. Sharp a watershed is omega and either side, no life.

Then there's Λ - anti-gravity (↑). Einstein thought the universe was static so he introduced a fiddle-factor to stabilise against gravitational collapse and keep the world in business. This he called the cosmological constant (Λ). But Einstein's equations could predict expansion too; and cosmos seems to be dynamically expanding. The expansion rate offsets gravity. The cosmological constant not only balances against collapse but seems to drive (or pull) space outwards. Thus its value is determined from

---

[82] see Chapter 1, Glossary and Index.

the dark energy/ cosmic repulsive force/ levity required to make the big bang model fit with observation. Such observation includes, as well as redshifts and *CMBR* (cosmic microwave background radiation), the apparent acceleration of distant galaxies away from us. Such levity, a vacuum power that pulls apart, is small but not, of course, precisely zero. Too great and gravity were overwhelmed; too little and its force would quench the antigravity. Thus, as far as we observers are concerned, the strength of the cosmological constant seems to have initialised exactly 'right'.

We've fathomed no 'deep formula', no way of calculating the root ratios whence all creation is derived. Nor, apart from intellectual speculation, has anyone detected levity. Its dark energy's another ghost. Indeed, if 95% or more of physics is a phantom who can claim omniscience? With known (luminous or baryonic) matter only 4.5% and dark matter 22% more, 73% of cosmic mass-energy needs be supplied by Λ in its aspect of dark, repulsive energy. This yields a cosmological constant of 0.73 and translates into a non-zero but minuscule energy-mass density of 10^{-15} joules/cm³; but although quantum physics requires a Λ value its uncertainty principle calculates a value of 10^{105} joules/cm³ - vast discrepancy to the wrecking tune of 10^{120} !!

That's no discrepancy to boast about. Dark matter and dark energy fill gaping holes in theory. In fact, by definition they're insensible; you need to sense them by not doing so! Thus, illogically or logically, the quandary demands that you *infer* their presence; you *interpret* how the unseen you believe in must influence what you can see. *But there is no direct experimental evidence for machos, sterile neutrinos, wimps, chameleons or any other kind of omnipresent, unseen interactor - let alone inflation's inflatons.* **Such entities do not, effectively, exist outside the cosmological arenas for which they were invented and by which they are believed in; so, in order to bridge the gap between the predictions of theory and actual observation, gravity, expansion and big bang need fiddle factors as a tottering Ptolemaic theory (of earth at the centre of cosmos) once needed layers of epicyclical adjustment.**

So, are Einstein's mathematics incomplete? Is exotic matter real or a figment of equations? Theories proliferate but *CERN* collider upgrades, the Euclid spacecraft and powerful spectroscopes might help discern and thus decide the nature of dark, guessed-at ghosts. However, will such revolution in perspective, if it comes to anything, ever come to countenance the immaterial, marshalling influence of archetype? Maybe, maybe not, but either way, you bet, an intense race is on to couple theory with reality! Please pass a very costly dose of paracetamol.

Labours of an Empty Womb

There's no business like show business! What force of creativity invigorated and invigorates a latent void's virginity? Are 'things' themselves the fruit? Can human thinking gather how an empty womb gave birth, how matter breaks from nothing?

Something for nothing (or from nothing) is, for physics, basic nonsense. Thus, for starters, let's realise that first appearance of the physical dimension and subsequent description is no more an explanation of its zero-point than explanation of your own development from its egg (Chapter 24). *In both cases what kind of 'nonsense' pre-initialised? On what metaphysical priority does each case depend?*

Silence concerning the crucial, pre-initial condition of physical nothingness is so profound that the instance needs rephrasing. Thus, before we take to zero absolute, let's reverse. Could immaterial ever rise from a material base? **How can non-physical (or metaphysical) appear from physical?** Yet, metaphysically active and physically passive, immaterial information is at every level of creation a reality. Therefore, whence emerged this subtle aspect of our universe? **If matter can't produce what's not itself then big bang or any other naturalistic explanation is insufficient, by itself, to account for everything.** Evolution, we shall see, can't even organise a single cell. Information is a critical ingredient. *This is physics' missing factor, its hidden and invariable element.* **It's missing since it's immaterial and, for this reason, science in exclusively materialistic mode can never find (but may infer) its own centrality.** Perhaps materialistic myths of origin aren't, therefore, what is wholly what. Perhaps they're a dimension out of true.

You hadn't thought of that? Re-frame the proposition using fundamental physics. The Second Law of Thermodynamics states that our one-way world runs only down, becomes exhausted and wears out. Since stars still shine such slack is incomplete - cosmos can't have been around forever. The First Law, again without exception, states that energy (and matter) cannot be created or destroyed; thus cosmos can't create itself. A starting-point non-physical (since there was nothing physical about) is metaphysical. Call this metaphysical projection a transcendent missive; and its cosmic miracle a revelation of which you're a natural part.

Transcendent projection from a metaphysical dimension might be lawfully correct but, according to materialistic logic that excludes the metaphysical, big-bang-out-of-nothing-physical is a non-starter. Cross it off your cosmic list but (since otherwise you're left with nothing) at the same time generously allow that nothing's moment once arrived. Could there be, just before epiphany or post-apocalypse, such a 'thing' in all of nature as a zero absolute? At the foot of what rainbow, at the edge of which ever-receding horizon or from what character of less-than-dot-in-space emerged Point Alpha? How thence did Alpha's absence crack asunder and the world stream out? Nothing couldn't even crack a joke. How, therefore, might *Ad Hoc,* a masseur who will oil your process through its problems, first rub space and knead the cosmic session in?

And if the intellectual fingers of *Ad Hoc's* massage don't press deep and muscular enough, we'll take a scalpel to Big Accident, dissect proposals for the stages of development and then perform anatomy upon the body of the universe. Such surgical analysis is called cosmology. *One of its wondrous, post-initial descriptions will eventually turn out right.*

In Stalinist Russia the principle behind all scientific work was atheism. Such a world has to include, in principle, some mechanism of material 'start-me-up'. In the venerable tradition of a Russian fairy-tale or, nowadays, of international pseudoscience George Gamov obliged. While comrade Alexander Oparin took care of the Little Accident[83] Gamov took The Big One. His explosive if not revolutionary conjecture of the 1920's was that 'nothing banged out everything'. Such startling wind-up seemed to blow the 'everlasting theory' of a steady state to smithereens and so Fred Hoyle mocked and super-glued it with a misleading label that has stuck - 'big bang'. Gamov's noiseless noise survives, however, as the dominant, modern version of cosmogony - big-bang cosmology. Why should his suggestion be construed as real? Because it is reasonably derived from what is assumed to be a straightforward extrapolation back from here and now in the expansive process of our universe to there and then, a point of naked singularity. Shelves of books retail the 'burden of proof' for this astonishing process; its causative event is based on inference and standard, plausible interpretation drawn from two observations.

The first is the Doppler red shift as applied to astronomical objects.

Suppose, in spite of some objections, a red-shift *does* indicate recession due to, say, the universal expansion of space. From shift an entity's inferred velocity is measured; this, related to an estimate of galactic distance and assuming a constant rate of universal expansion, permits an approximate estimate of age. It is thus possible to extrapolate, by theoretically contracting backwards in time and space, and reach an alpha point of hypothetical beginning.

The second is CMBR (cosmic microwave background radiation).

If cosmos *is* expanding there must be an outer limit, one defined by distance of expansion. This boundary must, by definition, be the limit of background radiation because such 'fossil light' is practically the 'bones' of Bang. Has not the rubric of universe been 'set in stone' this way? Gamov had construed the action of his naked singularity to be a cosmic, thermonuclear 'explosion'. As its remnant he predicted 'hot' *CMBR* (cosmic microwave background radiation) at 50°K. According to Einstein's General Theory of Relativity, however, the expansion of space would have 'stretched' or red-shifted the initial fireball's radiation by a factor of a thousand down to longer wavelengths. In 1965 Arno Penzias

---

[83] see Chapters 20 and 21.

and Robert Wilson discovered a background radio 'hiss', a spherical shell of microwave radiation interpreted according to theory as 'the faint afterglow of genesis'. Wherever astronomers point their telescopes they do indeed record such invisible glow. Thus, hail first light; we are observing cosmic dawn! We register, in effect, the patterns on a bubble whose surface has from its commencement been the limit of the universe and now is swollen to the absolute cut-off point for our inspections 14 billion years from point of origin. The temperature of this 'fossil radiation', *CMBR*, is ~ 2.726°K, that is, less than a freezing 2.8 degrees above absolute zero. This value concurs with Einstein's prediction. *CMBR* is taken as a proof of material creation causing Penzias to note (in the book *Cosmos, Bios and Theos*) that, 'Astronomy leads us to a unique event, a universe which was created out of nothing, one with the very delicate balance needed to provide exactly the conditions required to permit life, and one which has an underlying (one might say 'supernatural') plan'. Not an unnatural plan; one so natural it substantiates all nature. He meant, of course, 'super-materialistic'. In terms of cosmos this is what transcendent means.

The WMAP Picture

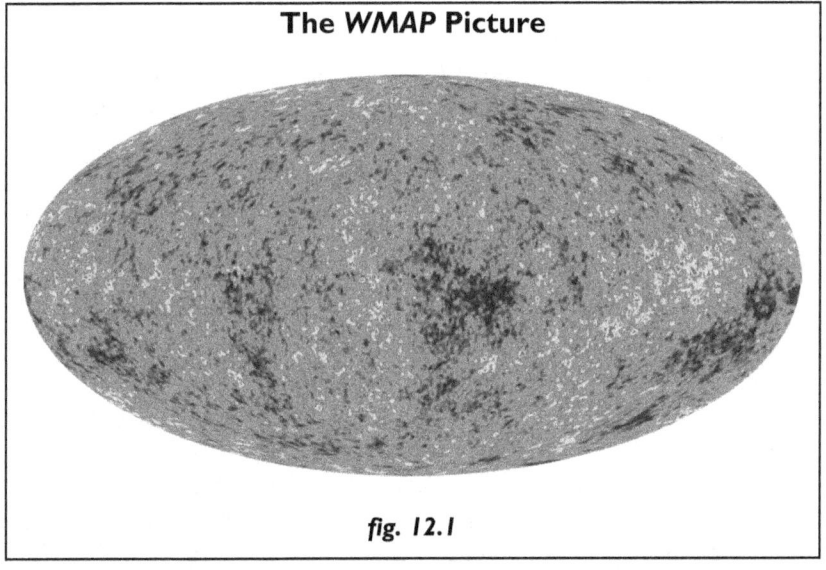

fig. 12.1

Several space-borne telescopes have probed the 'homogeneous' and 'isotropic' *CMBR*. Although it seemed smoother than a baby's bottom a satellite (*COBE*) spotted puckers on the cosmic infant's rear. Later (in 2001) 'precision cosmology' in the form of *WMAP* (Wilkinson microwave anisotropy probe) flew to check. Computers were engaged to enhance the signal to noise ratio until minuscule variations were indeed detected. In other words, although you can barely distinguish between signal and noise, statistical evidence was supposed to demonstrate hot and cold 'puckers' that differ by less than a 'most delicate' one hundred-thousandths (30µK) of a degree. These are as slight as close examination of a shellac finish with a microscope might conceivably disclose. Nevertheless, after such significant enhancements of data over a decade, variations on the cosmic flatness and homogeneity were announced. Planck telescope confirms that, according to present assumptions, these 'wrinkles on the face of time' signify youth not age upon the cosmic brow. The micro-fluctuations have now been enhanced by contrasting colours; and, it is easily suggested, from such cosmetic corrugations on the god-less face of space everything inflated and evolved. There is beauty, with or without make-up, in deep history's simplicity.

So, irregularities in quantum fluctuations once delivered, with inflation, micro-fluctuations of the *WMAP* picture. Why inflation should, in just a single micro-time and micro-space, have happened is unclear but, by this interpretation of the mathematics, cosmos must be 'quantum-shaped'; the universe evolved from vacuum; the vacuum is not empty (Chapter 9: New Vacuums) but brimful of virtual action. In this way 'nothing' has shaped everything! Except that matter's origin is transferred back to virtually nothing and the presence of this 'virtual' something in the first place has not been explained. Eternal matter just becomes eternal virtuality. Non-conscious space forever!

Except it's not so simple. Everything is best-guess speculation which may, if metaphysic is excluded by the faith of a religion, miss a vital clue. The actual as opposed to average distribution of material is not homogeneous. Massive super-clusters made of galaxies reel amid great voids of space. There are more than two hundred in the Milky Way alone. You don't find galaxies outside them and within the galaxies themselves globular clusters of stars revolve. What, since the origin of dust is unclear, started all the 'lumpiness'? What caused such grains to clump and grow? Perhaps space rippled but, if the world was quantum then, what collapsed its probability so that irregularities of micro-fluctuation, once 'inflated', did the perfect trick.

Moreover, as opposed to the axiom of homogeneity, *CMBR*'s irregularities appear aligned in a 'preferred direction' or 'along an axis'. If such 'preference' involved a polarised, global universe, perhaps one that revolves, it were for bangers strangely unexpected news. Indeed, the whole business of homogeneity is, for the cosmological principle, a problem. If Point Bang was chaotic and if its energetic storm has flown in all directions since, how could oppositely separate regions be 'causally connected'? Yet if this connection wasn't ever there, why is cosmos 'homogeneous' and 'isotropic'? The *ad hoc* solution proffered to solve this 'horizon problem' is again the unseen, absolutely unverifiable but currently useful mechanism of 'inflation'.

Ad hoc cosmogony is labour incomplete. What Casual Father impregnated Luck? You claim loose Luck disorderly delivered Happenstance to space and you're a bastard born of Chance! No doubt, the instinct for creation is invisible but from conception you can definitely try to measure the development of universal body. Measurement of world's nativity is not the same, however, as an explanation of the origin of seed or the explicit program of development that you post-natally observe. Perhaps, embedded in the Mother, there is systematic calm behind apparent chaos of her child. Was its fine-tuned, healthy cosmic body really sparked from ovoid emptiness? Clean nothingness; then suddenly a miracle - a version of immaculate conception!? Did it burst from a pointless singularity, a codeless space much smaller than the dot beneath this question mark - especially when you're told this dot was nowhere and yet, as paradoxically as holy spirit, everywhere at once? Perhaps even genetic, archetypal code was in prior place. Calculation cannot shroud the mystery.

Babies are, however, certainly conceived from small but highly-programmed eggs. Thus don't say 'big-bang blow-up' but 'an orderly projection' or (as p.177) 'a *coded* cosmic egg'.[84] *Codes involve a different level of perspective - one including immaterial, metaphysical conception.* **Have you grasped parameters this bang ejects are, to produce our current state of cosmos, so exact that they resemble very, very fussy engineering - precise and tightly interlocked fine-tuning of constraints? Couldn't physic have been coded from before its genesis?**

Order Below a Physical Start

Cosmos may expand from nothing physical; if, however, it is really everything then what is it expanding to? Consider, too, that we observe from *'inside'* physicality. *CMBR* and humans are 'inside' the space and time a big-bang may have made. Eyes can't peer outside the universe and yet, in principle, you can. The idea that you need to 'go beyond' by crossing light years to the 'outside' is understandable but incorrect. Mind, whose dimensions are hierarchically *above* matter, can in principle transcend material cosmos without travelling an inch - except that body, bodily sensations and attachments shackle focus to the earth.

Mind's *above* so never mind. Let's dip *below* the start. Three, Two, One, Zero... Emergent physicality! Now it's coming! Can you feel a cosmos coming on?

Bang goes nothing. Boom, like that! What a paradox! A singularity of zero size is held to hold, implicitly, a universe of mass. 'Prior to creation's space and time there's absolutely nothing', sophists cry - not even quantum rules you'd lift a ripple from. Neither rules nor game. So whence and how, before a vacuum full of weightless *ZPR*, did one extraordinary ripple self-ignite, grow unimaginably hot yet at the same time super-finely ordered and, thence, break by steps into a wide and merry cosmos-

---

[84] see Index: alphabet and cosmo-logical language.

building dance? A transparent, silent cloud of gases swells through space. If you need to make a space inside of nothing then the swelling makes space for itself. Why, though, you might ask, should any 'fragment of the void' clench itself as tightly as a density a trillion times as great as water's before it took relief? Indeed, can *nothing* ever have a density or heat? The truth is the reverse; and thus it seems that nothing never was! Nor was the bang a bang! Acceleration was luxurious as automatic gearing in a car. But how did absolutely nothing in a black box bulge into creation's air? Was it reasonless eruption? Nothing stretched and warped? That a cosmos-worth of radiation issued from a non-existent gun might well turn out to be by far the tallest, broadest story that mathematics ever told!

A Miraculous Projection

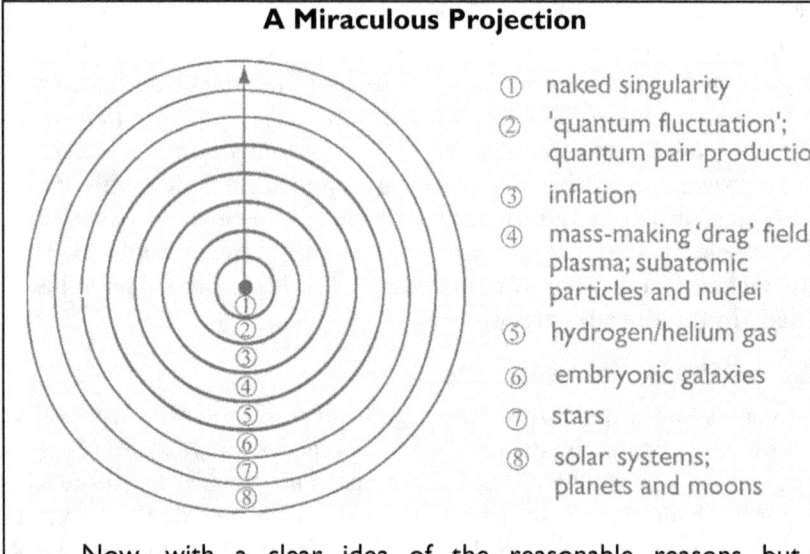

① naked singularity
② 'quantum fluctuation'; quantum pair production
③ inflation
④ mass-making 'drag' field; plasma; subatomic particles and nuclei
⑤ hydrogen/helium gas
⑥ embryonic galaxies
⑦ stars
⑧ solar systems; planets and moons

Now, with a clear idea of the reasonable reasons but nonetheless limitations of scientific method, inference and interpretation, we can turn to the ways in which science struggles to understand how, as a physical Point Alpha, something must have sparked and then developed out of nothing. Remember, though, materialism's miracle knows nothing of creation's gradient. It is a purely one-tiered job. How do you try and squeeze three tiers to one? Like this.

fig. 12.2

Look at your thumbnail. Imagine once again that, for no reason, force for a universe jumped out from nowhere (at least a point far smaller than the nail). Nothing ambushed destiny and plundered endlessly! Walk on the wild side. Check the massive swag. Gaze upon cliffs, gorges, chains of ragged mountains and the other treasures of this earth; the slabs of rock you slap are but a microscopic fraction of galactic clusters that, *Ad Hoc* is claiming seriously, erupted from a microscopic, smaller-than-a-pinhead fraction of your thumb - unless he paradoxically means the primal dot in nothing's everywhere!

Snap your fingers twice. Already *Ad Hoc* has massaged an answer to your question. Following the drift of Lady Luck allow the labours of her empty womb. Set your stopwatch. Press! After naked speculation of a singularity assume, from an abyss of nothingness, things magically as from a wizard's topper start to rise. The greatest trick, the luckiest break of all time! Nothing has been somehow fertilised! From such 'electric' touch an ovoid fluctuation in false vacuum has uncurled; this minuscule and trembling instability is due to be Creation's Seed. Assume *ad hoc* expansion and in only billionths of a second swirling particles emerging from the Generator's heat.

Aha! A second, bigger whoosh occurs. Bang on bang on time. Because the first exertion would have been in danger of collapsing back inside itself you need an even bigger miracle to free it up and drive the show onto the road. 'Cool expansion' grew until the universe was cabbage-large; the fading gong of midnight ushers in a day that's pumpkin-sized. Then 'hot inflation' had to blast things to the skies. This magical phenomenon, this inspiration *Ad Hoc* kneaded was devised to blow the first bang up a trillion times in billionths of a second. It exploded much, much faster than the speed of light then, like the first one, braked just right. Thus an assumptive boost of cosmic after-burner, inflation with inflaton fields, has ironed out all irregularities and, most unusual for a double bomb, set things bang on track for how they are today. Or has mathematics dreamed a damned fine myth?

Bang! Boom! Simplicity itself. What headache, even to a physicist, could such a grand yet simple vision of creation present? Sadly, pass more paracetamol.

Did the beginning really start or not? If so then is its singularity a state of *physical* affairs or not? If it is then what rules regulate material 'nothingness'? If not, then atheism places faith in an unknown, lifeless absence. It contemplates a gaping, godless gap. This sounds like religion, albeit of inverted kind. Before 'creation' any naked singularity would, since there wasn't any matter yet, have been beyond material law. Therefore neither its origin nor its first 'creative motion' would have been based on such law. *If, on this basis, creation by design's excluded from consideration then so equally must be a naturalistic bang and any consequential theory concerning origins.* **Projection from an immaterial source cannot be based on natural law. But unnatural conception is, by naturalists, outlawed.**

A 'miracle' is an event nobody understands; and faith, unless the miracle in question has without doubt been perceived, is the conviction of events unseen - like cosmogonic bangs, divinities and cells from mud and water. Even though it seems as good as tracked to lair by theory, the miracle of cosmos formed *ex nihilo* is clearly unclear. The crucial template of initial conditions is cached inside a Planck-sized black box. At this point, where micro- and macro-worlds both converge upon infinity, physical description breaks down and exponential improbability breaks loose.

Yet orderly projection from transcendent metaphysic certainly occurred. It produced a cosmos-worth of push-and-shove where you, lucky winner, live. You have conjured Lady Luck yet what could Luck produce without some prior form to work her magic on? A barren ovary is seedless; its womb cannot create a thing. If mindless spontaneity is chance then Luck alone would constitute an unexpected, unpredictable and yet primordial sway whose principality comprises empty impotence. No rule at all. Her second name is Anarchy. You gaze upon an empress of abstraction, a pose of photogenic nakedness whose main attraction is her lack of all intelligence - Lady Luck. Luck's baby was an Accidental Throw.

You Ain't Seen Nothin' Yet

Assumptions pile upon assumptions. Faithful *Ad Hoc* sweats. No doubt, we're here because the universe is as it is. That's no surprise. How came it so? Was its initial condition chance or not? The universal body is sharply defined by a precise set of over thirty interdependent settings.[85] *Their values combine to generate a universal pin-code that was either preordained or at least intrinsic in primordial projection. Indeed, the dials are set for the sun, earth, you and me to an accuracy computed by Oxford mathematician Roger Penrose at 10^{10} to power 123!*[86] **If true, that cuts chance completely out.** Erasure of coincidence. The probability of your bullet hitting a nail-head at the other end of cosmos first shot is vastly greater. Mathematicians consider odds longer than 'only' 10^{50} against to be zero. That is to say, there is statistically no chance whatsoever that cosmos and its dependent life are accidental. **The Penrose computation, if valid, indicates that odds against the observed, law-abiding universe appearing by chance are stupendously astronomical; and the facts appear to support his calculation. The consequences of such statistical annihilation of chance (and therefore a purely materialistic explanation of life) are developed later.**[87]

The knobs on the cosmic generator are fine-tuned but not by evolution. There's no time in billionths of a second for exact combinations to accidentally evolve. Thus, suffice it here to note that prior nothingness cannot evolve; that, post-start, basic forces haven't changed or natural laws evolved; and entropy does not affect this framework of our universe. **Such**

---

[85] see also Chap. 7: Can Math Help Us?; and this Chapter: Fit; and Index: physical constants.
[86] https://evolutionnews.org/2010/04/roger_penrose_on_cosmic_finetu/
[87] Chapters 17 and 19 to 26.

preordained fine-tuning is both critical and integral to cosmos. *It might well be argued, therefore, that numerous co-incidents, acting in cooperation, hint at more than a coincidence.*

Tuned is 'fixed'. Of course non-conscious cosmos, once behaviour's fixed, will ever after fix itself. Never mind the fixer, do you want to be electrified? Electrified by awe and wonder at the place we live? Where galaxies like silver raindrops drift across an everlasting sky and, in that night, the light of billions of suns composing every drop showers down on you? Go to a planetarium, be overwhelmed, sing in the stellar rain!

While you're singing you're not thinking! *Don't think modern theory's infelicities evaporate with vacuums, particles, big bangs and exotic zoos of matter.* What is the universe apart from what you see? Stars, suns, moons, clouds and galaxies compose its regularities. Why did these specific structures come to be? How did a primordial spray of light or sub-atomic particles condense to atoms, molecules and nebulae? Perhaps expansion was the coolant. Was it cooling that, with superfine precision, precipitated protons - only 0.2% heavier would have generated instability ensuring no atomic nuclei, atoms, molecules and therefore stars and galaxies could have fallen from the primal, legendary mist? Similarly, whence 'condensed' an exactly complementary drizzle of electrons? The word 'gas' derives from χάος (chaos) meaning space; what a cosmic mass of hot gas poured from nothing's maw!

How do you inflict a 'shock' upon a nebula sufficient to compress it to the point where gravity can start to act? After all, the only force you have is gravity. Ionizing radiation will induce electric currents and magnetic fields. The turbulence of these will, even at low temperatures, work against contraction. How, therefore, can you 'prime' your cloud until, against expansion, turbulence and any other kind of levity, it's ready to contract? You see the problem. It just doesn't want to happen. Even less so early on when 'great walls', super-clusters and plain, single galaxies were thought to form but gas was even hotter. Why didn't levity trump gravity? This is called a 'priming problem'.

Perhaps the contractor was a vortex, fluctuation or a massive swirl that 'something' set in motion. Perhaps hypothetical axions or nugatory neutrinos constitute 'something' called dark matter; perhaps swarms of exotic substance seed our haloed galaxies. Or gravity, against expansive pressure, spun gas-galaxies like candy-floss, stapled super-massive black-holes at their core and crushed smaller clouds into the stars. Next how did hydrogen solidify to stardust and then other, larger objects wheeling round in space? How are the objects that astronomy observes initially forged? The truth is that we do not know; but, definitely, we plausibly best-guess.

Ah, so! Puzzlement. A smile from glistening heaven heavy with inscrutability. From particles to galaxies still stretches mystery but *Ad Hoc*'s eyes are twinkling as he waves his wand. 'Fluctuation' is his magic word. Could the answer, cries the masseur, beat like gong and simply 'fluctuations' do the trick? This man must interpret vastness that's fantastically and beautifully photographed from probing satellites and, such as the Hubble Telescope, other brilliant technology. From nebulosity through galaxies to stars, planets, comets, asteroids and opaque clouds of dust *Ad Hoc* must massage theories of creation. It's not clear how cosmos formed. In our own stable system how were planets made? Are orbits fixed; did revolutionary turbulence shake up its past? Fluke is explanation's order. In this vein our cosmological facilitator airs a stream of plausible ideas. But even if, eventually, some super-theory seemed to 'fix' each aspect of all astronomical constructions and thus universal history, how sure is sure? Where random rules how indisputable is certainty?

The nature of an object's or a mechanism's origin is different from its current operation. And an equation that omits its immaterial potential - up-front information - always winds up in the arms of serendipity; its speculative history supplants originator with an unsubstantiated tale of flukes.[88] So, inevitably, hunch and fluke comprise the logic that substantiates a modern, naturalistic tale of origins. Yet if *Ad Hoc* is, to the astronomical extent we've seen, allowed to airbrush ignorance while sharpening lines round any passing cloud of speculation, why not speculate, *ad hoc*, upon an immaterial factor; *why not* allow an extra, metaphysical, informative dimension that 'decides' material facts? The answer is, '*Ad Hoc* is not allowed to'. It's simply that the lad's from *PAM*'s exclusive school; and materialism's discipline, naturalistic methodology, denies the possibility. Respectable science is solely materialistic. Disrespectful boys, stepped out of bounds, are summarily expelled. And so, of nothing-physical, you ain't seen nothing yet.

Approaching a Magnificent Mythology

From great to unseen small, a dislocation from reality's abroad. No doubt that basic, quantum characters (electron, proton, photon and so on) remain the same - fixed fit-for-purpose even if they act with wobbly 'probability'. A complex manifold of finely-tuned constraints contain the cosmic pattern well but, *bottom-up*, how far does a causal explanation have to lead?

We laughed at bishops 'counting' angels on a medieval pinhead; ironically, we don't at mathematicians 'counting out' the universes

---

[88] as well as cosmological see also (Book 3) biological evolution.

modern secularity requires! Thus let us pile, until *Ad Hoc* is wild with abandonment, more assumptions on assumptions. Quantum theory helps; and M-conjecture lends dimensional plurality. Maybe one could cloak these extra speculations with invisibility inside topologies unappealingly tagged Calabi-Yau spaces. 10^{500} sorts of these minutiae hypothetically determine the possible number of super-string vibrations and, with each theoretically inflatable into a universe, your imagination could obtain 10^{500} types. No empirical reality at all - but never mind, what fairy-story has? Emboldened by symbolic numbers you assume that those invisible dimensions harbour aspects of materiality called 'branes'. If 'branes' can interact across dimensions you might expect titanic, other-worldly clashes and, perhaps, repeated cosmic bangs. Mythic worlds made up! Chance galore! No chance that chance could fail to generate our lucky break. Doubtless, naturally selected, our universe evolved! And (what's to stop it?) still evolves!

Ad Hoc par excellence! Modern mythology, now of scientific class. Esoteric mathematics has collapsed wavefunctions every way; it seems that many worlds inevitably appear. Perhaps they're inflated quantum fluctuations forever bubbling in a vast creation field. Such a cauldron of imagination might be fun but naturalistic methodology's been well and truly dropped - all in the name of mindlessly constructing everything. Thus we've stepped to epic climax. **The wand that conjures vast odds into none against, the sword with which to, full and final, cut Deliberate down is a climactic throw of speculation**. It's an idea phrased as 'atheism's last refuge'.

Atheism's Last Refuge

What a trump! What a game! Cosmology comes in a packet with a warning to your mental health - an addiction to speculation can damage your actuality! When all is said and done hard physics is a neutral judge that neither precludes nor suggests necessity for G-words such as 'God'. Hard-line materialism, on the other hand, is driven by agendum. No doubt the fine-tuned facts of cosmos could imply a tuner and, therefore, archetypal shapes of resonance through which, like air though organ pipes, harmonic notes of cosmic symphony appear. With her agendum in such parlous state, Lady Luck must gird her loins and see off Lord Deliberate's advance. To gather strength a strong puff on the pipe of rational imagination is a soothing subterfuge!

Therefore, in order to avoid the vast improbability that an accidental universe should be precisely life-aligned and to sidestep the obvious consequent implication of archetype or a creator some naturalists have grasped a prodigal, most inefficient, multiplex resolution of their philosophical dilemma. *It's not quite in the AVP but, instead of space-time's four, it seems convenient to proliferate dimensions; instead of simplicity to read luxuriant complexity; and in place of a single universe to imagine many. Infinitely many.* This is Genesis. **Welcome to the idea of a multiverse.**

Sir, who is fooled? Phantasmagorical cosmology! **Such multiverse is an invention simply to avoid the idea of transcendent metaphysic and its pressing implications of design. Conjuring one is atheism's last resort.**

Let's repeat the charge. *Atheism's sophisticated last refuge is a figment of imagination.* Not testable, not falsifiable, it is science fiction in the naturalistic genre and, as such, unscientific. In fact, magnificent mythology, it thrives by faith alone!

Eternal, omnipresent matter's where you root such very clever, naturalistic faith. It's not a new idea but has been wrapped afresh. You consider a plurality of worlds; you imagine multiplicities of galaxies might 'bubble up' eternally inflating from creation's spring. Such an unseen cornucopia would not be hierarchically constructed; and would be material alone. Has, however, atheism's cosmological preference, eternal matter, any substance? The *fact* is that by observation we have only lighted on a single uni-verse; the *fact* is that, although abstruse hypotheses may argue cats and dogs, we'll never know if its material presence or its absence (space) is infinite because the search for confirmation would take endless time.

There burns a need to fix fine-tuning's vast improbability. *Bottom-up*, the tactic is to flee, at cosmic lengths and any cost, the idea of a Certain Fixer. The tactic's trick is to transform, by an infinity of tries, impossibility to utter certainty. It's not only crafted, by the way, with cosmology. The *AVS* (Authorised Version of Science) includes biology (life evolved from inorganic chemicals and, afterwards, random mutations). And, through an incomprehensible complexity composed of nervous, electronic switches, consciousness is floated for psychology. Probability is blithely plied until it seems, plausibly and even easily, anything can happen mindlessly! With Luck involved rigour can't resist. Such outright, brazen speculation casts theology's shy miracles in shade and shame. Chance and time are the magician's hands. The trick is truth. If there might seem to lurk the reason of a cosmic, immaterial intelligence then to hell with Ockham, parsimonious principles and elegance. Flout his rule. And, if the red light flashes urgently enough, then flog the flouting shamelessly. What contrivance is too great to rescue rationalism's naturalistic outlook and its consort, atheism? **Thus an industrial-scale exercise in speculation is invoked to roll, without a shred of evidence, an endless suite of bangs and crashes called a multiverse.** Such transfiguration could and, in scientism's mind-set, should for all eternity delete Deliberate.

Can clever cats grasp astrophysics? Do even wiser monkeys have the brain to understand, in its entirety, the screen of world that life's film finds them in? Law, chance, chance-killer, doubts and certainties - the argument flips to and fro. Scales of debate have swung this chapter up and down between a Theory of Intelligence and None. Intelligence, ignored by scientism's prejudice, is in fact the crucial issue. An imaginary multiverse by chance is certainly designed - by atheologists in order to replace design by 'the appearance of design by chance'! **But is the cosmic infrastructure really just an accident? Or do its principles and rules derive from a priority, a precedent called archetypal memory in universal mind?**

That's the point of this debate. After all, observational interpretations that converge on big-bang theory may be right. *Big-bang may or may not well describe, from 'inside' physicality, how cosmos was initially projected.* *Nor will cosmology observe a 'hidden', immaterial factor. It may describe initial random motion but its surges, swirls and particles subscribe (as do explosion, storm or natural uncertainties today) to 'law'. Crucially, therefore, what* **caused** *material projection?*

Advance in physics demonstrates extreme precision of the natural constants and constraints; it has discovered that the finest and most intricate of 'tuning' underwrites the universal 'swing'. *If an impression of coordinate design is overwhelming it is reasonable, against a prejudice of chancers to the contrary, to presume an architect.* Such presumption precedes any thought of animated forms; but it anticipates an inundation by the revelations of biology. Advances here illuminate design of coded information, programmed hierarchies and coherent harmony of complex biochemistry in life's most basic, microscopic unit, cells. And what about the shapes incarnate purpose, multicellular development, exquisitely constructs from them? Like you. **Therefore, well before fierce but rational interrogations of the most dramatic, cosmic chasm parting life from non-life and the theory of evolution have begun, bring up the lights. Let Lord Deliberate step forward upon a stage inanimate to take a prefatory bow.**

Points Omega

Initiation, action, end: this is the omnipresent dialectic order that pervades creation. Things were originally wound up but how at last, having wound down, will they wind up? The business of annihilation hangs upon your standpoint. Was there steady-state, eternal matter, an uninitiated world without an end? Or a cosmic missile issued from a non-existent gun? Was it mythic, multiversal phantoms or titanic branes whose clashing sparked off worlds?

And, if there's an end, will expansion cycle with contraction in a yoyo-go - perhaps even differentially as various local masses crunch **by**

(↓) **gravity** at different rates (and therefore times) into their so-called 'infinite black densities'; or will all vanish down One Great Black Hole? On the other hand, could (↑) **levity** define the nature of point omega? Indications are that, 'sucked' into lambda's vacuum energy, expansion of the universe is speeding up. So will creation's empty womb accelerate its spawn to death; must the lights go out inside a side-less, barren and expanding, everlasting tomb?

	existence	*Essence*
	created events	*No Thing*
	end/ action	*Initiation*
	sink/ process	*Archetypal Source*
	polar expression	*Potential*
	alphabet	*Alpha = Omega*
↓	*informed*	*informant* ↑
	entropy	*stimulation*
	end/ death	*process/ lifetime*
	exhausted void	*action*
	omega	*alphabet*

Hierarchy changes everything. If latent archetype's vibrated into action what would happen if the excitation ceased? A universal power cut;[89] instant collapse through stages whence the grid had been initially devolved? Dissolution of existence means that only Essence stays. Alpha and Omega are thus, for potential matter or for mind, the same - s*ine qua non*, be-all and end-all, beginning and the end of all created worlds - nor just in time but Presence, Archetypal Presence.

The first leg of this philosophical odyssey has now docked. Alpha is, for Natural Dialectic, Omega. The end is the beginning. The volume ends but in the next we'll see how its system applies to you personally - to your mind, body, friends and neighbourhood.

---

[89] see also Chapter 11: Matter's Holy Ghost.

Book 2: Psychology
Active Information (Psychological)
Chapter 13: Consciousness

It was declared on the first page of the Preface that a scientific world-view not profoundly and completely coming to terms with the nature of subjective mind can have no serious pretension of completeness. Now we approach, beyond mathematical mythology, the temple of psychology. Its establishment relates that consciousness is made of nerves and thoughts *are* aspects of dynamic brain. Mind, it's inferred, is down to molecules and atoms.

Thus, if subjective mind *is* immaterial, its study can't be, ever, 'scientific'; but, materialism demarcates, objective study of *brain* is allowed. Despite such fiat is it possible to generate a frame of reference including metaphysic? Base-line needs underlining. ***If the axiom that informative mind and energetic matter are two different kinds of element is true the logic of this volume is entirely unassailable***. Thus we include and not, as materialistic neuroscience might intend, diminish or exclude the primal factor of psychology, life's immaterial states of consciousness.

Psyche and Psychology

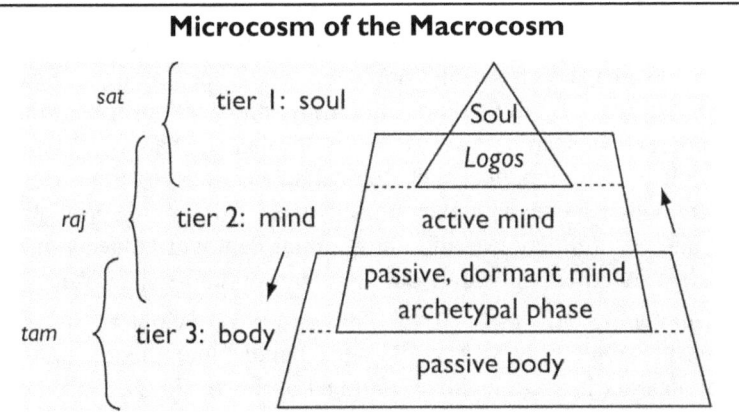

Do you reflect creation as a whole? Is your constitution in the image of a three-tiered universe? In this analogy the anti-parallels run *in* (for knowledge), *out* (for action); input, output - their two tracks lead to and from an information-centre at the head. In this respect they form an image representing you. This is how, in body's tier-3, passive zone, your nervous system is arranged; *indeed, as we'll see, it represents the informative way that your whole body is disposed, that is, it represents the structure of biology.*

At this point it may also be noted that, from a *top-down* point of view, created mind-with-body is seen as a 'soul vessel'. Thus every different pot (or organism) contains the same pure water. In this manner a human may be seen as soul having a physical experience rather than a body erroneously imagining its metaphysical soul.

fig. 13.1

	tier 3/ tier 2	Tier 1
	body/ mind	Psyche/ Soul
	lesser selves	Transcendent Self
	conscio-material range	Conciousness
↓	physic	metaphysic ↑
	energetic	informative
	material	immaterial
	non-conscious	states/ grades of mind
	body/ tier 3	mind/ tier 2

Microcosm of the macrocosm - a simple, three-tiered cosmic ziggurat represents the overview.

The word *'psychology'* means study of *psyche* (from a Greek word generally translated 'soul'). But current 'psychology' were better termed *'noology'*. This is because the discipline studies mind, for which the Greek is *noos* (from the verb 'to notice' or 'to think'), and not *psyche*. Or does it? Perhaps not even 'noology' goes far enough. **Scientific materialism can, by definition, only allow physical composition.** Thus contemporary research swings even further towards a hands-on, neurological commitment. Psychological science has, to this extent, lost patience with psyche and externalised itself from immaterial mind to study nervous reflections; and thus, embroiled with objects made of atoms, has now died. In death's oblivion lives psycho-senselessness.

In metaphysical fact, psychology (and psychiatry) deal with immaterial subjectivity. After all, what handle into the mental life of communicators has a telephone engineer's clip reading electronic blips on the line? Lines, even a whole country's network, are non-conscious. Material science is able to read, at most, such blips on nervous wires. It may be argued, therefore, that neurological discipline may affect the physiological condition of the network but that this reduction is, intrinsically, unable to tap or treat the metaphysical part, communication of meaning, that is, the real, subjective point of the system.

Materialism thus sells a soul-less psychiatric line. Holism can't subscribe.

Soul to Sell?

If your soul's for sale to whom will you sell out? If to Mephistopheles then what is he? The devil, sure, but what about amoral pact with yet another buyer - Mammon? Do you believe, excessively materialistically, that cell or cells 'evolved' a 'soul' whatever 'soul' might be? If you believe it's just imagination by material mind, a figment of the brain, then you've sold out to soullessness.

According to this *modus cognoscendi* so-called 'soul' evolved from cells. Since cell evolved mind must have too. Brain evolved and brain is mind and so whatever you call soul is at root cellular - and is not cellular the same as chemical? A chorus from molecular biology agrees. That's therefore what brain scientists have to think. Embodied 'soul' might verbally exist but only squeezed beyond the actual world's periphery. **Materialism thus thoroughly degrades Creation's Central Factor into vagueness whose reality is now diminished and dissolved among mysterious properties of hormones, genes and neurons in grey jelly.** Souls are not for test tubes; thus decline in use and understanding of the 'soul' word correlates with the ascendancy of scientism. Concept of soul, once materially dissolved away, leaves but the physicality of brain.

The Neurological Delusion

Could mind live apart from brain? Dead bodies aren't, for sure, alive; a mindless corpse would then imply that, in a universe of matter, immaterial mind exists and constitutes the core of our terrestrial experience.

Such substitution, metaphysical for physical, is, however, scientifically 'unacceptable'. 'Impossible' - whatever future studies might disclose. Death is just 'parts failure', a dislocation of molecular mechanics whence once arose the corpse's mind. Don't informative computers also terminally crash? Won't you? 'Immaterial' is a word too far; it breaks the mind-set's basic rules; it kicks a prop that's critical to sustenance of naturalistic faith. So, creed decrees, thought's entirely a nervous matter.[90] *Thus, naturally, psychology emerges as the study of neurological phenomena.* Well-educated and evolved professors tilt their weights towards neuroscience as the guru. Carbon, oxygen and more - soul is the activity (incredibly, incomprehensibly complex, mind you) of particles. Consciousness 'emerges' from non-conscious molecules grouped in some special way. Isn't thinking generated by the soft-wired workings of a brain? Just sling sufficient atoms, in the form of nerves, together - they'll become no less than self-aware!

That mind is an illusion is the neurological delusion. To mistake its neural correlate (as, say, registered in brain scans) for experience itself is error prime as, aforesaid, taking electronic pulses in a wire for all there is to telephonic conversation. It is a

---

[90] see Chapter 0: Scientific Delusions; also Chapters 5, 6 and 18.

prime, elementary fault, a first category philosophical mistake; it might be termed full-blown, psychological mythology. To identify consciousness as an illusion is itself, denying the reality of one's own experience, a pernicious - even dangerous - delusion. Who, however, cares for error when the guesses of materialistic faith cap all?

Does Brain Originate or Mediate?

One party, it is clear, believes life is a phantom of the atoms. Brain *causes* subjectivity. Thought (therefore belief and all the purposive effects of will and faith) is part and parcel of nerve chemistry. And what is the *experience* of consciousness? Thought (therefore belief and all the purposive effects of will and faith) is part and parcel of nerve chemistry. And what is the *experience* of consciousness? The essentially robotic view of neuroscience holds that nerves *are* consciousness. We just don't yet understand, the faithful purr, how brain's 'emergent properties' can squeeze experience out or how the juice that's 'you' must be exuded from its molecules. A revelation is, however, prophesied. Materialism's scientific certainty decrees that life will be reduced to chemistry and mind experimentally identified as simply due to complicated ionicity! You are a product of your physiology and so, at root, your genes alone. Life has, hasn't it, to be an electronic after-thought?

Nervous particles and atoms aren't, like atoms anywhere, alive. Therefore, if life is made of them it shouldn't be alive! *Thus the other line suggests, conversely, that brain isn't an originator but a mediator.* A filter. It is a **transducer device** that, like any mediation network (e.g. radio), must be sufficiently well-constructed to handle large volumes of two-way traffic.[91] It accepts environmental signals and translates them (↑) 'upward' into mind's experience; and issues orders (↓) 'downward' into body chemistry. As an organ of 'cockpit control' its 'dashboard' accurately connects an immaterial mind to a material body and, thereby, physical conditions. Of course, young pilots (babies) have to learn to fly; thence we and other kinds of creature navigate, in the vehicle of body, various sagas on the senseless stage of matter. In this view mind and brain, although compounded, are quite separate entities - the former metaphysical and latter physical. Brain chemistry's identified as a design that expedites exchange of information. Your head is thus a medium!

Although we recognise mind's products materialism cannot think how it projects designs on matter, that is, how thought affects a brain. Psychosomasis is neither recognised nor understood. The 'cop-out' is decree, by naturalistic bull, that the whole lot's aimless matter.

Top-down, however, immaterial cause transcends its outcome. Physical behaviour is the consequence of thought; mind's the 'hidden variable' whose projection specifies arrangements of material objects (brain to body and, thence, environment); and, *vice versa*, understands

---

[91] see Chapters 0: Opposite Directions of Mind and 16: Signal Translation.

demands made by its circumstance. Transcendent conceptual influence continually moves your world. Metaphysic unto physic, did transcendence through conceptual archetype inform projection of this cosmos?

Making no material difference by adding immateriality, the Dialectic simply reconstructs creation on the basis of a 'conscio-material' duality. In short, perhaps brain neither does nor ever did enjoy a seamless, subjective experience. The implications of this seminal idea are so extensive that this whole book explores them.

Consciousness

This is what it's all about. Without it you are nothing. The star of every play is mind; the kingpin of psychology is consciousness. What is the 'thing'?

First and foremost, consciousness unifies; its very nature, both in experience and intellectual endeavour, is to unify. To make overall sense. Isn't your own conscious mind what unifies, as you, the world around. This is life. Consciousness is the Great Unifier, the Great Connector.

Non-conscious matter, on the other hand, is a special case of its subjective absence - gases, streams and rocks don't know a thing. Their oblivion's polar opposite is total wakefulness. Of what, you ask, does this consist? As matter's pure non-consciousness exists could not a concentrate of immaterial information - pure consciousness - have being too? The last two centuries of psychology and fifty prior to them have given distinct answers when you ask the nature of the beast. Is it simple or in fact a very complicated thing? Is it a separate entity or 'just' a product of the brain? Myriad experiments involve non-conscious chemistry of nerves; thousands more devices prod and probe the way that consciousness is modulated as it operates a human body (and other bodies that are more a vet's affair). Brain is an intellectual's mirror house whose light is bounced around to various effects. Ingenious experiments test this and that response. Theories slap as thick and fast as theses on an academic table but, when all is said and done, though *ATP* and ions run the brain, what *is* the 'electric' that drives mind? What *is* consciousness that underlies a quagmire of confabulation and, equally, the clear and seamless 'knowing' we call life?

Energetic cosmos is non-conscious. It is *pure non-consciousness*. Its variability involves (as also does that mix of form and information, mind) such properties as concentration, state and reactivity. Thus lo! a senseless clump of atoms first became aware! From darkness somehow through subconsciousness the light of mind evolved unless, contrarily, creation's root turns out to be oblivion's antipode - *Pure Knowledge, Consciousness*.

Materialism doesn't like this phrase at all. *Top-down*, if non-conscious matter forms the cosmic sink its source is consciousness. **Uncreated Consciousness is primary; existential forms are secondary.** The grades of form constraining formless consciousness involve such mental qualities as mood, intensity of focus and intelligence. Could, like that transcendent burst of motive energy we call 'the primal singularity', a highest purity of consciousness project all lower grades of mind and, when non-consciousness is reached, solely energetic patterns of behaviour? **What light is to physics Transcendence is to psychology; in other words, the only absolute measure of consciousness, and therefore psychology, is Transcendent Super-Consciousness**. From mind holistic distillation therefore reaches high; but materialism recognises only base degree, the lifeless one.

Open Up the Brainbox!

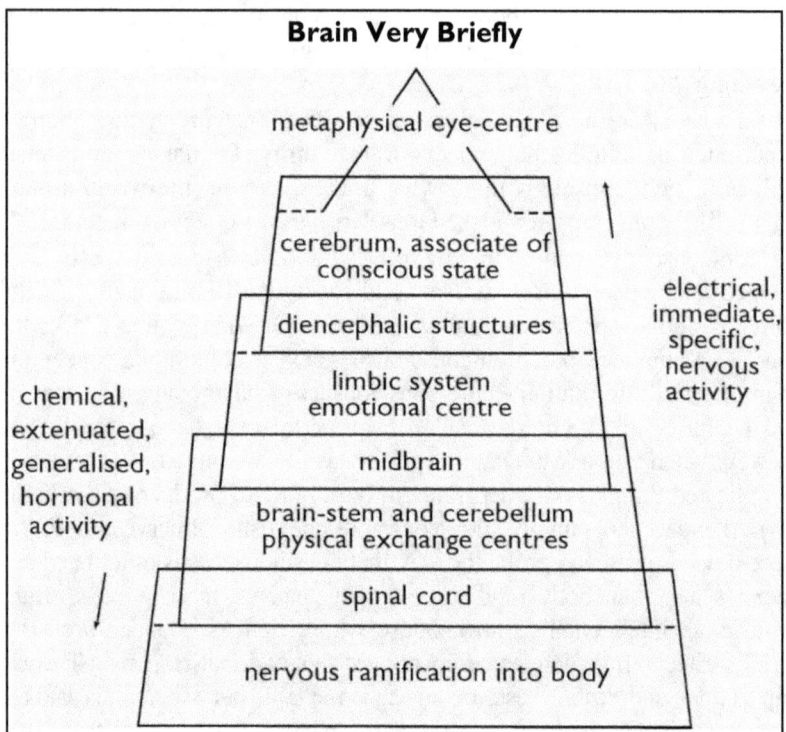

Can it be demonstrated that a nervous system reflects *top-down* order; and that various types of brain implicate degrees of flowering of capacity? In principle you always start top-centre with your personal cosmological axis, pivot, or eye-centre. In this subjective, metaphysical domain you would, as the psychology of conscious mind describes, actively experience the process of information. From here you would expect to drop back, down and out through more objective, reflex centres of control towards physical detail. You expect to fall through instinctive level and its sub-conscious, psychosomatic interface (or *PSI*) to a physical arena, that is, to the energetic, bilateral domain of body. **The ziggurat suggests that brain shows more than simple, physical, bilateral polarity. Its**

hierarchical descent along an information axis called construction also, more fundamentally, reflects the conscio-material gradient from positive metaphysical to negative, physically-embodied pole. Such descent marks the lower order of psychology.

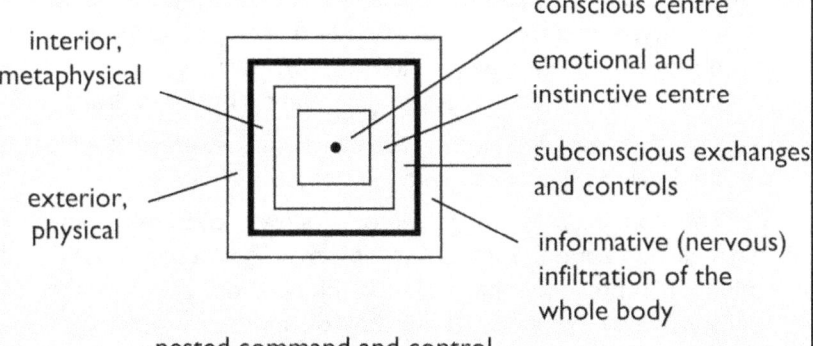

nested command and control

From subtle, psychological interior to gross, physical exterior there exists a clear motor-materialisation of conscious cause as it affects bodily behaviours and activities; and, in the opposite, sensory direction, a drawing together of multiplex environment into a single, coherent, unified whole - conscious experience. The tiers of a ziggurat are nested. **The human information centre is mind but this nest describes the dialectical construction of its interconnection with the physical world, the 'dashboard' through which it pilots the vehicle of body. Mind drives through its physical environment using a panel of control called brain.**

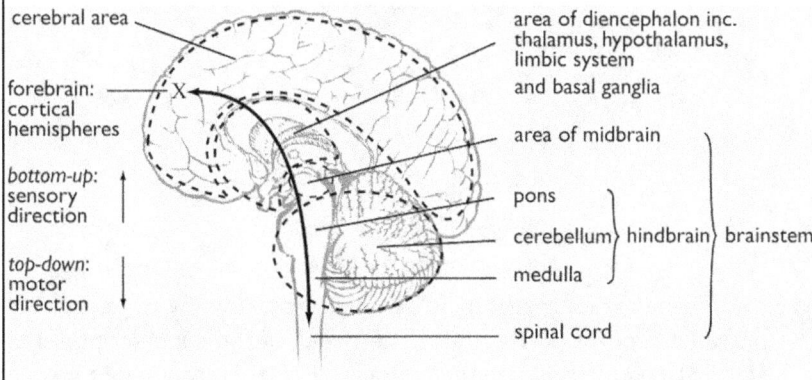

This illustration shows the main divisions of the brain that run from X in *top-down* order following the order of psychology.

The *top-down* 'motor vector' follows, inside out, the order of creation. Such an act informs the exterior world (including, mostly, one's own body). It translates into the way we execute decisions and rearrange materials; as such, it represents 'information out'. It also represents, dialectically, the way that information stored in such material potential as an archetypal memory, the *DNA* code or an egg is expressed in biological development. And, finally, it represents the original order of a brain's design (see 'Build Yourself a Brain').

Two worlds interlock. *Top-down* and *bottom-up* are meshed; mind (↑) is linked with (↓) body.

The *bottom-up* 'sensory vector' follows, outside in, the path of perception. Such an act informs the interior world, mind. It represents the path of 'information in' from the exterior side, that is, the path of a return to HQ - headquarters of the brain and mind. It marks an ascent from non-conscious through sub-conscious to conscious condition.

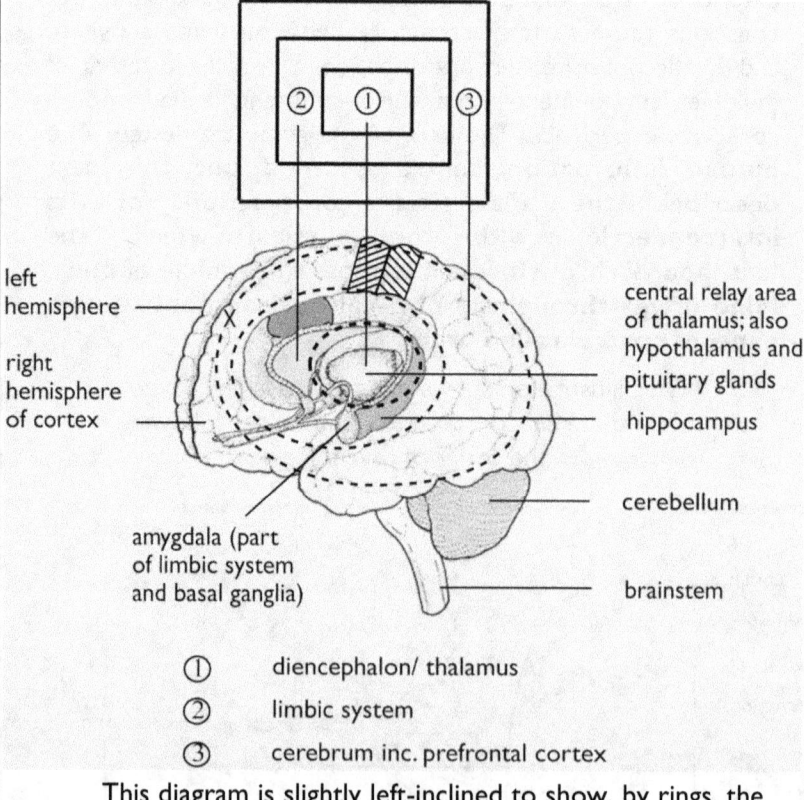

① diencephalon/ thalamus
② limbic system
③ cerebrum inc. prefrontal cortex

This diagram is slightly left-inclined to show, by rings, the nested structure of the brain. The order of the superscript is (1 to 3) taken from a *bottom-up* perspective. Embryonic brain develops (↑) to engage sub-conscious, autonomic and then conscious states; it blossoms as the flower of **cerebrum**. Cerebrum marks a 'maturity' of consciousness. X marks the spot. Front-centre of its cortex at the full unfolding of the flower is the place you think and therefore are. In other words, you inhabit a metaphysical point of reference, an axis or a link through brain and body to the material world.

Let's trace *top-down* order from your conscious place, your thinking space called metaphysically '**third eye**' or 'eye-centre' and, as its physical co-location, the **prefrontal cortex**. Travelling back and down you pass the motor, sensory and other media centres of enfolding cerebrum; here nerves from sense receptors or to muscles interface electric impulse with the faculties of conscious mind. Below you fall upon a second and interior 'cradle'. On each side of this cradle's outer edge find a **hippocampus** next to an **amygdala**. These organs are, it seems, involved with data storage and retrieval; they log experience in the manner of a record/ playback head; they catch or release a moment that, in fixity, is called a memory. The whole cradle is called the **limbic 'system'**. Along with a further, nested structure called the **basal ganglia** this area involves basic instinct. Nested inside the basal ganglia you find a further 'Russian doll', a central relay station that translates electrical messages to and from the right place in the cortex. It is called the **thalamus** and, within and below its compass, is revealed a triplex nest of master command-and-control glands. These electro-physiological lynch pins, that 'mate' nervous with chemical, hormonal *IT* programs, are called **pineal, hypothalamic** and **pituitary glands**. Below this 'arch' a connection is established with control units in charge of reflex, physiological functions; the 'primordial issue' of a **brain stem** links conscious and sub-conscious to non-conscious dimensions; in company with the master glands it ties metaphysic to physic, it bonds 'heaven' to the body's earth. This stem comprises a **midbrain** and a **hindbrain**. The latter is composed of the **pons** and **medulla oblongata**. These factors are linked internally by a 'slot' called the **reticular formation** which embeds a switch involved with choice-less sleep and choiceful waking. Pons-with-medulla (site of automated regulation of thermal, cardiovascular and respiratory systems) is also connected with a rear-disposed unit called the **cerebellum**. This sub-cortical workstation, sometimes called the 'auto-pilot',

> receives copies of all traffic to and from the body. Its dedicated function is fast processing and refinement of the sensory/ motor responses that govern physical orientation and motion. From the medulla neural structure falls along a spinal trunk-route; from this trunk nerves ramify, as 'branches from an upturned tree with roots in heaven', into the body's earth.
>
> The whole nervous system is material. Its physiology is as intrinsically reflex, inanimate and chemically 'dead' as a vehicle's dashboard or its wiring, dials and linkage systems.
>
> **fig. 13.2**

Perhaps you're still unsure. Why doubt that consciousness evolved with cerebral development? The scientific answer's elementary, dear chap! Quarks, electrons and four forces - at base life, science and life science all depend on these. Subjectivity at root is earth and what is brain but complicated earth? Therefore unhinge the cranium! Open up the brain's box! What is there to see?

Is there no logic in the way a brain is built; or is it just an aggregate of happy accidents? *It is logically surprising that, for no reason, non-conscious and illogical chaos should have constructed order to a very highly systematic climax in the most complex working system of the cosmos, an information processor whose whole, sole, negentropic business is order - a central nervous system and associated brain.* Did matter, getting far more than it didn't bargain for, perchance 'evolve' a brain? Materialistically speaking, it must have. *In reality did it?*

Open up the box. Can you find a thought or single memory? We think of thought but remember, for a thinker, memory is critical to hold the world together. Could, somehow, molecular circuitry 'hold' a memory, thought or seamless process of electrical perception so that, in effect, your subjective experience is reduced to nets of special physicality - synaptic neurons? Are you persuaded 'cognizance', essentially what you are, is electricity-in-motion?

Or is there more to life than nerves and, bundled by the billion, a mass of master-connectivity called brain? Does no concept underwrite development and specifically programmed structure of organic calculation? Is there no metaphysic to the mind? At least, none of such founding fathers of neurophysiology as Wilder Penfield, Roger Sperry and Sir John Eccles believed that mind was brain; and for Natural Dialectic mind including memory is, both in origin and operation, metaphysical. *Brain is a medium.* Of course, embodied mind must have connection with specific input (called sensation) and control the neuromuscular output. **However, to confuse electromagnetic** pulsations with the subjective message they carry is, from a *top-down* perspective, the standard error of materialistic perspective. *It is to mistake message-bearer for message, thoroughly confuse priorities and take effect for cause.*

Build Yourself a Brain!

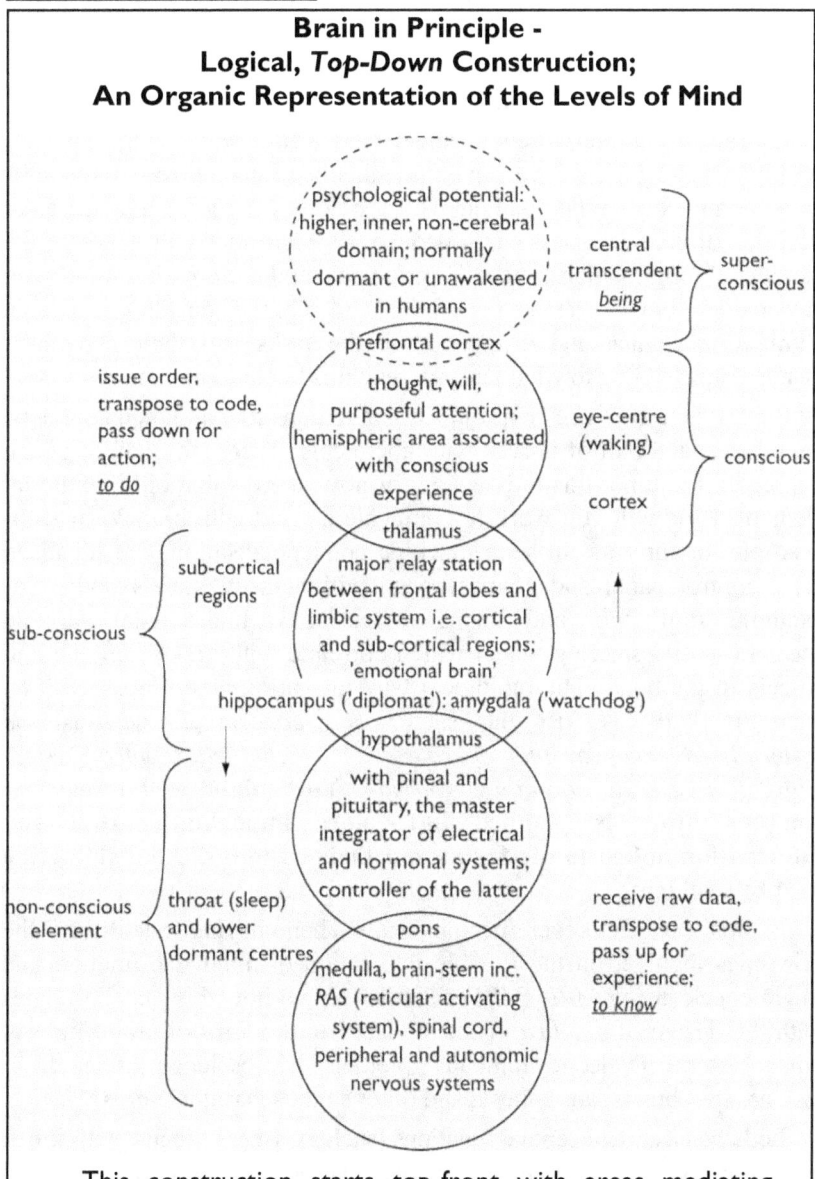

This construction starts top-front with areas mediating physical awareness and falls back down a conscio-material gradient to those representing sub-consciousness and non-conscious areas of homeostatic control. The structure and

> mechanisms of brain reflect the conceptual principles and information of mind; brain itself is composed entirely of non-conscious, chemical elements.
>
> *fig. 13.3*

Experience might seem simply seamless but its medium, brain, is not. Does anything in cosmos rival its complexity? How, therefore, mustering every ounce of your intelligence, would you invent an equal? What would your three-pounder look like?

A brain so simple we could understand would be so simple that we couldn't; but, *top-down*, we can enumerate the principles round which its great complexity might gather. Thus, before examining the different qualities of consciousness, let's deduce a system that will generate the faculties through which to feed its principals and thereby link mind with physicality. Let's sketch a definite and natural development and, eliminating chance and error, outline the tasks and their associated mechanisms leading to what lets you read this book - a human brain.

First, remember, you're not just going to construct and then connect a complex coded ball of nerves with chemicals. There's a body to attach but, even if you put this to one side for now, there's that old immaterial element. How will you work to conjugate a mind with brain? Take, for example, motor cars. In such machine conceptualisations are localised on a control panel and connected up within the vehicle's body. For example 'stop', 'go', and 'change course' have brake pedal, variable accelerator and steering-wheel which correlate with brakes, engine and wheels in the body. The biological way that information locks with an energetic body is just the same. *The conceptual 'wish-list' is materialised in a way that operates like a cybernetic control system with, if necessary, conscious override.* **You might well infer that complex brain is planned but, more importantly, that any information linked to life-forms and their cosmos did not shape the dust by accident.**

In this case, let's get metaphysical! Then, having established the principles involved in the flexible interaction of mind with matter, one might check (using *MRI/ fMRI* or similar scans) how this plan correlates with the design of a *deluxe* 'control panel' in the form of brain. Indeed, what Natural Dialectic logically predicts neuroscience often, from lesions, split brains, anaesthesia and other studies, has uncovered.

Let's be clear. Conceptual functions (such as 'I-ness', focus, will-power and so on) need to differentiate, qualify, interpret, organise, prioritise and so forth. Then you'd need, by reason or by reflex (programmed reason) to respond to all events. The list builds up. This outline sketch denotes some factors you'd mandate and thus need to mediate with physics through invention of sufficient mechanism and machine.

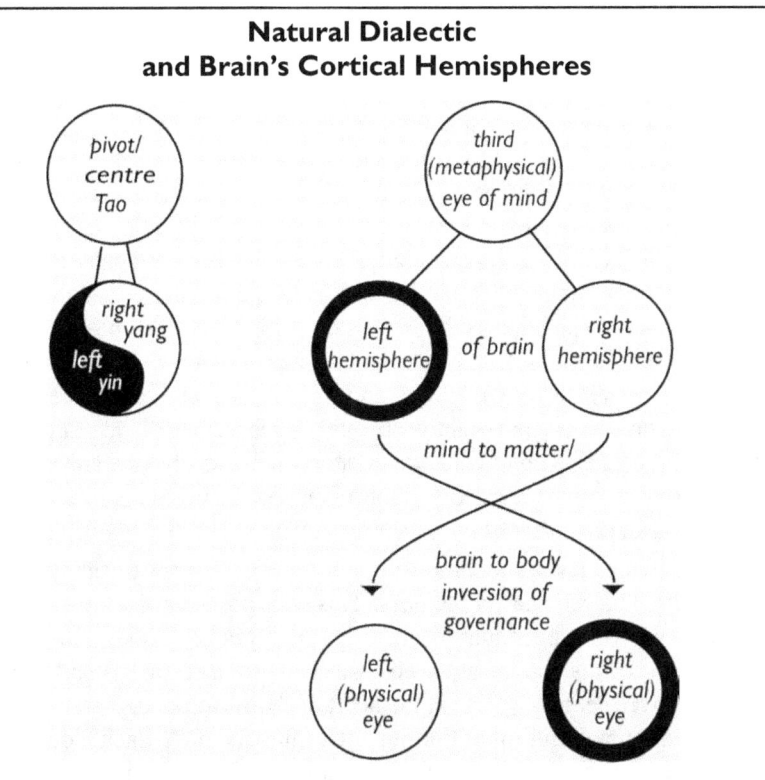

fig. 13.4 (see also 1.1, 1.2 and 3.1)

↓ left hemisphere	right hemisphere ↑
detail	context
analysis	integration
local interest	global interest
processing/ classification	exploration
assistant	leader

You'd need, as with any plan, things prearranged because all bio-informative systems must develop from genetic code encapsulated in a single egg or seed. However, did development from zygote into brain evolve; did inversion (reflective asymmetry) come about by chance? As we'll see, it's easy and logical to return that story to its seller. 'I don't buy it.'

Check this *top-down* construction; bear in mind that each conceptual item on the list must correlate coherently with mechanisms and the body as a whole.

One might reasonably deduce that brains did not evolve by way of semi-working craziness, hemiplegic discord or, except according to imaginative guesswork, mutant pointlessness. While nerves then brains developed what a bedlam would have been! Sparse sanity at every level as life's senselessness couldn't even try to get a grip! For even trilobite or early worm any imprecision in its chemistry would, drug-like, induce bizarre, unfit behaviour; mutant madness cannot build organic logic but, left to its own devices, mighty soon destroys. On the contrary, compartmental systems integrate towards an anticipated goal. Brains work bio-logically; they pack purpose and their shapes reflect it. Such shape impacts the way its owner - spider, bird, bee, you and me - sees the same environment. One might deduce, *top-down*, this is the way that we're each *meant* to see the world; and thus, in our own terms, are not half-hinged or quarter-cut but rightly operate. **Brain serves, it is argued, as a medium.** It connects mind with matter; it channels archetypal architecture and, in our human, hemispheric case, consciousness in ways appropriate to the asymmetrical, polar modes by which we pay attention to the world.

The hemispheres are a composite, two-in-one, that act together in a way that Natural Dialectic easily includes. This does not mean each working may not have its own agendum or that implicit gradient does not apply to operations of the brain as well. Balance (homeostasis) and gradient (hierarchy) are central aspects at the heart of cosmic order and control. What, after all, are brain and mind but our own mini-cosmos? Each side plays its complementary part. Information's shuttled to-fro for interpretation and decision. *Such psychological homeostasis or dynamic resolution in cerebral action amounts to cycling of dialectical synthesis.* **Indeed, such cycle might be termed a basic 'unit of embodied mind' - in turn reflected as a 'unit of cerebral action'.** The nerves involved summate; they are not, but they compose the basis for, a physical experience. They serve, like electronic chips, to unify all matters on a single, conscious screen.

Why has radically conserved bilateralism, from earliest organic times, been brought to bear upon the information organ, brain, itself? Is it all by accident, or does construction purposely reflect a sliding scale of balance in between the fundamental, polar constants of our universe? Clear signs of a rational division of information-processing at all levels emerge from close inspection of the metaphysical necessities of life married to the hemispheric architecture of material brain. Such conjugation represents paradoxically complementary opposition that is ubiquitous in nature's operations. Indeed, common sense and scientific scans now both confirm the dialectical polarity of brain-function, mind and, which both reflects mind and of which mind is the paramount reflector, cosmos.

We might note that, should one want to build a brain, then its conceptual architecture might well be designed to best reflect the inner architecture of the universe itself. You could build a set of stacks involving the normal polar characteristics with respect to order, mode of operation, creativity, community, morality, emotion and so forth. Polar ways of thinking would combine - with every nuance catered for - within the whole, most intimate but temporary conjugation of a human mind with matter.

Wait, though! This isn't all by any means! **There is much more to brain than just conceptual development.** Principle to practice, plans need be materially realised. Observe, therefore, the generation of a mindless miracle! Brain's factory is automated; it's completely codified. Nerves are mostly made of water but their architecture is atomically precise. About 5000 neurons per intrauterine second and five times that number of ancillaries, such as glial cells, swarm to craft an embryonic brain. Their journeys ramify great distances to pre-set destinations forging links that work. Such prodigious activity runs as smoothly as well-oiled Victorian or silicious Elizabethan machinations towards, at maturity, 100 billion informers mounted in a three-dimensional, pinkish, kilo-and-a-quarter jelly. Observe (from primitive streak through neural groove, tube, divisions into forebrain, midbrain, hindbrain and its first brain waves - at seven weeks - to the flowering of a foetal, infantile then adult masterpiece) beauty, power and, jewel in the cosmic crown, unsurpassed complexity.

An adult's 90 billion nerves are nowhere near enough!! And each nerve will communicate with between 100 and 250,000 others summing to perhaps a hundred trillion links; your brain's connectors would, unraveled, wrap four times around the world! Now add plasticity. Each link involves capacity to modulate intensity of its transmission, change its shape and redefine its connectivity. Indeed, the network's like a 3-D switchboard whose jack plugs (cells such as 'place cells') are switched about according to the calls of sense. Don't think connection is so simple either. Every synapse, as such relays are called, trades in nothing less than the cyclic cooperation of many complex, deliberately-fashioned chemicals. Each cycle's purposive metabolism can modulate the passage of a nervous impulse in just thousandths of a second! In each full second, using only 30 watts to support perhaps 1000 trillion computational steps, minimal electrical drives maximal calculating power - power far transcending human wit. No doubt, furthermore, that digitally coded transmissions within a system of electrical impulses form the physical basis for subjective elements of thought, perception and personal memory; they link mind with its body and they do so very subtly. How fine is your experience? Its dynamism does not tolerate mismatch, mistake or accident. Phenomenal accuracy has developed into healthy brains as different as their owners, that is, much biologically the same.

Tap your skull. You've got it all in there but we've not finished even yet!!! You need an integrated blood supply, *CNS*, *PNS*, digestive tract with *ENS* or 'second brain' (perhaps 500 million nerves) embedded in its tubes, ventilation system, sense organs and muscular exactitude - all the working gig. Parts are irreducibly a part of one another. Every part of every body, bound to its instruction manual, genome, is so connected and constrained.

Eyes, for example, only work in context of much greater whole. No wonder, Darwin shuddered, blinked and winked at what his mirror saw. It takes a lot to raise a smile! How, even when you've stripped the eyeball, nerves and brain to parts, is a focused image then translated into something seen? What *is* sensitivity? What biochemical steps led up to granting photosensitivity a gossamer relationship with mind? Who is the subjective seer? *Is not the clear intention that you, endowed with powerful equipment, see?*

Electricians understand the problems faulty wiring brings. Mutations in a circuit can't improve its capability. Bugs are defects: they are information's losers. Could *you* make mind? Or codify a brain with all the rest? **Then, after all, it's only left to automate the whole lot's fabrication!** The sum of man's intelligence can't even start to build a topknot yet, the mind-is-meat brigade insist, a mindless girl whose brilliance must far surpass our own 'conceived' and blindly built creation's prodigy - the human mind!!!! Mortar board and first first-class honours scroll. *Maxima cum laude*. Didn't Lady Luck do well?

Towards a Unified Theory of Science and Psychology

If mind is brain, thought is nervous and experience a kind of electronic motion then psychology's a science. The pair is unified by common physicality! Materialism's line, with theory and its practice well advanced, is without doubt the only reasonable one. Surely, as a naturalist, you'll claim such thrust as 'towards' implies is as redundant as it's ill-conceived.

Do genes and biochemistry explain the lot? Do chemical arrangements by-produce sensations that an atheist calls thought - so that he cannot trust his arguments are true? If this book changes how you think is that only down to synapse-reconnections; does it even matter that some nerves have changed as well? Indeed, is change of world-view open to mathematical description or can scientific language made of numbers and equations fully satisfy a study of psychology? No doubt that in the case of measurement and observation of the various chemo-physical expressions of behaviour (nervous, hormonal, muscular and so on) maths can help. Statistics can apply to brain and subconscious, automatic reflex of the mind, that is, to instinct; reflex reaction is calculable. If, however, the internal genius of a human mind is changed by *choosing* a response, this innovative 'override' is mathematically incalculable. The less conditioned, freer is a choice the less amenable it is to sums. Nor do numbers ever catch a feeling. Have you ever phrased equations for experience, emotion, symphony or dream? Or even colour green?

Is thought an object you can analyse, like nerves, underneath a microscope? If the key element of mind, your subjectivity, has simply been reduced to ghostliness and the central, immaterial dimension of psychology's reality electrified then how far past a neurological glass-ceiling has science scaled?

If, however, metaphysic is conceded how is it reconciled, how holistically combined with physic? Why should the rules of physic even cover it? Mind's rules are different. Its dimensions aren't the same. Mental motion is a product of desire; and knowledge isn't, any more than information or imagination, physical. How, therefore, do conscious mind and its unconscious body meet? How are the elements of two worlds linked? If, as opposed to naturalistic axiom, mind is not a sort of matter we will try and see how it could be. Certainly, if there is metaphysic to subjective feeling, the title 'towards a theory of science and psychology' were never ill-conceived!

Firstly, though, let's sketch the form of an essential psychology.

do/ know	*Be*
objective/ subjective effects	*Subjective First Cause*
lesser levels of consciousness	*Pure Consciousness*
body/ mind	*Psyche/ Soul*
passive/ active expressions	*Potential*
↓ *passive information*	*active information* ↑
programmed	*programming*
non-conscious/ objective	*subjective/ conscious*
reflex	*voluntary*
physical action	*psychological oversight*

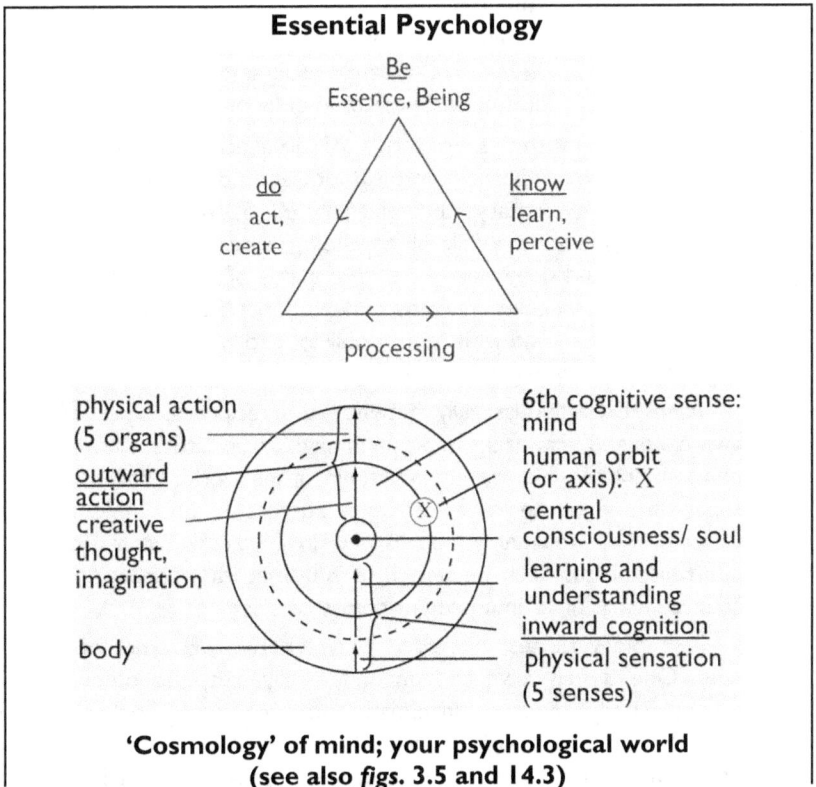

'Cosmology' of mind; your psychological world
(see also *figs*. 3.5 and 14.3)

Do you get the ups and downs, that is, the swing of it?
Balance of Mind: Natural Dialectic and the Five States of Mind

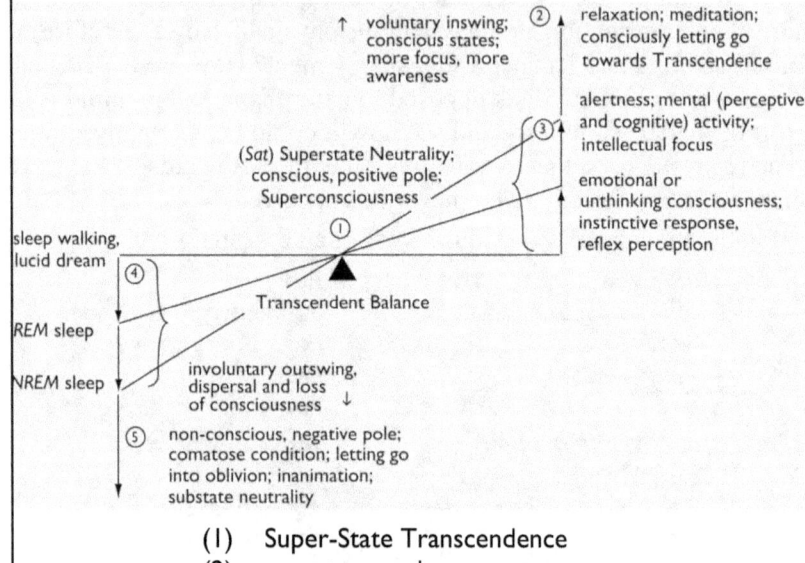

(1) Super-State Transcendence
(2) ascent towards super-state
(3) aminergic waking focus
(4) cholinergic involuntary phase
(5) shut-down, 'flatness', sub-state condition of material oblivion

This diagram shows, in terms of dialectical gradient, an oscillation between poles. Nervous modulation occurs between waking (2, 3; voluntary) and sleeping (4, 5; involuntary or dormant) conditions. While all psychologies recognise waking, dream-sleep and deep sleep as altered states of awareness, the Dialectic also includes a Transcendent Super-State (1). Four levels of mind ascend from none (oblivion) towards the Primal Super-State. Do you get the swing of it?

Minds oscillate elastically between the reciprocal poles of awareness and unawareness. At mid-range 'hallucinatory' dreams complement sensible, waking awareness of the world. While the super-state extreme of Pure Consciousness is uncommonly attained it is voluntarily approached in meditative states of internalised focus. Such 'deep thought' is diametrically opposed by an involuntary 'drop' into deep or comatose sleep.

Psychology ranges, therefore, between two extremes of void. One is fully alive and unconditionally free, the other psychologically impotent, inanimate and permanently locked away. Human psychology involves a brain and allied nervous of wakefulness but could some (e.g. coral, sponges and unicellular

> amoeba, paramecium or euglena) live permanently dormant - as apparently comatose as plants, fungi or bacteria?
>
> *fig.* 13.5 (see also *fig.*14.2)

It is also useful, in order to obtain a cyclical perspective, to view psychology in terms of polar oscillation.

	range of consciousness	*Super-State*
	normal states	*Ultra-/ Super-Conscious*
	degrees of obscurity	*Clear Mind/ Transparency*
	partial waking	*Fully Awake*
↓	*inattention*	*focus* ↑
	blinder mind	*clearer mind*
	unconscious/ oblivious	*aware*
	infra-/ sub-conscious	*conscious*
	involuntary	*voluntary*
	sub-state/ unawake	*partial waking*

Representation of the world by reciprocal processes seems to fit psychology. This stack illustrates that, while all psychologies recognise at least three states of mind (waking, dream and deep-sleep), there exist two more. The first of these, Transcendence or Enlightenment, is well-known and accorded pole position (both micro- and macrocosmically) by all except materialistic versions of the subject. The second, material oblivion, is understandably ignored. From a dialectical point of view, however, sub-state matter (of which brain is made) is located at the base of the conscio-material gradient; it forms the 'coccyx' of a universal spine; and the enervated, inanimate psychology of non-consciousness is called physical science.

Top-down, therefore, any attempt to reconcile science and psychology must logically start at the top (Conscious Soul) rather than bottom (matter). It must start at Axis, Centre, Source or Pivot rather than with the effects of subsequent informative or energetic motions. For this reason Chapter 14 deals with the two conscious states, super-state transcendent and our ordinary partial-waking. Next, Chapter 15 falls to cover, at the sub-conscious end of mental balance, the conditions of dormant mind, dreaming and deep-sleep; finally Chapters 16 and 17 cover the psychosomatic domain of instinct, memory and archetype; they account for the psychosomatic connection between mind and its substate, non-conscious material body. Description then falls outside the actor's mind (Chapters 19 - 25) to his gross body or biological phenotype. Still further from the Centre, outside his own physical periphery, there lives a community of other actors. These comprise, set in the local scene, a cast of interactive organisms networked on earth's eco-stage (Chapter 26). The whole flexible set includes, in principle and in practice, any player at any time anywhere in the universal drama.

Chapter 14: Five States of Mind

Top-down, Hierarchical Psychology

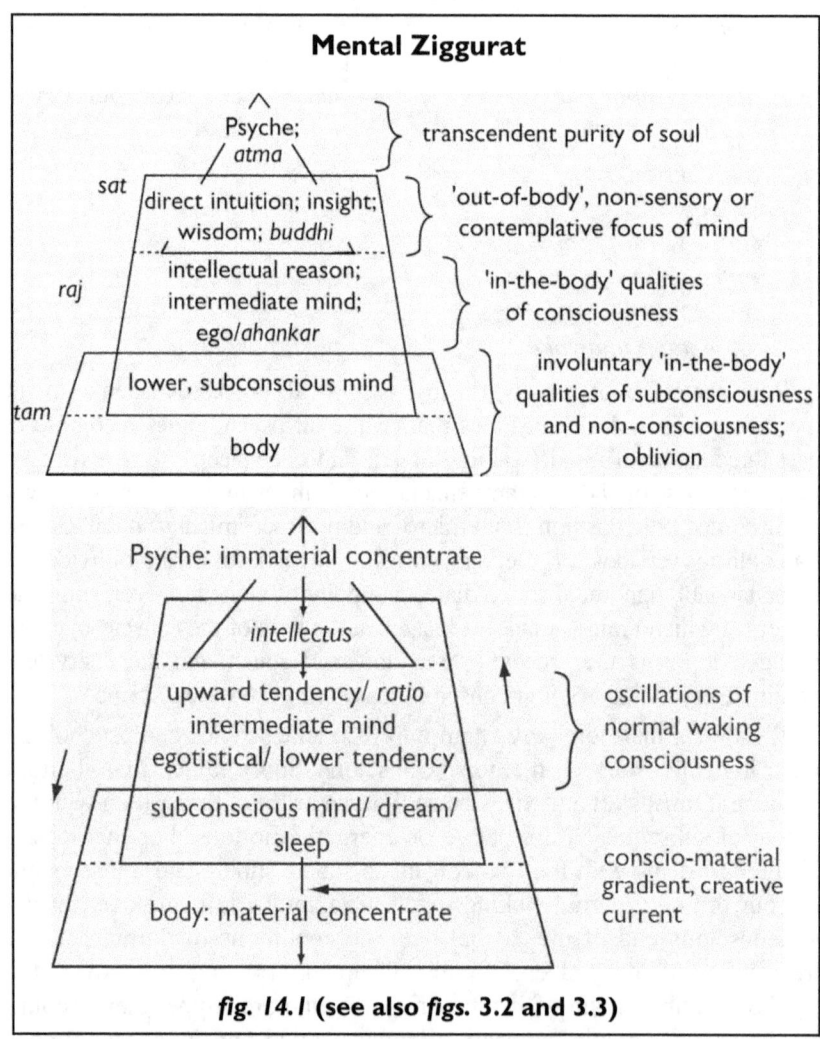

fig. 14.1 (see also figs. 3.2 and 3.3)

Steps; but the grades of hierarchy are not set in space. They are vectored in accord with energetic power and comprehensive scope of information and control. *Top-down*, *buddhi* and *intellectus* might be termed higher mind or *mind-in-principle*. *Ratio* and *ego* are lower or commonly experienced aspects of *mind-in-practice*. Such hierarchical psychology explains the levels and directions to which operation of the mind is geared; and, with oscillations in between these levels, nervous correlation in the human brain. In which gears does your life predominantly run?

info. out/ info. in	Information Centre
lower/ higher	Third Eye
↓ *lower*	*higher* ↑
sensory bias	*contemplative bias*
spiritual anaesthesia	*material anaesthesia*
diversification/ detail	*unification/ principle*

States of mind are, in effect, conditioned consciousness; they are a dynamic function of concentration, purity of focus and the object of attention. The more diffuse is concentration the lower (more restricted) its state until, eventually, the light's phased out. Over the next two chapters let's examine these 'mind-bands'. Check *fig.* 14.2

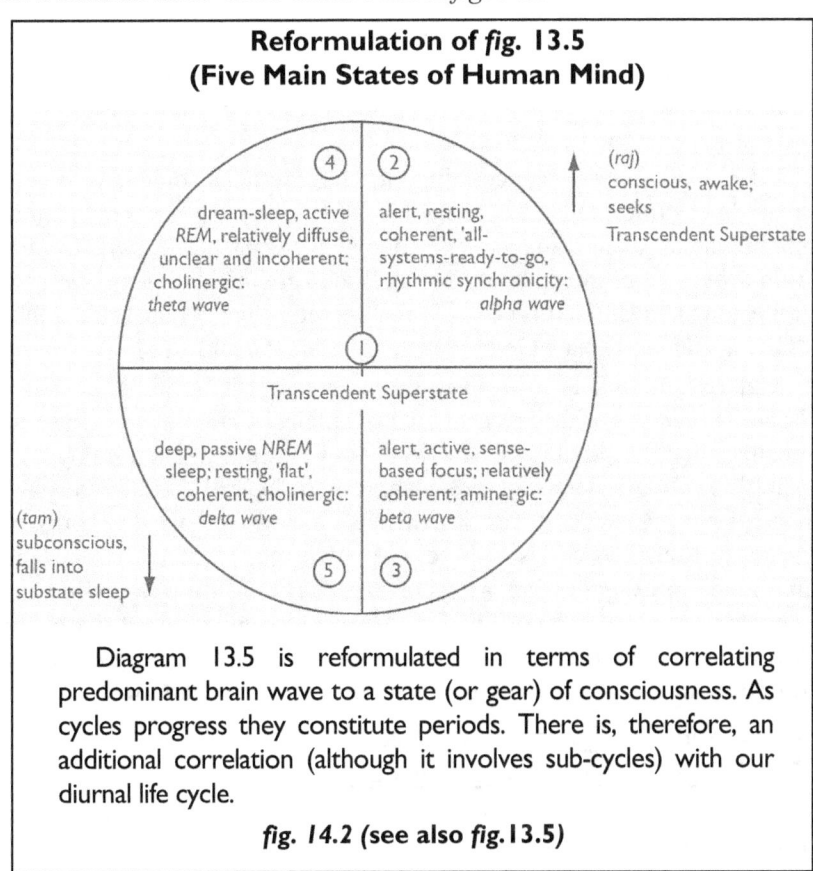

fig. 14.2 (see also fig.13.5)

Top-down, as you'd expect, we start at grade 1 and continue with its phasing down into grade 2, the highest state of normal human mind. Such normalcy, your waking state, includes within grades 2 and 3 the operating sub-divisions we call higher, intermediate and lower mind. Levels 4 and 5 fade to sub-consciousness. Let's investigate.

spectrum of consciousness	*Super-Conscious*
relativity of forms	*Centred Meditation*
(2),(3),(4),(5)	*(1)*
↓ *downward tendency*	*upward tendency* ↑
passive/ exhausted	*active/ alert*
involuntary	*voluntary*
non-conscious/ subconscious	*conscious*
busy-in-sleep/ dreaming/ theta (4)	*busy awake beta (3)*
deep sleep/ coma/ delta (5)	*contemplative/ alpha (2)*

First State of (Super-)Consciousness - The Psychology of Transcendence

Life's real secret rests, according to the zealous atheists who in 1953 discovered how it coiled, in a chemical called *DNA*. In their view mystic truth is actually a 'God Delusion' that's evolved from genes, their proteins and electric fields of charge.

Life's *real* secret is, according to the mystics, something we've already met - Informative Top Teleology.[92] It is an immaterial moment of enlightenment. Nanak analysed no followers' genes but was his Central Truth a figment of uneducated lies? Is Buddhist enlightenment, *Nirvana*, an illusion? Did Christ miss Crick's real secret? What is Lao Tzu's Alpha Moment, Ramakrishna's Great Idea, the world's Great Love?

Materialism stutters to explain Tzu's Moment. Its reflex is denial. It discredits, charges with deceit or plain explains it right away. Erase all immateriality! Yet what is love but living union? Sexual union is physical; friendship is dependent; but Friendship, metaphysical communion with Single, Living, Cosmic Source, is a mystic's holy grail. This, the 'Unifying Theory of Metaphysics', is realised as Knowledge, Truth and Love. **It is, therefore, no exaggeration whatsoever to say that human civilisation is constructed from and around a supra-religious tryst with the eternal moment. Human faith, hope and ideals are logically derived from the materially meaningless experience of transcendence.**

Second and Third States of Consciousness - The Psychology of Waking Normalcy

From a *bottom-up* perspective consciousness is a question of nervous and biochemical activity alone.

Top-down a holist agrees that bio-psychological elements permit the transfer of perception to and from the body but not that this is the whole story. In this view *metaphysical mind* spans two phases and interlinks

---

[92] Chapter 5: (*Sat*) Potential Information and Top Teleology; *fig.* 8.3 see *GUE*; also Chapter 1 *figs*. 1.1-3 Axis or Original Potential.

with a third. These are *conscious*, *sub-conscious* (psychosomatic) and dependent *physical* (or non-conscious, biological).[93]

Therefore we'll now deal with the scale of subjectivity, relativity of consciousness or, in short, prismatic spectrum of our waking mind. The engines of this mind are urges and desires; and its quality, from high to low, is scaled into three bands. In which directions and how strongly does the changing concentration of your focus weave?

Higher Normality involves inward, contemplative focus. Such focus tends (↑) 'upwards' towards a growing comprehension of general or universal principles. In seeking release from apparent disarray, confusion and limitation-in-specific-detail this higher 'mind-in-principle' more approaches and therefore reflects the characteristics of its highest grades - *Logos* and Transcendence. These include, as well as unification, continuity, coherence, integration, relationship and communication. *Such positive, right-hand principles educe a well-balanced, focused and attractive personality. Such a side is idealistic and strives for wisdom, beauty and love.* In dialectical terms, an increase in the predominance of (*raj* ↑) right-hand characteristics '*expands*' consciousness.

The next band is *Intermediate Normality*. Ego is self's intellectual executive and weaves, from sense to thought and back, with serial in-out reasoning to achieve its ends. As a wilful, manipulative schemer 'mind-in-practicality' is a driving instrument of problem-solving and achieving goals. Like any middleman, however, *ego* has to balance both sides of equations and in so doing cuts both ways. Its centaur-like operation may involve higher (↑) or lower (↓) moods, tendencies and desires; and it employs, at different times and in various persons, varying focus of intelligence.

Lower Normality. If there's an upside there's a (↓) downside too. Now comes a dark day. The other side of mind is a *separator*. The main static or downward principle of our psychology is *confinement* with its correlates of structural discontinuity, differentiation and individuality. *Negative, left-hand principles educe disruptive, chaotic and unattractive expression.* The body-focused isolate is selfish, restricted in scope, easily distracted, aggressive, turbulent and as demanding as unpopular. The lowest cast of conscious mind shows fear, depression, laziness and stultification. A human in such common but '*constricted*' consciousness is 'animal' inasmuch as an animal is most involved with unreasoning, sense-driven reflex. Continually noisy sensation is consumed by lusts and physical survival; there 'bubbles', in addition, a continual, diffuse chatter, day-dreaming sometimes called 'default mode'. 'Lower mind' lacks focus; its negativity seethes with passionate attachments or else with fear, hate and jealousy. Its base-pole ignorance is dimness and a dead-weight drag.

---

[93] Check *fig.* 14.1. Chapter 14 deals with conscious mind, Chapters 15-17 with subconscious and 19-25 with physical body.

Norms and Loops

Psychological Homeostasis

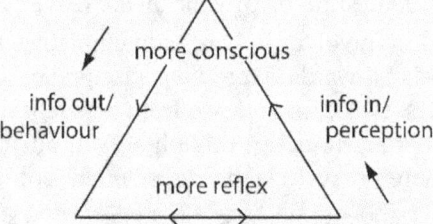

control centre,
equilibrator,
more or less conscious/ reflex

info out/ behaviour   more conscious   info in/ perception

more reflex

Buoyant animation tends towards dynamic equilibrium. Conscious desires and subconscious urges act, for their duration, as goals. They act, in terms of homeostasis, as an axis, norm or point of balance to be met and, maybe, cycled round repeatedly. See also *figs.* 1.4, 3.5 and 13.5 and Chapter 19 and Glossary.

Door of Perception

'contemplative door'
inwards towards Centre

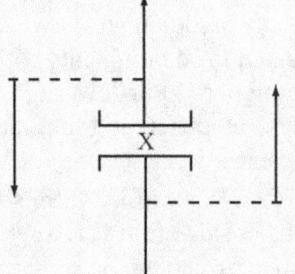

detachment from Centre; scatter of attention towards periphery i.e. worldly business

detachment from bodily business; contemplative concentration of focus towards Centre

X marks the door of your psychological house, the third eye or seat of mind. These arrows correspond to information loops.

Loops of Perception (local and in universal context)

an information loop

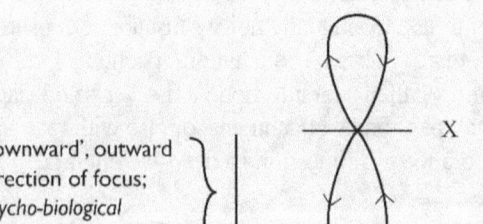

'upward,' inward direction of focus; psychological homeostasis

X is the eye centre direction of focus; it represents the pivot between inward and outward aspects of mind.

'downward', outward direction of focus; psycho-biological homeostasis

To an end, called balanced mind, two vectored loops engage; linked at our cosmological axis, the eye-centre, they radiate both 'up' and 'down'.

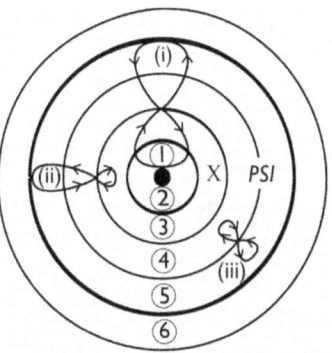

① Centre; Cosmological Axis
② superconscious state
③ conscious state/
④ personal cosmological axis/ third eye (X)
   subconscious state/
   PSI (psychosomatic interface)
⑤ physical body
⑥ external world

inner and outer psychological loops:
(i) complete
(ii) normal - emphasis on physical sensation and action
(iii) psychosomatic loop (see also fig. 17.3)

fig. 14.3 see also figs 1.4, 3.5, 13.5

If mind shows gradient it also works in cycles. Mind, regulator and equilibrator, keeps balance and decides direction. It seeks to navigate an ocean swelling with desires. These waves of urge are whipped by wishes. Knowledge, creativity and instinct need be satisfied. In this case two types of *norm* and *loop* arise.

Homeostasis[94] is regulation according to *norms*. Circumstances are continually monitored. Information is compared with a norm. Does it measure up? OK, do nothing. Does it deviate? OK, correct the deviation. The process is simple and fundamentally dependent on pre-set, programmed norms. In biological terms it involves deviation from a norm that is in consequence corrected; such correction, leaving the organism again 'feeling right', is an answer. Psycho-biological instincts involve the same kind of homeostasis; the problem is a need or urge (sometimes, by frustration, translated to a conscious problem) and its satisfaction, the answer. Accordance with physiological norms is a stability called health.

Now we turn to *loops*. Check *fig.* 14.3iii; and in *fig.* 14.3iv loops within the circle X are upper, metaphysical and those outside lower, physical.

Upper, Metaphysical Loop

In conscious mind our norms, like wind and weather, often change. In this case changeable desires to understand and problem-solve seek different answers; satisfaction of desire yields equilibrium. *The 'upward-facing' loop of internalisation is contemplative.* It is involved with internal,

---

[94] see Chapter 19; also Glossary and Index.

do/ know	*Be*
info. out/ info. in	*Information Centre*
subordinate principles	*First Principle*
logical expression	*Primal Latency*
↓ *to do*	*to know* ↑
info. out/ output	*info. in/ input*
differentiation	*integration*
creation	*knowledge*
behaviour/ response	*understanding*
actor/ action	*preceptor/ perception*
downward loop	*upward loop*
outward focus	*inward focus*
down from centre	*up to centre*
externalisation/mind to matter	*internalisation/matter to mind*

psychological homeostasis, that is, the achievement of balance through resolution and reasonable solutions to various problems. Such *voluntary* focus involves no physical organ; it uses neither sense nor muscle but, as ships pass, you might spot nervous waves relating navigation of a thought to body's brain. Its course engages the application of negative feedback to plans until the goal - an attractant object of desire - is attained. Such desires, flexible and ever-changing, compose the human mind. Their goals may include emotional satisfaction or relief from ignorance and pain. The purpose of this loop is, therefore, informative exchange and rearrangements within the metaphysical domain; and, therefore, increasing grasp of principles by which details of the world are understood and classified. The more general a principle the more powerful is its exercise.

The *purpose* of engaging this loop is to improve the quality of experience. Its course therefore blossoms with degrees of intellect; its fruit is every kind of scheme. It may, however, in deeper contemplation, ascend through higher mind towards First Principle. Upper mind's loop may achieve the full realisation of human potential called Return. ***Thus it spans from earthly possibilities to Central Potential.*** This is the bulls-eye of a wise man's aim.

Lower, Physical Loop

On the other, dialectical hand an outward, 'downward-facing' loop of externalisation concerns the domain of body and environment. It thus involves external, **psycho-biological homeostasis**, that is, the achievement of satisfactory balance through reasonable buffers to various bodily urges. *Involuntary* exchange of messages involves conscious, contextual response to local pressures, psychosomatic instinct and, lower down, biological homeostasis. Input, regulation, output. Sensor, processor, effector. This dynamic program of equilibration deploys physically informative nervous, hormonal and energetic muscular systems. Its course thereby engages the application of negative feedback for specific, codified behaviour until the goal - a pre-set biological normality called health - is achieved.

Inwardly, sense organs pass coded input for translation into psychological perception, a dynamic symbolism called the image of experience. Where anti-parallel motor and sensory nervous systems are, in humans, controlled by a central processor called brain the hormonal 'brains' are glands. These two systems, electrical and chemical, are linked by a triumvirate of glands (pineal, hypothalamus and pituitary) cradled at the heart of brain right in the middle of your very well constructed head. Communication also includes psycho-biological exchange with subconscious files (called instinct); it thus involves, according to Natural Dialectic, metaphysical psychosomasis by way of a typical mnemone.[95]

After *processing*, output is relayed in the opposite, *outward* direction. Instruction of encephalic mediation is translated into action *via* motor nerves (using 'on-off' depolarisation/ polarisation), antagonistic hormones (using chemicals) and muscles. Thus, in conjunction, norms of body are maintained.

The physical loop is of inflexible, automatic and repetitive function. Survival of a body gives no choice. Subconscious mind, systems of control and metabolic pathways synergise to promote health, survival and quality of biological condition; their reflex feedback stabilizes a body in its particular form of dynamic equilibrium. ***Such mechanism spans earth-life wherever bodies grow and go but is, like any machine, eventually going nowhere but demise.***

Quality of Mind

Reason cuts both ways. The loops aren't mutually exclusive. The upper loop, whose target is fulfilment of a wish, shows imaginative, internal focus that is most intense in contemplation. The lower, whose purpose is fulfilment of instinctive norms, shows reflex absorption; its external focus is most intense in fascinated observation. We oscillate within both loops. Their extremes, Truth and body-centred interests, are poles apart; but may, as Love and animal relationships, entwine.

Health and satisfaction of desire both 'feel right'. Feeling right is certainly a quality of mind. Is, though, individual 'feeling right' the same as tendency to preferential use of upper loop? And does it ally with universal 'good'? Does it secure morality? If quality is judged according to the criteria on which mind makes decisions, psychology immediately finds itself transported to a hallowed philosophical arena. Nor is this area a

---

[95] Chapters 15 and 16: Personal and Typical Mnemones; also Glossary.

scientific one. What is 'good'? What is 'best'? *What is true; of what nature is **top information** and the key to its disclosure?* **You ask, in other words, in what or whom exists the Highest Truth; and what, therefore, are the best criteria or is The Best Criterion on which to base our lives.**[96]

Put in the simplest terms, if higher self (second state of consciousness) approaches (↑) Truth then, by inversion (*fig.* 2.4), lower mind involves itself in the reverse - falsehood and illusion. Engrossed in body the third state's material desires (↓) distort the Truth until, approaching full distortion, mind must enter devilry. The ancient battle's fought in humans; benevolence and evil vie in you; which way will you tend? **It's clear the quality of psychology's central 'object' of research, mind, involves character and, in turn, a clear moral stance.**

Quality of Information

	lesser truth	*Truth*
	shades	*Light/ Consciousness*
	grades of knowledge	*Knowledge*
	grades of priority	*Importance*
↓	*negative*	*positive* ↑
	error	*truth*
	dim	*bright*
	oblivious/ ignorant	*knowledgeable*
	increasingly misinformed	*increasingly informed*
	incorrect/ wrong	*right/ correct*
	trivial	*important*
	inefficient	*efficient*
	useless	*useful*

Information is often called by its users 'intelligence'. *In fact, intelligence is a function of attentive focus and ability to spot the principle behind a practice.* It is, therefore, not passive data but an active quality of mind inferred from the relative ability to absorb, process and create information. Life is intelligent, always goal-oriented (teleological) and orders things according to its designs. *The greater an intelligence the faster, more accurately, intricately, thoroughly and efficiently are its problems solved and complex goals achieved. In short, the more teleological is its nature.*

If quality of mind is a function of intelligence and mind itself an information-centre, then what about the quality of information that it filters and then operates according to? By what criteria can we judge its value?

The importance or triviality of information is rated relative to various

---

[96] see also Chapter 4: *Is* There an Absolute Morality?; and Chapter 5: (*Sat*) Potential and Transcendent Information.

desires and sets of priorities. Such priorities and the wishes that sequenced them change. Instinct, education, advice and cultural experience help shape such changes and, in life's relative muddle, provide answers and directions. What are religions and political creeds but attempts to encourage coherence, develop *esprit de corps*, solve problems and emphasise idealistic priorities? Their purpose is to orient, create a certain quality of mind and thereby improve the way it handles information.

Everybody operates according to their own variable, personal set of priorities but do there also exist permanent, natural and universal priorities? Is there a hierarchy of information, a set of principles? If so, what is nature's 'top set'? Is there Prime Information? Is there anything of Absolute Importance?

The Ascent of Man

	existence	*Essence*
	lesser being	*Supreme Being*
	expression	*Potential*
	body/ mind	*Soul*
	periphery	*Centre*
↓	*descent*	*ascent* ↑
	compress/ diminish	*grow/ liberate*
	passive/ involuntary	*active/ voluntary*
	created form	*creativity*
	materialisation	*dematerialisation*
	entropy	*negentropy*
	devolution	*evolution*
	body	*mind*

Not from four to two legs but from sole towards soul - improve your quality of information. Rise (↑) towards order; now deliberately evolve towards the highest kind of information, the most powerful knowledge in the universe! When the waves of fear, *ego* and the passions settle then the peaceful 'residue', the central concentrate, is closed upon. *Such union with Natural Essence reigns Supreme; and the ascent of man, in seeking such Supremacy, is voluntary.* A seeker purposely engages the return slope of creation's gradient and ascends the right-hand path to its Extremity, Communion with Truth. *This is so-called science of the soul.* Psyche means soul.[97] Supreme Psychology is therefore taught by a 'soul man', an enlightened mystic. **This saint's healing constitutes Supreme Psychiatry; the practice lifts (↑) you to the origin and apex of the world. What might you find, Top Value, at this peak?**

---

[97] Chapter 13: Psyche and Psychology.

Passive Information (Psychological)
Chapter 15: Subconsciousness

For a materialist, who guesses informative consciousness is an 'outcrop' of brain chemistry, the next four chapters are irrelevant. Nor will you have met their content in the context of a modern scientific course. It requires no genius to figure their logic but the work has not been done simply because the head of such a course is not turned wholly in the right direction. By materialistic paradigm information and energy, mind and matter are of wholly the same quality. And if this quality is physical it is non-conscious. Consciousness is therefore a peculiar effect of certain formulations of non-consciousness. How strange are mind and subjectivity! How queer 'scientific' speculations made by quarks and leptons of a brain!

The Sub-conscious, Psychosomatic Sandwich

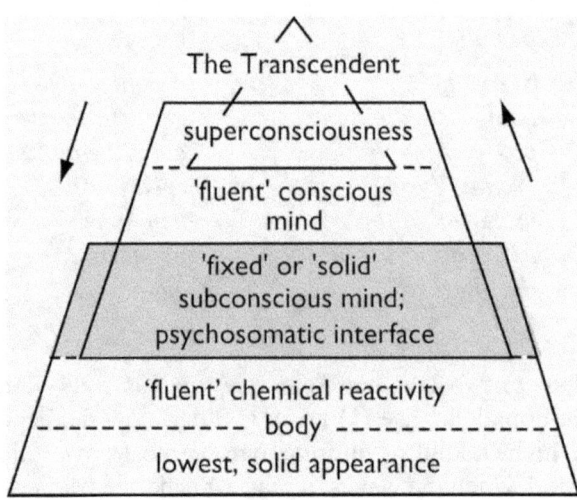

fig. 15.1 (see also fig. 14.1)

However, from a *top-down*, hierarchical perspective the subject matter of the next three chapters, sub-consciousness, is sandwiched between the conscious grounds of cognitive psychology above and non-conscious biology below.

A Black Box

Material bodies are hard, scientific ground; what, however, is mind's mystery about? Physical science is uncomfortable with conscious mind because, by definition, it is *not* non-consciously robotic. Its flexibilities involve, even at the base levels of involuntary instinct, memory and sleep, some degree of unpredictability, uncontrollability and unrepeatability. Such relative automatism decreases in proportion to the advent of wakefulness and voluntary thought. Of course, mind's natural enough but if it is insisted 'natural is material alone' then unnatural immateriality is expelled conceptually out-of-bounds. Whatever science-as-we-know-it can't explain becomes 'abnormal', 'paranormal', supernatural' or just plain impossible. **Thus, when only brain is treated, mind becomes an obvious, pseudoscientific mystery - a psychological black box.**

Mind includes sub-consciousness; in subjective practice, this is difficult to investigate because you can enter it neither voluntarily from above (you lose awareness) nor involuntarily from below (it is non-material). Indeed, your brain is in your white box, skull; but mind's whole business is a 'black box' - perhaps because it isn't physical at all. **Both sub-consciousness and the proposed evolution of that state are, as with consciousness, subsidiary black boxes.**

Sleepy Head

	imbalance	*Balance*
↓	*loss of balance*	*equilibration* ↑
	accrued instabilities	*homeostasis*
	descent into darkness	*ascent into light*
	towards malfunction	*towards healthy norm*
	psycho-physical exhaustion	*resurrection*
	entropic slump	*negentropic freshness*
	down to sleep	*waking up*

Holist and materialist agree, sleep's a slide towards oblivion but then, from 'little death', oscillatory ascension up again. This diurnal oscillation rules our lives but the mechanisms that underlie its 'simple' homeostasis are no simple proposition. No drowsy engineer could craft its composition nor doctorate reveal its elegant complexity. We have not yet bared the secrets of our own mammalian nether world nor is the restfulness of many other kinds of organism understood.

From a *bottom-up* perspective mind is simply a 'functional state of brain'. It is not that informative consciousness, as a separate entity from matter, is 'squeezed out' of this picture. It never existed. Therefore the difference between waking and sleeping is really one of chemistry. Just as more or less of the hormone called testosterone affects the subjective as well as objective sides of sexuality, so chemical instructions program the global states of brain and, therefore, mind - sleeping and waking are 'simply' biochemical effects. Thus, as knowledge is a fragile result dependent on the dynamics of codified molecular configurations so is its extinction, sleep. Indeed, since 'little death' and conscious life are both made of atomic forces, what is either but a formulation of oblivion?

An 'off-on' toggle for sleep-waking involves several major, interlocking switches. These regulators engage negative feedback using detectors and effectors in the form of a complex mix of specialised neurons and codified, correctly-triggered neurochemical modulators. They exert control mainly from the brain's centrally 'enthroned' hypothalamic master-gland (fig. 13.2). The *SCN* (suprachiasmatic nucleus) is a light-dark recognition system that receives information from specialised retinal cells. Acting as a clock it helps entrain your flip-flop with earth's night and day. *VLPO* (ventrolateral preoptic nucleus) is a 'sleep switch' flipped by the chemicals adenosine and serotonin - the latter also found, curiously, in plants and fungi. It makes chemicals that inhibit the 'arousal' areas of *RAS* (reticular activating system stretching from hypothalamus to lower brain stem). Finally orexin, also connected with basic 'house-keeping' such as thermoregulation and appetite, is projected by specific neurons in the hypothalamus throughout the central nervous system. This protein promotes the yield of serotonin in brain stem nuclei and thus, unblocking the voluntary system, acts as a wake-up call. Conversely, a lack of orexin induces sleepiness or, in some cases, chronic narcolepsy. You could see orexin as a 'toggle' between the two fundamental conditions of embodied mind, wakefulness and its lack. No orexin, no wakefulness. Too little orexin and a tendency to narcolepsy.

Thus living brain, always 'switched on', operates in two main modes. Your 'light', knowledgeable mode of life involves a conscious sense of purpose, attention, logical sequencing and general cognitive control. The 'dark' side blocks sensory and motor systems at thalamo-cortical and spinal levels respectively. Thus its activity is internalised and no 'waking behaviour' normally occurs. *VLPO* is aroused and issues chemical orders. Mind is now swept passively. Slumber may overwhelm its consciously directed thoughts and wash them into a psychotic stream, a hallucinatory course of dreams. It would amount, in this context, to a clearing and a cleansing process - reason's busted flush, psychological evacuation; and, physically, it represents release of nervous tension and a purge of toxins (such as excess build-up of β-amyloid proteins) accumulated from brain's waking use.

The point of this abbreviation is to show that, in principle, sleep-waking oscillation performs like all biological homeostasis (see Glossary) and thus, in practice, is a complicated mechanism. **Top-down notes that the operation of cybernetic switching systems and homeostatic goals of all kinds need coordinate components, calibration and a definite objective.** The dialectical oscillation of homeostasis revolves around a central target. It pursues a pre-set, balanced swing. **But coded, well-informed and closely interlocking systems are, like working algorithms laid into computer chips, not the gift of mindless energy.** How, piecemeal and planlessly, could coma have evolved into sleep-waking's regulated rhythms? Do you think conscious knowing and unconscious ignorance occurred by sheer chance?

Nor is sleep only cyclic but elastic. You fall deep asleep and, from this nadir, 'bounce' up into dreams. Thus, carefully controlled by many switches, each of four or five 'bounces' is in two parts. At last you 'surface', that is, wake. And between extremes its 'period' involves a scale, a conscio-material measure of reciprocity between 'off-line' sleeping and 'on-line' waking poles of consciousness. You can, for example, day-dream while awake and wake (in dream) while asleep; and, when it's not focused on an interest or task, the 'bubbling' mind diffuses into 'default mode'. This mode roams context in terms of emotive memories, desires and schemes. Indeed, the point of meditation is to quieten such continual mental chatter (an effervescence that's reflected by the brain). Meditative anaesthesia is conscious, voluntary and ascends towards greater awareness. By contrast, medical anaesthetists delve down. They deliberately induce coma in another individual; their dimmer switch, using chemicals such as propofol or xenon, approaches total darkness, irretrievable disengagement. They, for life's sake, manipulate the death-zone.

In short, psychology is a question of tendency and predominance. For the sake of simplicity we divide it into waking awareness (states 1 to 3 of *fig.* 14.2) and sub-conscious sleep (states 4 and 5). Sleep with uncoordinated dreams is as biologically programmed as wakefulness with its coordinated reasons. Natural Dialectic logically suggests that, in a structure built for polar reciprocity and equilibration, sleep supports the maintenance of higher wakefulness. Our life's homeostatic cycle sustains an otherwise unsustainable condition of unstressed, balanced health, both psychological and physical. *How, though, could a non-conscious death-zone aimlessly evolve through anaesthetic coma into wakeful self-awareness? In scientific detail, without just-so story, how did chemical 'mutations' wake you stepwise over aeons from material oblivion's abyss?* **It is reasonable to assert that the hypnotic routine (whose archetypal form appears designed in a binary, dialectical way rather than cobbled up by accident) is critically, precisely programmed**; but the irrational Darwinian explanation constitutes a **deep black box**.

The Fourth State of Consciousness -
The Psychology of Dreaming

Are you ready for a fall to the diffuse conclusion of psychological entropy, for a subjective drop into the labyrinth of underworld? You know what it is to be mentally as well as physically exhausted. You've often dropped off into mind's flat, dark condition we might call inertial equilibrium. 'Little death' is not the world's end so let's take a snooze cruise; let us simply fall asleep.

For dreamers dreams are real enough; but the experience is untrammeled by either external events or the ability to reason. Waves wash equally on what is in their path; a torch shone randomly around picks out disconnected or illogically connected objects and events. The files are scattered, narrative is blurred. Although dreams (the uncontrolled, internal generation of psychological events) can occur at any stage in sleep, their illogical disconnections stream in vivid profusion at periods of rapid eye movement (*REM*) sleep. Movements in *REM* periods seem to want to correspond with a dream's drama but are, for critical security, inhibited by induced and natural paralysis at spinal level. Successive 'bounces' become less pronounced as the moment of reawakening approaches. You then break surface into a relatively logical, voluntary association with physical phenomena.

In this apparent chaos is there reason? Dreams serve, like waking mind but lacking sensory restraint, to relieve circumstantial pressures in the form of dangers, problems and anxieties. Subjective equilibration is the game. For relief's solution, relaxation, dreams roam uncontrolled through memories. Of what is lunar dreamscape made but these? We'll soon explore the subconscious world of personal and typical (instinctive) mnemones, that is, major files of memories. Such world is not *per se* irrational but, carried passive on an incoherent dream-stream, an observer's various slumberous visions of it is. Sometimes, waking, he recalls his lucid travel or more jumbled narrative. There are even, in the loop, memories of dreams.

The Fifth State of Consciousness - The Psychology of Deep Sleep

In deep, non-*REM* (or *NREM*) sleep the 'upper', voluntary structures of brain are cut from the loop. A sleeper's movements, including eye movement, are much restricted; sensation is dull or absent. Brainwaves, the overall coordinators of the central nervous system, slow to between 0.6 and 3 hertz. These are so-called delta waves. Maybe deltas drop to zero. Brain death. If, by head injury, stroke, tumour or poison, the sleep/ wake toggles fail or signals cannot reach the forebrain then the patient drops into oblivion. The curtain falls but drama does not start again. Coma is an open tomb, an unpinned shroud or wake-less sleep - though in its stillness deeper grooves of mind (archetypal constructions but also profound personal impressions and rote such as language) stay frozen yet intact.

Organisms sleep in different ways and yet, not dropping off into their 'little deaths' at all, they'd die. Sleep is vital. Sleep refreshes and it heals. Survival is insistent, for example, that a brain is regularly cleaned (sleep's neural shrinkage lets in spinal fluid to wash toxins out). How, though, did genes evolve the physiology for 'off-line' maintenance or *REM* and *NREM* sleep? And, as well as chemical complexity, dormancy's a metaphysical affair; it knits up *mind's* ravelled sleeve of care. A just-so story for such wonder, best beloved, is that elevation from the 'sixth state' into coma must have woken life from its primordial lifelessness; and that, before you sleep on it, evolution really woke up when, atomically, the chemistry of slumber re-arranged itself as conscious beings such as you!

The Sixth State of Consciousness - The 'Psychology' of Physic

There's a rider tagged upon creation's tail. The sixth state of consciousness, if you could call it one, is its special case of total absence. Zero on the vertical, informative coordinate (see *fig. 2.5*). Absolute mindlessness. Non-consciousness is the wholly frosted, jet-black opacity of our body-chemistry and its phenomenal circumstance, the earth and starry universe.[98] *Its physical dimension is the only one that naturalism can interrogate directly.*

Frozen Time

You sleep but your past does not disappear. You wake and your past has not disappeared. You think you have forgotten, you may even suffer amnesia but memories remain. They are how we freeze time. **A memory is frozen time. It symbolically encodes the past.** *A memory is a thought object and, as such, has no life of its own.* A disc encodes music once recorded 'live'; it's a memory that, when replayed, affects the present and, from this, the future. Thus mental memory is an encoded image; it is a record and, on conscious recollection, becomes a presence of past action that may affect the future. What applies to music may apply to plan or any form of creativity. Could you call memory indefinitely suspended animation?

Bottom-up, mind is materialistically thought to be some aspect of dynamic brain and, from this neuroscientific but still philosophical angle, memory is a part of grey matter. Its storage bit, a hypothetical straw sometimes called an 'engram', is grasped as a 'nervous trace', 'a packet of proteins and lipids' or some other pattern of representation. But what configuration of proteins means 'virtue', what ionic pattern *is* a feeling of devotion? The mechanism by which an experience is first encrypted, later accurately located and then decoded back to the remembrance of a loved-one's face is unknown. The gradual evolution of remembering, continuously survival-critical for knowing what to think or do, is also **a pitch-black box.**

Top-down, **neither memory nor knowledge are inherently physical.** No doubt, correlated nervous circuitry acts as a storage-and-retrieval system that, by association, allows the immaterial library of remembrance its efficient, selective interaction with an innervated body. Thus 'engrams' (are they sited at synapses, inside a nerve cell's body or elsewhere?) may, if they exist, indeed *relate* to physical experience; they may act as a recognition trigger, reference point or body's resonance with an experience. And organs (such as hippocampus and amygdala) certainly seem to log experience in the manner of a record/ playback head; they catch or release a moment that, in fixity, is called a memory. But if such 'storage' or 'playback' button fails the system's compromised. Either

---

[98] see Glossary: conscio-material (c-m) dipole and Chapter 2: The Basic Existential Dipole.

records are not made in the first place or the connection becomes impaired or irretrievable. *But the 'disc' of memory itself is metaphysical.*

Recollection is one thing, endurance another. Memories, whose species include short-term, personal and long-term, archetypal forms, endure.[99] The latter, it may be argued, are innate, read-only files that are laid into the fabric of a being; they are like carrier waves of an archetypal broadcast. And the former are like individual modulations on the waves of such channels; they are read-write overlays, programs, records of varying endurance and intensity ranging from the ephemeral scratch-pad that supports moment-to-moment orientation to sharp, deep emotional impressions that colour a lifetime's responses. Either way, the past is present. Life can't forever start from scratch!

Once made, however, a metaphysical memory is, as a mind-object, physically indestructible. This does not, firstly, mean it can't be reinforced by repeated similar experience or 'learning by heart'; or that, conversely, it can't fade or, by retrieval and re-storage, be gradually falsified. Who can fully trust their memory - not least when, by retrieval's absence, impressions fade and are lost anyway? Secondly, it does not mean the instrument of brain itself, as well as interfacing personally constructed memories, is not constructed as a medium for archetypal instinct and related, type-specific records. Indeed, you might definitely expect it was.

<u>It needs be re-emphasised that an archetype is a memory and, as such, a thought object</u>. If sub-conscious mind is memory, neither its personal nor archetypal aspects think or feel anything. *An archetype, as lifeless as a stone, has no life of its own. A memory is, whether universal or individual, an informative structure that of itself no more contains consciousness than a test tube or a church.* It is, simply, like the record of a song, a frozen piece of time - one that is played or, if you will, accessed incessantly by as many users as there are physical operations or cells of its type. You might conceive its archetype as any organism's brainless brain.

Memory, the only form of metaphysical information storage, is the shape of infra- or sub-consciousness. *Indeed, it <u>is</u> sub-consciousness*. **Sub-consciousness is made of memories**. It is an organism's library of precondition and conditions that comprised its past, that is, its past experience. The precondition is its archetype, the basis of its sort. We might call this sort of memory '*typical*' while '*personal*' experience includes both active (created and transmitted) and passive (received) information. Files are 'written' to the personal library with a sharpness of imprint proportional to their impact or importance to the owner. And they are filed according to their association with other data in the owner's bank. Whilst focused, deliberate access is called *remembering*, unfocused and involuntary

---

[99] see Chapters 15 and 16: Personal and Typical Mnemones.

awareness of memory is called *dreaming*. The latter includes physical response to dreams such as sleep-talking or -walking. Below these dreamy shallows there plunges imageless and inconceivable abyss - to levels of deep sleep and even sub-sleep, coma. The latter are still metaphysical as opposed to physical non-conscious conditions. At this level reside general, archetypal or 'read-only' impressions, that is, the body-linked archetype explored in the following section and, especially, Chapter 16.

Memory is not necessarily a medium 'frozen' like a photograph. Nor, although it comprises a concept or basic expression of an idea, is an archetypal memory. Memories may operate like movies and store programs that, once triggered, can unfold like stories in a sequence to their completion. Any stored plan is, like a computer program, such a memory. Programs are, although dynamic, still a frozen form of mind; and they're replete with information. They specify the most efficient means to a well-defined end. **You might argue in this vein that biological structures are codification incarnate; and that the concept they express is an archetypal program.**

In this case memory, including biologically ubiquitous archetypal memory, represents another Darwinian black box.

Psychosomatic Linkage

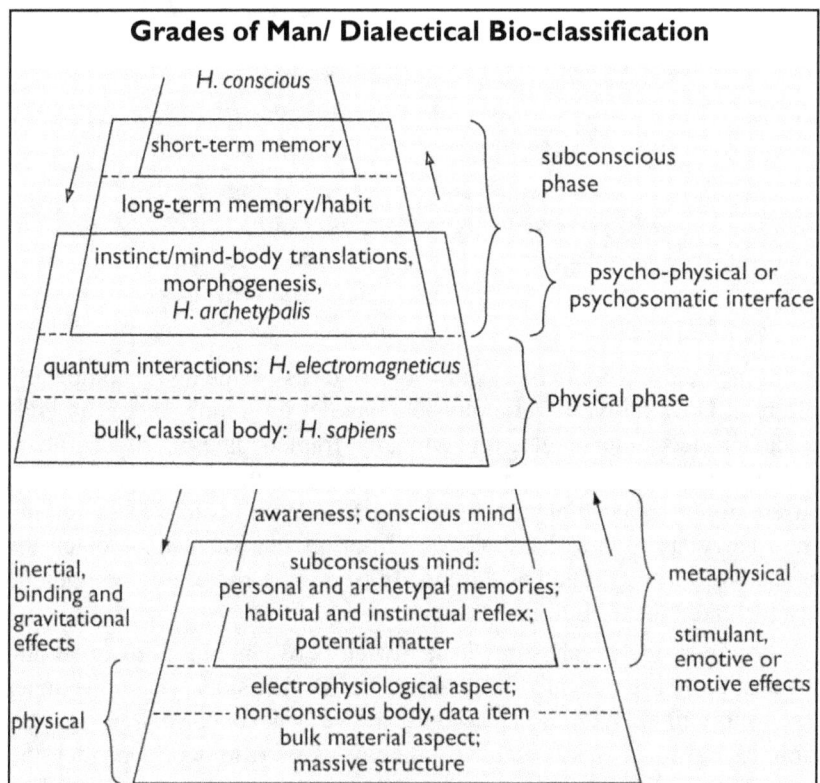

These ziggurats comprise holistic/ dialectical bio-classification. This involves the 'vertical', informative coordinate of creation. While we experience the link between conscious mind and non-conscious body as the condition of sleep, sub-conscious mind is also the repository of memories; *and these diagrams introduce a logical but revolutionary, top-down derivative from the conscio-material spectrum - the 'dormant' position of archetypal memory.* **Such channel of order is further identified with both the passive 'infrastructure of information' - codes including archetype - and with potential matter whence issue the various basic patterns of physics, chemistry and, with respect to complex 'bundles-of-laws' or 'programmed routines' called organisms, biology.** Such perspective will be elaborated over the next three chapters.

fig. 15.2 (see also 15.3).

physical/ objective correlates	*metaphysical governor*
non-conscious effects	*sub-conscious cause*
classical/ quantum matter	*potential matter*
material expression	*archetypal memory*
physical cause/ effect	*psychosomatic medium*
sensation/ action	*psychosomasis*
physical body	*morphogene*
H. sapiens/ H. electromagneticus	*H. archetypalis*
↓ *informed classical level*	*informant quantum level* ↑
chemical receiver	*radiant influence*
anatomical shape	*electromagnetic/ biochemical patterns*
H. sapiens	*H. electromagneticus*

From below (*bottom-up*) an objective assessment is completely different. How can you 'get a physical handle' on mind? Access ascends as far as electronic or electrochemically framed patterns of charge in motion. We can peer from outside through a glass darkly. Voltage, current and associated electrochemical effects carry us to a glass ceiling. They carry us to the threshold of sensation but we cannot enter the subjective house of another's mind. Indeed, to a materialistic rationale any such attempt were futile.

Top-down, the psychosomatic sandwich between conscious mind and physical body is practically inaccessible from either side. It is, in the upper case, subliminal and in the second, above whose body it is ranged, sublime. However, just because something is hard to access or measure scientifically does not mean it neither exists nor impacts our reality.

Natural Dialectic, noting the profound explanatory shortcomings of one-tiered materialism, simply and reasonably proposes an addition to physical electrochemistry - non-material, metaphysical mind. What greater inherent improbability, wrote the founder of modern neurophysiology, Sir Charles Sherrington, than that our being should rest on two fundamental elements than on one alone?

Physic ignores, denies or is confused by the immaterial aspect of information. The instruments of science cannot register and therefore science can't conceive of a 'gap' which separates psychological from physical worlds. If, however, what is termed illusory exists, this 'gap' is crossed at every waking moment. The elusive transition (*psychosomasis*) occurs with perfect ease and without discernible lapse in time. Hand-in-glove. If mind and body synchromesh like gears together it is a prime necessity to narrow the 'gap' between our *top-down* and *bottom-up* versions of their interaction. In brief, we need to identify and clarify the nature of the point at which 'I' exchange information with the physical syntax of a nervous system - where mind meets matter. As with quantum/ large-scale rift (needing at least quantum gravity to heal) so mind/ matter 'disconnection' needs its meeting-point. *What might heal the latter rift; what might be the structure of a psychosomatic interface?*

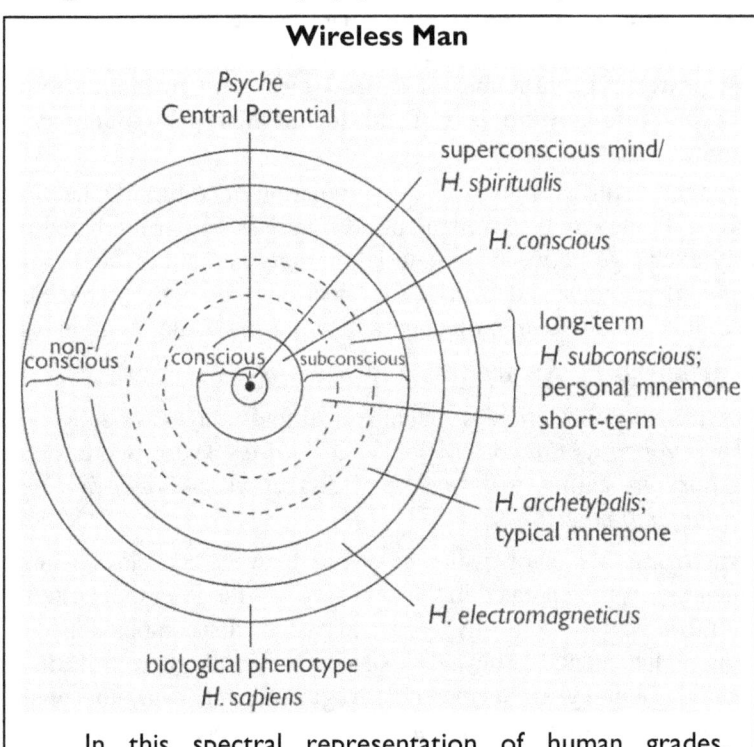

Wireless Man

Psyche
Central Potential

superconscious mind/ H. spiritualis

H. conscious

non-conscious / conscious / subconscious

long-term
H. subconscious;
personal mnemone
short-term

H. archetypalis;
typical mnemone

H. electromagneticus

biological phenotype
H. sapiens

In this spectral representation of human grades, sheaths or bodies only *H. sapiens* is wired, fixed or made of classical matter. The others are wireless or radiant.

> Thus you might visualise yourself (and any other organism) as both wireless and a wired anatomy, that is, as a composite of fixed and fluid sheaths. As discrete electron orbitals surround a nucleus these sheaths surround the Nuclear Psyche. Separated by 'an exclusion principle' their phased 'energies' materialise towards the periphery of creation. There the most dense, wired level - called your body - is obtained.
>
> *fig. 15.3 (see also 16.1)*

The basic principles of psychosomasis are clear. *Mind (at most gross, subconscious level) conjoins with matter (at subtlest, least-massive/ almost-immaterial level); elusive quantum probabilities pinned-down substantiate, it seems, certain processes; photon and electron precede, in the sense of underwrite, molecular and bulk reactivities; and, where electrodynamics describes the effect of moving electric charges and their interaction with electric and magnetic fields, biological electrodynamics precedes all bio-molecular considerations.* **Every biological process is electrical; and the flow of endogenous currents is the primary and not secondary feature of physical life. Not only biochemistry but quantum biochemistry heave to the fore. Natural Dialectic lifts perspective from molecular to a vibratory, field perspective. It is thus suggested that, at electrical and wireless levels, patterns of subconscious mind meet and influence matter; archetypal information is relayed to chemistry by polar charge and light.**

It amounts, furthermore, to a key prediction of Natural Dialectic that science will one day discover the mind-mindless border mediated by an orderly exchange between immaterial, archetypal patterns and material patterns of photons and charge (as electrons and ions). The electric couple link, of course, to molecular and supra-molecular aggregates.

Synchromesh 1 - Awareness and Memory

Top-down, information is metaphysical and body physical. *Linkage divides into two sections - an upper and a lower.* These two basic sorts of memory are called, for the sake of dissection, *personal and typical (or archetypal) mnemones.*

Synchromesh 1 involves the *upper link* between conscious awareness and memory. It is elaborated in the next section - the **Personal Mnemone**. This mnemone is specifically 'you and yours'. Its composition derives from individual mental or physical experience. **This higher, more inward cog gears memory, in terms of storage and retrieval, to conscious mind.** It includes short, medium and long-term records of mental activity. Its conjunction is therefore with your cosmological axis, point X, third eye or 'power-point' at which you think.

Psychosomatic Linkage by Domain

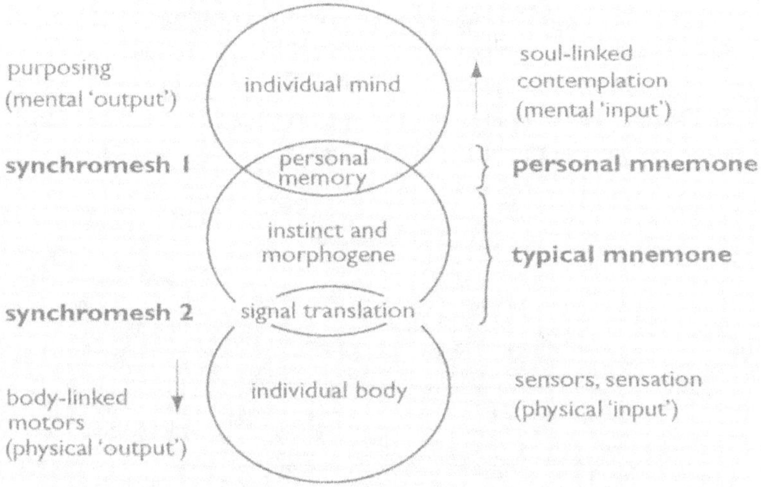

purposing (mental 'output')

individual mind

soul-linked contemplation (mental 'input')

synchromesh 1

personal memory

} personal mnemone

instinct and morphogene

} typical mnemone

synchromesh 2

signal translation

body-linked motors (physical 'output')

individual body

sensors, sensation (physical 'input')

Three domains of human being: conscious, sub-conscious and non-conscious. This diagram indicates the order of linkage, through sub-consciousness, between mind and body. Various *personal memories* (such as language, habitual behaviour, familiar faces and objects) are each inactive until reactivated (brought to consciousness) to provide context for conscious mind; such memories can be short-term or long. On the other hand the three aspects of *typical (or archetypal) mind* are sub-consciously active; they are, like genes in a genome, permanently associated with every cell, all development and all multicellular bodies. 'Signal translation' is the sub-routine that, where appropriate, provides for an exchange of information between conscious mind and body; it comprises the channel that translates information (for example, nervous information) into perception or, conversely, decision into the morse that culminates in specific muscular action. *Instinct* works upwards into conscious mind; or downwards through the faculty of signal translation and/ or morphogenic sub-routines to engage with cell-constructed body.

fig. 15.4

Suggested Architecture of the Subconscious

Two analogies are useful when thinking of the psychosomatic interface. Using the analogy of a mind machine, the computer, its '**conscious** *CPU*" (central processing unit) is not the machine itself. It is the inventor or user who employs the machine to serve his purposes. A computer as sold (or born) already involves immutable logic expressed in the form of operating system and various chip circuitries. **The algorithms of subconsciousness are likened to the programs of an operating system.** This informative software, varying with each type of organism, constitutes the typical mnemone; instinct is physically expressed by typical behaviour and morphogene by the typical hardware of a body-system. On the other hand, personal files (of the memory cache called personal mnemone) accrue through individual usage.

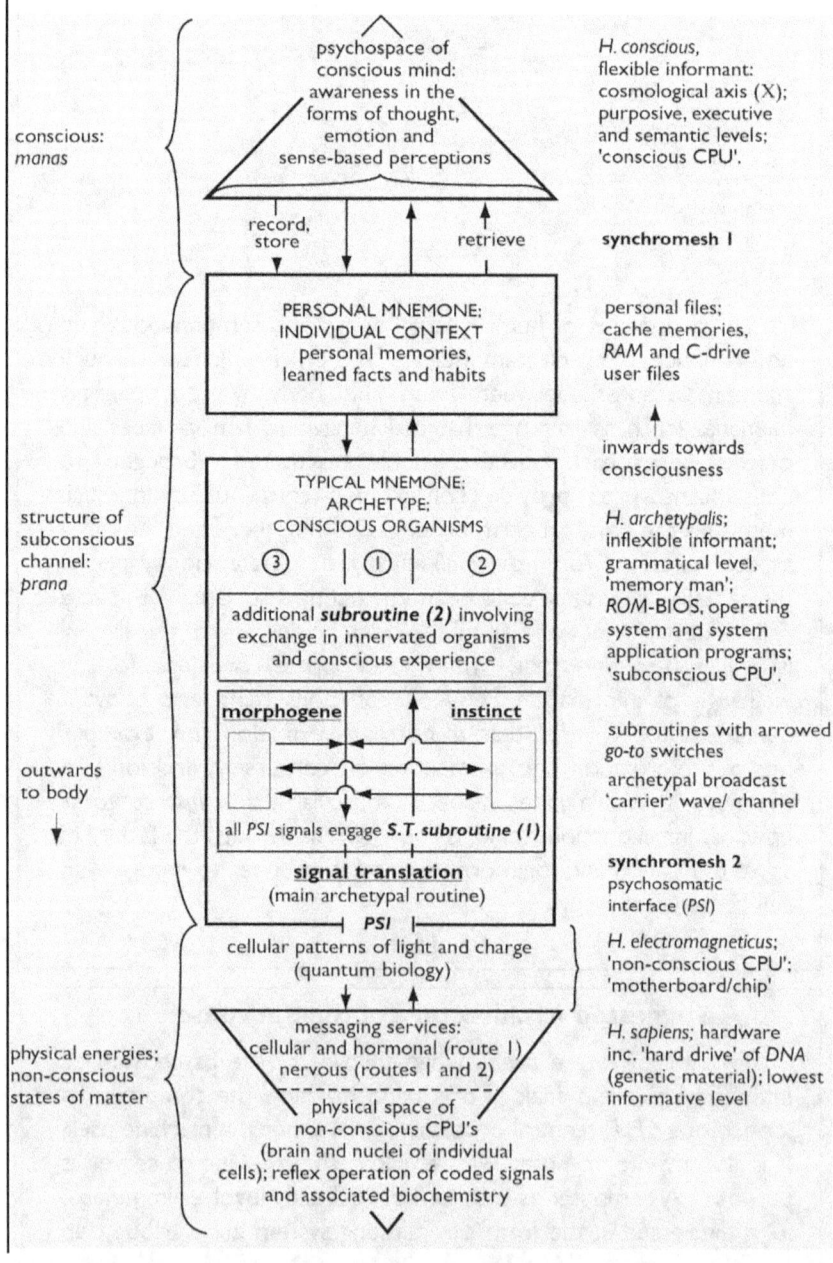

Mind is an information-centre. Although it may be represented by material vehicles such as nervous pulsation, ions, hormones and so on, information itself is a metaphysical arrangement. And the natural mode of metaphysical storage is memory. The two main divisions of memory are called mnemones. The flow-diagram above illustrates a possible 'synchromeshing' interaction of such mnemones with awareness, body and each other. It draws together the 'upper' personal and 'lower' typical characteristics of man. The upper 'personal mnemone' is an aspect of memory where images specific to an individual are stored. Such images provide a context continually and sub-consciously referenced by association. In fact, a personal mnemone might be viewed as records built up by an individual. Personal mnemone can therefore be viewed as a lens, filter or context built from a person's experience which 'colours' the way he or she thinks.

You can also view a second layer of sub-consciousness, the lower 'typical mnemone' or 'archetype' in this way - except this lens is, no more than a body, built from personal experience. If personal is 'weakly bonded' to typical mnemone, the latter is strongly bonded to every cell in every body that it represents. This 'typical mnemone', called in your case *H. archetypalis* (Chapters 16 and 17), is an archetypal memory. It can be visualised, metaphorically, as a dynamic blueprint, plan-stored-as-a-video or read-only input/output operating software. It comprises, in the form of instinct, the natural, dormant sub-structure of mind; and the 'read-only' architectural component, called morphogene, is responsible for the patterns of development, maintenance and repair of the physical body. In this capacity it interacts at electronic or electromagnetic level with chemical messenger services and, thereby, body. You might liken these physical services to 'buses' transferring information on a wired, material 'motherboard'.

Or, using the analogy of radio, a typical mnemone is the immutable 'carrier' frequency or broadcast wavelength (e.g. *H. archetypalis* or *Drosophila archetypalis*). As particular programs are transmitted on a specific channel so personal (which comprises a unique store of records for any particular organism with sensibility) rides on the back of typical mnemone.

Look at the 'circuitry' implied by these two diagrams. **The typical mnemone is an unconscious attractor, a sphere of influence or archetype in universal mind.** It is useful to think of it in terms of three major components, interlinked associates called (1) signal translation, (2) instinct and (3) morphogene (also called the morphogenetic sub-routine or biological archetype).

The combination of personal and typical mnemones amounts to a complete psychological and, implicitly, physical record of the organism it represents. Files are for storage but also communication. *H. archetypalis* is, sandwiched between conscious and non-conscious, physical domains, a transmitter.

It acts, in other words, as a signal translator, code formulator or transducer in the exchange between psychological and physical events. In this exchange instinct reaches up to consciousness and the morphogene down to non-consciousness. Both parties are referenced; response is circumstantial but their 'set' remains unchanged.

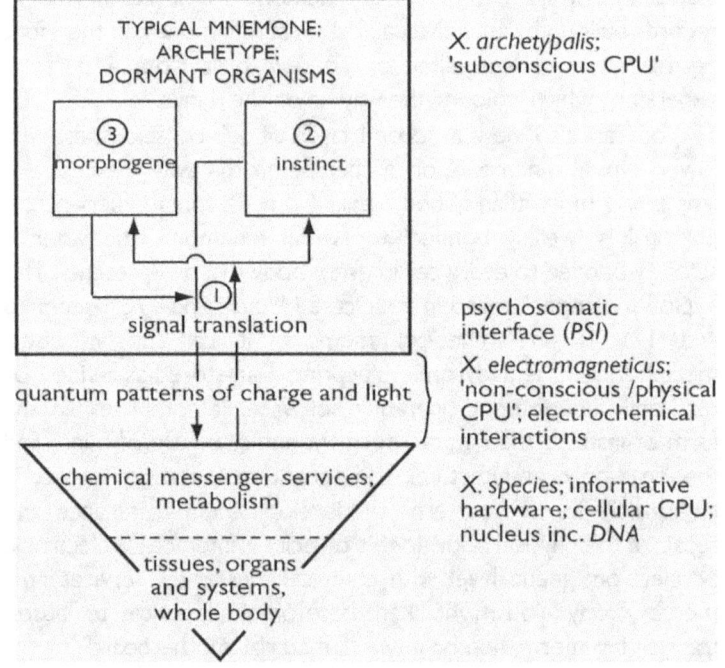

If, as in the case of animals, there is nervous sensitivity, then signal translation interfaces between conscious mind and body.

What about a biological operating system for organisms without nervous sensibility? In the case of plants, fungi, bacteria and so on, the 'circuitry' is reduced as shown; only dormant, typical mnemone and cell (or multicellular body) is involved.

Either way, this mnemone is a psychosomatic communicator whose sub-conscious mode of operation is association. *The method by which its bio-patterns are continually referenced and employed, either psychologically from above or physically from below, is association by resonance.*

fig. 15.5

In dialectical terms the *lower cog* is archetype; this form of memory is, with respect to an organism's psychology, called its **Typical Mnemone**. Such a mnemone includes three major sub-routines that will be explained in due course - instinct, signal translation and the morphogene. It gears mind to matter at the *PSI*.[100] This smooth, psychosomatic conjunction, called **Synchromesh 2**, is dealt with in the next chapter.

The Personal Mnemone

The analogy with radio is a channel, carrier-wave or frequency; with a computer it's the chips and operating system. These are likened to an organism's archetypal memory or *typical mnemone*.

Radio frequencies are, however, personalised by messages, programs or various broadcast schedules and a computer operating system individualised by chosen applications through which work is done. Coded transmissions need a local context to perform what minds that use them want; and so, to the generality of archetypal memory, specific context needs be added; local environment is integrated in a universal framework so that each different body's life makes sense. Contextual files, called *personal mnemone*, constitute an individual's history. They are the sum of what he has, to any point, experienced and learnt; they constitute the other, personal half of that subconscious sea of influence on which we sail our waking hours. After all, if nothing were remembered, how could anything be recognised or acted on? Sub-conscious whispers influence present wakefulness continually. Files of memory compose sub-consciousness and, in remembrance, let you move beyond step one. Without them learning could not be increased; in their absence nothing could be done.

Top-down, embodiment involves the *association* of both metaphysical and physical parts - mind and body. The material fraction of memory involves sensory experience. *Record is struck in association with a particular configuration of neural pattern. Such pattern will naturally cross-link with other such patterns, each of which acts as a key circuit, 'tag' or physical address for its particular memory.*

Such classified association is, of course, facilitated by physically different 'reception desks' (areas of the brain) for information from different senses (e.g. eyes, ears) and homeostatic detectors (e.g. hypothalamus, medulla, pineal); and, conversely, by points of origin for the transmission of chemical and motor response to different parts of the body. In short, an individual memory is prompted by or prompts physical response; its key circuit acts as the interface by which immaterial memory either ($\downarrow$ motor) acts on or ($\uparrow$ sensory) is acted on by body and its world. Clearly, if such neural link fades, is blocked or degraded there occurs

---

[100] *figs.* 3.5 and 15.5.

relative loss of access to the associated memory. Circumstances such as dementia are a case in point. It does not mean the metaphysical impression is impaired. Indeed, it would follow that if the neural address were repaired or mind left the body altogether the memory would become retrievable. Call memory a library, a database or an information-centre. Whichever simile you use mind works, both in its informative and physical components, as parts cooperative within a single whole and towards a single goal - the satisfied survival of its user.

The latter's focus of search, access and retrieval also works by association. **Association,**[101] **whether psychological or bioelectrical, involves resonance**. On the physical side a composite address of causal nervous stimulations would suffice. Such original pattern, on re-stimulation, would address the immaterial memory; and an incalculable search formula for a memory states that *ease of access is proportional to original 'depth'* (of imprint), *quality* (positive, negative or neutral emotional 'charge'), *intensity of seeker's interest and association factors* (according to the owner's perspective) *divided by fade* (the time between imprint and retrieval). Brain is an algorithmic point of linkage, a medium between two worlds.

There are *two* main divisions of personal mnemone. The *first* is an 'historic present'. This scratchpad of 'short-term' or 'working' memory is an ever-changing read-write 'dynamic' that allows you to function from moment to moment without losing track of time, location and the course of your plans. It might be likened to the cache or working memory of a computer; it is some reverberating circuit telling me what just went leading into what's now going and may well go on.

Salient features of short-term focus are, in accordance with their perceived importance, automatically filed to a *second* division, the 'historic past' or longer-term '*RAM*' storage. Whatever pulls attention to its centre is, for that moment, relatively 'important'. The key positive factors that direct attention and therefore, equally, remembrance, are purpose, interest and pleasure; and the key negatives are shock, pain and fear. Their log *is* personal memory. Of course, as well as one-off 'hits', sheer repetitious grind impresses deeply too. In either case no particular 'long-term memory switch' is needed to translate short-term into long-term '*RAM*' files. You do not need to quote frontal cortex (the physical correlate of mind's 'third eye'), hippocampal bulge or anywhere else as an agency of transfer - simple depth of emotional intensity (including the desire to learn) or repetition does the trick. It is noted that between its birth and death a living organism's 'computer' and, therefore, its 'software/ hardware complex' is never switched off. Therefore *RAM* becomes, effectively its database of personal or 'learned' files - the record of its life experience.

---

[101] Chapter 16: How Does the Association Work?

Chapter 16: Archetype

Firstly, check the Glossary and Index for archetype.

Secondly, note the failure of Darwinism's primary mechanism, chance (on whose product natural selection secondarily acts), to create sufficient initial genetic conditions for any organism (Chapters 20 and 21). This is compounded by a complete lack of the foresight intrinsic in all codification but, especially, that involving body-plans, reproduction, sex and development (Chapters 22 to 25).

Thirdly, note the pre- and post-modern use of the term archetype. This involves (*sat*) potential identified by Natural Dialectic as memory in universal mind (*figs*. 2.6, 3.2, 3.3). Given the failure of material mechanism such immaterial mechanism, natural archetype, is reasonably proposed as the provider/ controller of *all* basic forms, letters or notes of physical creation (see Index: alphabet and music; also Chapters 11: Matter's Holy Ghost and 19: Conceptual Biology).

Finally, this book states its position distinctly. It does not, therefore, entertain the large, confusing and confused philosophical discussion that surrounds what might be construed as parallels of archetype. These include Platonic ideas (εἴδη or ἰδέαι) and his 'finite set of changeless, natural forms'; also Aristotelian 'potential' (δύναμις), energetic process (ἐνέργεια) and end-products called entelechies (ἐντελέχειαι).

Natural Dialectic calls psychosomatic *archetype* in its physical aspect *potential matter* and in its psychological *typical mnemone*.

The Typical Mnemone

If, on the one hand, you believe neural networks substantiate sensation, memory, thought and consciousness then 'typical mnemone'[102] is a brisk irrelevance. If you equate morphogenesis with molecular biology alone then 'memory man' is superfluous nonsense. You might well, if *vis medicatrix* is other than chemical, deny its clinical influence. What cannot be physically explored does not medically exist. You state that the wrong questions are being asked on the wrong premises and obviously, therefore, the answers will be senseless; medical school will award and industry, in terms of pharmaceuticals, certainly applaud you.

Materialism ridicules the idea of natural informative structure, a metaphysical 'cloud' or mnemonic database called archetype. Where, however, does its absence leave us? Mind not molecule anticipates. No

---

[102] synonym, psychologically phrased, of biological archetype and, physically phrased, of potential matter.

doubt, automatic chemistry in cells preserves, repairs, reproduces and precisely orchestrates developmental patterns. For example, cells 'sense' gradients of chemicals like safety valves 'sense' steam - but what molecular 'instinct' for configurative survival sets such a bio-system up? Why should any coherent system of construction self-construct from incoherent molecules? **How could encoded protocols, each step irrelevant without the rest, gradually appear?** For what reason might a step, any step, emerge at random and then wait - since only with full muster registered can any step coordinate in profitable work? Don't stop at metabolic pathways leading to a target molecule. What reason is there for each stage of a development inexorably leading to anticipated adult, reproductive form?[103] Is such intensely informative program squeezed from mindless twists of serendipity? If chance is an irrational answer it's still the best materialism will have concerning source of information. *Might, however, immaterial elements conspire with the material components of biology; might rational mind conceive the mechanisms that can serve its purposes; and subconscious control (the exercise of a mnemonic archetype) inform all forms of life?*

Thus, as regards this issue, is not the fundamental error of naturalistic methodology to deny any element or force dissociated from matter? Perhaps the indulgence that it's 'all in genes' or 'simply chemical' is a professional deformation and a philosophical delusion too. If you consider information, reason and mind's other faculties to be material-only entities when, in fact despite consideration, they are not - what of materialism then? Metaphysic that cannot be tested physically lies at the root of materialistic (and therefore scientific) denial of an archetypal mind-matter linkage; but if, say, immaterial subjectivity *is* real and if your interest is finding out the actual truth, then tunes will need to change. No nonsense! What follows makes sense, in this light, of your internal metaphysic.

From a *top-down* point of view the 'general field' of cosmos is composed of many spheres of influence each with its kind of excitation and results. Indeed, field theory constitutes the summit of materialism. Each type preordains a pattern of activity; thus, controlling character of particle and force, it organizes matter. Modern physics (but not yet biology) explains its world in terms of them. Moreover non-conscious, energetic nuclei, called suns or stars, radiate great fields of influence. Why, therefore, should a nucleus of mind not also radiate its influence consciously? Why shouldn't even memories unconsciously exert their immaterial force? Body is your personal abode but, as body's elements are chemical, what element is psychologically natural in you? Is there anything of universal mind; and, if so, how do its files of programs link with an organism's 'psycho-chemistry'?

---

[103] Chapter 24: The Reproductive Archetype etc

We sail, it was noted, 'an ocean of context, gist, detail, currents and association'. Surface consciousness is tossed although, of course, the captain tries to keep course on an even keel. This elemental ocean is composed of memories of shallower and deep effect. Shallow rips and deeper tows; undercurrents that affect the transient, surface waves of circumstance. One more analogy illustrates the two major kinds of memory. Carrier wave with input modulation. That is, for a given channel (e.g. human, oak or bee) a specific supplement (such as yours, mine or each bee's) accretes by memorised events. Thus a personal program is developed; a unique drama is composed. General, archetypal instinct (part of the broadcast called 'typical mnemone') is, although innate, modulated by habits learned throughout a personal lifetime. The involuntary 'will' of such subconscious composite is known as 'force of habit'.

It bears emphasis that subconscious memory is uncreative. Its record is, of itself, passive, purposeless and fixed. *And yet its reflex is, like that of any technology, originally the product of creative purpose; and such purpose is the motive behind the otherwise motiveless government of biological bodies.* **It is the origin of meaning and rationale in biology.** Let's take a peep, therefore, at the distinct rationale behind the psychosomatic nature of *H. sapiens* - you. It has already been identified (*fig.* 15.5) as a typical mnemone called *H. archetypalis*.

H. archetypalis, the Image of Man

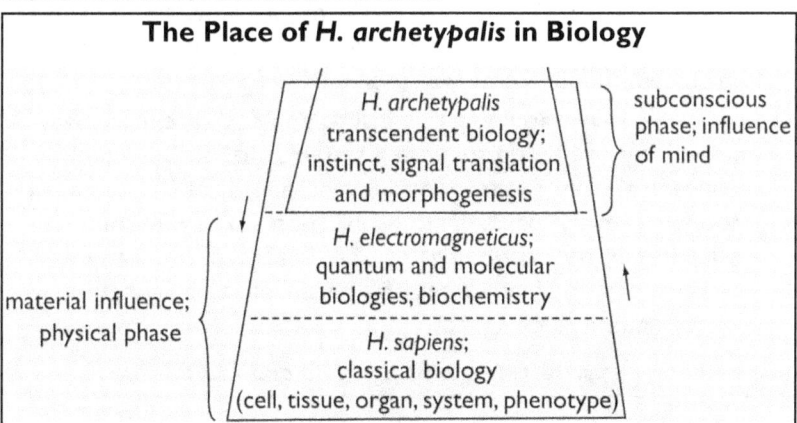

Check *fig.* 15.3. Metaphysical archetypes such as *H. archetypalis*, 'the image of man' or 'memory man' are *not* (because they involve the immaterial, informative coordinate) the same as any taxonomic definition of species. In the dialectical view information is broadcast from archetypal level to its biological correlate. Such local receivers are cells or multicellular bodies. Generic program acts as a homeostatic regulator, a reference-point or 'norm' that integrates purposive developmental sequences, chemical algorithms (such as metabolic pathways or mitosis) and the morphogenetic disposition of pre-coded building supplies for growth, maintenance and repair. An archetype is, in effect, a lawful program. Each one represents a statute (the organism as a whole) with its bundle of clauses (sub-routines). Informative principles or laws derive from a level above. The level above matter is mind - in this case a subconscious store of logic called an archetypal memory. As your own behavioural patterns derive from a context of memories, so do those of nature. Indeed, are not your instincts and their body part and parcel of a very natural pattern, the physical expression of a frame of reference, a metaphysical criterion called **deep and natural memory?**

Mind naturally encodes information. Subconscious mind consists of coded files that we call memories. It is suggested that passive, archetypal files inform, at the lowest level of existence, material energy. Every cell partakes, as an antenna, of archetypal as well as electromagnetic guidance. This couple acts as a regime within which correlated sub-routines correspond to, entrain and control the form and function of physical counterparts. Their program, in conjunction with 'read-only' material script called *DNA*, generates and maintains physical form; but why, any more than a program's binary code resembles what it generates on screen, should archetype resemble bio-form?

If you think 'archetype' is just a 'cop-out' in explaining how things are then I suppose a systems analyst must think conceptual plans for any working mechanism are 'cop-outs' too. However, philosophical objection to a program's purpose in no way mitigates the impact of its natural possibility and, if an immaterial element of information exists, natural fact.

fig. 16.1. (see also figs. 15.3 and 15.5)

A typical mnemone is composed of *three* main sub-routines; or, if a protocol is a standard procedure for regulating the transmission of data between two end-points, three linked protocols. Together translational, instinctive and morphogenetic programs comprise an individual's archetype. If, as in the case of plants, fungi etc., there is no conscious component then the translator element is reduced from nervous to chemical (e.g. hormonal) messaging alone. *H. electromagneticus* has, in the human case, been identified as the physical side of message-exchange and *H. archetypalis*, with its routines, the metaphysical correlate.

Two important aspects of archetype are marked. **The *first* is that, just as the genetic 'book of life' is found as a nuclear genome in every cell, so every cell accompanies its typical mnemone.** The *second* is that such generic 'broadcast' survives the death of any individual cell or body. Genetic material is passed from generation to generation with its accompaniment, a cell; the physical medium of transmission is, of course, a single fertilized egg or product of asexual division. The metaphysical mnemone is not thus passed; you might liken it to a generic permanence of which any ephemeral cell is the attuned receiver; destruction of local apparatus does not affect the broadcast. Nor can archetype become extinct.

In this view typical mnemone is the metaphysical blueprint for a body; it is body-linked at the level of a cell. Cell not soul has biological memory. Each one also contains genetic code whose switching systems produce appropriate materials for its own sustenance and, where applicable, communication with the rest of the multicellular operation called a body. In the case of a hologram different pixels appear differently, suppress or lock out parts of the whole picture but still contain its whole potential; similarly, in the mnemonic case, 'a pixel of the archetypal hologram' interacts to excite or suppress an appropriate cellular routine from the whole program of its organism's life-cycle. Possible mechanisms of mnemonic/ genetic interaction are suggested later in this chapter.

Thus a cell or body's typical mnemone is seen as morphogenetic 'intelligence' that moulds substance and thereby coordinates the production of functional shape - shape whose function relates specifically and accurately to its purpose. **That is to say, all cells partake of memory.** *DNA*'s **material book of life is correlated with a book of immaterial information in the form of archetype. Each cell responds, as well as to its circumstance, to its subconscious archetype.** For example, human cells each resonate with *H. archetypalis* and each cat cell with *Felis archetypalis*. So, within the whole picture, does each tissue or organ made of cells. Each pixel partakes of and corresponds to the whole. Just as a cell can express only its own relevant fraction of the genome, so it resonates with its relevant fraction of archetype. Individual sub-routines or linked groups of them are accessed within the framework of a coherent whole. In this way the archetype is a complete context, a governing template whose different sub-routines are accessed as parts of a coherent master program. Thus body cells and unicellular organisms alike 'know' their behavioural patterns. **This knowledge is not conscious; it is unwitting as sub-conscious means and is.** *But all cells have <u>dormant</u> mind that interacts with their chemistry, chemistry that includes physical correlates of their 'dormant intelligence' in the form of information systems (cell signalling, hormones, nerves etc.) and a database of DNA-written coils (or solenoids?).* Nor is this passive form of mind less separate from atomic action than a dye from its cloth or a solute from solvent. It is closer, close as broadcast is to the picture on its TV screen. Closer than a nanometre, mind substantiates material expression. It is, although materialism misses its dimension, a control centre.

Signal Translation

Life's key aspect is exchange of information. A great deal of such exchange occurs in our bodies without us knowing about it. It is possible that, except in the case of some animals, no exchange of information is ever consummated in awareness. But this is not to deny any biological unit, including a 'dormant' plant, fungus or bacterium, sub-consciousness. *Each single or multi-celled type of organism with its appropriate natural protocols of information exchange senses the world, and responds to it, through the structure of its own chemistry and archetypal memory, that is, in its own typically programmed, automatic way.*

It is reasonable, although the exact systems flow-charts are unknown, to suggest that this memory is programmed, like a computer (also known as a mind machine), to achieve specific ends. And, like a physical counterpart known as *DNA*, to suggest that its size, complexity, reasons and logic inhabit every cell of every body. It has been has been suggested (*figs*. 15.5 and 6) that this subconscious databank comprises three cooperative subroutines. All inward traffic across the psychosomatic border (*PSI* or Synchromesh 2) passes from quantum level (patterns of charge and light) to a metaphysical **signal translation** routine. This modulates the input for issue to either or both of two subroutines called **instinct** and the **morphogene**. In the case of innervated organisms sensory data identified for translation into conscious experience is also passed through an extension, **an additional subroutine of signal translation** for passage across the interface called Synchromesh 1. And, of course, for outward motor traffic the reverse flow occurs.

We know that translation from nervous impulse to conscious experience occurs but experimental neuroscience has not considered a hierarchical process that involves metaphysial mechanisms. We do not know how but the process certainly involves governance, that is, a codified program of physical and physical cooperation. It therefore demands corresponding structural and functional unity. Only in this way can the coherent propagation of hierarchically organised signals occur.

Thus the first sub-routine of archetypal mind involves the translation of signal information between mind and body.

Instinct

Default mode. The *second* subconscious faculty involves routine - *instinct*. Instinct is a term that covers, in broad terms, reflex actions that are neither taught nor acquired. Such 'auto-pilot' is always purposive in scope and involves behavioural strategies of varying complexity.

Purpose, strategy and tactic *anticipate* an outcome. Whether it is timely switching that supplies a metabolic pathway, an organ's cyclic function or whole body's integrative purpose to 'live on', all ingredients of life involve, as well as the requisite biological tools, conceptual know-how and, therefore, mind-behind. *The issue is not only operation but also the origin of programs of behaviour, that is, of information.*

Bottom-up genes make and modify all instinct. Genes alone make a spider spin its web, generate complex courtship rituals and, at root, will even let you think! Maybe, however, you don't buy half-truths. Maybe thought and instinct aren't just functions of your *DNA* and a magnificent but as-yet unexplained development, the brain; maybe that, *top-down*, there's a rational, prior place for immaterial mind. Cell, organ and organism are like aerials that resonate with information. And, if archetypal resonance rings possible, why not two or three mnemonic channels (*fig.* 15.5)? **DNA sequences are already known as a chemical data storage system that is incredibly responsive, specific and precise. Could not an immediate, radio system with its conceptual broadcast, archetype, be prior, proactive and equally precise?** Doesn't an essential link, mind-body psychosomasis, engage communication of a subtle, rapid, highly-programmed kind?

A *top-down* point already made needs hammering home. Instinct is innate and unlearned ingenuity; it is problem-solving often in sophisticated ways. Spiders' networks shout the world is not material alone. As well as web-life's tricks instinctual memories include call-signs, food-finding, hunting strategies, migratory navigation and complex, distinct breeding patterns such as courtship, home-building, nursing and rearing; and, when life bodes ill, what about a creature's instinct for effective natural medicines (zoopharmacognosy)? Instincts are crucial and ubiquitous. **Their evolution is, of course, a Darwinian black box because their aspirational exercise isn't physical alone.** Indeed, at root, they're immaterial. Metaphysical. You need a framework, a broader box that's outside but inclusive of the biochemistry. Within this box the instinctive aspect of mnemone works as an ancient, indelible and often complex inscription. **It is a psychological datum, immaterial software, an 'under-writing' of intelligence, a common theme whose various strands are woven in the warp of archetypal memory. Auto-pilot is, therefore, an aspect of a universal mind.**

Of course, any program's expedition involves physical kit. After all, the life's on earth. In case of instinct genetic, hormonal and other apparatus trips the 'job' - but is job derived from them or do they serve its plan? If so, whither did this plan, a pattern of behaviour, evolve? There is nowhere evidence that an instinct evolves through stages of inferior skill or inadequate function. **Natural Dialectic therefore proposes that instinct, because it is the cause of such an obvious**

effect as behavioural goals, is our closest sensation of the immanent, numinous presence of archetypal memory. In the case of gradual neo-Darwinian evolution, on the other hand, instinct and associated physiologies are, *as if yet another black box*, incoherently explained.

Morphogene

Is there any part of biology that does not exhibit breath-taking engineering, design and innovation in order to achieve specific ends? Nature (though perhaps not wholly material) is, in fact, a consummate chemist, physicist and engineer. Well-educated men, armed with 'inorganic' technologies, haves now come lately to the natural table and demonstrate appreciation with an exercise dubbed 'biomimicry'. They take inspiration from the codified ideas behind ubiquitous bio-technologies. Who in history suspected every cell of every body was codified? And that translation mechanisms were in place to transform a passive, 1-d chemical expression of code into the 3-d shapes of an automated factory; and all the other shapes that compose myriad intricate devices and efficient mechanisms every body's system as a whole employs? Whence, though, logically do *code* and *program* come from? Mind or chance? Chance, guess Nobel laureates. And yet, since mind's action cannot be directly observed, the notion of morphogenic, archetypal code has been ignored. So Natural Dialectic forces logical conclusions and suggest some answers that these pages have already started to propose. *Figs.* 15.5-6 act as a primer here.

DNA no more resembles bodies than airwaves resemble the picture on your TV screen. Receiver noise, buzz, whistles or an aggregate of purposeless 'interference' is unacceptable. You tune in. And, just as your TV's antenna picks up broadcast, what wireless code informs the operation of a cell's *DNA* and its morphogenetic (shape-making) capabilities? In other words, *DNA* certainly informs the development and sustenance of 3-d shapes but what 'oversees' the complicated, automated, cybernetic operations of its nano-biochemistry? It is suggested that, as regards the construction of complex bio-shapes, the broadcaster is a subroutine of archetype called morphogene.

Morphogene is a body's *vis medicatrix*. It is a type of organism's 'ideal health', a homeostatic norm against which aspects of a physical expression of that type are constantly monitored. Input crosses the *PSI* border, is routinely processed and reflex adjustment signalled in return. Morphogene is like a system's 'signalman' that triggers switches. It is also like *CAD* software that an engineer employs, an image-source able to project and respond to 3-d shape instructions. Though still relatively crudely, such engineers use not pixels but 'voxels' (3-d datapoints containing multiple instructions for colour, texture, material and so on) to 'print' planned objects from a 3-d printer or scan, using f*MRI* machines, specific activity throughout the brain. Could nature have again preceded man, this time producing holographic 'voxellated' bodies?

In this case, while *instinct is the behavioural/ functional subset* of archetypal memory, *morphogene is the 'cache' of structural order.* ***Morphogene for morphogenesis is realised as the third, base subroutine of a typical mnemone, an aspect of mind closely associated with electrochemical function and our earthly coil (if not DNA coil too).***

Any organising template exists independent of the content that it frames. A machine is made of parts but coherence, integrity and purpose are its frame. *The whole is greater than the sum of its parts.* And, as Chapters 5 and 6 show, that greater part is mind. Similarly, chemicals and their interactions constitute the divisible parts of an organism but they do not sum to its structural and functional whole - the coherence, integrity and purpose reflected in its shape. The role of a morphogene is the unification, coherence and thereby governance, in terms of resonance with electrical patterns, of a life-unit, an organism. Biological *H. sapiens* is, as well as the phenotype of a genotype, the gross expression of electrical interactions that constitute *H. electromagneticus*. Each factor is physical, inanimate and essentially insensitive. Their reflex patterns are a gross reflection of an independent organiser - the subtle, sub-conscious, metaphysical infrastructure called *H. archetypalis*.

A morphogene is the generic form of storage for a biological program. It is an organism's shape and working dynamic in principle; and genetic practice supplies the detailed, individual differences of phenotype-on-theme. Just as mind responds to sense perceptions from its environment so a cell responds, by way of electrodynamic signals, with its physico-chemical context. The archetypal morphogene operates like a 'control satellite' off which specific, local *DNA* and other molecular signals are bounced; it is the 'brain' whose subconscious map coordinates those references and thereby integrates the operational hardware of a unit called either cell or, cohesively, multicellular body. Your own 'brain', replete with sub-routines, orchestrates *H. sapiens* as a conceptual whole through the medium of *H. electromagneticus*. It therefore includes a developmental sequence with, as a final 'still', the adult form.

Archetypal program is a magnet, attractor or prior organiser. Circumstances push but goal-oriented program draws a process forward in time. Thus the real psychosomatic question asks precisely how connection is made between informative mind and forces, atoms and molecules of the phenotypic composition that confronts you in the mirror - your body. A simple magnetic field exerts influence; it organises iron filings round the magnet. A mind-field tries to organise the world to fit its own desires; and both conscious and sub-conscious patterns can exert direct influence over heart rate, breathing, body chemistry etc. In a similar way, as an electromagnetic broadcast is transduced into pictures by a TV set, it was suggested that a 'template of health' exerts influence over physical agents such as the genetic code, particular materials and their expression, phenotype. No single element will act within the format of a coherent program without this animating influence. *DNA* or protein outside the context of a living cell is impotent and inanimate. The bioelectric and morphogenetic constituents of its composition no longer resonate. The molecular level has lost guidance and succumbs to entropy; the organism's dead.

How, then, to sum this up?

Bottom-up, developmental/ morphogenetic information is construed in reductionist terms as the product of chemical activity accidentally coordinated piecemeal, aimlessly, over aeons.[104]

Top-down such view of information-generation is deficient. A *hierarchical* structure of control[105] would suggest a step *up* from the molecular structure which creates an aerial for the controls of an electrodynamic level. Such quantum biochemistry it names in humans, for convenience, *H. electromagneticus*. Such a wireless body with its endogenous flows of current, fields and associated photons, is in turn a medium - as are electromagnetic waves for radio - for the primary, archetypal broadcast. Conversely, if you like, steps *down* discharge a body's working shapes.

H. electromagneticus

A metaphysical point of reference, *H. archetypalis*, is physically realised by its electromagnetic correlate dubbed, dialectically, *H. electromagneticus*.

Hints have been around awhile. Faraday showed that a changing magnetic field is accompanied by an electrical field. Maxwell showed the reverse and that interaction between the two generates electromagnetic waves (and, therefore, particles). All electrical fields therefore have a magnetic component. And the fact is that all living cells generate their own fields complemented by ionic motions and charged interactions. It might therefore, given this electro-dynamism, be reasonable to describe a cell as an electromagnetic unit.

In 1935 Dr. H. S. Burr, Professor of Neuroanatomy at Yale University, and Dr. F. S. C. Northrop established that all 'living matter' from a slime-mould to an elephant, from a seed to a human being, is surrounded and controlled by electrical fields. Indeed, every cell pumps ions to maintain a healthy, electrical potential across its lipid membranes.

Electromagnetism involves photons. Is light not in fact the language of life's cells? Fritz Popp, founder of the International Institute of Biophysics at Neuss, Germany, proposes that low-level light emissions

---

[104] Both with respect to metabolic dynamics and (Chapter 25: *passim*) development.
[105] *fig.* 16.1.

are a common property of all cells. Such weak luminescence ranges from thermal (infra-red) to ultra-violet. Not only do electrochemical forces across cell membranes help control their permeability. Colleague Marc Bischof believes that weak, coherent e-m fields combine to regulate not only the cell's surface but its internal members. Thus, correlated with the positions, densities and movements of electrons, a 'signal web of light' might harmonise cooperation of organelles with each other and with chemicals throughout the cell. Moreover cellular cytoskeleton's coiled, semi-conducting filaments and tubules, whose other jobs include structural support and transport track-ways, compose a network for the conduction of charge and generation of electromagnetic fields. With this, as well as body heat, living organisms faintly glow.

These 'wires' of electrodynamic propagation may power up other structures such as protein alpha-helices and coiled/ solenoidal *DNA*. Indeed, Bischof postulates that *DNA* pulses as a 'light pump', that is, as if an aerial both emitting and absorbing light. As brain-waves control the *CNS* such a mobile web of light would, with its various frequencies, control cellular operations. Cannot radio carrier waves incorporate signals we tune into and call programs? Similarly, might light signals control not only internal operations but also employ cytoskeletal components, where they attach to the surface membrane, to form electrical and possibly 'fibre optic' channels to the exterior matrix? This extra-cellular matrix (*ECM*) provides a medium for body-wide bioelectrical linkage. It is, therefore, possible that Burr's L- (for life) fields coincide with the description of cells as electromagnetic units and so with the idea of a mass-free lattice identified in the last chapter (*fig.* 15.3) as a quantum-level sheath called, for organisms in general, *X. electromagneticus*.

To summarise, all matter, including biological matter, involves both structural/ particulate and dynamic/ vibratory elements. Quantum physicists concentrate on the latter; chemists intimately recognise them both. Quantum biology is, however, scarcely recognised. Nevertheless 'acoustic resonance' or 'resonant association' are already known to play a crucial part in such biological technology as photosynthesis and sensation (visual, olfactory and so on). According to *fig.* 16.1 this should not, dialectically, surprise us. **Simply, *H. electromagneticus* is, composed of billions of subatomic frequencies, the human shape of quantum biology.** And, as resonance is part of molecular dynamics, so its weightless waves and bosons are identified as a prime candidate for the physical side of the psychosomatic exchange of information. It is, therefore, suggested that psychosomatic exchange occurs between a typical mnemone (with its three parts of signal translation, instinct and morphogene) and molecular structures through the medium of highly organised electrical, electromagnetic and quantum vibrational patterns. These patterns are, as the active site of an enzyme is framed by the conformation of its protein chain, framed by gross matter; but, inwardly and equally, from archetypal program. **In other words, just as quantum effects are increasingly harnessed by industrial *IT* to perform specific tasks, so biological technology, including the expression of code and metabolism, will be found to employ them - ubiquitously.**

Thus the mode of mnemone/ body interaction is seen as a dynamic, coherent association of two anatomies, wired and wireless. As quantum world rules gross appearances so wireless orchestrates the wired. In your case fixed psychological harmonics called memories interact with body through the medium of photon, electron and atomic bond vibrations. In such association atoms and molecules of your visible body, *H. sapiens*, act as oscillating resonators. *Thus interlocked, Janus-like psychosomatic bodies can be seen as sub-conscious information, in the form of memory, acting in concert with complex electrical fields. Sub-conscious mind is the hand, electricity the glove.* **There is nothing in biology to suggest such concert does not exist, indeed there is much to support it.**

Synchromesh 2 - Psychosomasis

Synchromesh 1 dealt with personal and typical mnemones and the way they mesh with each other and with conscious awareness 'above'.

Synchromesh 2 deals with the way psychological meshes with physical 'below'. It concerns, in this case, the junction of all three aspects of typical mnemone with the physical side of the psychosomatic interface - an electrodynamic, resonant medium called variously your 'network of light' or, aforementioned, *H. electromagneticus*. Psychosomasis is simply a word that identifies the process of translation, a transduction of information between *H. archetypalis* and *H. electromagneticus*.

There is no doubt, the universe is in vibration. The cosmic transmission of information and energy is, at root, vibratory. Ordered oscillation is called harmony; harmony is the grammar of music and music is a universal language. The theory of music is implicit in any recital. Could it transpire that explicit order of a cosmic recital is the product of implicit notation? Could its excitement represent harmonic code?

Why is the Master Analogy of Natural Dialectic[106] music? Cymatics (the study of the effect of waves on matter) has found striking evidence of the patterns that sound can induce on large-scale, let alone quantum-sized, matter. For example, by 1787 Ernst Chladni had drawn a violin bow against a thin metal plate covered with dry particles of dry sand and reported the resonant effect of given frequencies of standing wave. He watched, in a direct and obvious example of morphogenesis, implicit energy inform explicit shape.

---

[106] Chapter 6: Music.

More recently Swiss Hans Jenny invented a tonoscope which, using crystal oscillators to precisely vibrate plates or membranes, converts vibrations to patterns. Thus sounds uttered into a microphone yield visual representations on a screen. When vowels of ancient Hebrew and Sanskrit (but not other languages) were pronounced the vibrated particles took the shape of their written symbols; and when the Hindu sacred syllable for primordial creation, *Om*, is correctly intoned it produces a circle (representative of the Infinite Void from which all things issue). This progressively fills with concentric (nested) squares and triangles. Finally, as the last humming traces of 'm' fade, a *yantra* or meditational *mandala* takes shape. A *mandala* is an archetypal symbol, a formal, geometrical expression of sacred harmony, structured vibration and the orderly, dynamic basis from which a solid-state, physical universe derives. The diagram called a *yantra mandala* symbolizes an entire, sound-drummed cosmos. It is a perfectly symmetrical picture radiated from the Sound of Silence, from the Nature of Nothing. **Such harmonic resonance is foundational**. It raises strong possibilities, always avowed by the mystic, that sounds and names have internal properties of their own. Such 'intonations' are a fascinating reminder of, at physical level, the connections between sound, geometry and the development of form.

Resonance, whose orderly aspect is characterised by an analogy with music, is the tendency of a body or system to oscillate with larger amplitude when disturbed by the same frequencies as its own natural ones. *It therefore intimately involves the vibratory transfer of energy.* Such transfer is an integral part of all vibratory systems wherein waves interfere with/ destroy or cohere/ amplify each other. Science is familiar, for example, with nuclear magnetic, electromagnetic, electron spin, acoustic and mechanical forms of resonance. Acoustic (musical) resonance involves the sympathetic vibration of stretched strings or air in pipes; thus it is easily understood that, as well as tuned circuits, quartz crystals and so on, musical instruments are resonators. Indeed, the motion of waveforms is closely associated with harmonics. Sound waves are governed by the law of harmonic relationship whereby notes of the right frequency combine into chords, and chords and notes in time into harmonies. Resonance occurs when one note or chord vibrates in harmony with components of notes or chords in a different octave. One body vibrates in sympathy with another. A similar sympathy can occur in the oscillations of electrical and mechanical phenomena. Energy is amplified and transferred by resonance and attunement. Common examples of electromagnetic resonance include tuning a transmitter/ receiver and photo-electric initiation of the photosynthetic process, that is, the first step in life's chemistry.

For quantum physics matter is certain vibratory frequencies of energy; and, from a dialectical point of view, it is simply stresses, strains or tensions in the medium of their absence, that is, nothingness. **There is, however, nothing random in a highly orderly creation derived from first acoustic principles.** The universe is, physic agrees, a kind of machine finely-tuned according to about twenty fundamental physical constants, archetypes or notes. Could these notes constitute the instruments that shape such chords as elicit an electron, proton, atom, molecule and thereby, vastly larger and more orchestrated, you? *Like organ pipes such instrumental archetypes would be the shapes of resonance through which the notes of cosmos can appear.* **The idea of musical archetypes as a source of physical order may alienate one-tiered materialism but its synergy dances to the tune of Natural Dialectic's Central Metaphor.**

What is being proposed is, in other words, vibration prescribed by standing waves. Such information is, essentially, musical. The resonance patterns that a fundamental string can support would give rise to the properties of an elementary particle such as an electron (*cf.* the harmonic equation for standing and probability waves). Motion, energy, mass, force, charge, and so on are determined, in effect, by the precise, vibrant events that a particle's internal strings execute in symphony. This 'un-struck music', a particular 'chime' or 'intonation', is its hallmark. Just as a spectroscopic analyst understands that each sort of molecule emits its own 'fingerprint', 'signature' or 'bar-coded *PIN*' by which it is uniquely recognised, so different 'chimes' show as the forces and particles of nature. Waves both initiate and lay their fingerprints on form; and, as with musical theory, the rules authorise permutations like chords rather than discords - a system fundamentally harmonic and not cacophonic.

Are atoms harmonic oscillators? Are quantum harmonics reflected in larger visible dimensions? Are crystals an example of this coordination? The answer in each case is yes; and the effects of harmonic vibration on, for example, jets of gas and flame in air are dramatic. You can also make 'electromagnetic plants' whose flowery patterns arise in a dish of ferrofluid placed over a single iron-core AC coil; the fluid accords with lines of the field. And it is significant that Chladni's figures often imitate familiar organic patterns that we see in nature and, especially, biological structures. Pattern clearly relates to frequency of cycle. The ancient architecture of leaf, fern, wood grain and many other kinds of morph appear. Music is frozen. Orphic sound and archetypal, Pythagorean geometry are thus combined.

In this view inner action naturally precedes and governs outer structure. Visible expression of energy is the inverse of vibratory pattern; visible mirrors invisible. *Between the 'dead' ground of material shapes course channels of energy - the invisible rules.* **Oscillation between polarities, cycles, vibratory rhythms, the interrelationship of waveforms and resonance are at the heart of Natural Dialectic.**

How Does the Connection Work?

What, therefore, is the psychosomatic grail? What earths mind to matter? How, mind > body/ body > mind, does the connective 'ligand' work?

Rhythmic beat, coherence, harmony - their influence draws the whole world into order. It chimes bells and dances. Music vibrates shapes and its songs, each in their own way, feel right. Thus natural music of our cosmos is expressed. *The heart of morphogenesis is buried deep in resonance.* **A key phrase in the suggested explanation for the wireless, psychosomatic transfer of information between subconscious structures of the mind and the physical plane is 'resonant association'.**

Internet, television, wireless - energetic frequencies can carry information; antennae are tuned for resonant association with specific frequencies; and, since the emission/ absorption of electromagnetic radiation by atoms and molecules peaks at each one's natural frequency, they act as aerials. **It is thus suggested that psychosomasis between mind and body is, with appropriate devices in place, common and continual.** On the *mind's side* of the interface the device is called a mnemone; on the *body's* it occurs when the appropriate molecular/ cellular bioelectrical configurations are in place. Synergetic resonance occurs whenever thought affects nervous/ muscular activity or sense data is transferred for symbolic appreciation in mind. The logical postulate for such informative transfer of harmonic codes is the most subtle, least physical substance - photons.

An argument was developed suggesting that the principle of psychosomasis[107] was, mind ⬄ body, resonant translation at an electrodynamic/ quantum level. **In this respect, regarding the relationship of nuclear archetype with electromagnetic configuration, resonance sources communication; and its agent, matter-in-principle, is vibrant, quantum forms.**

Thus, transduction is the game's name. A typical mnemone, using signal translation or the morphogene, mediates both conscious and subconscious transmission. Archetypal broadcast is 'grounded' by a multiplex composed of photons, electronic dispositions in transmembrane voltage fields, cytoskeleton or large, molecular configurations including *DNA* coils. As physical so also, given appropriate equipment, psycho-physical transduction occurs; it exchanges patterns between mind and body worlds. *The modus operandi of psychosomatic broadcast is therefore, in a word, attunement. Resonance.* It involves transduction between recorded information (memory) and physical energy.

In short, mind is linked to matter by a wireless anatomy whose instrument is resonant association.

---

[107] Chapter 8: Principles of a Unified Theory of Matter/ Chapter 15: Psychosomatic Linkage/ Chapter 16: Psychosomasis.

Chapter 17: Caduceus

Faced with an opinionated wall of 'no!', you have to keep on asking. Is there an immaterial element, information? Not material vehicles that carry information passively - speech, inscriptions, drawings and so on - but an immaterial driver. Not brain or nerves but metaphysic passive (memory) and active (consciousness, intelligence and thought). For atheism's secularity the idea is heretical. Philosophies, mind-sets and scientific paradigms depend on 'no!'

If the answer's 'yes!', however, then a link exists. Mind and material dimensions must be joined. A psychosomatic *PSI* needs be identified. Here, therefore, we try again.

The Logic of Embodiment

classical/ quantum matter	*Potential Matter*
classical/ quantum biologies	*Conceptual/ Archetypal Biology*
outworking	*Plan/ Morphogene*
informative agents	*Informative Source*
↓ *H. sapiens*	*H. electromagneticus* ↑
classical biology	*quantum biology*
gross constraint	*subtle motions*
fixture/ container	*biochemistry/ drive*
structure/ anatomy	*physiology/ function*
external appearance	*internal informants*
informed product	*informant messaging*
seeming sameness	*rapid (inter-)actions*

Is there logic of embodiment or is there the embodiment of logic, of functional, dialectical logic? What is the purpose of a body?

Bottom-up, there's none. No logic, no design. The business of bodies is a case of chance-originated configuration constrained by survivability.

Moreover, science seems as yet to have scarcely grasped the central importance of quantum physics in biology; but, no doubt, within five minutes of such realisation some quantum theory of that 'bio-fact', evolution, will have been evolved!

Top-down, however, the quantum biology for which *H. electromagneticus* stands both precedes and generates the level of gross forms that we study.[108] **Its 'phantom' will describe the ways, behind her staid and overt manners, that nature uses wild and subtle tricks; it will track the medium from which all classical appearances logically and,**

---

[108] *figs.* 2.5, 3.3, 5.1, 15.2, 15.3 and 16.1.

in ways which the last chapter began to describe, orderly emerge. Suppose, therefore, that hierarchical, *top-down* construction exists. Material order's crystallised around prior plan. Archetype informs a quantum level which in turn informs gross, physical projection; thus, within a reproductively enabled bio-form, an individual life can undergo adventure on earth's plane. From this angle an incarnation is not construed as an unusual, essentially purposeless arrangement of sub-atomic particles. *On the contrary, from molecule through cell and organ to whole construction it must work as intended.*

Are you not, in this view, an incarnation of supremely functional logic? Is your body not a brilliant birthright? Put this prosaically. Do you express ideas with symbols? Don't you use fluid, energetic code called speech or fix your language into written words? Biological ideas are, similarly, expressed through symbols. Their physical expression is genetic code. This language is fixed on 'ink and parchment' known as *DNA*. A 'book of life' is stored on a nuclear database. **This cellular database is, in short, a symbolic expression of ideas.**

Core Principles

The potential that precedes any bio-systematic construction or activity is its plan. Information. Messages inform, organs are informed. *The remit is survival.* Survival in the present involves energy metabolism (with associated organs and organelles) to 'finance' sensation and action. Bodies resist (by dynamic equilibrium called homeostasis) but eventually succumb to the universal process of entropy. *They die; thus survival into an indefinite future involves reproduction, development and their associated apparatus.* The whole business is thoroughly informed, coded and works to plan. No chance.

From a bottom-up, materialistic point of view this is nonsense. Mind, a material outcrop, is dependent on matter and not *vice versa*. Neither logic nor symbolism can appear in a structure thrown up piecemeal by chance. There is, therefore, clearly no logic in the construction of bio-logical bodies. Atoms have no mind. Where there is no mind there is no logic. Accident, in the form of evolutionary mutation and natural selection, is the best you can suppose. **Although any explanation of the evolution of protein synthesis, mitosis, sex and so on collects Darwinian black boxes, you've no alternative.** Therefore, a compendium of *interim* 'just-so' hypotheses has been devised. For example, developmental embodiment is supposed to follow a course of evolution that may have involved the disruption of genetic switches, neoteny, mathematical transformations (as per D'Arcy Thompson), mathematical attractors, arcane chaos theories and so on. *Uncertainty about the evolution of morphogenesis is, however, but a subset of overriding certainty that such 'progressive' process must have happened unintentionally.* Naturalistic methodology and, with it, scientific orthodoxy, thus inescapably demands no less. Perhaps it's perilous to break such barrier and disagree!

Top-down there is complete agreement that physical is physical. A body, including its intricate hierarchies of switching and control, is *per se* entirely physical - as physical as any technology. Its entire, informed structure will eventually be understood in terms of molecular, cellular and higher level interactions. It is, however, the *arrangement* quite as much as the substance of components that falls under scrutiny; and the logic of Natural Dialectic suggests embodiment reflects a gradient of creation as it falls, from top to bottom, through its basic dipoles - information and energy. **The construction of human body, by reflecting the conscio-material gradient, also reflects the construction of cosmos. This is why it is said, accurately, that 'microcosm reflects macrocosm' and 'the universe is in man'**. Thus, recruiting the device of *H. archetypalis*, we can further probe how the logic of embodiment works.

Morphogenic Crystals

Top-down there's more to things than meets the eye - or any scientific instrument. Mind, even in its 'solid' form, is metaphysical. It is an 'information cloud', an 'immaterial field'. Memories are particular, persistent patterns/ constructions in that field; and archetypal memory as much a broadcast as an object which, like a film or computational sub-routine, communicates information. The focus here is on typical mnemone and, within it, morphogene. Morphogenic input, bio-logic's output; code (symbolic metaphysic), physical expression. Psycho-biological morphogene 'precipitates' material patterns of behaviour; nature's forms are morphogenic crystals; you are, in this sense, a crystal of the human archetype.

In this case, because it is part of ourselves, *H. archetypalis* and not, say, *Equus archetypalis* (horses), *Scyphomedusa archetypalis* (jellyfish) or *Hymenomycetes archetypalis* (gill-bearing toadstools) is used as the most relevant exemplar. Its expression will soon be traced in a pattern called the caduceus.

Biological bodies, which are in all cases at least chemically dynamic, are not seen as solid objects but, primarily, purpose-channelled, flowing events. In this case an archetypal key is the association and amplification (or malignant interference and reduction) of vibrations - *resonance*. In psychological terms 'resonant association' means 'feeling right'. Things chime. They click. We resonate with preferred energies (e.g. sound waves of particular music) and behaviours (e.g. ways of thinking expressed by concept, language and behaviour). This is how we feel what, for us, is right. Click 'n' chime are friends. In the case of biological form the resonance is (*fig.* 15.5) between psychological *H. archetypalis* and the physical set's radio broadcast, *H. electromagneticus*.

You are, physically, vastly more similar (about 99.9%) to every other human, especially your parents, than dissimilar.[109] Archetypal-sex-expressed delivers this minute dissimilarity; it spices minor-seeming-major variations on our theme. Sperm and egg involve two half-sets of chromosomes which, coalescent, form a zygote. It is as if, in fusion, two similar versions of the

**Vibrant Morphogene
Information Loop between Physical Body
and Morphogenetic Mnemone**

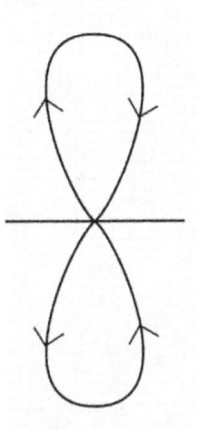

cosmological axis (X)

'upper subconscious',
personal mnemone;
'lower subconscious',
typical mnemone;
(H. archetypalis)

H. electromagneticus

genetic and other physical
information systems;
molecules, cells, organs etc.;

H. sapiens
biological phenotype

This psychosomatic loop engages rings 4 and 5 on *fig. 14.3*. Bio-physical input 'rebounds' from its particular morphogene. In other words, slightly different genetic make-up and surrounding circumstance (commonly called 'nature and nurture') generate physical variation on archetypal theme.

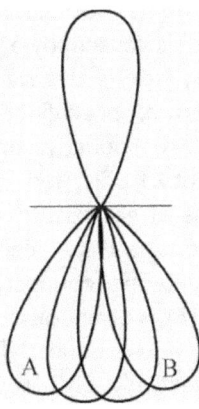

immaterial cause;
single, universal archetype

material effect;
local variations; individual
forms, races, hybrids
or (eg. A and B) species

---

[109] see Chaps 23: N-p-c *DNA*; and 24: Archetypal Sex Expressed.

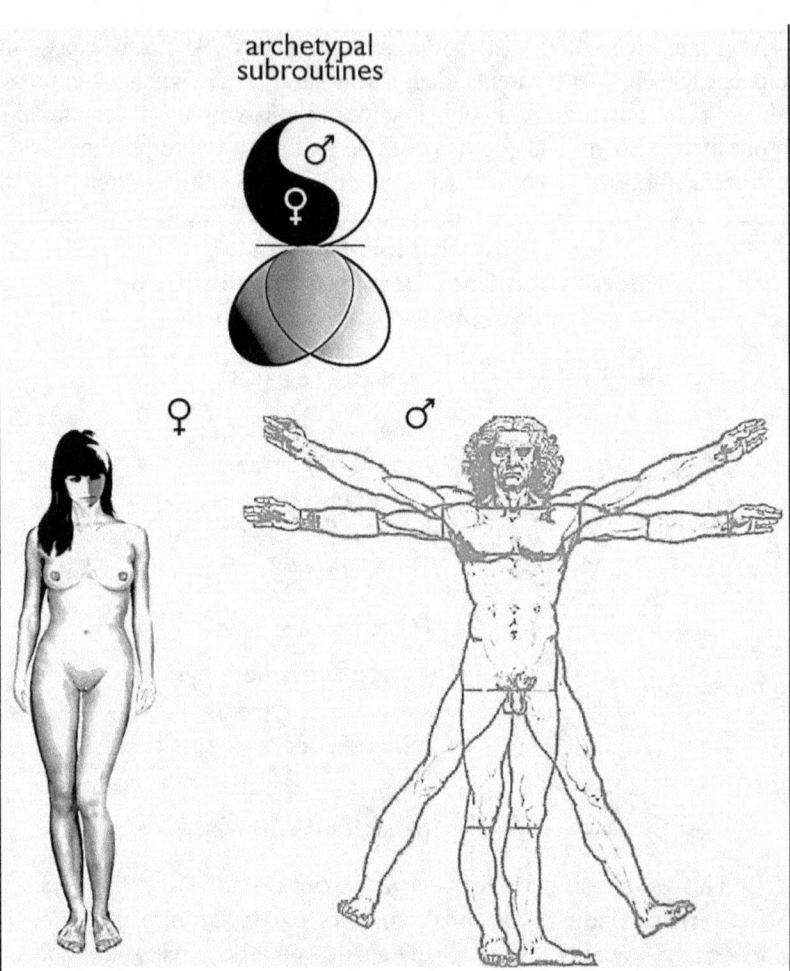

Sexual organisms are 'dimorphic', that is, are polarised into male and female morphs. While these represent complementary opposites of a single type, the issue is not simply one of emphasis. A sexless primordial 'gonad' is caused, by a pre-coded signal, to develop into male testis or, lacking this signal, female ovary. And further definite, deliberate, predefined switches trigger either male or female external reproductive organs from a 'bipolar precursor' - an embodiment of polarised information intended to create one of two forms of genitals. In short, bio-form is built from appropriate genetic, cellular and morphogenetic interaction. How does the connection work? Communication is (see Chapter 16) continuously bounced between physical and metaphysical sub-routines. The latter, parts of an 'instinctive' program, effectively orchestrate a 3-D physical reflection of either male or female archetype.

fig. 17.1

same record were played together, two singers combined to mix a single species of their song or two photographs of faces with slightly different features were superimposed. Is the product more like dad or mum? A new 'diploid' superimposition appears whose exact genetic expression, in conjunction with *H. electromagneticus* (or electrical form), represents the dominant traces from each parent. Such likenesses are reflections of physical not archetypal feedback. Generic morphogene is changeless; local contents ring specific changes. Of subtle differences between all humans you possess precisely, half from each, those of your parents. Thus the fine structure of a child's features 'resounds' with its parents, grandparents and, increasingly diluted in effect, further ancestors. Maternal, paternal or neither line may dominate. The outcome depends on the purely physical genetic recombination during parental meiosis and the differential strengths of resonance with an archetypal 'master copy'. The generic, archetypal template can accordingly be seen, both with respect to sexual and asexual reproduction, as a probability-generator. It is a changeless 'tympanum', 'habit' or 'instrument of form' whose physical reflections vary while at the same time it endures as a strongly conservative 'attractor'. Thus physical strains may vary with each throw of the sexual dice but always tend, if circumstance allows, to rebound towards their preordained and well-proportioned norm.

In general the more powerful a material reflection, due to pristine resonance with archetype, the stronger its routine, the more dominant a particular influence and the greater the probability of its expression; and, conversely, the more it 'recedes' from archetype, then the weaker its resonant association and the greater chance that 'off-key', erroneous transfer will occur. Over generations some biological characteristics are locally amplified (become dominant in a population) and others recede. It may even happen that a particular physical resonance 'breaks away' to form a recognisable but distinct version of the same basic song. It certainly happens that suites of the same musical genre (say, blues, operatic, Bach etc.) sport their own signature of style. They fall into distinct orders, classes, families; but though they may share notation patterns, cadences or other commonalities one style does not transform itself into the next; and *DNA* mutations don't as much transform as jar like out-of-tune performances. Intra-specific variation within genre is called, in biological terms, a new race or species. Various genres (genera) will make up different types; perhaps even distinct and separate families can radiate within a type. Where does a classifier draw his line? Does nature set impregnable a bar; no change beyond a definite extent; no unlimited plasticity? No, cries evolution, though its music lacks composer. Yes, retaliates the rational notion of design.

The rational idea of a biological archetype used to be the norm. The naturalistic notion of transformation from type to type of organism through long chains of small but 'progressive' distortions ('progressive' evolution) has so distorted this original mode of thought that it is worth rephrasing. General characteristics (e.g. facial features, organs, body proportions) occur throughout a biological type. A single, sub-conscious archetype (which includes, where appropriate, male *and* female sexual programs) is a kind of 'resonance structure' inflected in physical practice by genetic composition and circumstance. Its metaphysical form (or image, if you like) is no more shaped like a mature organism than an egg or the electromagnetic frequencies of program that will be transduced into 'mature' end-product on a screen. Different bodies work the changes round this median, this source of 'original perfection'. There is, in other words, detailed variation driven from the physical side and which appears as different individuals, races or even species. *Such secondary variation is sometimes called micro-evolution but no primary archetype is morphed into another.* Mixed offspring, reversing reproductive isolation of parental groups, tend to reduce peculiarities of race and, to that extent, emphasize the norm. And the 'vibrant criterion' of type will tend to correct even such excessive stress or deformity of its field as might result from severe genetic or physical damage. *Such effective 'elasticity' promotes the so-called 'wild-type'. Such type is strongly conservative, selected for and anti-evolutionary - as we find in practice.*

Genes play, no doubt, a crucial physical half. They 'express' specific proteins; many of these operate biochemistry's automated production lines. *Top-down*, they are part of the hardware employed to express software but are not the *source* of morphogenetic information. In short, the integrity and accuracy of a genetic database affects the application programs that use it and the physical materials, specially processed to construct hardware, that derive from it. **However, the primary, all-powerful gene is a materialistic myth**.

The Logic of Development

Could you infer the presence of a morphogene from the bio-logic of development?

As a packed parachute is a different shape from its expanded use, so the potential of an egg is different both from the unpacking process and its full expansion into target product - finished, adult form. *In each case the whole construction and its operations are preconceived, preordained and, from the very first, work perfectly.* They had better!

In the previous chapter it was noted that all cells and bodies are permeated and controlled by electrical fields. Salamanders, for example, possess such a matrix complete with positive and negative pole arranged along the longitudinal axis of the body. When biologists trace the development of such field back through the growth of an embryo they find that it exists even in the unfertilised egg. They mark with blue dye the pole of the egg where there is a noticeable drop in voltage and note that the head of the salamander always grows opposite that point. Eggs, which also involve chemical gradients (or differentials), are electrically polar; and embryonic cells were arranging themselves according to the pattern of an electrical field that was present before the individual came into existence. Yet though some startling discoveries concerning the morphogenetic properties of electrical fields have been made, no general theory of electrophysiology has been developed. *There arises, moreover, when the complex shape of any mechanism is built, the question of what moulded it according to what. In a biological case the question is what moulded, moulds and controls the electromagnetic matrix itself - since its purposeful and systematic shapes cannot be the product of chemistry or physics left to their own device.* **The linkage of causal bio-electrical patterns with genetic, physiological and anatomical effects will be further confirmed and pursued in Chapter 25.**

Biological morphogenesis always involves a precise and intricate plan. Equally clearly, it evokes variation-on-theme. Did its complex mechanisms, undoubtedly physical and thereby susceptible to scientific resolution, occur by accidents?

Top-down, they never did. Information rules coherent integration of forward-thinking plan (is not an adult *target* of an egg?). How, in this case, does *H. archetypalis*, through the agency of *H. electromagneticus*, impress its top-to-tail sequence on the adult logic of *H. sapiens*?

Caduceus: the Human Morphogene

The *caduceus*[110] is an ancient symbol. It was the staff of Hermes (or Mercury) whose power was thought and who was, thereby, the wingèd messenger (or information-exchanger) between the gods. Although any metaphysical import has been forgotten and it has become confused with the rod of Asclepius, Greek god of healing, caduceus remains the totem of many scientific medicine men. From its central bulb spread two wings which represent the higher, flying capacity of conscious thought as expressed in the form of cerebral hemispheres of the brain. In biological terms the bulb itself is said to represent sub-cortical, central forebrain. This structure, called the diencephalon, links conscious to sub-conscious and totally automatic, physical regulation of the body. The diencephalon contains both the 'seat of emotion' (limbic system) and a triplet of neuro-hormonal governors in the form of pineal, hypothalamic and pituitary glands; from their 'throne' at the spatial centre of the brain, these masters regulate all autonomic and hormonal information and, thereby, the body. The bulb, representing sub-conscious, psychosomatic aspects of mind-brain, also includes the cerebellum, core processing units of the brain stem and reticular system, midbrain, pons and medulla. In this region

---

[110] see also Glossary.

reside the orchestration or government of balance, an on-off toggle for waking-state or sleep, the coordination of autonomic functions and a kind of switchboard, point of exchange or crossing-over for the routes of sensory and motor traffic between brain and lower body. The staff (central, spinal cord) and the double helix that surrounds it represent the informative distribution of archetypal (i.e. sub-conscious) energy.

How can matter set up systems? How can it, 'any which way', ever *think*? Yet isn't it, materialism claims, doing just this at this moment in your brain? Aren't you 'thinking matter'? Conversely, from a holistic point of view, there must be a psychosomatic linkage of which nervous or other systems form the somatic side. The other side of this informative linkage, one 'sandwiched' between conscious mind and non-conscious physicality, is sub-conscious mind. A persistent mental shape or record is called a memory; memories, whether personal or archetypal, are the 'stuff' of sub-conscious mind. An archetype is, in these terms, a complex mental shape; if it is created personally it is saved as our idea, if universally

Caduceus - a Schematic Shape of Man

The caduceus is a representation of basic human structure. It is an outline-in-principle, a matrix, a part of typical or archetypal mnemone also known as *vis medicatrix* or morphogene of *H. archetypalis*. Its logic represents the conscious element as wings, made physical by the two cortical lobes of the brain; the knob or bulb represents sub-conscious elements manifest as diencephalon, cerebellum and brainstem; staff and snakes represent three major channels of information exchange - central, inward (sensory) and outward (motor) - whose physical correlate is the nervous system. As glands distribute chemical information, so the caducean crossovers represent central nodes for the distribution of sub-conscious energy.

Caducean Structure Reflected in a Cohesive Protein

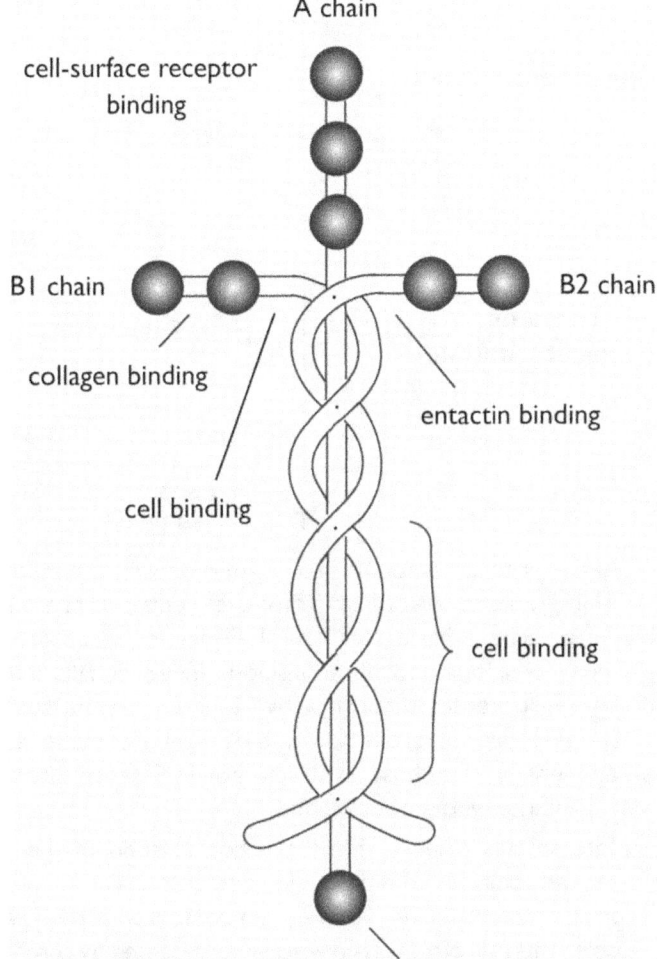

Curiously, the archetypal structure of our physical being, caduceus, is reflected in a cross-shaped cell-cell adhesion molecule called laminin. Found in basement membrane this glycoprotein forms a networked foundation for structural (as opposed to informative) cohesion. It binds to cell membrane and, at the same time, extra-cellular matrix (*ECM*). Without it the tissues of most animal cells and organs would fall apart. No glue, no you - without it you'd disintegrate! Caducean coils and crosses are thus crucial!

fig. 17.2

Human Extent: the Conscio-material Gradient

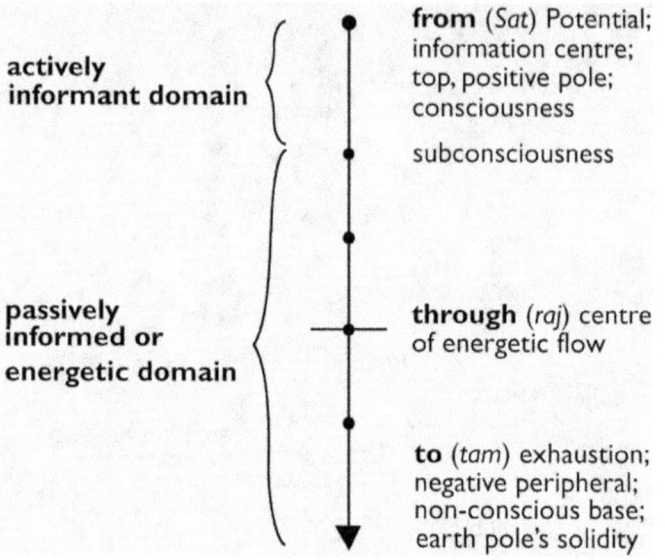

Human radius amounts to a microcosmic conscio-material gradient. A radius from the centre represents cosmic polarity - basic to Natural Dialectic - that drops between conscious and non-conscious levels. It falls from informative potential through active function to the static, structural phase of gross form, that is, bulk shape and containment. In this sense, with its root in Essence, human form spans the whole of existence.

Note, in this respect, that the major function of the (sat) informative domain is command and control to maintain balance. Conscious command is flexible, sub-conscious archetype is inflexible. This pair of zones is physically represented by cerebrum and lower brain with thyroid metabolic control.

The informed or energetic domain also involves two divisions. (Raj) energy's centre of balance is at the solar plexus. In a human the systems of the upper trunk involved with energy (ATP) production include, lungs, digestive system, heart and blood transport system, pancreas and liver. These break down nutrients to 'loose' chemicals that are irradiated to support each cell's metabolism, growth and maintenance.

Solidity swings low. Exhaust drops from organs of the lower trunk (colon, kidneys and the urogenital system). From reproductive organs is expelled a fresh but earthbound, solid form. No nerves branch into base coccygeal and pelvic supports.

> A limited but simple analogy of your own gradient is with a TV. The set.'s predesigned body is plugged into a power supply. When switched on, it receives its 'screen-life' of psychologically fascinating episodes broadcast by programmers. The point is informative communication; this business, from preparatory starts to finishes, is always initiated by mind; mind controls the whole show.
>
> *fig. 17.3 (see also figs. 5.1 and 19.1)*

it is part of the blueprint for physical creation, a file of archetypes called universal memory. In ours, the human blueprint, it is a 'virtual' matrix whose 'archetypal program' constitutes psychological behaviour (instinct) and governs its own reflection in the localised shapes (both in development and maturity) of an associated body. In other words, you represent the projection of a plan upon an earthly screen.

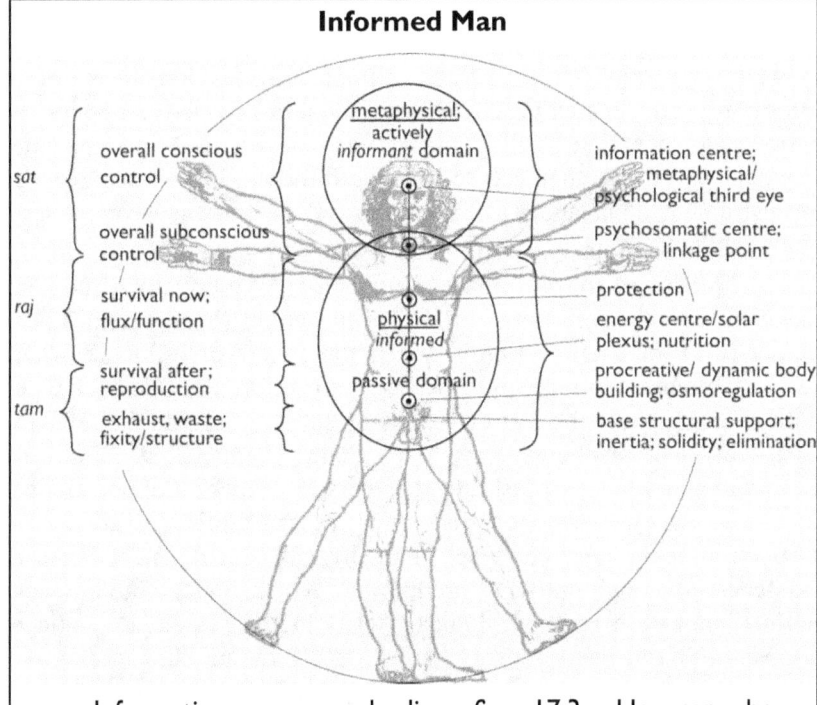

Information man embodies *fig.* 17.3. He can be represented in the form of two domains, the *informative* and *informed* or *energetic* domains. These domains involve active and passive information respectively.

Thus, as well as bilateral, humans also show vertical, top > toe reflective asymmetry of complementary opposites.[111]

fig. 17.4

---

[111] see *fig.* 2.4 and Chapters 2 and 3.

Now consider (*fig.* 17.3) 'the human extent' as a *wireless anatomy* with conscious mind as its positive pole and pelvic foundation as the negative.

Information man is archetypically reflected in the so-called esoteric oriental systems that involve psychosomatic nodes with associated glands, physiology and anatomy. There is, from this perspective, no argument with neurological mapping. It is simply a question of broadening the psychosomatic context in which the transmission and storage biological of information occurs.

The functional logic of man is now inspected. The caduceus is analysed in terms of two domains of influence, the informant and energetic (or informed), that involve the component basis of its logic.

The Informant Domain

You, of either male or female polarity, express universal humanity. Your archetype, of all biological types, best represents unrolling of dialectical phases down a cosmic scale.

At the top mind impresses and is impressed. Its domain is one of information, understanding and desire. Conscious mind generates the current and negotiates fulfilment of wish or whim. Mind generates imbalance by desire then acts to cancel out the tilt. It is, in this respect, an instrument of balance. Mind's terrestrial instrument is brain; head glands (hypothalamus, pituitary and pineal) control the rest.

At the base of the informant domain it is, symbolically, the neck that connects 'heaven with earth'. It links mind (head section) with matter (mid- and tail-sections); and is the sphere of influence where sub-conscious mind plugs into body, where information is exchanged with energy, where the mirror-worlds mesh. Not only do the left- and right-hand hemispheres of the brain 'shine' at different, complementary sorts of task but they reflect events on opposite sides of the body. In terms of dialectical polarity this 'inversion' or 'reflective asymmetry' between metaphysical and physical poles of existence is logical; but why should blind evolution generate the well-defined inversion or 'neurological twist' in the systematic logic of almost all motor and sensory pathways? Why do they cross over (decussate) in this psychosomatic region of brain stem and upper spinal cord and thus reflect, in the channel between heaven and earth called a neck, such a mind/ matter mirror?

Below consciousness the system drops from the eye-centre into sub-conscious influence. What I do not know I cannot disturb or be disturbed about. It would be impossible to have to think about every physical need so that lower operations are mostly automated. Such automation is both sub-conscious (although a conscious override is sometimes possible) and physical. Therefore in keeping with its role as psychosomatic link the second node binds upward with the conscious phase and, downwards, with the body. The physical expressions associated with this zone are post-cerebral areas of the brain (including cerebellum, medulla and the head glands) and, as controlling agent of the rate and quality of cell metabolism, the thyroid-parathyroid complex. Its management, sited just above its energetic zone, involves rate of metabolism; and also the dynamic, integrated cooperation of a specific class of molecules.

Chemists identify a molecule by its unique spectrographic signature. It is important to conceive of molecules as vibrant energy. Each sort is 'bar-coded' by the measure of its spectral lines of light; it emits a unique vibratory pattern, kinetic 'fingerprint', 'chime' or 'ring-tone'. Such tones act as subtle switches. In the thyroid system, for example, they are characterised by different configurations of iodine and calcium atoms.

Thus associative 'chime' is seen as the dynamic address system. Such electromagnetic resonance constitutes the body-side of linkage between mental, emotional and physical energies. Electro-physiology, neurology and endocrinology are seen as studies of a rapid interaction between archetypal as well as physical influences. In this view psychosomatic interactions are tightly controlled by the system of 'ring-tones'. Social molecules will ring each other up! The order of the instant is overall linkage, coordination and cooperation whose objective is the maintenance of homeostatic balances specified by informative archetype in the role of 'health-plan' for its physical correlate.

The Informed, Energetic Domain

The hierarchical order of human state descends from the level of informative mind to informed energy. If you *know* with your head you *do* with the body below. *Below head and neck subtends the province of energy, of torso, limbs and, of refined agency, their hands and feet.* In the descent of any hierarchy power and influence is gradually restricted. The central zone's for energy; lungs, heart, liver and digestive system cooperate to radiate nutrients to every cell and thus promote, by respiration, *ATP* production. This, sunlight's sliver, drives most biochemistry.

Below the belly-button, at the end of business, falls exhaust. In front liquids, solids to base rear - this is expulsion's zone. Osmoregulation is located in the kidney, egestion in the colon; and, where a finished job (maturity) recycles, there's earthbound reproductive apparatus binding life unto its wheel. A body's logically devolved along the dialectic gradient from information (the potential for orderly behaviour) through energetic business down to bodily expulsion out into a careless world! All, supposedly, in order that's evolved by mutant serendipity!

Do we therefore suppose a cell's or body's perfect health evolved through grades of imperfection, that is, millions of sicker stages? Or prefer that archetypal vigour radiates original, dynamically fine-tuned health for every kind? *If your body's health is indeed the result of interaction between physical and archetypal (sub-conscious) patterns, it did not evolve by accident.* The tendency of every system is to bounce back to its archetypal norm, its *vis medicatrix*. **This metaphysical form is the very basis of homeostasis, physical life's overriding principle. Health is the norm**. Wounds heal, infections are fought and cell debris cleared. Indeed, the central nervous and autonomic, endocrine and immune systems are integrated in such a holistic way that experts sometimes use the phrase 'psycho-neuro-immunological system' to describe their cohesion. Mind, as every doctor knows, affects matter. *The channels of 'psycho-somatic interaction' are precisely what this chapter is describing - unless you still neuro-scientifically believe that mind is brain and psychological equates, essentially, with chemical.*

Swing life round to the other side. Instead of reason seek, as rational Darwinians do, its absence. Is, as Lady Luck avers, the irrational cause of reason chance? Yet reason floods biology. Metabolic pathways, cells and code-specific systems all express, in different organisms, common physiological functions. These functions represent, effectively, 'reasons' subordinate to the overall 'reason' that drives a biological body. This 'reason' is to live more life, that is, survival. Homeostasis buffers changes in and outside bodies. Such equilibration promotes balanced working, that is, survival. 'Sub-reasons' therefore include sensitivity, biochemical command and control, digestion, respiration, osmoregulation, reproduction etc. *Which of these functions is, where found, superfluous?* Which of its associated organs, metabolic factors or coded molecules is, lacking a reason for existence, therefore redundant? *Which can survive without the others?* **Calculate the minimum number of functions necessary to support life, any life at all.**

Reason's nowhere absent but, with chance, like chalk and cheese. Reason is meaningful, chance meaningless. *Why should a highly rational system irrationally 'self-organise'?* Chapters 19-25 will demonstrate the fallacy of a belief that any chance whatsoever could invade the initial, entirely purposeful construction of a cell, a human or any other type of biological body. <u>Your own human microcosm underlines the utter feebleness of evolutionary explanation</u>. If materialism's rational it spotlights how, backing chance as its creator, irrational rationality's become. This is 'counter-intuition' for you!

Isn't it, on the other hand, inconceivable that such a logical, integrated, self-consistent embodiment, constructed with highly specific complexity in accordance with grades of the conscio-material gradient of creation itself, occurred by chance? **If reason wins, chance and time's grand theory crumble back to their home ground - they bite their progenitor, the dust.**

Chapter 18: Death

One Thing Is Certain…

Will you never die? Do you suppress the thought or, as society around you seems to, live in death-denial and suffer from delusion? The end is nigh. The end is, naturally, upon us - in maybe fifty years or, perhaps, a single hour. We are tenants of a body's lease whose life is short. This whole world will disappear. It will be no more. Yet commentators turn the sense upon its head, evade the issue and thus think it's claimed the starry universe's time is up, that cosmic death is nigh! **For cosmos it is far away; but for each life on earth one thing is absolutely certain - death.** Therefore cut cant and superstition. Approach its moment in a steady, lucid, balanced way. Death is, in Natural Dialectic's terms, a very lively subject.

D-Day

Are you ready for your D-Day? Is it not the world's way bodies turn to dust? So are we, you and I, prepared for our respective D-Days? All-change at the end of our 'normality'? Why did the hermit contemplate a skull? What about the other end of birth-day, death-day, and its hour of passing?

Of course, if everything's material then death would seem, where life's brief flicker guttered out, oblivion regained.

What is non-conscious matter but oblivion's tomb? Chemicals can't die nor is a body other than unconscious. Thus, locked in a biological construction, soul becomes immersed awhile in lifeless, physical phenomena. It becomes an actor on the world's stage but if immaterial mind and energy of body are two different qualities then they can part. *What meshes can detach. When the pot cracks life may, as from tomb, fly. When the psychosomatic connection is broken you part with your pretty (or not so pretty) arrangement of chemicals; you part with oblivion and the incarnate experience.* The Bible, Koran, Kabbalah, Zend Avesta, Mahabharata, Ramayan, Granth Sahib, Buddhist Sutras and all other human inspirations except materialism have distinguished, in sophisticated or simple language, between the sacred and profane, between soul and flesh, life and death. **Indeed, a main point of mystic practice (and therefore tradition) is to experience and thereby understand the ways of dying and of death.** In this view physical materials of an outer sheath are shed and, incoherent, are recycled ecologically; but, equally, the inner sheath persists. Mind is recycled in a different but coherent sort of way. *If the objective, physical side goes but the subjective, metaphysical side remains then where, death, art thy sting?*

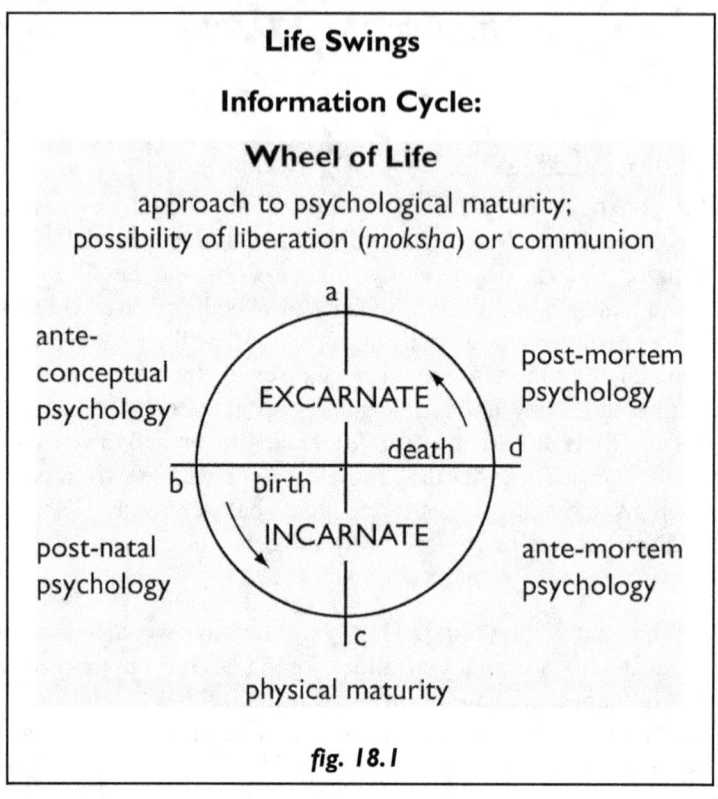

fig. 18.1

Psychological recycling? This, if we consider mind and its attention metaphysical, re-begs the hackneyed question of psychosomatic interface and the nature of the conjunction (at out-swing or birth) or disjunction (at in-swing or death) of body and mind. *How does subjective recycling work?*

It is necessary, in order to understand the human life cycle, to consider both incarnate and excarnate phases of its pendulous swing.

The four phases of this cycle are, from a subjective point of view, transitions called birth (incarnation), (incarnate) biological life, dying (excarnation) and (excarnate) metaphysical life. To describe any cycle you have to decide where it begins. *Here the logical, top-down starting-point ('a' in fig. 18.1) is in-swing at its zenith, the moment of equilibrium before out-swing commences.* This moment occurs, however, in an excarnate condition; as western psychology operates within a framework of materialism it excoriates the existence of such a state.

The Logic of Disembodiment

It will become clear that, contrary to the previous chapter's logic of incarnation, the *top-down* logic of disembodiment is dominated by the upward, right-hand path of Natural Dialectic.

The Logic of Disembodiment

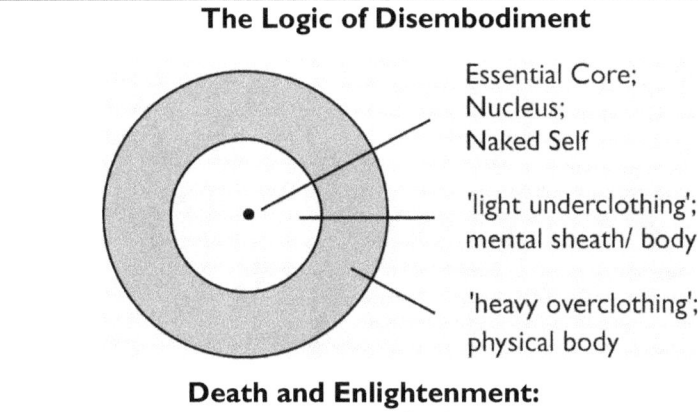

Essential Core;
Nucleus;
Naked Self

'light underclothing';
mental sheath/ body

'heavy overclothing';
physical body

Death and Enlightenment: Excarnate Cycle

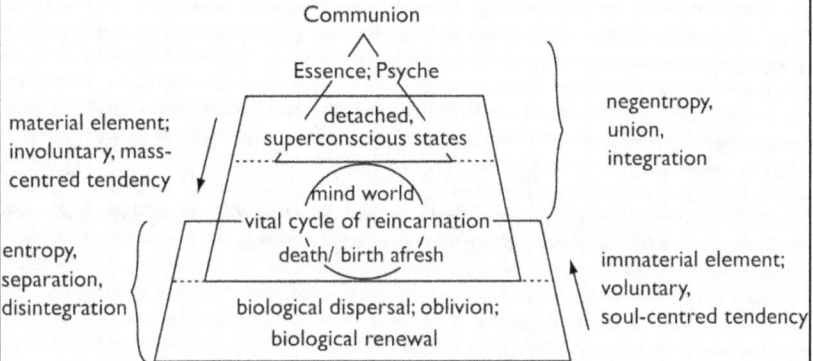

Death cracks the pot. Dust to dust the outer, lower case disintegrates.

Mind-brain identity is crucial for the sustenance of atheistic materialism. No brain, no mind. Mind reconvenes with dust.

The separate identity of immaterial, informative mind and material brain is, however, basic to all variations of 'expanded science'. Thus, according to such *top-down* holism, death of physic is but birth into the mind-world. The 'great sleep' may be physically unaware but psychological experience persists. Animation's factor tends, according to its purity, to rise. You *are* your mind. If mind is clear Communion is guaranteed. If not, what rises falls. Gravity propels the bundle of accounts (mind weighted with the balance of past actions and desires) towards an incarnation commensurate with the incoming character of mind. New clothes for old - death of body is the final cut but mind will live to fight its next embodied day. What neurological nonsense!

fig. 18.2

tam/ raj	*Sat*
existential episodes	*Life*
body/ mind	*Soul*
oscillation/ revolution	*Axis*
↓ *tam*	*raj* ↑
negative/ outward	*inward/ positive*
out-swing/ descent	*in-swing/ ascent*
entropy	*negentropy*
creation	*dissolution*
incarnation/ embodiment	*disembodiment/ excarnation*
body	*mind*

Towards a Unified Theory of Life and Death

You say, perhaps, there's only matter and therefore after-life and reincarnation are impossible. Science has progressed beyond such superstition.

That is nonsense. How can material science 'progress beyond' the immaterial? Scientism simply ridicules its mention; but if immateriality is real what does ridiculing or denying it avail? Denial is a guess but not a scientific one. *Indeed, in its relative maturity science has now reached a couple of interesting limits - mind and death.*

shells	*Nucleus*
lifetimes	*Essential Life*
↓ *objective*	*subjective* ↑
involuntary	*voluntary*
death	*life-time*

Death is a closed box; it is a critical unknown, the last horizon of ignorance. What is the nature of its inevitable 'black hole'? There are two views:

(1) the *objective* which looks, from the outside, at another's

(2) the *subjective* which, crossing the event horizon, experiences death for itself

Bottom-up only the former, physical view is possible. Objective science can't include subjective views of conscious life or death. There is no logic of disembodiment save that the dynamo of respiration fails. Collapse of proton gradients across cell membranes marks, invisibly, full stop. Psychology post-mortem never starts because, since I am only body, death's the end. Disintegration of atomic aggregation is the death of life's illusion, that is, of a soul-less life that never really lived. And faith, remember and recite, decrees that mind is somehow a reflection *caused* by a molecular arrangement. Brainwaves are life; their flatness signifies the end. Atoms of cerebral molecules survive but their arrangement falls, along with cells that they compose, apart. Ashes to ashes, dust to dust - that is the entropic, earth-bound kind of end.

Unless, perhaps, 'cryonically suspended' by materialism's type of hope and faith your refrigerated form awaits the day white-coated angels judge is right for resurrection (of as yet unspecified a species) then 'reload' you into some unfrozen form of paradise!

Even more sophisticated than deep-freeze, why not simulate the active 'connectome' of brain? If this vast, sparkling, 3-D web of interaction could be downloaded into storage space - enough of which we don't yet have - you might have lifted thought onto a massive chip. Edit, copy and then paste yourself into a million more. Next, as if playing programmed games, you could be released on myriad platforms each 'living' an *ersatz* reality. You could do this while your life was still embodied but, when that was dead, someone else could press the button for luxuriant revival and machine-dependent immortality. You can sense the moral bomb all this implies but never mind. These boys, members of the mind-is-meat brigade, seem to crave at any cost construction of a resurrection kit!

If that were possible. But if 'mind-meat' is a delusion so is 'lifting' mind into another kind of clay. The exercise would sum to an advanced expression of AI (artificial intelligence).[112] Playing God, could anyone create the spark of life from death in an electronic, e-m or atomic form? No doubt, the IT robot would be subtle but material. Its brain-chipped program of behaviours would, like any computer, simulate the calculations of its initial programmer - in this case through the same quantum level as that of H. electromagneticus.[113] However, if the element of subjectivity did not materialise then such experiment would fail. An illusion of life would not be alive. How, then, could it yield victory over death?

Thus, from a *top-down* perspective, to assert that mind can never leave or dwell apart from body is posturing by materialistic creed. True, there occurs disintegration, dispersal or disposal of a body. Mind is not, however, construed as the by-product of grey matter. Thus to seek a material explanation of immaterial mind is as demented as physically and chemically analysing a piano in order to discover hidden melodies or a composer's mind; and destruction of a keyboard with its loss of sound is not the end-all of a melody. That is a delusion.

While mind-brain identity is crucial for the sustenance of atheistic materialism, the separate identity of immaterial, informative mind and material brain is basic to all variations of holism. No doubt, life-transforming revelation can't be scientifically verified; it's anecdotal as a dream or any personal experience.

---

[112] see Chapters 6: Mind Machines and 13: Neurological Delusion.
[113] see Chapter 16: *H. electromagneticus* and Psychosomasis.

No doubt, equally, child, poet, scientist and farmer differently describe a tree. Various cultural metaphors describe a lone emotion; this does not detract from its reality. How do you describe an inward vision such as *OBE* (out-of-body experience) or *NDE* (near-death experience)? By referencing, in your language, your own cultural imagery? Does such description undermine the metaphysical reality? To inexperienced sceptics, perhaps, naturalistic methodology can't treat an immaterial entity; nor does it deal in metaphor. But if a single *OBE* were ever inferred beyond doubt, it would demonstrate brain is not mind, mind not matter. No doubt that doubt, a sceptic's tool, keeps materialism sane but if, equally, a single *NDE* or other 'paranormal' event were caught within its methodology (and is this possible?) materialism logically falls and with it any part of a synthetic world-view that comforts atheistic interpretations of origins, mind, human purposes and capabilities, death and the nature of life and consciousness. <u>A radical, broadening re-appraisal would become necessary</u>. **No less than a transforming shift of paradigm would have to overtake scientific materialism.**

Thus, *bottom-up*, incessant emphasis is placed on neurochemistry's illusions and 'the visions of a dying brain'.

Post-mortem Psychology

You can think that, on your death, consciousness will disappear. This does not mean perforce it will.

Death is, by 'type-2' subjective/ metaphysical token, a release from bodily incarceration. *However, in a physical sense nothing 'enters', 'leaves' or 'goes' anywhere.* Immaterial mind, having quit its connection with the material universe, a connection called its body, is no longer in it. Nothing has flipped except some biological switch. Nothing has changed except material disconnection and a shattering. Membrane voltage disappears, body's currents cease to flow and the atoms of a corpse are re-arranged. Once 'unplugged' from its physical medium dematerialised mind is left in its own metaphysical place - the so-called information field. Its program is still broadcast; and, as filings round a magnet, its coherence round the soul remains. In this sense an absolute finality of death is, as all the manuals of 'truth metaphysical' agree, an illusion born of ignorance.

Out from a womb; birth into the body-world is the opposite of leaving it. From physical to metaphysical, death is birth-day 'back' into the mind-world. Implicit in almost every funeral rite (when you're the centre of attention without knowing it) is the assumption death is not an end but a transition. *A change of state. Therefore is not the greatest lesson in life to learn how to die properly, to use death's window of opportunity successfully*? Vacuums suck. Will death's subjective void suck 'up' or 'down'? Have you prepared so that, in your own case, you are drawn, unconstrained by body's space and time, to happiness not psycho-hells? Have you learnt how, post-mortem, to avoid a rebirth into shocks and pains of physical embodiment; how, at death or before birth, to avoid involuntary expulsion into lower qualities of life; and thus how to voluntarily attune with suction that will draw you towards enlightenment and even to communion with the Infinite Heart of Nature? Books of the Dead accompany Books of Life. Guides and travelogues on the art of dying and excarnate adventures have always been popular. Well-known eschatologies include the European '*Ars Moriendi*' and Egyptian, Mayan and Tibetan Books of the Dead. All holistic faiths exhort clean living to die well. This is the perennial theme - to get a grip on death's reality. Success at D-Day! Successful victory post-mortem!

Anathema

Before post-mortem psychology we encountered materialistic anathema - post-mortem psychology is denied. Any kind of life after body-death is denied. Before ante-natal psychology we encounter another kind of anathema. *No doubt reincarnation is not compulsory but orthodox Semitic faith denies its cycle altogether.* The final, official source of Christian denial proceeds from Anathemas pronounced by the Fifth Ecumenical Council convened in Constantinople for political motives by the Emperor Justinian on 2-6-553 AD. This synod was a dubious publicity exercise designed to drum up solidarity within the Eastern Church and to condemn anti-monophysite schisms that threatened a united religious and political front. Monophysites believed that the body of Christ was wholly divine and never for a moment combined human and divine attributes. Whatever the obscure theological import of this dogma, it was held by the Empress Theodora as she challenged western doctrines and especially those of Origen. On this rigged Council only 6 out of 165 bishops were from the western church. Through its Anathemas the teachings of Origen, which included the pre-existence and transmigration of souls, were condemned. The idea of reincarnation was expurgated.

Nevertheless although it was anathematised, other early church fathers believed in the theory. The Gnostics, neo-Platonists, the Alexandrine School and others expressed sympathy with the idea. Nevertheless, for one and a half thousand years orthodox officialdom has emphasised Salvation but, due to a council meeting, excised and thereby forgotten the existential wheel of life, one whose cycle includes transmigration. The idea of 'metempsychosis', as the transmigration of souls was called, rumbled on in various sects until its last known 'death by anathema' in 1439 by the Council of Lyons. Now dogma does not recognise its buoyant view. It only lectures individual eternity in various styles of heaven or hell.

Reincarnation is, by comparison, a dynamic possibility whose metaphysical episodes are as natural as those physical. It harmoniously incorporates the fundamental principles of cause, effect, motivation, free will, the law of psychological and physical *karma* (action and reaction), fate, destiny, justice, equilibrium, oscillatory motion, reciprocity, hierarchy, recycling, gravity, levity, life, death and relativity set against a Central Absolute. In this view heaven and hell are textures of mind; they vary in degree but are intrinsic qualities of the mental spectrum. Mind's engine, on the other hand, is motivation; it is focus of attention and willpower concentrated by desire. An individual's texture of mind fluctuates according to the quality of his strong or habitual desires, consequent actions and their results. *His or her mental projector creates, in its flux, destiny which is realised, at the appropriate moment, as fate. Therefore kinetic fate, which is a result, can include the result of a desire to 'try again' as a human - rebirth.* The theory of reincarnation is reasonable. It does not preclude Salvation. The reverse. <u>Salvation is included as the special, highest, absolutely anti-gravitational, immaterial case - the positive goal of non-reincarnation.</u>

Immortalities

Immortality is packed in three main brands - the unconscious (subconscious and non-conscious), conscious and super-conscious. In two of these (*fig. 18.3: 1 and 2*) immateriality and immortality combine. And two cases (2 and 3) are of existential or apparent immortality whilst (1) is quintessentially real.

Bodies come and go; forms do not last. Such passing transformations include the senescence of animate bodies. In this case 'immortality', in its material aspect, resides in the databank of each individual life. This bank is codified on chemical 'paper and ink' called *DNA*. Language is therefore the heart of an intricate mechanism whose *purpose* is not only to support individual life but postpone its death and the extinction of its life's type as

Brands of Immortality

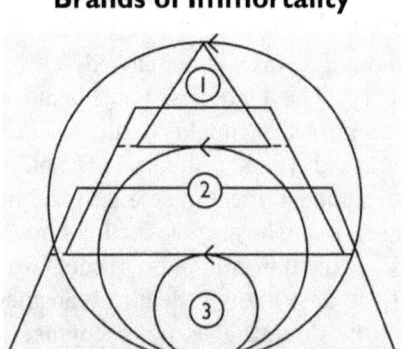

Immortality seen in terms of cycles of Salvation (1), mind (2) and material cosmos (3).

It can also be viewed (below) in terms of cosmic ziggurat.

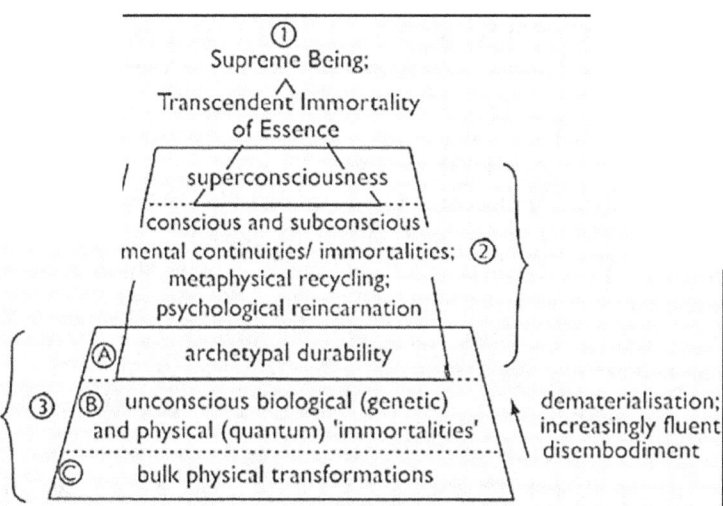

Three brands of immortality are visualised
in terms of cosmic ziggurat.

The first (1) transcends existence;
it is called Essential Salvation.

The second (2) involves metaphysical recycling.

The lowest (3) is material - in turn subdivided into three parts. These are archetype/ potential matter (A); quantum 'immortalities' (such as proton and electron) and genomic 'immortality' (B); and an endless transmigration of bulk chemical structures (C) through formulations based on (B) and (A).

As regards 3B:

Genetic 'Immortality' through Offspring

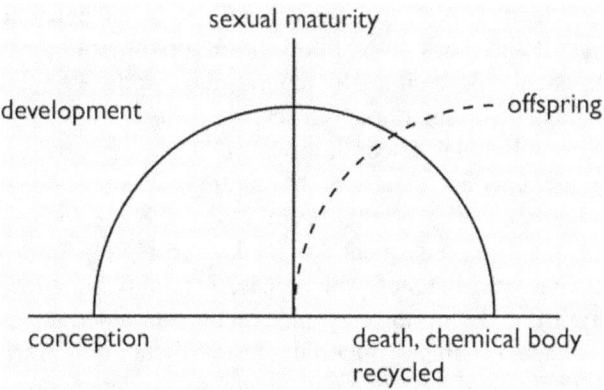

This illustration of genetic 'immortality' links with *fig. 18.4*.

fig. 18.3

a whole. What might be called 'the computerised operation of digital bio-information' is a program for survival. It serves to 'outwit' the destruction of progressive, straight-line aging by recycling the physical libraries of information (genomes) required to reproduce 'photocopies' of the various forms of life on earth. In this sense asexual reproduction, oscillation between polyp and medusoid phases of one kind of jellyfish life-cycle or aggressive Hela (cancer) cells might be construed as quasi-immortal - until, for one or all, disaster strikes. Similarly, every birth or germination represents a kind of resurrection. Life-cycles are repeated. Each organism might presume *vicarious 'immortality'* through its descendants. Body dies except ancestral genes live on.

	grades B/C	*grade A*
	expression	*potential*
	physical expression	*information/ program*
	cycles/ changes	*axis*
	concentric shells	*nucleus*
	lifetime	*relative immortality*
↓	*grade C*	*grade B* ↑
	bulk aggregate	*inner constituent*
	inflexible phenotype	*flexible switching*
	discontinuous-in-time	*continuous-through-time*
	outcome/ conclusion	*nuclear text/ DNA*
	informed	*informant*
	adult form	*development*
	run-down/ entropy	*build-up/ stimulus*
	aging/ oldness	*reproductive youth*
	death	*vitality*

Let's be clear. This is the 'least real' yet 'most apparent' kind of immortality. It involves symbolic, informative continuity, although constituent chemicals continually churn; it reproduces a theoretically endless series of similar but different bodies. General continuity complements individual discontinuity. Life's line is a thread. Death is cheated by a connective strand as thin as tiny chromosomal strings in a single cell that's called a zygote. From a zygote bodies of a genome are regenerated and the consequence of lack of offspring or extinction is postponed. For you, but not your genes, it is survival *in absentia*. In short, mortality ends immortality unless, by reproduction, genes make perpetual escape. In this sequential sense 'digital bio-information' is, subject to a congenial environment or lack of its destruction, deathless; with fit copy and without extinction there'd theoretically exist an animate eternity, an 'immortality of genes'. Brand 3. Such creed is, indeed, the socio-biological essence of material religion.

High-level Death is Life

Materialism's anathemas include metaphysical recycling (brand 2) and, to which we now turn, brand 1.

Evolution is a word that, with a sense of progress and development, you can use in many ways. Revise the Glossary. Although the Darwinian theory of biological evolution implies 'target-less improvement', the sense of Natural Dialectic certainly includes a natural goal. Can you, therefore, mentally evolve? What kind of education stimulates such lightening? What kind of message personally evolved by you is broadcast just by 'being you' on earth's great stage? That is to say, what is the quality of your own theatre, 'soap' or psychodrama within the family of life on earth? Is its predomination to the right or left-hand column of the stack? Towards what final sun could your life, like a flower, unfold? Is there, indeed an evolution of the soul?

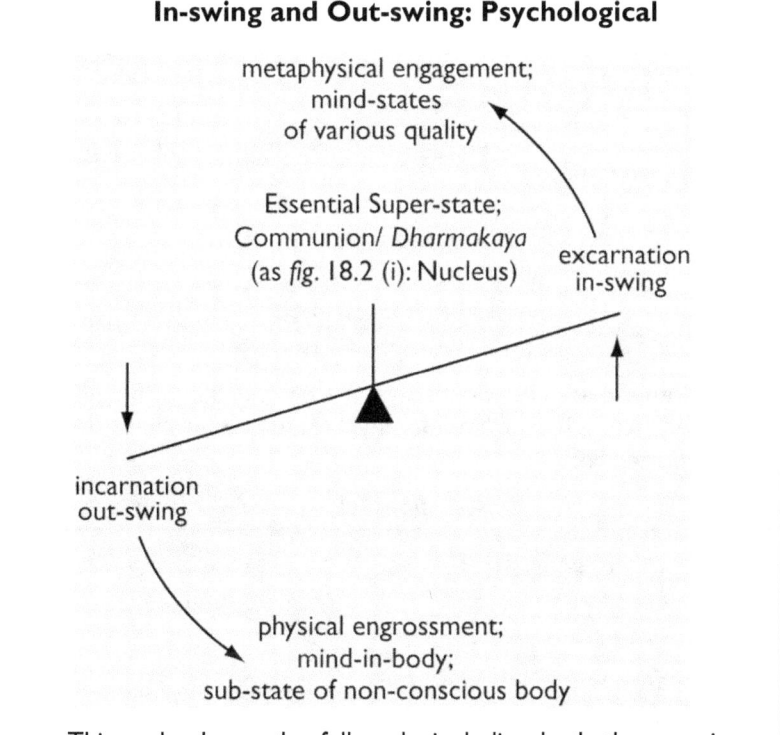

This scale shows the full cycle including both the cosmic poles between which mind can swing. It shows Essential Super-State, the *Dharmakaya* of Enlightenment, Singularity of Self, Source or Father; and it shows its anti-pole, the sub-state oblivion of matter. This sub-state includes your material body.

…s are ecologically recycled; minds, if not centred …nce in Transcendent Essence, are also recycled ac… …eir constitution.

Full Wheel: Psychological and Biological

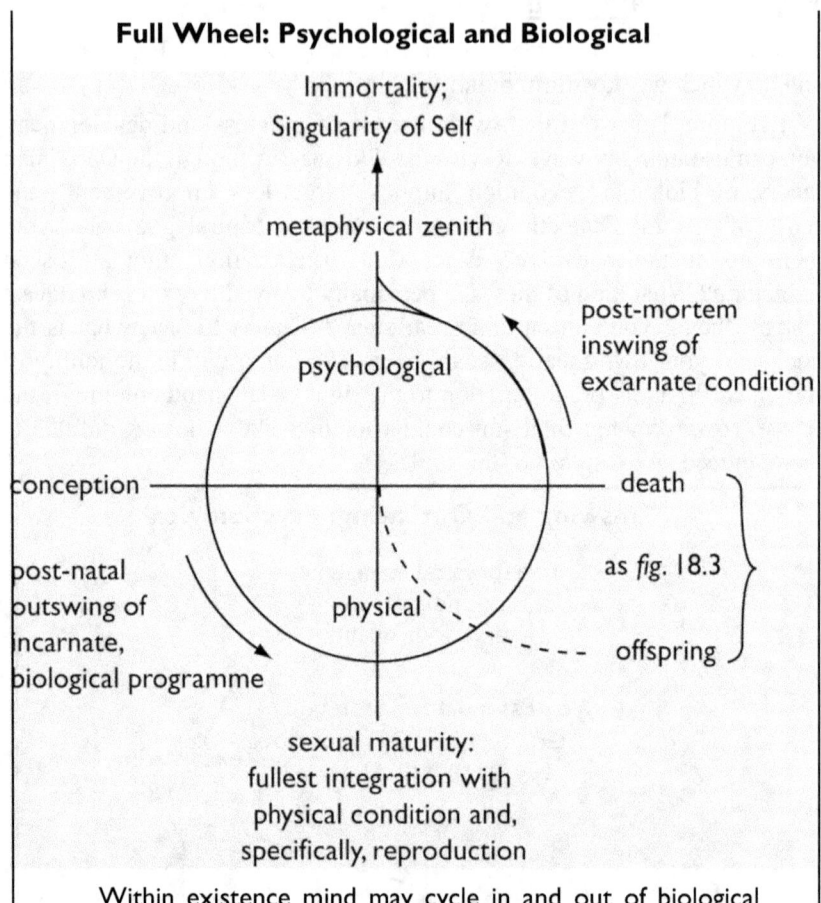

Within existence mind may cycle in and out of biological embodiment. Genetic 'immortality' is illustrated (as in *fig*. 18.3) by a dotted line.

fig. 18.4

existence below grades of error	*Essential Super-State Transcendent Illumination*
↓ descent degradation outward perspective physical attraction body-centred devolution	*ascent* ↑ *improvement inward perspective metaphysical attraction soul-centred evolution*

Body's death is death of outer shell. Mind's death is death of inner shell. **Death of *ego*, that is, death of ever-bubbling-up desires is the second, higher level birth-day; such liberation from both shells is where crusading mystics aim to be.** Materialism's world believes that no-one has returned who has been dead for long yet meditators, shedding mind and body in their deepest exercise, can die at will. You can die each day. And if you die before you die then you will not die when you die. You will be released, unshackled, into Your (and Everyone's) Communion. Where mind and body represent existence you will experience Essence. Thus, if they are not deluded and an immaterial entity is central to creation, the mystics want Transcendent Immortality. They have come to know Life Absolute. Where in this could you find transience, relativity of change or death? Look round this extraordinary universe. To reach its Living Source were awesome as a goal and far past words when once achieved. *If, therefore, a mystic is materialism's clown still there is method in his madness. Might a universal fool not, breaking out of cycles of existence, come to consciously rest in Final yet Original, Eternal Truth?*[114] **High death is thus, in metamorphosis, the Highest Form of Life.**

Ante-natal Psychology

If not brand 1 you get brand 2. This sort of existential immortality is mind recycled. Your reincarnated form will, however, definitely not be you - at least as you identify with present body and its years. How could it be if, somewhere, you will take fresh parents and another body's circumstance? What, therefore, passed over and what happens now?

What is your 'vibration'? How did you sing life's song? What about you and within you inexorably counts? Is it details or mood, inclinations and emotional experience that, stripped of physical constraint, will carry through? This sheaf, no doubt, sources fantasies. The gravity or levity of these will steer, like fates, to destinies. Quality of mind determines where it falls.

If mind 'blinked first' and shrank from inmost climax then, the moment lost, its tide has turned. *Out-swing* has begun again. There follow episodes, either short or extenuated according to the driving strength of earth-linked pulls-and-pushes. Such motivations are, in effect, the process of 'falling out' towards another birth. Psychological precipitation. In other words, what goes up must, unless it escapes gravity, come down. Having failed to escape the gravity of existence, having failed to obtain the enlightenment of type-1 Immortality, where

---

[114] *fig.* 18.3 (1).

will it be carried? A mind gravitates according to its quality of thought, deed, interests, memory and habit. These compose the amalgam of character. They are the sheaf of inner context that predicates, in turn, the outer. Outer context shows as fate and destiny. Fate is what you're undergoing. It is what you have brought on yourself. Destiny is what, according to response to fate, you will undergo. A mind's re-entry into the physical zone proceeds in proportion to its moral imperfections and material bias. The descent is buffeted by winds in mental air, currents of both fear and desire. At some stage, like anything that falls from the sky, it will have to land. By choice or otherwise metaphysical mind will assume another container. You will dress, according to the character of your debts and desires, for another physical adventure. According to this process you will 'drop through the dark tunnel' into physicality.

If reincarnation is a fact you'll never prove it by experiment. Subjective metaphysic is not open to objective, physical investigation except with respect to its lower linkage with electromagnetism of a body's nerves and brain. This link is broken when you die; there is no link between a past physical body and its follow-up. Dead meat decays; a corpse is not reanimated nor will a last trump ever more than scatter dust. Reincarnation is not the resurrection of old ash or bones. Exhumation or lab tests are meaningless. What is any ride in mind but intangible, invisible and anecdotal? What else, therefore, could you call a claim for life after death, life preceding birth (that happen in the metaphysic of a mind-world) or life after after-life (birth yielding physical reincarnation)? What proof of life on life? Yet an anecdote involving usual or unusual experience may, though 'unscientific', still reflect the truth.

Now, by the end of the ante-natal phase, morphogenic and genetic instruction have created a foetus; and, by resonant association, there is linked with its typical mnemone the personal mnemone of an incarnating individual. This hierarchical process of materialisation has produced the heart-beat of a child. At the end of the psychology of reincarnation we have therefore pulsed down to the subject of the next five chapters, bodily biology. It is time to be (re-)born.

Book 3: Biology
Passive Information (Biological)
Chapter 19: Unified Biology

The first assumption for an *AVS* (Authorised Version of Science) is metaphysical. Materialism. Naturalistic methodology involves a working assumption that all features of the physical universe can be explained by material causes. Any immaterial factor is proscribed. It is, by decision but not necessarily by fact, excluded from consideration of present operations (problematic for psychology) and past events (problematic for initial projection of the universe and, we'll soon see, for every form of life).

To try and solve these problems *AVP* has piled assumption on assumption to create magnificent mythology. Now, after the transcendent projection of cosmic energy, we turn to explosions of specific, complex information, that is, the transcendent projection of biological form. Of course, the *AVB* avers, such informative projection never happened since, by simple to complex formulation, life evolved. The trajectory of this 'progressive' mind-set follows its inexorable logic - including assumptions of chemical evolution, a universal common ancestor, an ancestral tree of life and unlimited, gradual transformations, form to form. **If, however, from first step this trajectory is incorrect biology may outclass even physics' fabulous inventions!**

What, after all, *does* life consist of? Is it invested wholly in atomic bio-form or, like all technology, does it imply there's more than only particles and forces? In which case let's presume materialism's metaphysic's incomplete and there exists an element of immaterial information as well as of material energy. **The basis of biology is information so the basic bio-issue is its origin and nature**.[115] Over the course of the next seven chapters we'll highlight distortions (often cleverly conceived) that arise if dialectic metaphysic's existential dipole - energy *and* information - is ignored; and if facts are clustered round a paradigm packed tightly with materialistic suppositions we'll de-cluster and assiduously rebuild. Chapters 20 to 25 will demonstrate, in fact, that the interpretations of the popular Darwinian paradigm are, due to false assumptions and half-truths, distortive perceptions of the whole truth. *The evolutionary mind-set of the AVS has generated even*

---

[115] see Chapter 0: Scientific Delusions, Chapters 5 and 6 *passim* and Chapter 20: Alchemy; also Glossary: Transcendent Projections.

more magnificent mythology - not involving only cosmos but biology of life on earth.

The Principles of a Unified Theory of Biology

Naturalism is tenaciously, religiously defended. The metaphysic of its mind-set is sacrosanct; Darwinism is cemented into its world-walls; trespass on that holy space is charged with secular anathema. Thus, in order to sensibly discuss an inclusive, holistic alternative, let's first suggest a brief but broadened theory of biology.

Such a unified theory considers both conceptual and physically expressed elements of biology. It combines informative with energetic dimensions. Can bio-reason automatically arise from randomness, does (code-)specified complexity occur by chance or not?

In addition it reflects the three cosmic fundamentals and their universal order of events - potential, action and exhaustion;[116] and, remember, the potential for systematic action is information. As molecular biology well understands, information constitutes the top priority.

tam/ raj	*Sat*
exhausted/ kinetic phases	*Potential Phase*
subsequent order	*Archetype*
codified outworking	*Message*
consequent effects	*Prior Information*
↓ *tam*	*raj* ↑
exhausted phase	*kinetic phase*
informed structure	*informant function*
cell/ organ	*metabolism*
individual part	*systemic communication*
external/ gross	*internal/ subtle*
end-product	*process*

In the diagrams below you have it. Fundamental principles instruct the functions and encasing structures of biology; and complex information systems, hierarchical communication, homeostasis, energy metabolism and systematic structure (including the mechanisms of exhaust and reproduction) are basics that compose the outline of all organisms. *How, when these parameters are needed from the start of any life, could nature creep towards order out of accidents?* **It will be clearly demonstrated that the panoply of life's fully informed operations is**

---

[116] see Chapter 1: cosmic fundamentals - link from physics.

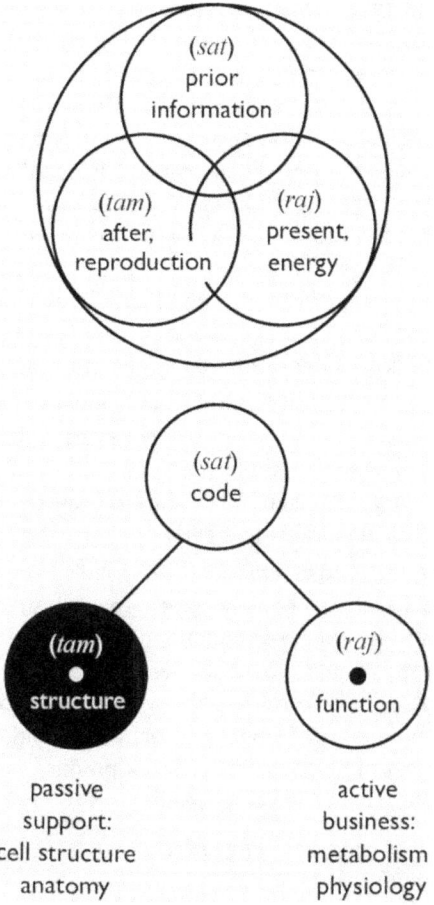

A Dialectical Plan of the Way Life Works

(sat) prior information

(tam) after, reproduction

(raj) present, energy

(sat) code

(tam) structure

(raj) function

passive
support:
cell structure
anatomy

active
business:
metabolism
physiology

What is the fundamental order of creation?[117] (*Sat*) informative potential governs (*raj*) energetic behaviour whose action lapses to (*tam*) exhaustion. Biology, we'll see, is permeated by such order.

Information precedes. It is prior and anticipates. Information is the potential and *sine qua non* for the action that consequently issues orderly.

Information comes, as previously explained in Chapter 6, in two forms - active/ conscious and passive/ unconscious.

The **conscious, informant case** involves sensation, intention and knowledge; it employs, in order to experience the body-world, a voluntary nervous system.

---

[117] see Chapters 1, 3, 5, 6 and Index: order and cosmic fundamentals.

> The **unconscious, informed case** involves both psychological and biological components. These include archetype (instinct and morphogene); reflex balance, called homeostasis, by way of nervous, hormonal and other systems; cybernetic metabolism controlled by preordination in the form of genetic code carried chemically by *DNA*; and muscular organs of action and response. Life's process is one of dynamic equilibrium. Equilibration. Its goal is balance in accord with pre-set norms. Metabolism, being totally information-dependent, works with reference to precise, incoming messages and equally precise genetic response. Such response is indexed, switched and flexibly monitored by non-protein-coding and epigenetic factors. Life is, in this way, an incarnate flux of order due to information.
>
> **Energy** provides for survival now. It involves, metabolically, photosynthesis and respiration. It promotes cell biochemistry, trans-membrane voltages, physiological processes and, on the large-scale, (nervous) sensation and (muscular) motion. The character of all function is energetic.
>
> **Structure**, whose character is solidity, represents the outermost, fixed (or flexibly fixed) realisation of shape. Its base domain is energy's container, a fixed expression of internal, orderly flux. In other words, a 'phenotype' (see Glossary) is a peripheral aspect whose body both reflects and fixes the shape of inward information and energy. The end-product of structural development, maturity, is reproductive. The cycle starts again.
>
> *fig. 19.1*

as entirely consistent with deliberate design as it is entirely inconsistent with neo-Darwinian evolution's core 'co-creators'. This couple comprise non-orderly (that is, chaotic) chance and non-creative selectivity by death. Such mindless co-creators can't, we'll find, create a single cell.

Natural Dialectic is one way to phrase biology. You could equally apply the reasoning of an engineer, that is, informed, intelligent construction. Systems generated randomly don't constitute a rational bio-treatment.

The Basis of Biology is Information

Biologist Theodosius Dobhzhansky is famous for coining a popular mantra: 'nothing makes sense in biology except in the light of evolution'. *This is utter nonsense.* **The actual, iconoclastic fact is: 'nothing makes sense in biology except in the light of information.'** You may drag evolution in on information's tail but it is simply a fashionable word (check carefully in the Glossary) used in biology to mean several different things. *In reality, codes and signals run the bio-show.* **Not evolution but information is the basis of biology. <u>To recapitulate: it is not, trivially, an origin of species but, fundamentally, the origin of information that is at issue.</u>**

> **Biology in Brief: Structured Energy Plugged into Information**
>
> **(i) The Cell**
>
>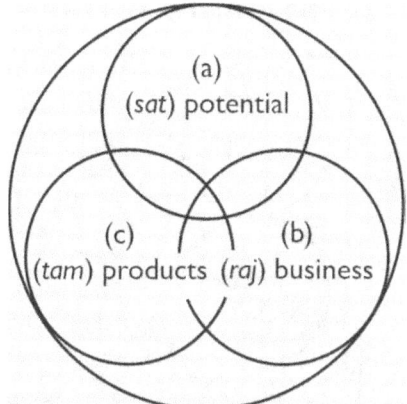
>
> Smaller than the eye can see cells embody marvellous technology. And cell construction reflects the dialectical way life works. The 'colours' of existence are seen as a conscio-material spectrum, that is, the hierarchical dispersion of consciousness-in-motion (active information or mind) and variously stored images (passive information set in memory or in codified matter). Biological 'colours' are seen as a latter, lower waveband of that overall spectrum. Just as energetic existence is shaped from Essence according to the three cosmic fundamentals (pre-active potential, action, end) so the graduated characteristics of biological plan are based on the metaphysical transparency of their potential matter, archetype. Cellular and multicellular composites are physical expressions of such recording; expression can be seen in terms of subtle, electromagnetic interactions and their gross effects viz. atomic, molecular and bulk biological structures.
>
> (a) Information is exchanged between nucleus and extra-nuclear zones (cytoplasm, membrane and extra-cellular components). A nucleus is a highly organised and sub-compartmentalised organelle. Its signalling agents are specifically informed biochemicals.

(b) Cytoplasm with energy-related organelles: the cell's business complex.

(c) Structural support: cytoskeleton and membranes involved in the stabilisation of intricate metabolism, packaging, intra- and extra-cellular import/ export duties, shape-making (morphogenetic) and boundary constructions.

(ii) **Link Scheme (with *fig. 19.1*)**

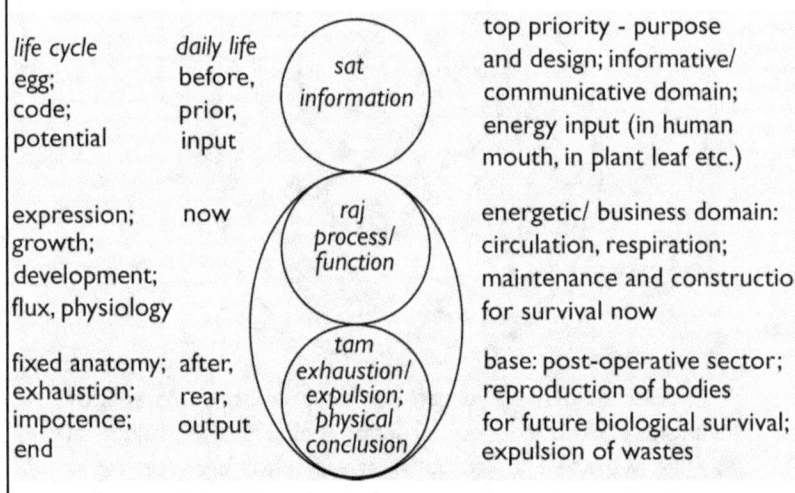

life cycle egg; code; potential	daily life before, prior, input	*sat* information	top priority - purpose and design; informative/ communicative domain; energy input (in human mouth, in plant leaf etc.)
expression; growth; development; flux, physiology	now	*raj* process/ function	energetic/ business domain: circulation, respiration; maintenance and construction for survival now
fixed anatomy; exhaustion; impotence; end	after, rear, output	*tam* exhaustion/ expulsion; physical conclusion	base: post-operative sector; reproduction of bodies for future biological survival; expulsion of wastes

Note, as in *figs*. 17.4 and 19.1, how biological expression involves cosmic fundamentals with respect to both life-cycle and day-to-day living through space-time. In the course of such dialectical sequence information precedes orderly outcome, energy supports current dynamic equilibrium and its sub-cycles are 'housed' in structures that, while exhausting, temporarily until the organism's death endure. Maintenance of such cycles, assured by whole-body reproduction, takes care of survival into the future.

Thus, in the manner of all natural action, information-in (rule) leads to process (or behaviour) and, finally, to information-out (completion). A chemical reaction could happen (has potential), in suitable circumstance reactants are fired and their process runs to fixed conclusion. So it is with life. Central potential (information reflected in a codified arrangement, *DNA*) is employed in metabolic business; the final result of either metabolic or phenotypic development is a target form. This target is a structural container, a fixed or stable form. At the end of its functional life a molecule and/

or molecular aggregate decays into waste. The end of process is, therefore, on the one hand maturity (programmed attainment of target) and on the other, worn-out (entropic) exhaustion from this target. An end of term involves fixity, expulsion or death.

(iii) Relate these representations to figs. 1.4 and 17.4

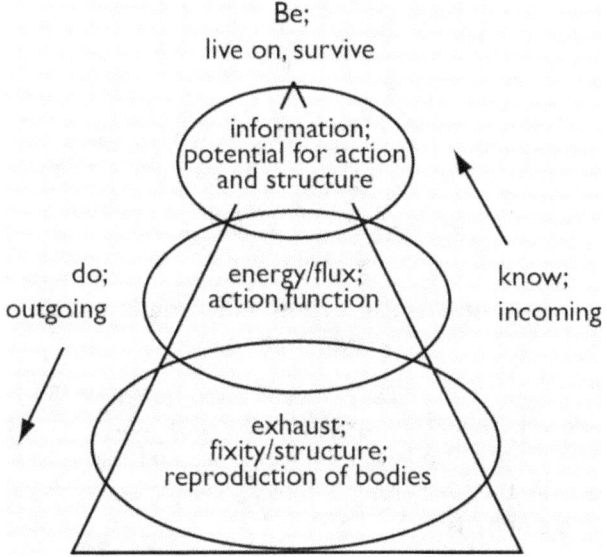

Natural Dialectic represents symbolically all cells and bodies. Yours embodies it particularly well. Top and centre, head and spine incorporate a concentration of information/ control systems. Its central portion (including trachea, lungs, oesophagus, stomach, pancreas, small intestine and liver) is involved with the uptake of materials for respiration; a 'warm' heart irradiates both gaseous and liquid elements. Energy means business. Arms are flexible extensions for hands that feel and manipulate; legs and feet are appendages that combine to give us balanced locomotion. Last and least but still a crucial part, the lower portion is involved with water regulation (kidney), the elimination of gaseous, liquid and solid exhaust (kidney, bladder and colon) and reproduction (by way of sexual apparatus) of fresh, fixed forms. Solid bodies are elaborated and drop out from here. The sweep of cosmic fundamentals clearly falls from skull to pelvis, brain to genitals.

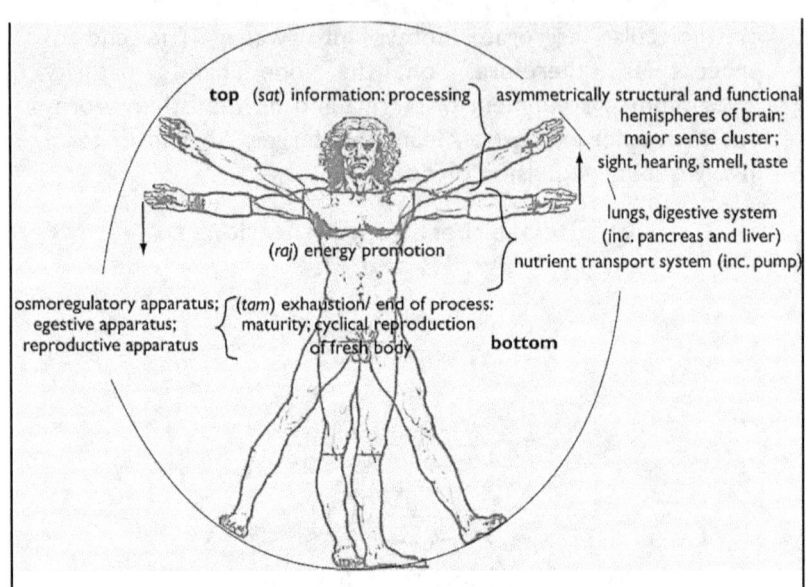

The Information Plug: Mammalian and Cellular Schemes

Information is plugged into energy/ matter in its active (physiological) and passive (structural) roles. The organic construction of bodies is organised around oscillators that reflect, in their different combinations, different shades of cosmic emphasis.

(a)

(b)

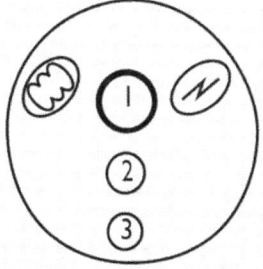

① nuclear, genetic information
② energy metabolism with chloroplast and mitochondrion
③ structural membranes inc. 'anatomical' cytoskeleton

Their 'set' reflects a 'cosmic' gradient from potential through kinetic to exhausted phase, from inward informative code through the outward expression of energetic functions to, as in all good engineering design, structure that most efficiently accommodates desired patterns of behaviour and their ends. The human case (*fig. 17.4*) incorporates a full set of such archetypal oscillators; the resulting balance is said to reflect the epitome of natural design not least in its top element, an information modulator called the brain. In a cell (1) nuclear 'brain' informs both (2) the biochemistry of cytoplasmic functions and (3) suitable structures for its industrial expedition.

(c)

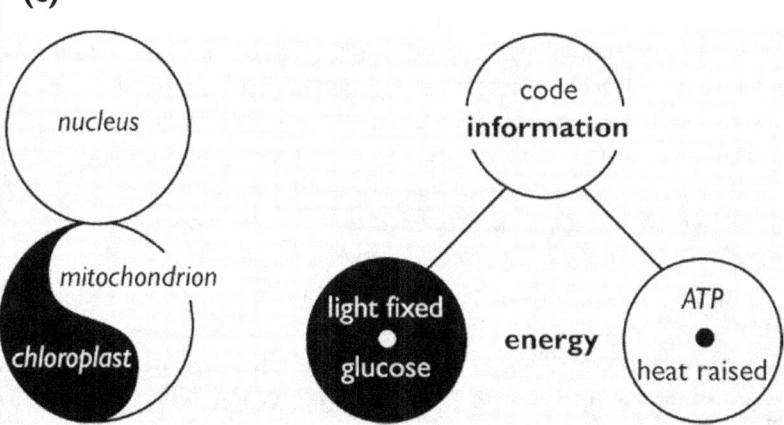

Compare this diagram with *figs*. 2.2 and 21.1. Information is the potential for orderly action; action depends on energy. Intriguingly, the correlation of this basic existential dipole is reflected in the substructure of a eukaryotic cell. A primary division of cell function is distinctly represented by organelles of information (nucleus) and polar energy (chloroplast for (↓) fixing light and mitochondrion for generating the agent - *ATP* - of biological 'steam', that is, of the orderly thermal (↑) release of energy-for-work).

> Indeed, if bodies are designed *top-down* a cell is *central outward*. Nuclear information leads the cytoplasm's business zone; its body is dynamically stabilised by membranes, walls and cytoskeleton.
>
> *fig. 19.2*

Thus the basis of biology's an immaterial entity. **The subject's root is metaphysical.** Yet modern science (with its concomitant philosophy) is locked into a restrictive, self-imposed self-definition - the philosophical materialism of naturalistic methodology. This leads logically, on its own predefined terms, to the unchallengeable conclusion that all bio-logical organisms (including ourselves) evolved as the product of chance events within the framework of natural law.

Such mind-set is, however, strangled by its own catch-22. No-one denies that information runs the bio-show; nor, as with machine, form and performance of bio-machinery appears 'as if' deliberate. We know (from Chapter 6) that all technology and printed literature passively incorporate intelligence. Mind (of their creator) lives invisibly and inextricably in them. But, by itself, can sheer matter (even grey stuff) comprehend, innovate and system-build? Or codify its aimless plans with script that turns into a purposeful reality?

Conversely, mind cuts straight to chases mindlessness can never see; and, always, coded plans instruct resultant bio-form. Cells, with their nucleic acid, clearly operate like any automated factory; they embody information 'by analogy'; it is 'as if' their core displays intent. Life, psychologically and biologically, is unequivocally information super-rich. *When does appearance of intent become so overwhelming you infer it's real; whence does what's inferred to be intentional arise; what's the nature and the source of information and, to wit, bio-information*?

Is randomness informant or informed? Rather, it degrades and swamps constructive logic. Faith that the basis of biology derives from fecklessness of chance is blind. **The following seven chapters will argue that mainstream science (and, in particular, molecular biology) has yet to fully grasp the clear, informative implications of its discoveries.**

More and more molecular biology shouts metaphysic since, as well as the chemical hardware, it increasingly illuminates the signals, switches, codes and carriers of complex operating software.

To rehearse, a book's as unintelligent as ink and paper; so are specific gigs of molecules. Thus, *bottom-up*, material information *is* its mindless agents, that is, its chemistry of carriage. It is passive; it's been gradually, obliviously 'organised', according to the *AVB*'s own *PCM*, by chance mutations acted on by variable constraints. Natural histories are promoted to support this evolutionary view; plausible narration presses to explain such mindless source of bio-information; science promenades with Lady Luck. *A post hoc story that articulates a chain of possible historical events is, however, not a scientific exercise in the same sense as operational physics, chemistry or applied biology.* Indeed, how do you define the subject's sphere of influence? Is validity exclusively conferred by a materialistic stance? Or, more rationally, is it by comparing competence of explanation to obtain best-inference or even, perhaps, a single, causal agency?

Archaeologists distinguish artefact from soil. Artefact's informed and it may help, dramatically, to face the fact that information is, at root, a natural and immaterial element. An archaeologist can also tell a bone or fossil from surrounding soil. Here lies metaphysic by the name of reason. In this case causal agency that's best inferred is life and life's biology, we'll see, is packed with coded information. Scientists with *SETI* (*NASA*'s search for extra-terrestrial intelligence) agree that they could spot intelligence apart from 'noise'. Still, naturalistic methodology and atheism team to resist redefinition of the *zeitgeist*; they resist admission that the source of immaterial information is, naturally, mind; and denigrate such inference as 'pseudoscience'. Thus evolution theory, materialism's version of life's truth, is clutched.

Top-down, however, information's sprung in mind and mind recruits some kind of symbolism to materially express itself. Symbolic code includes formatting rules (such as language with syntax and grammar), carriage (using sound, light or chemicals) and storage (memory, tape and so on). It includes communication whether through inanimate or animate an agent. Organisms involve these informative agents in abundance. Mostly their intelligence is reflected passively, as in machines; but active, creative, conscious response occurs in some as well.

From this point of view matter has no chance of ever generating mind's core *sine qua non*, information. As Chapter 6 describes in detail active precedes passive state. Mind's creativity rules matter's lack of it; mind's message gives rise to material patterns which, according to its purpose, other minds can understand. This frame of reference treats biological organisms as passive or 'rigidified' expressions of information. **Again, the basis of biology is therefore**[118] **metaphysical**. Metaphysical leads, physical follows. The correct angle from which to view biology, whereby its parts fall into place, is therefore from the angle of immaterial information. Code, syntax and semantics - translation to and from a purpose - are the operators of information. The origin of life is abstract code. Whence appears, in all the universe, by agency of wind or heat, dust or rain, a set of abstract symbols? **A cell is seen as information first, chemicals second.**

---

[118] see also Chapters 5, 6 and 13 – 17.

Yet current study, committed to scientific materialism, cannot or will not see genetic code or language of the genes this way. It treats, upside-down, chemical as primary. It is therefore forced to bet on matter, energy and chance alone. It relegates the cause of code to randomness and thereby incorrectly raises evolution theory and its random generator to a pedestal of total rectitude! Teach new generations, teach your children incorrectitude!

Biology is Hierarchical and Cyclical

Bottom-up, chance-born genes rule the roost. Genes are factors that store, in the manner of a database, coded information that can be translated, subject to cyclical feedback by various messengers, using a 'staff' of *RNA* and protein executors that are themselves the product of that code.

From a *top-down* point of view, however, information is invisible; its immaterial meaning and purpose intangibly but recognisably inhabit material shapes and behaviours. Inner precedes, substantiates and supports outer. Invisible creates visible. *Information precedes and orders energy/ matter; its creations are symbols reflecting mental formulations.*

Top, therefore, in the hierarchy of biology (*fig.* 19.2ii) presides its own potential - information. Integrated forms of homeostasis control the expression of plans; and networks of control are, by their nature, bureaucratically hierarchical and cyclical. Genetic and phenotypic complexities involve specific chains of cause that order preordained effects. Code anticipates. Signals communicate. Information operates with feedbacks which ensure dynamic systems 'stay in line'. Therefore, to find the causes of 'downstream' effects you have to travel to their 'upstream' cause. This cause is not natural selection. Nor is it mutant lesion of a gene or chromosome.

Hierarchical information-structures are conceptual. They are archetypal. Such imprimatur is clearly expressed in physical product; coded information (in the form of *DNA* and chemical signals) is the recipe for subtle, bio-molecular patterns from which are derived cells and cell interactions. On this basis, in turn, are constructed the large-scale organs and operations of biology. In other words, visible characteristics called phenotype depend on molecular action that, in turn, depends on genotypic and other pre-programmed information. <u>Life is so obviously purposeful in character that genetic chemicals are ascribed all kinds of animistic properties.</u> They 'compete', 'organise', 'express', 'program', 'adapt', 'select', 'create form', 'engage in evolutionary arms races' and even, lyrically, 'aspire to immortality'. In inverted commas, of course, because like every other kind of chemical genes are non-conscious and have to leave any panegyric to their PR men. No more than letters of an alphabet desire to make meaningful sentences do base-pairs 'author' *DNA* or genes 'conceive' of organisation as complex as cells or your body.

The origin of hierarchical order, cyclical flows of information and integrated function is always purpose. The origin of purpose and its attribute, meaning, is mind. *We might, therefore, reasonably infer that the basis of meaningfully informative, functional and structural biological hierarchies is mind.* What holds any scheme in mind but memory? A memory is a passive store and so, therefore, are subconscious archetypal memories. These archetypes are not products of the imagination of philosophers or, where conceptual design precedes their prototypes, of inventors. But they govern biology. They inform it from above. They are life's records not fixed on paper, disc or other chemical but naturally archived in memory. You might argue that body is in mind as much as mind in body. *Bodies are envelopes; they are the atomic outskirts of mind.*

Down the dialectical hierarchy business follows its directive. **After informative, biology's second set of hierarchies is energetic.** On cue, there radiates reflection of divinity - prismatic rainbow or a wall of light that boundlessly supports the house of life on earth. Hence, first in line stands photosynthesis. Nutrients are 'frozen' out; a rain of nectar and a snow of sugar crystals is precipitated from the leaves of plants. Fixed sunlight is next joined by energetic gas; oxygen fuels, like a bellows, respiration. This regenerates a bonded 'safety-match', a 'battery' called *ATP*. From this universal currency of 'grounded light' is struck a flame to drive metabolism and the chemistry of everybody's body. Energy metabolism is a key operation whose two aspects, anabolic photosynthesis and catabolic respiration, underwrite the 'buzz' of energy we know as life.

States of organic matter range in phase from liquid crystal matrix to hard shell and bone. Organisms involve all kinds of molecule comprising gels, membranes, organelles and various qualities of tissue, skin and hard materials. These constitute the board on which life's programmed circuits run; they provide containers for action, forms to facilitate function, anatomy to complement physiology. **Thus a third set of structural hierarchies completes the purposeful range of dialectical logic.**

Symbolically-coded information's orderly exchanged with matter; it unconsciously works, like a machine, according to its pre-set plan. Life's 'purpose' is equilibration round construction-and-survival schemes; such communicative cycles of equilibration are discharged by vibrant dynamism known to biology as homeostasis.

The Central Executive is Homeostasis

Life, the integrator, draws together; death, disintegrator, splits apart. Biological form waives, temporarily, the tax of death. The instrument of this avoidance, which allows a body time on earth, is homeostasis.

Executive homeostasis is informed, hierarchical and cyclical. Whether psychological or biological, it means adherence to a norm. Such norm is its target. This may be psychologically flexible (as with changing desires) or biologically fixed (as human thermoregulatory control pre-set at 37°C). Either way, in maintenance or development, procedure cycles cybernetically around a stabilising norm. Cycle, wave, vibration - call it what you will. Such dynamic equilibrium keeps (*sat*) balance and stability that we call health. Its vital control involves a switching system called *negative feedback*. Your home's central heating system is a simple example of such feedback. *It needs, along with associated plumbing and water supply, a minimum of three integrated mechanisms to work.* Firstly, a *sensor* tells the temperature. Secondly, a *regulator* decides if it is equal to, above or below a pre-set 'norm' and, finally, an *effector* accords (switches the boiler off, on or does nothing). Such a system, allowing the temperature to fluctuate, is, in effect, oscillatory.

It is worth emphasising that homeostatic regulation involving input, output and a balancing or regulatory processor has no use whatsoever on its own. It does not work in a vacuum but involves an irreducible minimum of complex, associated components and, in metabolic terms, pathways. Sub-systems have purposes and are, in cooperation, interlinked in a hierarchical fashion within an overall scheme or, in this case, biological organism. All codified factors, whether at molecular, cellular or bodily level, must therefore necessarily be fixed together in the right relationships at the same time in order to work.

Tight-rope balance is the key. Each individual, homeostatic cog cooperates in balance with its overall scheme. Where health is balance disharmony does not feel right, discord calls the doctor and cacophony cuts out health's song. Death's the trump of Old Man Entropy - except for a purposeful and advantageous form of death called apoptosis. Such organised cell-death promotes, like scaffolding, construction. How, you ask, could death-and-destruction harnessed as a developmental tool survive a piecemeal evolution of its highly integrated steps? More broadly how, as check-mate to extinction, could any codified development have cycled accidentally unto maturity?[119] Vital balance dances on death's tomb. It holds the end at bay so why, if life's end was unforeseen, did there evolve genetic patterns of senescence (such as the Hayflick limit) and, for every type of organism, preordained life-cycles? Finally, if homeostasis is the key, why hasn't bio-generalised extension of survival-time progressed towards the evolutionary apotheosis of a mortal's immortality? Or was life's tango pre-composed?

In short, so central and fundamental is the informed balance of homeostasis to all life on earth that a visible life-form, whether uni-

---

[119] see Chapters 24 and 25.

or multi-cellular, should be understood as, first and foremost, an invisible but preordained pattern of informative control.

Nuclear Super-Computing

A computer is a mind-machine entirely dependent on the memory of its programs, chips and databanks. Its construction and encoded programs are teleology objectified. Computer operations are based on conceptual Turing machines. These consist, like homeostatic systems, of input, processing controlled according to a set of rules and output. Scientists (such as Leonard Adleman) have realised that *RNA* and proteins responsible for *DNA* manipulation act, with the *DNA* itself, as conceptual machines - input *DNA*, process, output specifically required expressions of stored code. Their biochemistry operates as a Turing machine; bio-information is computerised. From this realisation the scientific development of artificial *DNA* computing is being developed.

No ifs or buts. Biological computing and its issue, biological machinery, are not 'as if'. They're not illusions or appearances but real. Here natural's as conceptual as anything that's artificially engineered. In this case the excuse of analogy, beloved of evolutionists, is swept away. Call them alphabetical or digital (as machine code) but *DNA* molecules store digital information; and the nucleic acid/ protein system *is* a calculating machine. Information, abstract and metaphysical, is carried on biochemical 'speech'. This 'speech' is subject to the linguistic architecture of alphabet, syntax and grammar; its expression is tightly regulated. Sentences, phrases and whole routines - metabolic modules - are fashioned to most efficiently exact desired effects, that is, requisite end-products for their system to survive. No doubt, as chemicals, genes are hardware but just like hardware called a c-drive or a book they carry immaterial meaning, reason, purpose and program. Informative message is carried on the concrete arrangement of materials. **Genetic code is as designed as sentences you think and speak.**

Bottom-up, the problem's to convince yourself that, granted a vastly complex starting-point, the computation of cell chemistry haphazardly, mindlessly 'improved itself'. What mindful systems flow-chart could you build to demonstrate this possibility?

Top-down, programmers know that, from a main routine, switches branch to sub-routines and, when a sub-routine is done the process cycles back to start again. **These routines are modules**. This conceptual character of algorithm, this repetitious use of switches and blocks of modular code is just what coded bio-systems show. It is how nature's life-forms, full of reason, always work.

DNA and *RNA* are nucleic acids; and, actually, these acids compose a **genomic super-computer** whose storage component (*DNA*) is operated on (for constructive, comparative and regulatory purposes) by

RNA and protein. Whereas man-made computers work to base-2 (0 and 1) bio-computing works to base-4 (the nucleotides C, G, A and T). It also works in four dimensions.

Its storage string of *DNA* is, like any line, one-dimensional.

The second dimension, where parts of the string affect each other directly or through the abovementioned *RNA* and protein proxies, is regulatory. The latter act as repressors, activators, transcription factors and so on.

Third dimensional computation involves genomic shape. Genes are not randomly scattered but clustered according to need. Even if not neighbours on a chromosome, cooperative genes are collected to proximity by the way embedded code causes their relevant *DNA* to fold.

In the fourth dimension, time, nuclear conformations may change according to a stage of an organism's development. Epigenetic factors are, we'll see, involved. Mobile elements called transposons and retrotransposons may also be involved in the process. The genome you end up with will not, at least in sequence or in sections used, be the one you started from.

Such complexity amounts to a teleological symphony. The bio-super-computer is a cosmic wonder. Only self-deception, a symptom of severe philosophical malaise, could believe any fully integrated, multi-layer control panel - least of all this one - ever built itself for no reason by chance! **One would, therefore, predict research will more and more reveal signs of bio-logic to the point that, in 'live' computing, such complex permutations, integrated combinations and hierarchical sets of regulation will further squeeze then nullify the notion that celled systems ever cropped up accidentally, that is, evolved.**

A computer is a mind machine. *On this basis it is established that the cybernetic operations of a cell, an object as thoroughly material as a computer, superbly meet the criteria that pass it as a mind machine.* **A cell is a mind machine. Its instruction manual is a program written up as 'genome'; and its systems hold semantic meaning as modules or entities of bio-form like, for example, you.**

If this is right then naturalism's explanation badly needs to twist interpretation of the facts - for mind read randomness. Henceforth we'll check the evidence accordingly.

Conceptual Biology

Is archetype a dangerous idea? For materialism it is fatal.

Bottom-up thus trenchantly denies it while increasingly the latest bio-science points 'upstream' towards a chanceless source. This source is, from a *top-down* point of view, derived from software called an archetype.

Software always issues from the purposes of prior intelligence; and archetype (or 'typical mnemone') **may be defined as a conceptual template or, in dynamic terms, a program.**

How, though, can frameworks that are 'logical', 'goal-oriented', 'reasonable' or 'purposeful' be implanted in biology? You may remember (from Chapter 1: Causality) *energetic* and *informative causation*. The former pushes you from past to future. **The latter is goal-oriented causation; and goals are in the future pulling you their way. They pull you *from* the future; they are metaphysical attractors, guides that govern your behaviour as they lead you through the world.** Of archetypal structure biology is generally concerned with instinct and the morphogene.[120] Morphogene means a specific, developmental program in universal mind, a metaphysical but also morphological attractor. Attractors pull things forward the way that, in response to pressing circumstance, they have to go. Push and pull cooperate. **Energetic and informative causation, running anti-parallel, are the way the world proceeds; and, in the context of biology, primary archetypes pull while secondary genes, body and its physical environment all push.**

The basis of a conceptual approach to biology *is* archetype. It should be crystal clear by now that an unconscious archetype is, like instinct or memory, real but not physical. Indeed it *is* a memory, a natural record; it is a sub-conscious mental, metaphysical construct, an idea or rationale physically realised in a variety of similar forms. Call such template biological (since it's expressed as bodies) or (since it's in mind and includes instinct) psycho-biological the same.

For the information theorist 'archetype' is an idea realised as a main, core routine (or type of organism). Around this core are linked sub-routines - some universal, others more specific - which are appropriately tailored to integrate with each other under the control of any particular main routine. These main routines and sub-routines serve as modules; and they are recognised as homologies. Such a central *top-down* concept is worth rephrasing. **Various tailored sub-routines are called from a Main Routine.** Suites of such modules combine as permutations around which different bodies are, under the coordination of such an Archetypal Master Routine, expressed.

In the *top-down*, hierarchical view of Natural Dialectic, tiered sets of modules compose, at top level, the range of life-forms and, lower down, their various coded and coherent parts. Again, the clearest example of this is demonstrated by the way that biological development is controlled. In such a scheme distribution of modules across the living world is, in varying degree, mosaic. For example, one that is critical (like respiration

---

[120] *see* Chapter 16.

or nuclear operations) may appear, appropriately attuned, in every organism. Other sub-routines may occur in scattered and sometimes unexpectedly different locations (as for example, haemoglobin in humans and various kinds of plant; or luciferins, bioluminescent compounds each complete with complementary synthetic pathway and luciferase switch, dotted about in various kinds of insect, bacteria, protist, snail, jellyfish, ostracod, fish, squid and shrimp). While evolutionists term such apparently unconnected, mosaic appearance 'convergence',[121] from the perspective of archetypal computation it is called 'modular programming'.

Molecular biology's great strides are also heading straight towards the notion of an archetype. A similarity of genetic instructions is found to extend across a great variety of forms. For example, a high-level developmental complex of modules (*see* Glossary: 'homeotic gene') may code for outline body-plan in very different organisms such as human, duck or fly. Or it may call subroutines for the construction of completely different kinds of eye - a normal gene from a mouse can replace its mutant counterpart in the fly; now an eye, a fly's eye not a mouse's, is produced. In other words, such genes act as 'go-to' switches in an archetypal program of development.

The power of shape-making is equally evident in the operation of hierarchically integrated circuits of developmental regulatory networks.[122] Experiments as well as logic (for example, the engineering principle of constraints) indicate the impossibility of any kind of mutation gradually creating the tight-knit nucleic acid and protein agencies of such computer-like functionality. In fact, non-destructive mutations for latterly-expressed, minor, micro-evolutionary variations may occur but never for early, top-level genetic expression affecting body-plan. Macro-evolution of this sort is out.[123] Furthermore, the newly-discovered layer of complexity, epigenetic programming, is observed to play a major role in morphogenesis. Its informative cooperation still more radically contradicts the historical hypothesis of gradual, randomly-generated adaptive 'solutions-by-mutations' to ecological challenges. *At this point neo-Darwinian explanation reduces, simply in reality, to wilful materialism, a faith of naturalistic wistfulness and a misleading, occasionally polemicised mythology.*

And you rejected, philosophically out-of-hand, conceptual archetype? Variations-on-thematic-archetype are wrung, by breeders

---

[121] see Glossary and Index: convergent evolution.
[122] see Chapter 25: first three sections.
[123] natural selection and mutation are inadequate mechanisms by which to realise the transformist hypothesis (see Chapters 22 and 23); modern genetics and molecular biology also increasingly militate against it; see Science 210 (4472): ps. 883-887, 1980.

and mutation, on the forms of dog, horse and many other organisms. Yet the direction of natural selection, though supporting trivial differentiation, can't *create* a metabolic pathway, tissue, organ or a body-plan. It operates the wrong way - (↓) down towards *loss* of alleles and genetic information. Variants are, as even pedigrees may show, mainly due to bred-out loss of information and not evolution's necessary gain; then death always freezes up a limit to plasticity. **In fact, by artificially and fiercely driving such selection to its breaking point you'd soon empirically discover whether nature could snap barriers of type and *innovate*. <u>There is no evidence, not even nascent, rudimentary evidence, it can</u>**. By contrast deliberate not random changes rung upon the archetypal sub-routines is how Natural Dialectic would interpret facts. It would informatively explain convergence, co-evolution or mosaic evolution as the engineering application of a theme to local detail, of a principle routine to different practices. *It would be found no accident that a mosaic conservation of both form and function pervaded somehow, somewhere, every genetic/ phenotypic form of life. These general routines and sub-routines - locally reflected by material genes, molecular conformities and thence the larger structures that entirely constitute biology - would reside in immaterial mind.*

In summary *top-down*, hierarchical perspective rediscovers a primary bio-structural agent - archetype. Round this 'axis' secondary variation (due to adaptive potential,[124] sexual recombination and, to a lesser extent, genetic mutation) continually occurs. *In other words, so-called micro-evolution is a secondary process which, far from progressing to macro-evolution, is dependent on the expression of a primary agent, conceptual archetype.* **Such a view, with concept preceding material chance, is logical, reasonable and data-compliant.** Its order reflects what we find everywhere. <u>However, because it includes an 'unnatural', immaterial element - intrinsic information - it implies the antithesis of a naturalistic, evolutionary account of origins</u>. For this reason the strength of the archetypal against the weakness of the evolutionary idea will be thoroughly scrutinised over the course of the next five chapters.

---

[124] Chapter 23: Super-codes and Adaptive Potential.

Chapter 20: Alchemy

Chemical Evolution?

The issue is not one of religion or opinion but science and logic.

One explanation for our beginning is that we were deliberately created. The other, which is the neo-Darwinian theory of evolution, is that we were not. The latter starts with a process called chemical evolution. This phrase implies that lifeless chemicals 'evolved' to the point whence they could 'self-construct' the primary unit of life, a reproductive cell; it means the generation, perhaps gradually over a long period of time, of life from non-living components by physical means alone. This process[125] is integrally part of, strictly not the same as, Darwin's consequent evolution. After all, it casts no role for natural selection, variation or mutation; Darwin merely hinted, hopefully, that once upon a warm pond....

E cellula omnis cellula. Only from a parent cell does daughter come. **No exception has ever been found to this rule so that it is called The Law of Biogenesis**. In 1860, a year after the publication of the Origin of Species, creationist Louis Pasteur decisively debunked the medieval notion of abiogenesis (then called spontaneous generation) by demonstrating that broth in sterile flasks did not spoil. This experiment, which underwrites the use of sterile equipment in medicine, therefore showed that life does not spontaneously arise. It confirmed the Law of Biogenesis. *Nor in modern times does any process of abiogenesis, whether quick or slow, man-made or natural occur.*

Darwin: Half Right, Wholly Wrong?

Is not Darwinism mostly faith in the unseen? It is not Darwin's facts but his hints, suggestions, interpretations and extrapolations one might take to task. In so doing we'll check examples but, remember, this abbreviated version has lots more lined up behind.

Non-Darwinian seed of a Darwinian tree of life must have evolved from barren pools or clays; from mud or (either warm or cool) saline solution. Could this be physically possible?

Which inference, yes or no, do the facts (most of which Victorian science did not know) support? If they support a positive then Lady Luck has won the toss. If they support a negative then 'life-from-non-life' flips to 'modern alchemy'. *An axe is laid upon phylogeny; the root is sliced; Darwinian seedlings never sprang to sprout a tree.* **No seed, no tree.**

---

[125] also called abiogenesis, biopoesis or prebiosis.

Perspectives on Three Central Tenets of neo-Darwinism - a Tabulation		Bottom-up	Top-down
✓ true ✗ false			
①	Abiogenesis/ chemical evolution	✓	✗
②	Variation (so-called microevolution) by mutation, adaptive potential and natural selection	✓	✓
③	Transformism (macroevolution) by mutation, natural selection or any other suggested means	✓	✗

Aren't half-truths the hardest ones to disentangle; and ones with greatest tendency to lead astray? **Indeed, neo-Darwinian evolution (as opposed to biological variation) may be viewed in the light of a logical fallacy, a trick called equivocation.** *This semantic trick is to conflate two entirely different matters - firstly, variation on existing features (tenet 2) and, secondly, tenet 3's <u>addition</u> of complex, highly informed fresh features (such as coherent organs, systems and so on) in the first place. You might even try to slip in tenet 1!*

Over the next couple of chapters we shall ask:

1. if bio-monomers (such as amino acids, nucleotides, sugars, lipids and so on) could spontaneously form in a single place in sufficient quantities to build a cell.

2. whether interconnected bio-polymers (such as proteins and nucleic acids) could form from these in water.

3. whether the integrated, codified metabolic pathways that a cell requires could appear in operational readiness from scratch.

4. whether a cell, operating in homeostatic, steady-state disequilibrium, could be constructed by chance in a continually changing external environment.

5. even if it were, could such dead-ringer of a living organism (an extremely well-informed, complex and huge conglomerate of atoms) be 'raised from the dead'? Could the pre-life corpse be rapidly primed to a dynamic, steady-state condition before chance decomposition or fatal degradation of any part occurred? Could life accidentally 'go live'?

> The holistic answer is, of course, no. On what evidence is naturalistic speculation, strongly theory-driven, based?
>
> Next Chapters 22-26 will question the mechanical possibility of macro-evolution. It will find that, on the ground, there is practically no evidence for this crucial extrapolation (from variation or so-called micro-evolution). Surely the evolutionary scenario is not composed by only hope's determination and a flair for rational story-telling?

fig. 20.1

Modern Alchemy

Transmutation of base chemical to gold; transmutation of base elements to life - scientific orthodoxy has, since naturalistic explanation lacks all choice, resurrected a fresh version of spontaneous generation. Thus it invests heavily in chemical evolution because, as such materialism can't too optimistically emphasise, we are alive due to a series of fortunate accidents. Therefore, life *must* have 'emerged' from non-life. You build it up as you can break it down (although, engineers aver, to build a working mechanistic system up from scratch is not at all the same as just to deconstruct it!). Life's reduced to atomic bits and blobs - except its basis, information, inconspicuously disappears. How, re-building atoms up, could the lost spec. be retrieved? It must, since how life sprang from non-life is materialism's holy grail.

It happened randomly, of course. *But although such alchemy is the foundation stone on which the materialistic explanation of life rests, it needs equally be emphasised that its process is purely speculative, very highly unlikely and, in practical fact, impossible.* **Impossible or, at best, practically impossible - on such thin nicety the cult of atheism stands or falls.** Such very-very-long-shot represents the first of many black-boxes ranged to protect a world-view vulnerable to the question 'whence came you?'; and on the offensive, to promote a plausible, naturalistic, evolutionary inference of your body's origin. *It may well have nothing in it and, if so, such black box represents a coffin for secular faith.* It can be verified neither by observation (no-one was there) nor experimentation. Even if intelligent, highly-directed contrivances produced 'life' from raw chemicals, this would not prove but only jam ajar the door of doubt - a possibility that uncontrived circumstance might mimic such intelligence. Or, door indubitably shut, it might equally imply a sheer impossibility; and, due to the incredible geochemical complexity of such a mission, lend credence to the necessity of mental guidance - by investigating scientists at least. **What, then, are the facts? How could you, knowing all that's so far known and yet despite scientific *zeitgeist*, rest confident no cell chemically evolved - and if no primal cell no multicellular construction 'organised' by chance and natural selection? We shall see.**

Over the next chapters (Chapter 20 from a *bottom-up* perspective, 21 *top-down*) we'll rehearse over thirty major, cumulative reasons why, despite protestations to the contrary, the speculation of terrestrial abiogenesis, whether in extreme or benign conditions, is a species of wishful thinking and, indeed, an unbelievable form of modern alchemy. Within this framework we'll note plenty of black boxes, contradictions and chicken-before-the-egg scenarios. Is it science or philosophy you want? *The abiogenetic canon is, we reiterate, zealously guarded because the alternative (which includes an informative dimension the current AVB lacks) causes the official doctrine - naturalistic materialism - to collapse. Its standard cloth is fuzzy: it is woven from indefinite time with, by definition, unpredictable and therefore incalculable chance.* Anything *might* happen. **Blurred, invisible, indefinitely flexible yet infinitesimally slender is the thread whence 'modern', atheistic *zeitgeist* is suspended.**

Not a Great Start

The origin of the earth has been dated at perhaps 4.57 billion years. A 'Hadean' period up until 3.8 billion years ago was inhospitable.

Atmospherics

It's enough to make you cough! Our atmosphere's not right. Pasteur showed spontaneous generation does not occur under current oxygenated atmospheric conditions. Therefore Alexander Oparin's resurrection had, of necessity, to hypothesise a strongly anoxic, reducing atmosphere rich in, say, ammonia, molecular hydrogen, methane, nitrogen, oxides of carbon, water vapour and perhaps formaldehyde and cyanide. Without this lethal assumption the evolutionary scenario fails because even simple organic compounds, the smallest biological building blocks, fail to form in weakly reducing air (say, oxides of carbon, water and nitrogen) and would crumble as soon as they formed if oxygen were present.

Such politically red-starred speculation was soon propagated in England by Marxist fellow-traveller 'JBS' (Jack Haldane). The Oparin-Haldane textbook model includes the notion of a dilute, marine 'soup' in which floated precursor molecules of amino acids, nucleic acids and so on. There is, though, no evidence for such a soup which anyway, being of high entropy and structureless is a worst case scenario for generating structure.

However, neither soup *nor* sky was likely to have been the case. Current speculation prefers a non-reducing, neutral atmosphere composed of nitrogen, water vapour, carbon dioxide and possibly carbon monoxide. UV-radiation could break the strong bond of the latter apart; and its atoms could react with the products of light-split water to give carbon dioxide. Meanwhile photolysis of water could generate considerable amounts of oxygen. While hostile to chemical evolution this latter molecule, in the form of ozone, protects life from life-destructive UV-rays. Its pre-biotic quantity is disputed but, anyway, oxidation is reduction's opposite. If, for example, appreciable quantities of oxygen were present they would be incompatible with the coexistence of reducing ammonia. Geological evidence can be interpreted, according to what you are seeking to find, both for and against various atmospheric compositions - although in the presence of an appreciable quantity of 'fresh air' such basic bio-compounds as amino acids and sugars are unstable. So was oxygen first evolved from organisms that could photosynthesise? Or, with photolysis of water by sunny ultra-violet radiation, was it always here? The evidence now 'waters down' reducing skies but has forensic rain washed theory right away?

The so-called 'oxygen-ultraviolet' paradox is, anyway, acute. If present, oxygen inhibits the formation of pre-biotic chemical reactions; if absent, then no ozone layer could have formed to shield fledgling products from destructive, short-wave radiation. Such radiation is lethal for bacteria and, we now know, almost 100% of viruses and other kinds of microbe too. Either with or without oxygen chemical evolution could not have happened. *Such is the lose-lose dilemma for a 'life-is-only-matter' point of view.*

Another popular scenario invokes primeval volcanoes. They belched, their lightning flashed and some kind of life-juice was evolved. Lo! This is how life happened! There is, however, little to indicate that such exhalations differed from today's - nitrogen, oxides of nitrogen and sulphur, carbon dioxide, water vapour and a little oxygen, itself swiftly lost in conversions to oxides.

In short, reducing atmosphere is a *sine qua non* for chemical evolution but likely never existed. Nor, if it had, would its presence have lent inevitability to life's unlikely start. It may improve upon a resort that attributes start-up biochemistry to crashing meteorites but could either device lift the odds minimally above nil? Let us, however, generously allow the possibility and at least the reasonable production of organic precursors such as methane, carbon dioxide, ammonia and so on.

Unnatural Interference

The classic laboratory attempt to simulate chemical evolution (by Harold Urey and his student, Stanley Miller, in the early 1950's) assumed an atmosphere based on the 'Oparin-Haldane model'. Such gases were circulated through glassware. Power sources assumed to have driven proto-syntheses exist the same today - cosmic rays, radioactivity, lightning and volcanoes. Instead a continuous spark discharge was used to simulate discontinuous bolts of lightning when, in reality, U-V light would have been the commoner source of energy. Such energy is biologically unhelpful - in its presence amino acids yields are low and *DNA* disrupted - but supporters argue that hydrogen sulphide in an early sky might have acted as an absorbent buffer.

This said, thermodynamic considerations predict the breakdown of any complex, improbable but ephemeral molecules. Natural conditions, especially harsh ones, are destructive of complexity. *They simplify and randomise.* Therefore carefully, intelligently-designed equipment employed an artificial device, a 'cold-trap', to isolate products from the deleterious influence of the abovementioned UV-radiation, electric discharge zone or raw heat. With this cold-trap a muddy-coloured product was found to contain some organic compounds including various kinds of uncoupled amino acids. Spirits soared! *But without the cold-trap chemicals would be destroyed as quickly as they were made; with it no more energy could enter to drive the hypothetical evolution of molecules forward.* **Such intelligent experiment, such unnatural interference is, we shall see, a hallmark of the laboratory search for pathways that might naturally produce organic chemicals.** It defeats the purpose but, if you believe the process *must* have happened, what else can you do?

Evaporated Soup

You need heat for the non-metabolic production of amino acids and so on. Therefore why not consider volcanoes belching in the sea? Newly discovered, lightless hydrothermal vents spew a superheated brew rich in metal ions, methane, hydrogen sulphide and other inorganic chemicals. Couldn't thermophiles (bacteria that inhabit the airless exteriors of these 'black smokers' at super-heated temperatures of up to 400° C) offer clues about the chemical evolution of life on a hot, early earth? Could hot, alkaline crustal water bubbling into acidic brine have conjured amino or fatty acids; or biochemistry have started in swirling, boiling and supposedly oxygen-free water - perhaps perched on or even down inside the crust itself? It's a pity, in that case, about amino acids whose half-life at 350° C is measured in minutes, small peptides in hours and sugars (at 250° C) in seconds. Watery furnaces that produce at the same time destroy. In addition there would be, at most, very little ammonia and, even so, *RNA* hydrolysis would occur in minutes; and dispersal into colder waters simply thins the soup while for long-term functional stability *outside* a cell you need to keep nucleic acids in a freezer!

If volcanic pitches or warm pools can't take the heat then why not pour cold water onto theory? Earth's oceans contain perhaps a billion cubic kilometres of water; thus for any bio-interesting substance to achieve even a minute fraction of molar concentration it would require billions of tons - after the great majority had been destroyed by ultra-violet radiation in the atmosphere before being washed down by rain. Even there it would not last long but, *ad hoc*, you assume it congregates in cool, fresh pools where water might evaporate and the concentration of its solute grow. You'd need strong evaporation since in this exceptional puddle the likely concentration of even the simplest amino acid (glycine) would be the equivalent of about 0.2 mg. in a swimming pool. Thin broth indeed!

Furthermore, water and its solutes are, within a cell, highly controlled so that uncontrolled water, with its destabilising motions and hydrolytic tendencies, seems an unlikely launching pad for life; yet equally, because of the need for a liquid 'reaction chamber', does the dry lack of it. Hot or cold, served in a bowl that's localised as pools or global as the oceans, primordial soup seems all in mind. It's evaporating as we speak. Evaporated soup.

Chained-Up Unchained

No biological polymer naturally occurs outside a living cell. It is not, therefore, easy to account for spontaneous, uninformed production - especially without enzymes and genetic protocol. The question of the origin of life always drives back to the question of the origin of orderly mechanisms such as metabolism.

Metabolism works by proteins; proteins operate the *DNA*; both are made of joined up units. Whether you obtain such polymers 'at random' in a lab or, as in biological practice, non-randomly as synthesized by order of a code, you have a problem. *DNA* nucleotides in water no more combine to make up polymers than dissolved amino acids spontaneously join into polypeptides. There is no such combination. This is because polymerisation is the product of a freely reversible 'condensation' reaction that expels water during the bonding process. By le Chatelier's principle of chemical equilibrium such polymerisation is, therefore, highly unlikely to buck the strong counter-gradient that water represents. In a cell enzymes are able to overcome this disability and build up chains, each to its biologically critical mass - but whence arose the enzymes that are needed to polymerise all types of bio-chain? *By what chicken-before-egg spectacular did cell metabolism first hatch into history?* **In short, a set of chains which can't be made without another set of chains which can't be naturally, chemically made at all is all you need to liberate a life-form from primeval sludge!** Even then, you don't need only chemicals but, critically, prior code. <u>**Any unguided production of biopolymers outside a cell, inside which metabolism is all strictly guided by instructive code, is fundamentally irrelevant to the question of abiogenesis**</u>.

Not So Sweet

Glucose is a product of photosynthesis whose process itself depends on the prior presence of sugars like ribulose bisphosphate - the commonest carbohydrate in the world - and its associated enzyme called rubisco. Single sugar molecules are classified by the carbon atoms they contain. **No five, six or seven-carbon members have so far been detected outside the biological sphere of influence**. However, very low levels of a three-carbon specimen were found in the Murchison meteorite (possibly due to contamination since high-up atmospheric entry temperatures would have been destructive).

How these and other life-essential sugars actually arose and dovetailed into a working cell's coded metabolism remains a total mystery. Indeed, realistic routes for the spontaneous production, simultaneous condensation and preservation of three-chemical, ribose-containing nucleotides on an uninhabited earth have not been identified. Vital sugars, only found historically encoded for in cells, could never have evolved. *No sugars, no ATP, no nucleic acids and no book of code.*

Bags of Life

Every cell has, as its boundary, a membrane that separates 'inside' from the world 'outside' and thus defines the basic unit of an incarnation. The ability to live depends, among numerous other critical factors, on the ability of this 'skin' to withstand disruption. It is made of molecules called phospholipids; and the fatty acids involved are constructed of chains between ten and twenty carbon atoms long - more would create molecules too insoluble to mobilise in water and less ones too soluble, slippery and disruptive. The phospholipid head-group includes, crucially, a negatively charged phosphate part.

In fact, natural, non-biological lipids are unknown. Nor have viable paths leading to them been identified.

Furthermore, enzymes create agents (such as pumps, specialised gates and signals) whose various duties are distributed throughout the membrane. Its bi-layer is thus, even in 'simple' bacteria, effectively asymmetrical, complex and functionally organised. The origin of such specific synthesis and disposition are problems that we'll soon discuss.

Simply put, biological lipids fall into evolution-busting, chicken-before-the-egg syndromes. Neither cell wall nor membrane's lipid bilayer can be made without proteins and nucleic acids; but no codified metabolism, such as protein or lipid synthesis, can occur except inside such phospholipids as it needs produce. **There can, in other words, be no molecular stability without a membrane to buffer and maintain it. So which came first - membrane or its makers? <u>Of course, all are needed simultaneously - with a minimum of another 300 or so proteins and other chemicals</u>. And all such chemicals depend, at the head of cellular bureaucracy, on informative, supervisory codes. So did symbolic, grammatical, anticipatory code or 'requisite' but meaningless chemical come first? Or, fully integrated, maybe both together?!**

Reflections

The left and right hand of a pair of gloves are mirror images. Such 'handedness', dubbed from the Greek 'chirality', is exhibited by chemicals called optical isomers. All non-metabolic synthesis of amino acid and sugar (including Miller's experimental products) shows a 50/50 or 'racemic' mixture of each hand. Miller thereby reflected death not life because life has 100%, obliterating poisonous 50/50 or any other odds, 'selected' by codification to control the segregation of its major biochemical components into *all* L-form (left-hand) amino acids and D-form (right-hand) sugars. Because the ability to do their jobs depends on 3-d shape and because 'handedness' is an integral part of such shape,

Mirror-image Chemistry (Chirality)

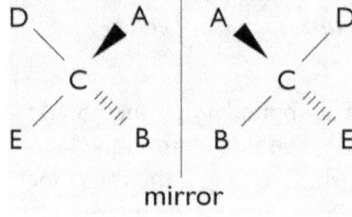

fig. 20.2

life's *DNA* and protein depend on rigid, metabolised segregation. Without such handy handedness metabolism couldn't happen. Life without metabolism?! *Metabolism neither occurs nor, speaking bio-logically, ever occurred without prior arrangement, without a purposive computational assemblage at whose heart is code.* Yet chiral rectitude depends on enzymes in the first place. **Thus another intractable, chicken-and-egg problem plagues wholly chemical, physical and even astronomical explanations of either single-handed, 'homochiral' isomer production or a *DNA* construction process.**

Join Up, Fold Up

A protein works according to a specifically-folded, biologically-active 3-d shape derived from multiple weak, electrostatic, hydrophobic and configurative interactions. These occur in the watery medium of its 'birth' and operation. A precise shape has a precise function and any deformation may dramatically decrease its operational efficiency. **In other words, incorrect folding means functional deficiency or destruction. No function, no survival; or, at very least, illness that a lack of proper protein generates.**[126] There are myriad ways in which each specific chain could theoretically fold into its 'minimum-energy structure'. However, in a cell *RNA*, having been edited by large and sophisticated molecular machines called spliceosomes and editosomes, travels to a similarly sophisticated translator, a ribosome. As its corresponding protein is extruded the latter always folds (or refolds after denaturation) in the 'correct' order leading to the same target shape within seconds. *It always must have because, as noted, the single correctly folded shape of any protein is critical to its operation.* How proteins fold to take their precisely required shape is therefore thought of as a 'grand problem' whose solution is well within Nobel Prize range. We already know that proteins don't just do it by themselves. They are helped by chaperones, signallers and, of course, all prior machinery. Whence co-evolved, in time for cell 1, ribosomes from the shadows of a primal marsh? A dialectical prediction: the answer will involve a coded (thus conceptual) mechanism much too complex to appear by chance.

---

[126] Chapter 23: Entropy of Information.

Number Games

A probability of less than 1 in 10^{50} is considered statistically zero, that is, impossible. **Many calculations have been made that illustrate, by this criterion, the practical impossibility of obtaining by accident a functional sequence of *DNA* and, therefore, protein - let alone a myriad in multiplex cooperation.**[127] In fact, no scientist has so far computed the probability of random searches finding, in the finite time of earth's history, the complex, codified systems ubiquitous in biology. Yet such absent computation is basic to the credibility of evolutionary theory. Nor does statistical imagination of remote a probability make up for physical impossibility - although materialistic faith and hope, transcending fact, will leap such bound.

Pristine Instruction

Instructive input is hierarchically prior to passive output. As ideas are translated into speech or script so purposive code precedes material representation. Machines are 'unnatural' configurations; minds created them. Similarly, meaning is encoded, stored and transmitted by the material vehicle of *DNA*. Its genetic and epigenetic grammar represents an unseen bio-logic. **Its specificity obeys the laws of chemistry but do not kid yourself that systematic, coded information's ever built from automatic, aggregated interactions that obey such law.** And if it isn't so constructed then each protein is, in the sense that it exists only as a direct consequence of cellular information, 'unnatural'. And although such natural artefact may service every cell, no protein is an island. Its work is cooperative with thousands more. It is an integral part of a larger, coherent but encoded and therefore 'unnatural' whole. Purposive wholes do not work if parts are misshapen, tolerance exceeded or the inventory is imprecise. *The irreducible sum of a machine is not the sum of its parts; the equation has to include information, arrangement and the exercise of purpose.* In both biological and technological cases these requirements are, of operational necessity composed at the same instant, expressed with astounding forethought and precision.

No doubt the molecular biology of the cell will eventually be understood fully in terms of the laws and processes of physics and chemistry. So are steam engines. *Such understanding does not, of itself, explain their origin - which was an idea.*

DNA

So, thus, to consideration of code's bio-carrier, a chemic elegance called *DNA*. Friedrich Miescher, who discovered it in 1869, did not realise its vital potential. Now we know that, as opposed to physic's

---

[127] Dembski W: The Design Inference (Cambridge University Press 1998).

dynamic code (Index: cosmo-logical language), this chemical fixes biological information; bodies are, effectively, its book expressed.

We have, if materialists, to ask how *DNA* was thermodynamically formed; and then, against this impossibility, how polynucleotides (in nature *never* found outside a cell) came to float, language intact, into composition of line 1, scene 1 of life's amazing play. Why *this* alphabet? How did *DNA* 'choose' such uncannily apt bases, exclude all other possibilities, find enzymes to polymerise and work it, devise a code, co-opt one systematic operation that linked it to ribosomal protein synthesis and, within the first cell's lifetime, another that allowed it to replicate? How, furthermore, did it adopt the intelligent-looking code sequences applicable to a second alphabet of twenty 'chosen' amino acids such that this second protein-language translates effectively from the first *DNA*-language to create a cooperative suite of structural and functional proteins that, in further turn, form metabolic pathways to create every other necessary (but no unnecessary) biochemical? And only with the breath-taking completion of such serendipity is *DNA* any more use than a floppy disc without contextual hardware. Therefore let us be mechanically pragmatic. *DNA* is a data storage facility; *RNA* is an agent of data transmission, comparison, translation and, in its ribosomal form (r-*RNA*), of catalytic function; and t-*RNA* modules constitute exact translation's vital key.

All kinds of nucleic acid are polymers composed of nucleotides. A nucleotide comprises a base, ribose or deoxyribose sugar and phosphate molecule. These are bound, although options are available, in a particular, always-the-same way. Linkage of single nucleotides into chains uses exclusively what is called 3-5 linkage between sugar and phosphate molecules. In pre-biotic simulation experiments predominant 2-5 is jumbled up in a 'racemic' mix with more stable 3-5 linkages. Therefore short, non-biological strands suffer the usual, predictable lack of sorting and selection. Without natural guidance (by agents such as polymerases) they have neither consistent 3-5 linkage nor base-pair matching. Any short, unguided 'starter-polymer', vanishingly remote in the pre-biotic first place, would as likely degrade as somehow construct a longer strand. Yet even 'simple' bacterial *DNA* is millions of specifically sequenced nucleotides long. Their doubly helical strands are required to wrap around each other in a very regular twine. This means a 100% non-racemic fit!

The 'chosen' alphabet of genetic code is in the form of nucleotides whose bases are guanine (G), cytosine (C), adenine (A), thymine (T) and, instead of thymine for *RNA*, uracil (U). Four of these have been synthesised in a laboratory using unnatural concentrations of pure (analar) components such as urea, hydrogen cyanide or cyanoacetylene. It is unlikely these chemicals ever existed naturally, at least in sufficient quantity and, if they had, interfering reactions and decomposition would have rendered pathways manipulated by chemists irrelevant. No doubt, the optical properties of bases minimize damage by U-V radiation - useful when incorporated into biological *DNA* because of buffering against mutation; but if any pre-nuclear molecule still pulled through then ambush by formaldehyde or a fluctuation in pH or heat would quickly finish off its game. Long-time is not on offer.

Another point - anti-parallel base pairing (G with C and A with T) lends stability to the very large and complex molecule. It is also fundamental to the fine structure and efficient function of *DNA*'s perfectly symmetric, doubly helical purveyance of life's code. The graceful stability is able, in order to easily undergo the transmission of information by transcription or replication, to precisely and rapidly unzip and re-zip.

Base pairs have been selected from all other possibilities for complementary fit. The weak bonds between each exclusive pair are, because of the correct angle between each hydrogen atom and its acceptor, optimal for this job. Such apt but essential association seems, out of trillions of unworkable possibilities, to exhibit a remarkably long series of quick and beneficial coincidences. Such codified fitness is the result both of 'correct' three-dimensional configurations, the right sizes to create a smoothly curvaceous double helix and a junction with sugar that produces the right weak bond connectivity and strength. If this 'perfect elegance' was absent at the start how could a cell begin to work, that is, transfer data and metabolise? **Indeed which, replicator or metabolism, issued first upon the vital stage?**

Any alphabetic letters go together but only certain sequences spell words. Similarly *DNA* bases/ nucleotides are chemically neutral with respect to each other. Crucially for the unrestricted expression of unpredictable, meaningful sequences no base-base affinity is preferred. **Such complete lack of affinity precludes biochemical predestination so how do you explain not operation but origination of functional genetic code?** In other words, where and how did such un-predestined, unpredictable yet (as mind-machines called computers seem 'intelligent') 'intelligent' code-for-programs first evolve?

Now let's focus for a moment on the natural, non-accidental, codified construction of the sort of nucleotide that has adenine as its base. It is called *AMP*. Its first form in the metabolic chain is a sugar called ribose-5-phosphate (itself the product of previous biochemistry). Its biosynthesis, under strict regulation that governs both method and amount of production, involves thirteen steps, twelve enzymes and some *ATP* (which, in a second chicken-before-the-egg situation, is itself constructed from *AMP*). Intermediates rarely have any use except, as in twists made with a Rubik cube, as stages towards a target molecule. Michael Behe summarises the general metabolic problem on p. 151 in his book 'Darwin's Black Box':

"The problem for Darwinian evolution is this: if only the end product of a complicated biosynthetic pathway is used in the cell, how did the pathway evolve in steps?"

Indeed natural selection, during eons spent waiting until every new recruit was 'correct and present', would not preserve each of the many intermediates, accessories and associated code through which useful chemicals are synthesised. The reverse. A part that doesn't have a system can't yet work or know that, if an as-yet-undefined new system were to invent itself, it might then gain a useful role. Lacking any target or agenda either entropy or natural selection would therefore eliminate each novelty as a non-integral irrelevance before it had a chance. Decomposition would occur. *To repeat, the longer a period of time the less likely is the survival of a complex molecule.* This problem is ubiquitous in biochemistry. *What applies to nucleotide metabolism applies to every other aspect of cellular (and therefore multicellular) biochemistry. Such biochemistry basically informs all 'larger' aspects of biology.* **Yet the central question of the gradual development of coded, polynucleotide *DNA* and metabolic pathways is answered with deafening silence.**

Supreme Elegance

DNA is a form of paper and ink that transmits a biological message. The paper, often called a 'backbone', is composed of a chain of phosphate and deoxyribose sugar molecules; and the 'shapes of ink' are, in this case, the abovementioned four bases (G, C, A, T). These carry, like letters of an alphabet, the sense of the biological world. They are *not* the message but its cast of type; they are just book-like purveyors of life's rationale. The actual code involves 4 letters (2 complementary pairs) and 64 3-letter words (called codons) with 21 'meanings' (20 amino acids and a 'stop'). This is the entire alphabet and dictionary for the whole diversity of life on earth and probably, if it were found to exist, beyond. The sequential order of these words commands the way protein is built. **Such code is often judged as optimal. It is supremely elegant. And if code is bio-universal then, since changes would be lethal to its sense, it must have been there from the very start.**

Supreme Density of Data Storage

If the code is conceptually optimal is it practically so? Thus, where *DNA* will write the future of computing, a quantitative description of its information-density beggars belief. Calculated at 10^{21} bits per cm^3, biological *DNA* embodies by far the highest known anywhere. A high-tech microchip might store 10^9 bits per cm^3 - which means *DNA* carriage is about a million, million times more efficient. All information stacked in the libraries of the world is estimated at about 10^{18} bits. If placed on superchips these would, piled up, stretch from earth to past the moon. Registered in *DNA* it would cover, like the *DNA* that informs every different human in the world, about five pinheads! A genetic copy of every kind of organism that has ever lived on earth would need a fraction of this space! Yet the capacity of *DNA* for compaction is such that your super-coiled chromosomes, an encyclopaedic genome that contains all biochemical information needed for your construction, are routinely packed to occupy a tiny fraction of a cell's volume. Most cells are much smaller than this full stop.

Compaction of space is matched by compression of data. For example, while each codon translates to a single amino acid, each such acid may be represented by several codons. Say, TAA, TAC and TAT all code for isoleucine; sixty-four codons stand for only twenty amino acids. Such 'redundancy' has been called 'degenerate'. However, as well as adding a measure of protection against unpredictable mutations such 'degeneracy' may foster compression in the form of overlapping meaning. In other words, it may promote multi-tasking; by using alternative codons different permutations of requirement are met without compromising protein design.[128] Such multi-tasking will, it is predicted, be found to impact splicing, *DNA* folding and epigenetic commands.

A straightforward letter needs wit to write but interweaving second or third level messages much more. **There is no entity in all of physics' known universe that deals in high-grade information as effectively as nucleic acids and their co-ordinated systems; even 'simple' cells *are* genetic super-computers.** You might well enquire how such multi-layered complexity of construction and such interwoven, informative mastery of support appeared by way of its complete antithesis - chance! I wouldn't bet on it!

Supreme Operation

DNA performs two critical and, in the sense that when it's doing one it's not doing the other, mutually exclusive operations. *They are protein synthesis and replication.* The function of protein synthesis is maintenance and preparation; that of replication duplication of information prior to cell division. Both are complex involving many co-operators and stages to completion. Every cell must perform them correctly.

Protein synthesis involves transcription and translation into protein using molecular industrial units, some composed of many subunits. Transcription is a multi-step process employing polymerase and other molecular machines. Translation, at the other end, is also multi-stage; many machines, including multiplex ribosomes, cooperate. **No theory has been proposed for the evolution of the critical, immediate and highly serendipitous linkage of protein with nucleic acid through the medium of such a suite of requisite, complexly-formed and yet precise t-*RNA* translators and corresponding synthetase tools.** Without this series intact nothing could start. So it goes on. *At least 100 different proteins, each coded for and synthesised by the very machine of which they are components, are used to convert DNA instruction into protein product-hardware.* Of the whole

---

[128] see Chapter 23: Hierarchical Language and Super-codes.

system the late Professor Malcolm Dixon, a founding father of enzymology (the study of enzymes) at Cambridge University, wrote, *"The ribosomal system under which we include DNA and the necessary co-factors, provides a mechanism ... for its own reproduction but not for its initial formation."*

Nor is full genome *replication,* employing replisomes, less a complex chip.

<u>*Codes and codification represent the opposite of chance; so does any creation devolved from prior coded instructions.*</u> Randomness in any code sequence destroys the code; and in any process of construction destroys its meaning, that is, its intended end-product. It is irrational to suggest that randomness could have spontaneously generated code sequences as super-specific as those of genetic code at the same moment as its integrated, self-consistent surrounding hard- and software operations. *No thread of DNA could function without myriad coordinates to operate upon it.* Indeed, such molecular machines perform the function of an operating system in a computer. Geared to perform co-efficiently they copy, splice, join, mend, regulate expression and, at micro-level, are the basic reason forms of life are able to exist. **Yet, chicken-and-egg again, the information for this crucial protein machinery is carried on the very *DNA* that it manipulates. So which came first - print factory or coding for it?**

Perplexity

Paradoxical perplexity in biochemistry! No chemical 'wants to survive', 'complicate itself' or 'get a life' let alone look to a reproductive future. You claim that *DNA* 'self-replicates'. **The problem is, however, that *DNA* cannot by itself replicate itself.** *It is an evolutionary* <u>myth</u>, *commonly propagated, that nucleic acids are self-replicating entities.* Man has laboured long and hard but failed to design a single 'self-replicator' so why should atoms blindly chuck one up? **The notion of SRM (self-replicating molecule) may be a white lie that helps keep materialism/ atheism afloat but the biological <u>fact</u> is that there exists no such thing.** Nucleic acid cannot copy in a vacuum. Genetic mechanisms cannot operate without 'satellite' operating systems. Just as a database needs surrounding soft and hardware systems, so life's metabolism needs satellite systems. <u>*Primary genetic material codes for protein but needs an operating system to make it work; the language of protein-coding DNA is, therefore, complemented by a second, interwoven non-protein-coding, regulatory language.*</u>[129] <u>*Such a multiplex system, comprising DNA, RNA and protein, is complicated, coded for by 'bi-lingual' DNA and only found in living (i.e. post-pre-biotic) organisations called cells.*</u>

So, again we have to ask which one came first - ***DNA, RNA*** **or all the proteins that cells need to make a protein? Or, indeed, the proteins needed to metabolise their own genes, that is, to synthesize their own essential templates made of *DNA*?** *The fact is that the machinery (or*

---

[129] see Chapter 22: Non-Protein-Coding *DNA*.

operating system) by which the cell transfers instructions or 'self-replicates' consists of about a hundred precise protein components which are themselves the product of the code. **This alone renders a materialistic theory of the first cell illogical and untenable; and thus abiogenesis a faith forlorn.**

R not DNA?

You *might*, we've seen, track down a nucleotide or two in all the ancients seas but polypeptides, polynucleotides or lipids? No. And, remember, natural selection cannot chemically evolve a single bio-molecule.

Nil desperandum. If you persuade yourself that single-stranded *RNA* alone could take on *DNA and* protein metabolic function (thus making a self-replicator that could catalyze its own synthesis) you might just convince yourself the deadlock's breakable. Pipettes in a lab must change the game. Pipette the pure and activated chemicals you need to synthesize a special, short and fragile strand of sequenced *RNA* called a homochiral ribozyme with very modest catalytic property. Chemical engineers have orchestrated their experimental choreographies; cross-contamination of the target products is averted at each algorithmic step. Lo! Couples conjugating have been caught in test tubes but, in such abnormal cases of 'self-catalysed replication', a prefabricated *RNA* molecule simply binds to limited extent with other fragments. Huge skill, effort and contrivance strive to upgrade such fractional to full, bio-mandatory replication.

A bit of snip-and-splice is most remote from triggering or running life on earth. There's far more to any cell than this. The research represents a hopeful and yet desperate apology for generating biological centrality - **code** *in order to* **metabolise**. The fact is *RNA* like *DNA* needs proteins to construct it in an independent cell. Yet even the *RNA* helices that inhabit various reproductively dependent viruses are not protein-free ribozymes; nor do ribozymes, though found in ribosomes, code for metabolic pathways. Could such an independent and yet inefficient *RNA*-based bio-prototype survive at all? But say it did. *What about the necessary transformation into current, universal DNA?* Having created *RNA*'s uracil chance would have to systematically replace it with thymine, replace ribose with a deoxygenated form and expunge all 'non-canonical' nucleotides, such as inosine, from any regular length of double helix. Then how did genes evolve with ribosomes for protein synthesis? Could, due to faulty but evolving text, early burly cells have survived an abundance of redundant, interfering proteins? None can now so it's unlikely then. And, when you've done with *DNA*'s transcription, what about the other non-redundant part, translation - purposive, conceptual, meaningful translation? Without the lynch-pin of translation, *RNA* to protein, life's basic, coded exercise just meaninglessly self-aborts. You need enzymes, metabolic pathways fully up-and-running straightaway. Then, as well as a transition phase of *RNA* to protein code-storing *DNA* must supersede unstable *RNA*. Where, meantime, has ancestral *RNA* gone? The molecules have disappeared but, apart from blowing in imagination's wind, there is no jot of evidence that they were ever there.

No problem with hypothesis. Dreaming's not against the law - especially when imperatives of theory drive. What, though, about reality? *In labs intelligence in swathes comes first.* The 'information-problem' of genetic replication has been attacked by carefully selective minds. Yet those directing ribozyme experiments paradoxically intend to show that, in a medium of *RNA*-world, chance must have animated lifeless chemicals![130]

Raw Energy Spawns Disarray

You'd think, when talking to some naturalists, that energy was everything. Doesn't immaterial information count?

Evolution's based on build-up; unusual, improbable constructions should increase in complexity. Its problem is run-down. Run-down, the way an energetic cosmos works, falls absolutely opposite to build-up's way. Time's arrow pierces Darwinism's heart with poison. Its dart is tipped, every moment of each day, with toxic entropy. Of course, raw energy can be injected into systems from outside but is that all an organism needs? Sunny weather's needed but is not sufficient unto animation. Crude radiation has to be refined before it can resuscitate, reinvigorate and flow through all the veins of earthly life. More than this, can energy alone initiate genetic code, metabolism and the systematic architectures that express vitality?

In short, in order to obtain constructive as opposed to a destructive biological effect raw energy must be specifically refined. Life forms need, indeed are themselves, open systems but they need specifically purified and not raw energy. They use sugar metabolically refined.[131] The intricacies of such refinery (which must have been instantiated round about life's crack of dawn) well exceed capacity of molecules 'without disturbance by intelligence' to purposelessly chance upon; but, without such orderly catalysis, life could not sustain its special, vital pistons of dynamic equilibrium, its clutch of orderly disequilibria defined as homeostasis.

Materialism won't accede. Ne'er say die since matter isn't dead but never lived - yet isn't 'living matter' what, exclusively, one claims a life-form is?

The problem is: mind organizes, life is intricately organized so how, philosophically deleting mind, could matter naturalistically (thus scientifically), intricately self-organise? Nor just self-organise but symbolically code itself? What, in short, is the material source of information?

---

[130] For a detailed demolition of such demonstration see Stephen Meyer's Signature in the Cell.
[131] Chapter 21: Energy Metabolism Perchance?

A mathematical model of almost any process can be made to work on paper as long as certain assumptions apply. Couldn't, hopefully, 'dissipative structures' such as vortices whip up a gene or gradually whirl up 'self-organised' complexity of cells - each comprising perhaps ten trillion atoms configured in a codified, bio-specific way? Ilya Prigogine understatedly writes: "Unfortunately, this (self-organisation) principle cannot explain the formation of biological structures. The probability that at ordinary temperatures a macroscopic number of molecules is assembled to give rise to the highly ordered structures and to the coordinated functions characterising living organisms is vanishingly small".[132] How, therefore, might entropy create negentropy?

OK. We need simple rules explaining how initial conditions might, delicately poised and then disturbed, yield unpredictable results. This sounds like the 'evolution' of uncodified, non-living systems. Might 'deterministic chaos' from a theory of catastrophe meet life's bill? Or, which describes the large-scale, repetitious and irregular generation of both animate and inanimate shapes, Mandelbrots' fractal geometry? The shapes of geology, physics, astronomy, chemistry and biology can all be viewed and described in terms, where self nests within self at every level, of 'self-similarity'. The 'Mandelbrot set' can, in theory, be infinitely magnified or diminished. While the real world seems to end at an indivisible duality of electrons and quarks the fractal is a geometrical structure that has, fantastically, detail at any size and level of perspective. There resides, at the heart of this 'mathematics of a natural infinity', cyclical feedback from a simple formula. But equations, formulae and m-sets, however much reiterated computationally, can't make a thing. They can describe and virtually explain the energetic patterns nature churns; they may describe self-organising operations of a wound-up wilderness as tension is released. But they do not account for innovation, creativity and origin of specified (in life also codified) complexity as found in every purposive machine.

Repeated attempts to explain the origins of programmed biological form in terms of an 'Auto-catalytic Self-organisation of Matter' or 'Principles of Evolutionary Innovation' are essential to uphold both the neo-Darwinian theory of evolution and, thus, vision in harmony with a by-default-naturalistic philosophy of science. *They are also, however, at root sophisticated, systematic attempts to ascribe the informative qualities of mind to matter and thus generate design without designer. Such flier amounts to so much huff, puff and bluff. It is an enterprise as wrong as it is misconceived.* Of course, creations of both chance and mind are constrained by physical law. **But life is never 'self-organised' by forces or external circumstance. It is controlled internally by very complex and specific codes.**

---

[132] Physics Today Vol. 25, No. 11, 1972 ps. 23-38.

Chapter 21: Cell Sell

Catalytic Philosophy

Disarray is not what determines the catalytic quality of the chemistry that *DNA* informs. Life's watery bodies are, intrinsically, impassive. Their temperature is low and pH as neutral as invites no unexpected chemical events. *Each cell is, effectively, a chemically blank page on which a 'writer', regulator or cybernetic program can control exactly what goes on.* Using a computer analogy the same exercise applies. The cell is hardware and its informative software controls, in a balanced, homeostatic way, the order of its chemistry. *In order that things run smoothly bugs are eliminated. 'Noise' is life's arch-enemy.*

Upon their background blank, however, cells write busily. They are 'control freaks' regimenting algorithmic text through cybernetic chemistry; their discipline is wielded by specific catalysts. Such catalysts, enzymes, are minuscule bio-machines. They are, to the three-dimensional position in space of an electronic charge (to fractions of billionths of a metre), extremely accurately constructed.[133] All life's processes, including *DNA*-replication, depend entirely on them.

Most biochemicals are sufficiently large and their conversions sufficiently complex to require suites of enzymes to guide initial input into a specifically required end-product. Rephrased, some starter chemical 'A' is fed through, say, ten steps but often many more during transformation to a target molecule (say, 'Q'). Each step involves a new substrate and another enzyme. Intermediate enzymes/ steps in the series are useless except for the critical part each plays in the chain. *Therefore some unspecified, currently unknown chemical selection would, without knowing what it was aiming for, have had to build up the 'correct' sequence in order for anything to work!* If a component enzyme were absent the chain could not proceed. 'Q' would not be reached. In fact, no enzyme (or its code) would have *any* use before Q was reached; and useless candidates are deselected. How thus, fully codified and operational, could life's pathways ever accidentally reach completion?

A metabolic pathway is, in other words, a molecular *developmental* pathway, one that develops towards a pre-arranged product. It is, with enzymes being like the lines in a highly structured computer program, *anticipatory*. Each codified and correctly-switchable step is just means to an end. 'Means to an end' means purpose. In a word you call it teleology. With this you switch to *top-down* mode.

---

[133] Chapter 20: Join-Up, Fold Up.

Otherwise explained, most proteins are enzymes and most enzymes useless except as specific intermediaries in a metabolic pathway. Therefore most proteins, present in the right quantities, time and place, must by definition constitute steps in such pathways; and, if such logistics were not satisfied, most genes would code for useless proteins. How come such great redundancy, prior to full complement and operation of a pathway, was retained while it was being built up step-by-step towards a naturally unknown metabolic target? Why does there remain no sign of any actual redundancy today? **Use arises only at the moment a full and correct suite of 'colleagues' for completion of a pathway falls into place.** *The fact is that every single metabolic circuit is, not only of itself but linked into its wider context, irreducibly complex.* <u>*Such pathways form the very foundation of life's continuity.*</u> **It might therefore be concluded that if biology knows nothing of their evolution then it knows nothing of evolution as a whole.**

It was noted that the first cell would have required at very least complete energetic and reproductive metabolisms, with associated genetic hardware and software, in operational place before it had any 'survival' value at all. *In short, multiple pathways including all intermediate enzymes, for which no 'selection' process could exist, must have arisen at the same time within the same ten millionths of a cubic metre of impossible puddle! Enzymes are themselves the product of code and so, with appropriate controls, must have been their genetic precursors.* **How, therefore, can you riskily sell cells as chancily evolved? Is the enormity of the problem for evolutionary theory dawning?**

Metabolism involves algorithmic chemistry. *Pathways, it needs be emphasised, depend on suites of intermediate enzymes precisely tailored to the step-by-step transformation of a raw material into a target molecule. They are entirely purposive and, moreover, each intermediate enzyme is useless except within the context of its pathway. It is all or nothing.* **Each route is an irreducibly complex system. Understandably, therefore, the silence that surrounds the evolution of metabolic pathways or, better defined, the circuitry of metabolic programs is deafening.**

Energy Metabolism Perchance?

Because it must have been there from the crack of life's dawn, fully-fledged, intact, let's examine energy metabolism. Such metabolism, a superb example of parsimony and elegance of design, pumps at the heart of life's engine. You can improve on chance; thus how, if you were asked to invent such a primary, conceptually beautiful system from scratch, might you devise its operation?

Life's energy comes in the form of *ATP*. Inherently unstable *ATP* itself depends, for fabrication, on complex and highly integrated programs called photosynthesis (except for a few bacteria with their own chemosynthetic complexities) and respiration. That's the rub. To make

The Light to Life Energy Conversion Chart

The parts of this conceptual chart illustrate how, like two inseparable sides of a coin, energy metabolism is a good example of dialectical polarity incorporated as an exquisite reflective asymmetry of complementary design.

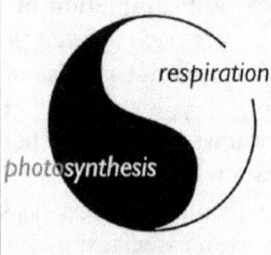
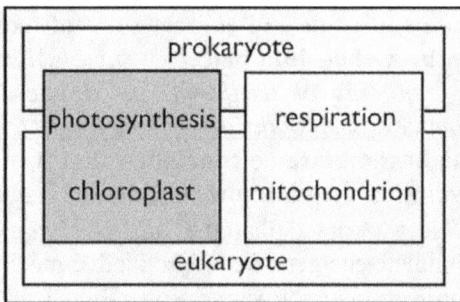

(i) Complementary energy units incorporated into life's phenomenally sophisticated engine express its dialectical polarity. Take the basic equation for photosynthesis (light + water + carbon dioxide > oxygen + sugar) and run it in reverse. You then have the basic chemical description of respiration - except that an input of light has been replaced by an output of 'frozen light' called *ATP*.

(ii) *Chloroplast* and *mitochondrion* are, like a dynamo and an electric motor, a mirror-imaged symmetry of opposites which interlock to drive life's engine. They are as complementary as right and left-hand gloves; and they amount to a fuel factory and a power station.

The conceptual point of a chloroplast, which is vested with its own *DNA* for the synthesis of crucial proteins, is to reduce carbon dioxide to glucose, that is, make life's fuel.

The reason for a mitochondrion, which is also vested with some genetic autonomy, is to oxidise glucose and other food materials i.e. to burn life's fuels in an efficient way. Where punctuated reproduction involves life cycles energy metabolism is a critical, continually spinning wheel.

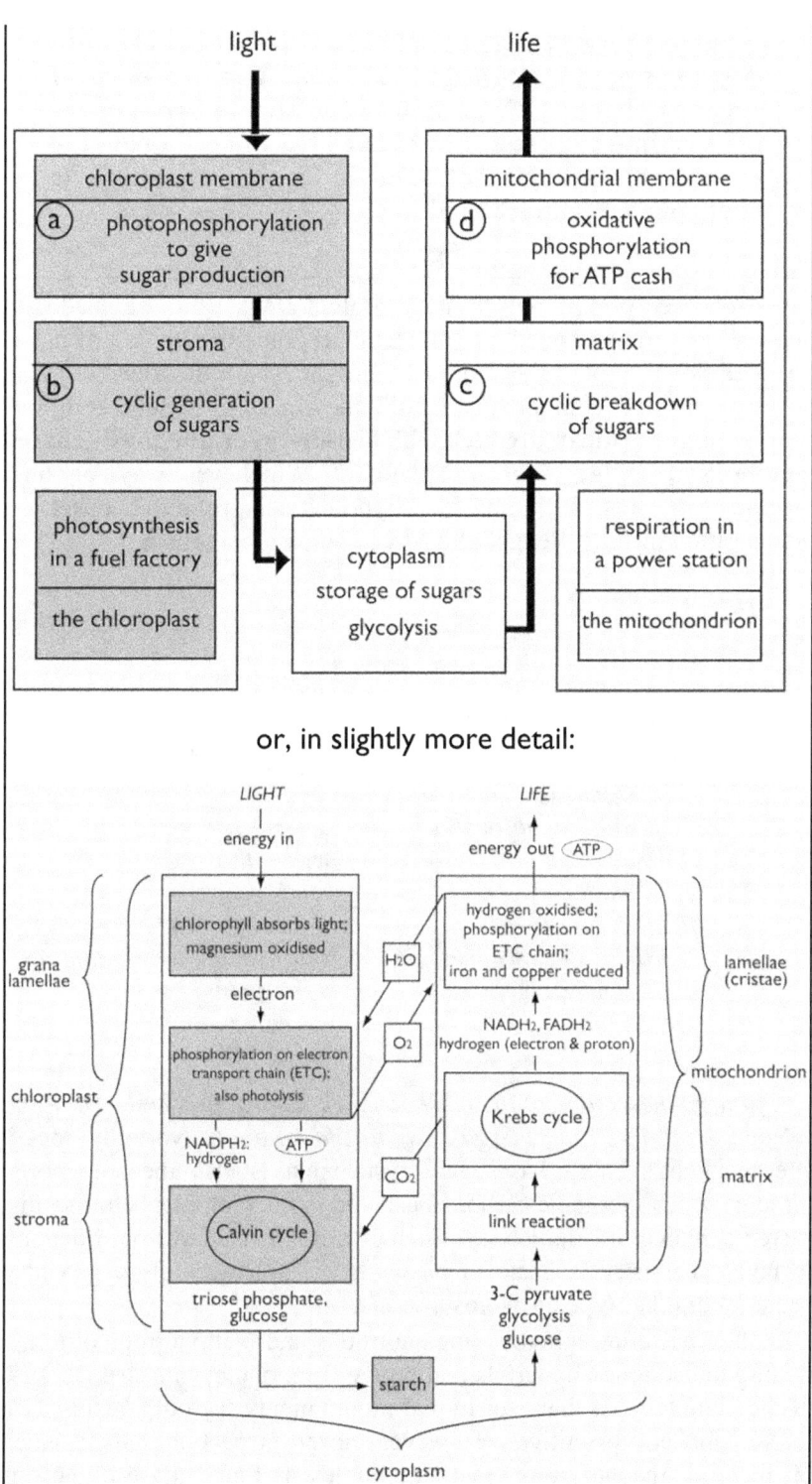

> As with suggested psychosomatic archetype (fig. 15.5) these diagrams suggest a bio-energetic archetype. **The presence of such an elegant, efficient solution to the problem of life's engine suggests that chance has been thoroughly outwitted. Indeed, not known for its ingenuity, it likely played no part at all.**
>
> *fig. 21.1*

DNA you need protein and to make protein *DNA*. Both are needed to make *ATP*; you even need that molecule to prime production of itself! Stripped-down, primal modules might be theorized. What mechanisms and components would they need precisely codified? **Primal or sophisticated which, chicken-or-egg, came first - energy metabolism, *DNA*, *ATP* or protein?** Whichever way or both together, *all* cells, including any hypothetical prototype, exact the precondition of fully-fledged, bug-free energy metabolism.

tam/ raj	*sat*
complex	simple
practices	principle
effects	cause
conversions/ transductions	light
↓ *tam/ yin*	*raj/ yang* ↑
anabolism/ small-to-large	*catabolism/ large-to-small*
light capture/ absorption	heat release/ emission
storage/ fixity	action/ flux
bond made	bond broken
light-to-sugar	thermal unit (ATP)
'earthing'/ materialisation	'quickening'/ dissolution
photosynthesis	respiration
fuel factory	power station
chloroplast	mitochondrion

Photosynthesis, one of the most complex natural processes known to science, is no less than the gateway to life on earth. Not only does it 'freeze' light's energy to sugar crystals but is also the sole portal through which inorganic carbon is ushered to the biological temple, that sacred container of life, a body. Such process of fixation from fluent air to body chemistry is basic to biosynthesis. What therefore preceded bacterial or other type of photosynthesis?

A few bacteria (called chemoautotrophs and methylotrophs) do not require the complex materials that derive from photosynthesis. Because of the simplicity of their nutritional requirements such organisms used to be labelled 'primitive'. However closer inspection has revealed biochemical machinery at least as complex as found those organisms that need to ingest carbohydrate. So-called *archaebacteria* involve a sodium gradient with phosphorylation no less sophisticated than our own. Even the simplest conceivable cell has phosphorylation and carbon fixation systems entirely dependent on suites of enzymes accurately produced by genetic hard and software.

Nor is glucose fabrication more than half the set-up that is critical for any moment full of life! *Respiration* is the other. How did this complicated, co-integrated but critical bio-mechanism 'pop up' in step to crystallise the working team? *While a cell can wait a few days or even weeks for the molecular biology of its reproductive apparatus to evolve, it needs energy metabolism immediately! That is, it needs mirror-imaged photosynthetic (or chemosynthetic) and respiratory systems coincidentally, immediately; or it needs, equally urgently, a digestive and a respiratory system.* This is too much for anyone to swallow.

Thus each aspect of this polar couple's physiologically-integrated structures and functions needs be as sharply explained in terms of evolution (which its coded profile does not fit) as bio-engineered design (which it does). The argument that earth is an 'open system' and daylight alone sustains life's processes is as weak and deceptive as claiming that codified, cellular mechanisms must have 'wriggled' out of water, warmth and mud. Whether for photosynthesis/ chemosynthesis or anaerobic/ aerobic respiration, no serious analysis of evolutionary possibilities is proffered. **In other words, the critical evolution of energy metabolism is (with that of general metabolism, the universal system of information-storage-and-transmission and the technique of protein construction) another 'just-so' black box.**

The Origin of Irreducibly Complex Mechanisms

To make a vehicle (as bodies and their parts are) mechanistic modules are conjoined; an automated factory makes and joins according to a program representing purpose, concept and its plan. Therefore switch Darwin's evolutionary problem round. *His puzzle does not involve an origin of species but the origin and existence of irreducibly complex and coordinated systems.*

Irreducibly complex* means, with respect to a system or a mechanism, one that must have several parts each made of the right material, the right positional relationships between parts (which will include quantum elements such as electrical charge) and, in the case of a biological mechanism, the right 'grammatical' instructions encoded in a manner that allows it to access data, perform tasks and repair and reproduce itself.** Such criteria must all be met. ***The parts must be simultaneously present and built to perform to the standard of minimal function.

Minimal Functionality

As noted in Chapter 16, structural molecules and quantum elements cooperate to execute biological function. Charge, photons and so on involve quantum effects as precise in timing and location as their molecular skeletons. Structural disjunction and discord of resonance combine to determine, by their extent, the full, partial or busted step of any particular bio-mechanical dynamic.

In this context, *minimal function* **means a mechanism must fulfil its promise to the extent that an efficient, acceptable level of performance is achieved.** Such codified, programmed performance includes, at each level of construction, not only the correct shape and interconnection of parts but the correct materials of which they are made. In the interests of efficiency technologists strive to combine minimum complexity of working parts with maximum effect at lowest energetic price. Such integration could, as well as applying to a single mechanism, also involve cooperation with other similarly irreducible mechanisms to engage a larger whole.

A bicycle and space shuttle involve different levels of complexity but both employ conceptual mechanisms and adequately perform their task. Likewise do bacterial and human forms. Thus, in all mechanisms, a single part at any level from molecular to systematic is no use by itself. For example, a television, although complex, is no use without each properly-integrated, working components; nor if it lacks programs, broadcasting centres, a power grid and an audience. Likewise, biological components need cooperate at every level with their environmental context. *Thus if, within a cell, it could be demonstrated that any serial metabolic process or organelle existed along with its informative infrastructure of coded instructions which could not possibly have been formed by numerous, successive, slight modifications Darwin's theory would, in his own words, 'absolutely break down'.* **In fact, it is multiply broken in every single organism.** And the more that is discovered about any cell or multicellular configuration the worse the problem becomes.

Complex systems don't start up spontaneously. Therefore, naturalistic scenarios place great emphasis upon simplicity. Primordial life *must* be simple. The assumption is that a last common universal ancestor (*LUCA*), with codes and basic metabolic functions now intact, gave rise to every other form of life. Thus keen interest is paid to the genomes of prokaryotes considered to be the oldest, ancestral representatives of our family. The question therefore becomes, how simple must 'simple' be?

You can't count parasites because you'd have to count their host. Free-livers, though, roam independently. The smallest known are an archaean called *Thermoplasma acidiphilum* (with its extraordinary membrane and about 1500 genes) and a bacterium called *Aquifex aeolicus* (a thermophile with 1512 genes from about one and a half million letters, bytes or base pairs). Tests lead to the consensus that the most hypothetical bareness of complexity, that is, absolute bio-simplicity may range between at least 250 and 400 different protein and *RNA* coding genes required to cooperate for a basic independent mode of life. This suggests that, from a *top-down* point of view, an absolutely minimal form of life would need between 30,000 and 50,000 correctly sequenced base pairs to constitute its book. This book would have to be, in micro-space and micro-time before least degradation of its parts began, encapsulated and enter into synergy. By synergy is meant cell functions organized and orchestrated (that is, energetically regulated) in the proto-cell. To produce over 250 inter-functional proteins simultaneously the lack of probability explodes to incomprehensible numbers such as make the amount of atoms in the universe or seconds since the world began look ludicrously small. To translate into the production of an *E. coli* bacterium, the size of whose genome (with 4288 genes) falls with the same order of magnitude as 'simple' microbes or the *LUCA*, you'd have to write trillions of zeroes (powers of ten, orders of magnitude) against the right code taking shape! You might have time but still need to crampon-scale the information cliffs. Piton-by-piton accounts of the separate, wind-blown origin of any component let alone whole integrated systems are absent. The heights of Mount Impossible are silent; and volumes of silence speak, when it comes to functional success, volumes.

This is fact. Abiogenesis is dead. No cellular mechanism, even the most 'primitive', could have been a 'homogeneous globule of slime' like Thomas Huxley's mythical *Bathybius haeckelii*. All that is 'primitive' is the understanding of that word's Victorian user. We now know that even the most 'primitive' cell is, marvellously miniaturised and of labyrinthine complexity, a computerised chemical factory.

The fact is not that Darwinism has a clue *how* 'irreducible complexities engaging minimal functionalities' evolved but only that, by scientism's incantation, they *somehow must have done. Innovation and not tinkering with innovation is, as Darwin himself recognised, the real problem.*

To rephrase the problem: the key factor in evolutionary theory is chance (acted on, in life, by death's selection). Yet the key activity rationally describing life is anti-randomisation - in other words, purposive command and control. Life has no positive role for error; chance degrades not creates biological forms. The implication is that, while one can dream that such irreducibly complex systems might have evolved, the hard facts bespeak a far stronger argument - highly rational, logical and intentional operations are, and always were, informed. *Information walks the walk with resolve. Chance, staggering and scaling nowhere clear, has no chance.*

Natural Nanotechnology

Technology involves machinery. A machine[134] is a body or assemblage of bodies used to transmit and modify force and motion, especially a construction in which several parts unite to produce intended results. Engineers develop, build and maintain machines. Professionals know perfectly well that a multipart machine must *start* its working life as a well-designed, functionally

---

[134] Chapter 6: Machines.

coherent unit; both then and thereafter, except perhaps for superficial faults including improvable performance, unless everything works nothing works. Such condition applies with equal force to nanotechnology, that is, the development of functional machines at ultra-small scale. Engineers are learning how to develop tiny devices such as molecule-sized wires, switches, valves, shuttles, rotary motors and, as they seek to mimic bio-materials, protein catalysts and *DNA* computers. In comparison with bio-machinery, however, their efforts are technically skilful but as yet unsophisticated. Natural nanotechnology is proving, with excellence, to have got there first.

How? 'How the gecko, without fail or fall, gradually got a grip' is one of myriad just-so stories evolutionism tries to spin. Not only gecko but shark, spider, octopus and many more, including parts of plants, inspire biomimicry. Biomimetic research is progressively revealing extraordinary concepts realised as codified, molecular machinery. In this case stark, bogey words to be ghost-busted loom - 'intelligent design'. Yet clear signs of this capacity is what, in specifically informed and integrated, operative complexity, researchers with intelligence are finding. In fact, three scientists were awarded a Nobel Prize (2016) for their 'exceptional insight', 'great skill' and undoubted intelligence in designing and developing molecular machines that are child's play compared to the codified, integrated suites which nature is supposed to have produced at random.

Happenstance and time completely lack intelligence. **Machines, moreover, work according to natural law but the latter cannot explain them.** This is because machines are informed by purpose and, in obtaining the fulfilment of that specification, are irreducibly complex in a way that purposeless forces and particles can never be. Molecular biology embodies, brilliantly, natural nanotechnology. When all components are dynamically in place you automatically find, at large-scale levels, equally brilliant technology revealed! 'Of course', scientism cries, 'it's worthless. Oblivion sends no message, nothing means a thing so don't get too impressed. Materialism's driver, energetic aimlessness, simply threshed and floundered into an illusion of design.' Just to inform you, in this case, information's sourced by lack of it. What paradox, evolutionary style!

Biosynthesis

Biologist Craig Venter defined a minimal genome by removing genes, one by one, from the already very small genome of a *Mycoplasma* bacterium. If the organism survived he discarded the gene until he was left with a minimal genome of 473 vital genes. But symbiotic *Carsonella rudii* has certainly lost the bare necessities of life since its present genome of 182 genes (~160000 'letters') is insufficient to replicate or transcribe and synthesize protein. A viable analogy for free-living *LUCA* might be ten times that amount. Both parasite and independent free-liver must also contain, as well as their irreducibly complex chunks of information inscribed on so-called self-replicating *RNA* or *DNA*, the correct quality and quantity of ions, proteins, carbohydrates, lipids and so on. Abiogenetic chance has absolutely no chance in real life; in spite of this, the dodgy hard-sell is to sell a lucky cell, to trade you natural biosynthesis.

What, however, if we add a dash of mind? What about unnatural scientific biosynthesis? Venter, thinking 'maximum impact with minimal genome', wants to engineer bacteria for medical and ecological purposes. So he transplanted one species' genome (with some machine-sequenced changes) into another species' body and so created, at the peak of a mountain of accumulated information and using great skill, tenacity and ingenuity, a novel form of 'primitive complexity'. For sure a highly intelligent achievement! *M. laboratorium*, a beauty of its kind, was nicknamed Synthia - but, although a biotechnical triumph, young 'spare-parts' Synthia was hardly synthesised from scratch. Just a tiny novel fraction came from an expensive sequence generator. Nor is Venter's mind at all the same as Lady Luck's. He honestly admits his cell is *not* an artificial form of life - just an innovative rearrangement of tweaked data.

Billions of *intelligent* man-hours have been invested by clever humans in acquiring sufficient knowledge to deliberately achieve even this very simple transformation in a lab. Millions more are needed to construct fresh chromosomes and, it's hoped, design economically useful, artificial forms of life. **Indeed, the whole thrust of deliberation is to achieve what chance does not**. Nor, in physical terms, are genes the whole game - you would have to copy no less than a whole cell. *To copy's not to innovate or to create. Nor, even then, would such deliberate facsimile of life-form demonstrate that one might ever come by chance.*

In Extremis

No bacterial constitution is a fraction as primitive as modern, humanly informed nanotechnology. *No contemporary organism is 'primitive'.* There exist only very complex or even more complex life-forms.

Even so, weren't proto-organisms on the hostile early earth hard-cell extremophiles? These are prokaryotes that thrive in very hot, cold, dry, acidic, alkaline or salty conditions. The likelihood of *any* large, specific molecular construction surviving the destructive effects of an extreme condition is minimal but, within the context of a membrane, hardy organisms thrive. So, did biosynthesis occur from scratch? Harsh inference from Synthia deflates the case. Hard-sell droops to impotence.

If terrestrial extremes won't do perhaps we'll glide off into freezing, airless space. Might a dust-cloud, if it harbours any carbon chemicals, throw life-on-earth a starting-line? **But 'dusting' earth with life in some 'panspermic' way evades not answers how it all began.** No more than showers of rain can random showers of organics 'germinate'. **And bacterial seeding simply moves the problem off to an unqualified, unquantifiable and unknown plot somewhere in space.** No trade here either. We are running out of options.

Tick Tock

Countdown. It's time for summary.

Michael Denton[135] summarises the whole problem admirably. He writes "The difficulty that is met in envisaging how the cell system could have originated gradually is essentially the same as that which is met in attempting to provide gradual evolutionary explanations of all the other complex adaptations in nature....The problem of the origin of life is not unique - it only represents the most dramatic example of the universal principle that complex systems cannot be approached gradually through functional intermediates because of the necessity of perfect co-adaptation of their components as a pre-condition of function....The origin of life problem lends further support to the notion that the divisions of nature arise out of the necessities of life rooted in the logic of the design of complex systems."

A Definite Flight from Science

All biological bodies are (as Chapter 19 told) entirely reliant on codified, integrated informative, energetic and exhaustive systems. Nor can any cell exist without its homeostatic and reproductive dynamics intact. Which is more miraculous, which more ludicrous - that such operational brilliance 'pop up' by chance or purposely? *Whose logic, in the face of science, is most strained?*

The facts are clear. **Chance and necessity engage, like natural selection, absolutely no creativity, anticipation, intent or intelligence - the latter being the only known generator,** *by its very nature*, **of specific, teleological information. These properties are, exclusively, the province of mind.**

The physical world is, on the other hand, completely mindless and, specifically or otherwise, no-one has a clue how life could mindlessly begin. *Instead you have, on pain of scientism's excommunication, to believe a wretchedly thin materialistic fable dressed with technical sounding jargon - chemical evolution.* Yes, it is true! **The myth of modern alchemy states that an unknown reacting mixture reacted in unknown reacting conditions to give unknown products by unknown mechanisms - against an accumulation of actual scientific evidence.** If this involves a flight from science what will you make of scientism's claims? If they wobble at a shaky altar of coincidence, by definition purpose and intelligent intent do not. The greater an intelligence, the more powerful in principle is its information and the less uncertain its designs. Knowledge and precision outwit accident. Chance is left no chance. **Life did not go live by accident**. Yet faith in science fiction, life-creation fables and a grand, irrational mythology persist.

---

[135] Evolution: A Theory in Crisis p. 269.

Chapter 22: Neo-Darwinism Isn't Fit

What's the Problem?

School taught you Darwinism works. The idea (random mutation acted on by natural selection) is simple. You received the *AVB* but nowadays its gloss is wearing very thin.

In short, can selection of mutations generate sufficient bio-novelty?

Such creativity needs to include the generation of a digital information system (grammar, molecular agents and operating systems) from scratch; the original generation of epigenetic co-systems and of hierarchical, computer-like routines of government whose tightly integrated circuits every cell displays; rapid innovation of new form (e.g. Cambrian body-plans); the origin of sex and developmental algorithms; the morphogenetic congruence of hierarchical layers of construction ranging from molecule through molecular coordination, organelle, cell, tissue, organ and system to target body-form; and the evocation of irreducibly-complex mechanisms at all these levels simultaneously (if things are to work).

The problem is, of course, not the origin of species but of concentrated information; and, in such magisterial dock, evolutionary theory repeatedly breaks down.

A systems engineer at ARM or INTEL would, if you announced his complex chips arose by chance, evince surprise; no less if you contended that the mode of his construction could, equally, have been by gradual accumulation of improvements mindlessly. Because denying information's source in mind denies its causal agency; and as we know, but Darwin didn't, symbolic information is the basis of biology. Though the discipline's dramatically progressed still evolutionary theorists with naturalistic goggles firmly pressed deny such agency. They seek inference and interpretation whose 'best explanation' categorically rejects an immaterial element - although we know that programs, books and speech are physical expressions always traceable to such an element.

'I remember at an early period of my own life showing to a man of high reputation as a teacher some matters which I happened to have observed. And I was very much struck and grieved to find that, while all the facts lay equally clear before him, only those which squared with his previous theories seemed to affect his organs of vision.'

Lord Joseph Lister (1827-1912).

Evolution is agendum-driven but its major props, The Big and Little Accidents, are both in quarantine. Can you start a cosmos out of nothing?

Maybe physical constraints were absent but surely it is reasonable[136] to argue that transcendent, metaphysical events and entities both were and are prior? Now, worse than quarantine, the Little One has, as the last two chapters showed, suffered a fatality. But you can't start without a start. *The problem is profound because the whole edifice of materialism demands primordial seed from which its 'family tree of life' can then branch up or, if you like, can commonly descend.* If, therefore, the hypothetical chemical evolution of a proto-cell *is* no more than sophisticated animism or an alchemical myth, then axe is laid to root. It becomes impossible to steam away to any evolutionary extravaganza.

Therefore this extravaganza, constantly portrayed as truth, needs investigation. Behind closed doors there lurk 'trade secrets'. A sign for public consumption reads 'Work in Progress' since, materialistically and therefore scientifically, evolution *must* work - but precisely how? Briefly, the next four chapters check the theory's manifest and deep deficiencies, its fundamental problem (chance-generated systematic novelty) and whether those solutions currently proffered cut the mustard. We'll subject the interpretations of its modern synthesis to an interrogation. Specifically we'll ask:

1. whether mutations and natural selection are sufficient mechanisms to generate more than variation called micro-evolution.
2. if they are not, and since no other mechanism has been found (with mind disqualified), what might generate the large influx of highly organised information needed to innovate a cooperative biological pathway, process, organ, system or organism? Neither innovation nor transformation has, on the large-scale required, ever been observed; upon close inspection not even nascent systems are found.
3. organisms involve complex editors to reduce rates of mutation and preserve the integrity of code; their effect actively minimises any chance of evolution. Moreover informative entropy also scrambles and deletes 'progress'; everywhere, both now and by extinction in the past, loss or degradation of genetic information is the norm. How, upstream against the torrents of this natural, anti-evolutionary deluge, could complex and correctly-sequenced order-building due to accidents occur?
4. does the fossil record substantiate the idea of macro-evolution?
5. how sex, development or any other agent of reproduction might have 'emerged' by chance.

---

[136] Chapters 5 – 12.

Forthright answers to such questions may well indicate that neo-Darwinism isn't fit and doesn't work; that promissory faith in future answers is, if not a disingenuous ploy, still insufficient to delay for long the theory's demise; and that, in the light of modern science, evolution's vital alchemy demands a radical revision.

Hold on! Research negotiates the problem thrashing in a void of vast improbability. A welter of suggestions wriggle (e.g. evo-devo, punctuated equilibrium or self-organizing entities); hypotheses proliferate (macro- or multiply-coordinated mutations, gene duplication, neutral evolution, epigenetic inheritance, natural genetic engineering and so on). But these, despite including elements of fact, evade the main, material issue. Assumptions pile upon assumptions in a *bottom-up* direction. Thus life's information-rich intensity *must* have been constructed, simple to complex, accidentally. Proposals vary in plausibility but *all* presuppose original instruction. They thereby beg the question by assuming prior information, both epigenetic and genetic, for tightly-coded innovation (not minor variation on such innovation). They don't explain how it's originally obtained and so two basic problems still aren't satisfied. **How is functional form created? And how was information, specifying great and orderly complexity, first injected into lives on earth?**

ARM and INTEL merchants tell you, where it comes to systematic, functional circuits, you must employ an engine to develop each and every integrated part. You need reason's piston to produce conceptual traction and create the informatics life requires. Detectives also know that, with more facts obtained, interpretations change. Life on earth is mostly history and, although contemporary agents are presumed at work, abductive reason chooses 'best' or 'only' candidate for its 'arrest'. Could evolution be the 'best' or 'only' candidate at interview? **The fact is Darwinism doesn't fit the facts.** Its icons - beak of finch, fruit fly, melanic moth, ancestral tree - brew a tale that's skewed with half-truths and, when interrogated, doesn't make the modern grade.[137] A scientific cauldron of conjecture regularly fails to bubble any form of life. *If first step failed*[138] *how did its 'successors' stumble into biological reality - unless progressive, evolutionary mind-set can devise explanatory animations and create a second, extra-grand mythology?*

The Editor: Natural Selection

Natural selection was the first agency proposed for evolution and thus we begin with it.

---

[137] In fact, a much better explanation for these and the majority of so-called evolutionary 'steps' is outlined in Chapter 23: Super-codes & Adaptive Potential.

[138] Chapters 20 and 21.

Edward Blyth (1810-73), a keen naturalist and 'Father of Indian Ornithology', published essays that first appeared in The Magazine of Natural History in 1835, 1836 and 1837. *These, which were read by Charles Darwin who afterwards corresponded with him, introduced the ideas of a struggle for existence, variation, natural selection and sexual selection.* These four are central Darwinian tenets and we might ask why, since in large part Darwin's work was based on Blyth's ideas, the former hardly acknowledged the latter's 'intellectual copyright'. Blyth made no more of his notion of natural selection than the facts warrant. He, like Darwin afterwards, drew attention to its passive, conservative function, using it not explain how new types arise from pre-existing ones but rather how a healthy population is conserved.

Alfred Wallace (1823-1913), co-founder of the theory of evolution and whose paper 'On the Tendency of Varieties to Depart Indefinitely from the Original Type'[139] preceded (1858) and precipitated publication of Darwin's 'Origin of Species' (1859), noted that natural selection not so much selects special variations as exterminates the most unfavourable. Such extermination is, in fact, a stabiliser; 'fit' traits blossom, 'unfit' wither as an organism's own ecology (its niche) dictates. Wallace, like Blyth, believed in original intelligent design.

The stakes were high. Darwin put his money on 'selection' but had no idea how it worked. Was it by the incorrect idea of Hippocratic pangenesis (Darwin's bet) or Lamarck's *IAC* (inheritance of acquired characteristics)? Nor, despite recombination, do fresh genes evolve from mating by two parents of a kind. Nevertheless, his bet replaced a 'supernatural designer' with a mindless kind of breeder that is central to the atheistic faith. Now, instead of materialism's apotheosis, let's treat editorial selection for what it actually is and, at the start of our inspection of its 'mechanism', post three *caveats*.

The first, Blyth's, is that natural selection is solely a process of elimination. It involves the disappearance of 'unfit' organisms. **But destruction is not creation. Natural selection originates absolutely nothing at all.** It is, simply, a fateful finger hovering above a genetic delete button, no more than a name given to the lucky survival or unfortunate death of an organism or group of organisms in a particular environment.

Secondly, the effect of natural selection is, in the case of speciation, to have *reduced* the genetic potential of an original gene pool. Alleles are deleted or a pool split into separate populations. Information is not gained but lost. *For evolution, which needs not information ($\downarrow$) loss but ($\uparrow$) gain, this is entirely the wrong direction.*

Thirdly, it is false to conceive of natural selection as a kind of 'ratchet' that holds on to 'an extrapolation of improvements' at any level

---

[139] For Dialectical connection see *fig.* 22.1 overleaf and Index: plasticity.

from nucleotide through protein to whole body shape. This is because without a prior plan to work towards there is no way for intelligence, let alone total lack of intelligence, to 'discriminate' a 'good' apart from 'bad' move as regards some novelty. *Indeed, it will breed out unwanted, nascent or non-functional characteristics.* **Only once an organ or a functional system (such as a beak or eye and associated factors) is complete and fitly working can natural selection act on trivial, accidental variation to that system. In a phrase, the 'mechanism' explains survival not arrival of the fittest.** It weeds the weaker but cannot create the fitter; it's as creative as a kitchen sieve.

In short, this mindless editor confers not change but stability by maintaining wild-type pedigree. Such **negative natural selection** deletes organisms damaged and disabled by chance events[140] so that, far from being a grand 'law of nature', it is just a trivial observation - an organism born defective does not reproduce. Its truism states the obvious. 'Weaker die, stronger live'. 'Who survives, survives'.

The Origin of Species

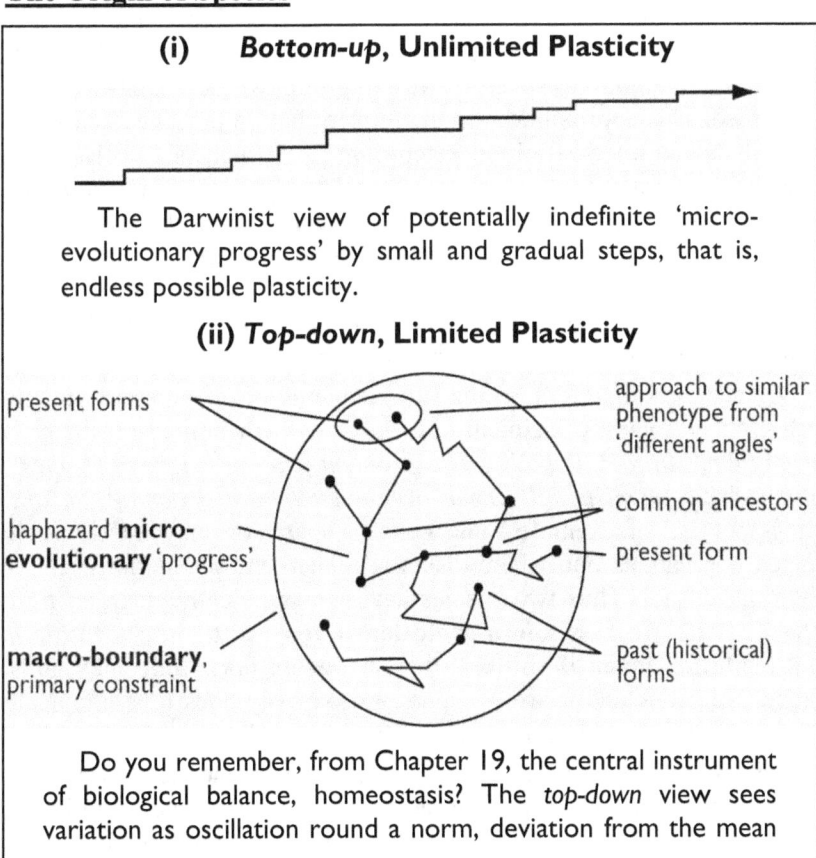

(i) *Bottom-up*, Unlimited Plasticity

The Darwinist view of potentially indefinite 'micro-evolutionary progress' by small and gradual steps, that is, endless possible plasticity.

(ii) *Top-down*, Limited Plasticity

present forms

haphazard **micro-evolutionary** 'progress'

macro-boundary, primary constraint

approach to similar phenotype from 'different angles'

common ancestors

present form

past (historical) forms

Do you remember, from Chapter 19, the central instrument of biological balance, homeostasis? The *top-down* view sees variation as oscillation round a norm, deviation from the mean

---

[140] see Chapter 23: Adaptive Potential for **positive natural selection.**

> or limited plasticity. **It proposes 'type' (see The Origin of Type) as the homeostatic norm around which limited differentiation may, within primary constraint, occur.**
>
> Consequently this diagram enlarges on the notion of random mutation and genetic drift within a typical boundary. It shows schematically how a number of species and sub-species can (as, for example, in the case of wolf, dingo, fox and abundant varieties of pet dog) exist simultaneously within the same constraint; how constrained common ancestry occurs; how a given species may be 'approached from another angle', that is, derive from a different lineage; and how it may, as such, be approximated at other times and places.
>
> In short, what is commonly called 'evolution' is dialectically described as 'principle-expressed-in-practice' or 'variation-on-theme' (see also Chapter 23: Evolution in Action?).
>
> *fig. 22.1 (see also fig. 5.1)*

What is the smallest category into which similar individuals can be grouped? Biology needs the concept of species, or something like it, to identify what it is talking about. Carl Linnaeus, the founder of modern taxonomy, thought a 'species' was a group whose members could interbreed. This approximates to the modern but, in practice, sometimes blurred taxonomic definition according to which a species is a population of organisms reproductively isolated (by various factors) from a similar group. Darwin, as opposed to Linnaeus, preferred to 'plasticise' his definition of species to what he saw as its cause - gradual change. He wrote *"I look at the term species as one arbitrarily given, for the sake of convenience, to a set of individuals closely resembling each other, and that it does not essentially differ from the term variety."*

A species is therefore, in the fluid dynamic of evolutionary time, a 'variety'; and variety incipient new species. Such limited plasticity is no big deal. *In fact, nobody has a problem with an origin of species - although taxonomists still argue over what the label actually means.* Ring-species, for example some gulls or sparrows, may interbreed in linked populations but, where distantly related end-populations 'close' the ring, do not. Thus two species have geographically emerged from one. **Call the process micro-evolution if you want; but variation is not evolution (even if you call it micro-evolution).** Gulls stay gulls, pigeons pigeons and sparrows sparrows. No way does this amount to macro-evolution, that is, evolution of fresh body-forms.

However, Darwin not only regarded varieties as incipient species but also proceeded to *extrapolate* on the hypothetical principle of unlimited plasticity. He *did* consider it a minor stage of macro-evolution. Such extrapolation, variation-*without*-theme, formed the very basis of his theory. Thus, because natural selection works on these varieties, it would seem that the origin of species is what, in its flexible gradualism, in the end denies them any durability. Species are simply moments in flux: snapshots of change. The iconic model of speciation derives from his observation of thirteen or so finches on the remote Galapagos Islands. It illustrates how, by various mechanisms, one group of descendants from a single population may diversify and perhaps cease to interbreed. The question is how far such variance extends. Is it simply elasticity within an archetypal theme or can it be *extrapolated* past macro-boundary to the point where, say, a mouse snaps out of 'mouseness' into something else? On the principle of unlimited plasticity does speciation accumulate so that this stretch and snap of boundaries keeps happening until the mouse has made a whale? Is the integrity of archetypal 'rings' paramount or can 'broad' *micro-evolution*, against all experimental evidence, proceed unchecked into *macro-evolution*? Can its unlimited plasticity transform bacteria into humans? This is the dogma. It is precisely why Darwin's disciples recoil at the anathema of 'archetype'.

Variation-unto-theme; small variation writ large is the Victorian sage's gamble and, if in the real world macro-theme runs unto micro-variation, his great, anti-parallel mistake. *Top-down*, on a genetically demonstrable principle of limited plasticity, typical variation might be thought of as different permutations of a single genetic potential expressed at different times and places.[141] Genetic potential constitutes the amount of variation that a type of organism can produce from operationally-organised genetic material already present and acting as a 'variation bank'. **In this way the definition of speciation might be seen as differentiation circling round a norm, wild-type or archetype. In other words, reversal of sagacious gamble. Theme-unto-variation; variation-on-theme.** In practice, when genetics decisively demonstrates the limits past which mutation does not survive it also demonstrates conclusively, through myriad experiments (e.g. with *E. coli* bacteria, fruit flies and mice), that the macro-evolutionary principle of unlimited plasticity (or unbridled extrapolation) is incorrect. Ask a dog-breeder if he has bred other than a dog. Ask him why.

Darwin saw the problem - even if you loan it bags of time. "The geological record is extremely imperfect and this fact will to a large extent explain why we do not find interminable varieties, connecting together all the extinct and existing forms of life by the finest graduated steps. He who rejects these views on the nature of the geological record will rightly reject my whole theory." Where Darwin staked his claim on continuity, *discontinuity* (the primary prediction of an archetypal model) is what we actually find. Nobody is more aware than a palaeontologist that Darwin's 'interminable varieties' are missing from life's petrified family album.

---

[141] see this chapter and chapters 23-25.

Galapagos and All That

Whistle up the dog again. Do you need proof selection doesn't work? Try Cruft's show. Over 400 breeds of the species *Canis lupus familiaris*, whose patriarch is a wolf, have been unnaturally bred (through the agency of breeders' minds) over the last 200 hundred years or so. Many different-looking hounds - boxer, whippet, sausage dog and poodle - are 'cat-walked'; but such inbred or otherwise artificially selected specialities actually show a loss of some 'bred-out' alleles, reduced frequency of others and, by concentration round the lower number left, in each case an overall diminution of canine information. Such 'selective pressure' means a *reduction* rather than increase in genomic diversity. 'Wild type' is diluted to 'recessive' oddities. Indeed, eugenic manipulation by the 'intelligent selection' of professional breeders has led to major suffering due to morphological deformities and other defects in many of the dogs concerned. *Reduced potential is, however, the exact opposite of what is required by evolution.* **Dogs will be dogs.** As with Cruft's so with Chelsea; you can 'evolve' a great variety of roses, dahlias or any other kind of plant - but such variation does not typologically stretch to transformation of a dahlia into rose and so forth. Breeders know intense selection or inbreeding *'devolves'* weakness in the novel stock; alleles are *lost* and there's a limit of plasticity past which you cannot press. **Despite its 'elasticity' a rose remains a rose.** So too do finches, mice, bacteria and flies. For example, studies of ornithologist David Lack noted such similarity between birds of some species of Galapagos finch that a hybrid might be indistinguishable; however, he did not observe interbreeding documented by Peter and Rosemary Grant, J. Weiner *et alii* between six of the thirteen species that were studied in the 1980's. Indeed, the next chapter discusses whether variation in beaks and feeding habits is better explained as a result of natural selection acting on random mutations or the expression of pre-coded adaptive potential for the beak sub-routine.

While variation - or micro-evolution - is accepted by all parties as fact the theory of macro-evolution is in a different category - unproven in fact, unacceptable by evidence and therefore, like chemical evolution, unacceptable in faith. **If that's the case then, in the mill of reason, where does this leave Darwin's central inspiration, the idea of a tree of life?**

The Tree of Life

Books may use the same language but this doesn't mean they can't, informatively, vary hugely; or that mindless editorial can transform one, by random cut-and-paste, into another. Could this be the case with books of life, organisms rich with information?

Phylogeny[142] means relationship based on the hypothetical assumption of an 'organic' tree of life. Branch-points (nodes) are, in

---

[142] see Glossary.

this vision, construed as 'common ancestors' on an evolutionary lineage; and similarities (homologies[143]) at every biological level from molecular to whole body are assumed to indicate evolutionary linkage. Once natural biosynthesis (Chapters 20 and 21) is over you can build trees/ lists, by likeness, of how life hypothetically evolved - from simple into complex forms with time. Their degree of difference is, equally, assumed to represent time elapsed since divergence from 'ancestral node' - an assumption not made by the more objective method of construction called cladistics.[144] *Such processional lists comprise the staple of all naturalistic explanation.*

But is the tree a tree? Or a forest of disparate trees and, like the nervous tree in you, upturned and sprung from soil of brain, from the conceptual roots of mind? For decades it has been recognised but tardily[145] admitted publicly. The theory of evolution is evolving and a revolution has occurred. Darwin's inspirational doodle isn't right. The central assumption of his tree of life is that homologous molecular or morphological patterns will, by comparison in different organisms, illustrate degrees of phylogenetic relationship. Has this idea been uprooted, sawn and cast as dust by an 'onslaught of negative evidence'? At least, if the assumption's not marked for the chop, some heavy grafting has to radically re-forest it.

One problem is that genomics, bio-informatics and other modern disciplines have revealed revolutionary non-linearities. Often genome sequences do not, as it was at first supposed they would, concur with suppositions drawn up in evolutionary trees. Visible characteristics appear similar but involve 'genetic discordance', anomaly zones' and 'anomalous gene trees'. In other words, disparities between molecular and morphological trees appear; parts of genetic make-up do not match a phenotype's[146] supposed phylogeny. And, if actual molecular pattern neither corresponds to expected homology nor to assumed developmental or adult anatomical similarity, then, *prima facie*, evolution has been compromised. Thus new models, such as coalescence model, try and patch the puncture up.

Moreover, trees constructed on the basis of biological abstracts (radial or bilateral form, protostomy and deuterostomy, mode of germ-cell formation, types of coelom and so on) conflict. Both developmental and anatomical similarities do not sprout congruent trees. This bodes ill since, historically, there should exist no blur of multiple, conflicting arboreal speculations but just, historically correct, a very single tree.

---

[143] see Glossary.
[144] see Glossary.
[145] for example, New Scientist 24-1-09 p.34.
[146] see Glossary.

As well as this kind of molecular/ morphological incongruence studies may omit a proportion of single gene comparisons from matrices because they produce 'odd' phylogenies.[147] Even different genes in the same organism (e.g. mitochondrial cytochromes c and b) can yield different phylogenies.[148] What's more, another species of nucleic acid (micro-*RNA*) adds substantially to 'arboreal disconnectedness'.[149] And, on top of this chaotic state of relationships, up to 20% of protein-coding *DNA* in each genome has been identified as 'orphan'. Hundreds of thousands of orphan sequences exhibit little similarity to any other. In other words, their *de novo* uniqueness negates traditional phylogeny. They seem bereft of ancestry.

If history can't be represented as a grand tree of life then Darwin's crucial metaphor has simply failed. *Ad hoc* excuses for such failure (e.g. horizontal gene transfer, varying rates of evolution, incomplete sampling etc.) multiply but only ever re-evaluate according to the kind of branching thoughts whose gaps are spliced up using dotted lines. Subtract these 'missing link-lines' (representing millions of absent fossils and screwed up by molecular/ morphological crossed wires). Thus transform 'tree of life' into a 'living forest'. No lone tree, no phylogeny. If twigs and leaves compose the living world then, Cheshire-cat-like, evolution's trunk and branches fade away. Instead *top-down,* archetypal image is much closer to the truth we find. From bird to beetle, frog to flowering plant and all the rest, the missing links seem nowhere but in human mind.

Homology: Common Descent or Common Design?

The theory of evolution is, we've seen, based on two main arguments - hereditary variation/ diversity by common descent and homology. A commonly used 'icon' of homology is the pentadactyl (five-digit) limb adapted as your hand for manipulation, as a whale's flipper, a bat's wing, a mole's trowel etc. Related form, different use. As opposed to disconnected form (say, avian, bat or insect wing) with analogous function (flight).

Homologous relationship implies common descent and, thereby, missing linkage - though you've been told a few such links (say, *Tiktaalik, Archaeopteryx* or *Pakicetids*) unarguably exist - at least, don't argue! And it is, furthermore, reiterated that some dinosaurs *must have* become birds; filamentous, hairy dino-fuzz therefore *must* represent 'primitive feathers'; and thus, in circular mode of argument,

---

[147] 'Bushes in the Tree of Life'; Rokas & Carroll; PLoS (Public Library of Science) Nov. 2006.
[148] Molecular Phylogenies Become Functional'; M. Yee; Trends in Ecology and Evolution Vol. 14 ps. 177-8, May 1999.
[149] 'Rewriting Evolution'; Elie Dolgin; Nature 486: 28-6-2012.

your claims constitute proof positive of your original assertion. Materialism can't interpret any other way.

Darwin defined homology as a "relationship between parts which results from their development from corresponding embryonic parts". It is, therefore, insufficient simply to compare end-phase, adult homologous structures. *These need to correlate with clear genetic and embryological homologies before any phylogenetic relationship can be considered. In other words, to be considered homologous a similar organ needs to develop in the same way from the same genes.* **In fact, this is not always the case. Homologous structures have been found specified by non-homologous genetic code and reached by different developmental pathways.** Homology of phenotype does not always mean homology of genotype. Code and developmental algorithms have found different ways of saying the same sort of thing. Such discontinuities bode ill for seamlessly connected parts of trees.

Phylogeny is Darwinism's lens. Its refractive error overlooks a modular, typological approach to biological classification. This is in spite of the fact that cladistic homologies interpreted according to *bottom-up*, phylogenetic, evolutionary principles can equally and systematically be approached in a *top-down*, typological way; and that, as breeders know, once selection is relaxed 'elastic' reversion soon draws towards resumption of the typological norm or, in geneticists' jargon, the 'wild type'. Denton[150] makes the non-Darwinian point:

"There is nothing more deceptive than an obvious fact. The same deep, homologous resemblance that serves to link all the members of one class together into a natural group also serves to distinguish that class unambiguously from all other classes. Similarly, the same hierarchic pattern that may be explained in terms of a theory of common descent also, by its very nature, implies the existence of deep divisions in the order of nature. The same facts of comparative anatomy which proclaim unity also proclaim division; while resemblance suggests evolution, division, especially where it appears profound, is counter-evidence against the whole notion of transmutation."

In short, phylogeny's interpretation from molecular homology is no more 'proof' of evolution than is such interpretation from a visible (phenotypic) counterpart. **Neither might reflect the truth.**

Further than homology a second problem is **analogy** and, with it, **convergence**. Here no links at all exist. They are completely missing. Form found in one organism appears in many unrelated ones as well. A duck-billed platypus shows what I mean. In fact, a wide-ranging roster of thousands of analogous and convergent forms have 'evolved independently many, sometimes hundreds, of times over'. The camera eye, for example, is

---

[150] Evolution, A Theory in Crisis p. 155.

found in such unrelated organisms as cephalopods (octopus & co.), annelids (worms), snails, jellyfish and vertebrates including us. What a puzzle. *How could chance-bound evolution innovate a very complex organ many different times?* **How blindly codify such ubiquitous convergence[151] as inhabits much biology so that all sorts of similar and optimally appropriate mechanisms arise in a mosaic and non-evolutionary way?**

Adaptation is described but the deep origin of form and function is not explained. So clear is it that randomness of mutation cannot generate such multiplicity of sophisticated designs (for flight, sight, echo-location, locomotion, mimicry etc. etc.) that the Darwinian model is in grave danger of shattering. In order to rescue it from certain death some other answer is imperative. For an evolution-saturated mind-set (such as professors of the faith all sport) the philosophical choice of response excludes what the facts cry out - **modular programming from which superbly engineered designs are developed.** This is the way that, *top-down*, programmers and systems analysts in the IT industry work; but it is metaphysical, essentially involving symbolic code. Instead, therefore, appeal is made to hypothetical, material 'principles of evolution', 'inevitability' forced by physical/ ecological constraints, 'bio-morphological laws' or speculative 'rules for naturalistic innovation'. Co-option (the use of existing genes for new proteins) and *HGT* (horizontal gene transfer from infections) are unlikely factors proposed as additional engines of functional innovation; and such biology certainly fails to explain the *origin*, let alone *mosaic origins*, of codified metabolic pathways, cells, tissues, organs, systems and organisms that develop from zygotes.

The problem is that 'old' biology treats *form* as primary and *symbolic form* (code) as secondary; *physical body (including its chemical DNA representation of information) is considered prior to symbolic, metaphysical codification.* Yet, conversely, biology actually depends on information expressed as form. Information, whose active source is mind,[152] controls the game. **Thus, in the 'new' biology, code is primary and its dependent, physical expression, secondary.**

In Dialectical terms, information precedes energy. In this sense you might expect the actual architecture of a building to reflect its plan; and individual components (bricks, tiles, sculptures) to reflect the context of both plan and the whole building that they represent. An architect or archaeologist will tell you that their basis for interpretation is the same. Do you, therefore, philosophically assume haphazard, mindless build-up or, conversely, that an architect has licence to commute his themes?

The *top-down*, conceptual aspect of biology is the same. Plan is foremost. **Typology interprets biological similarity as evidence of an**

---

[151] For a long list of convergences see Runes of Evolution (Conway Morris).
[152] Science and the Soul: Chapters 2, 5, 6, 13, 16 and 19.

efficient use of concept in design *and* accommodates exceptions to the planner's rule. Therefore, homology *and* analogy simply describe cases of a module 're-tuned' (or adapted) and spliced for the same or other purpose into the genome of similar or dissimilar organisms. Each is clearly the expression of programming whose hierarchical cascades of subroutine are impregnated with multiple adaptive switches. In this case the concept of heredity is, except as regards variation-on-theme, irrelevant; and archetype is, as the summary information of a type of organism, the general potential from which specific forms derive. It is, as real as any conceptual program, a metaphysical entity whose possibilities are 'printed out' in matter. *Biological truth therefore resides in the expression of modular routines and the distribution of conceptual parts appropriately tuned to the constraints of physically engineered design.* **Seen this way life certainly makes sense; and the appearance of convergent or mosaic assemblies is, far from being anomalous, a prediction of non-evolutionary design theory.**

The Origin of Type

For pre- and post-Darwinian biologists an archetype is the general, abstract representative of a systematic type. It comprises the idea, purpose and principles behind its physical expression, that is, it includes the metaphysical architecture of irreducibly complex, minimally functional systems. Blueprint is as good a word as ever. Doubly helical, symbolic *DNA* comprises an important reflection of such *informative potential*. Its code is charged with the storage and programmed (i.e. logical) transmission of instructions that comply with the commands for a cell's survival.

Ask the ARM and INTEL men again. Their chips' instructions constitute a deeply integrated, functionally consistent whole. Regulatory networks communicate the precise specifications of intention. Their work is no less teleological than the regulatory logic that governs bio-logical organisms. **A simple principle of constraint dictates that the greater integration the more likely that disturbance of any component will damage, destroy or 'disintegrate' overall function; and a simple principle of *top-down* computation exhorts the construction of purposeful compounds, that is, modules that can be refined to serve a goal in different environments.** From transistors to pipes, wheels, motors, screws and a million other archetypes engineers can redeploy a host of factors. Sometimes they use biomimetics and at others innovate from scratch. Where, in all this orderly and specified complexity, plays mindless chance a role? What sense in advocating, as creator, senselessness?

A macro-boundary is drawn round any archetype. Variation so far is permitted, plasticity is limited, the generality persists. *Such typological perception, allowing for innumerable yet constrained variation, is supported by three facts.*

Firstly, no-one has observed, either now or from the past, either abiogenesis or the myriad concatenations of intermediates that must chain class to class of organisms. **E cellula omnis cellula; this, firstly, is The Law of Biogenesis.** *And parents are observed, without exception, to reproduce their own kind; variation is limited not limitless;* **this, secondly, is The Law of Heredity.** *No evidence contradicts these most rigorously tested of all biological principles.* <u>But both contradict evolutionary theory.</u>

Secondly, the fossil record (next section) - not least in its Cambrian respect.

Thirdly, generations of extreme conditions inflicted by genetic engineers on *E. coli* bacteria, fruit flies etc. have produced genetic mutations and all sorts of deformities; new strains, races and, under the humanly imposed definition, species have also been observed *but no new types.* **E. coli remain *E. coli*. Fruit flies remain fruit flies.** Type *Drosophila* has existed for over 50 million years (since the Tertiary) and has undergone virtually no modification in that time; in this case both natural/ unintelligent and artificial/ intelligent selections entirely and without exception contradict the Darwinian hypothesis. The list goes on. Nor is diversification the same as evolution. **The question is not one of limited plasticity but the origin of a type in the first place.** *In this view the first law of genetics would be conservation of archetypal information.* There is no new injection of information only a kaleidoscope of original, resilient 'books' of code suffering, by now, a little wear and tear. There's no unlimited plasticity and no new types. Type remains true to type. As we find.

If, however, typological axioms such as immutability do not apply at the plastic level of species where is irreversible determination set? When does stem-variation have to stop; what exactly *is* its axial type? *Bottom-up*, the concept's unexpected. Unlimited plasticity must mean it can't exist. Thus, the question's not been asked nor research for answers been engaged.

Top-down, a type's an archetype; it is a metaphysical principle or, for any type of organism, its principal. Round such axis play eccentrically occurs; this practice is called variation-on-its-theme. Thus typical constraint *is* expected; macro-evolution's not. Such constraint would involve body-plan, coordinated systems and, possibly, the mosaic but well-tuned deployment of less crucial modules. *In this case one possible definition might be that a type is a group of organisms that possess a number of unique defining characteristics which occur in fundamentally invariant form in all the species of that class but which are not found even in rudimentary form in any species outside it.* Examples include birds with beaks, feathers etc. or mammals with hair, lactation, four-chambered heart etc. Types are, by this definition, exclusive and not approached gradually through a sequence of transitional forms.

Another possible definition is the complete incompatibility of alien-type male and female gametes. Fertilisation would be either incomplete or fail before division.

Thirdly, consider the possession of a unique set of characteristics whose distinction (in the form of functional homologies) was unapproachable by any gradual process. These characteristics are irreducibly complex mechanisms that include, as well as organic systems, specialised metabolic or developmental algorithms.

A fourth definition sets the integrity of type (as the basic unit of biological construction) at about the conventional classification level of family. In this respect you would expect to find distinct, archetypically adapted but physically unrelated types of fish, bird, flower, tree or mammal. You might, for example, expect to find hawk-types, heronoids, crows, waders, warblers, finches and so on, each scattered worldwide in its niche.

Denton[153] summarises: "All in all, the empirical pattern of existing nature conforms remarkably well to the typological model. *The basic typological axioms - that classes are absolutely distinct, that classes possess unique diagnostic characters and that these characters are present in fundamentally invariant form in all members of the class - apply almost universally throughout the entire realm of life*...(my italics). To refute typology and securely validate evolutionary claims would necessitate hundreds or even thousands of different species, all unambiguously intermediate in terms of their overall biology and in the physiology and anatomy of all their organ systems." Such intermediates exist, as Darwin admitted to Asa Gray, only in the imagination. If so, Darwinism is turned upside-down.

Types of Fossil

We come to nature's book of years, life's archive stacked within the international library of earth, the fossil record.

Science works by observation. Facts are gathered and theories proposed according to these facts. Is this the case with the theory of evolution? Is not true science reversed when facts, even contradictory anomalies, are squeezed by hook or crook to fit a theory? What if such distortion happens wholesale?

The fossil record is amenable to models that argue either evolution or design. Indeed, the superficial appearance of evolution has, we'll see, to some extent been imposed on it by dating rocks according to their fossils and their preconceived relationships. How confirmatory - or tautological - is that?

Fossils show variation in the past as now. They demonstrate neither

---

[153] Evolution: A Theory in Crisis p. 117.

how the variation happened nor that 'later' descended from 'earlier' fossil form; but because their evidence has been almost exclusively interpreted within the dogma of a Theory of No Intelligence a counterweight, a thorough reappraisal is needed. *Thus it needs be stated, in the interests of fairness, clarity and truth, that a major research project needs be established.* **Palaeontology, whose interpretative mind-set is exclusively evolutionistic, needs root and branch reassessment. Thorough unpicking and radical comparative debate is long overdue. Such international program would involve a thorough comparative analysis of each and every fossil - no doubt, a shaking of foundations.** *What does a top-down, archetypal view of palaeontology predict earth's graveyard will reveal?*

(1) *Since life's major routines and sub-routines will, at every level, concur with archetypal plan you might expect organisms to appear 'abruptly' and fully formed.*

(2) *The archetypes will persist.*

(3) *Systematic gaps (discontinuities) in the fossil record will underline the obvious yet intangible order of biological form as vested in a metaphysical archetype. Intermediates will not exist.* This does not mean that one type of organism may not resemble another or contain common modules such as eye, heart or echo-location. It does mean one kind never commutes into another. Missing links are neither expected nor, in fact, found. Contextual refinements are.

(4) *Continual variation on theme will occur due to intra-typical genetic exchange (asexual and sexual including hybridisation) and mutation.* Changes (called micro-evolution) will include degenerations, circumstantial improvements, changes in proportion and complexity but no new types.

(5) *Extinction may occur but because it represents loss and not origin of information is irrelevant.* You may argue that the loss of one type gives another a chance but what has this to do with macro-evolution?

The fossil record[154] displays, in fact, just these five predictions.

(1) Of 36 animal phyla (9 of which have no fossil record) 20 sprang, as spectacularly witnessed in the Burgess and Maotienshan shales, in a Cambrian 'explosion' of disparate body-forms lasting less than 10 million years. Other intense pulses of information characterise, for example, apparent 'explosions' of flowering

---

[154] There is only time, in Science and the Soul, to deal with fossils in principle. However, thought-provoking discoveries have been made in Canadian and Chinese shales (next page).

plant and lactating mammal forms.[155] Darwin himself initiated a series of 'artefact' hypotheses to explain such hard anomalies away but each one (for example, an insufficient fossil record) has been refuted.[156] However, the abrupt occurrence of disparate body-plans concurs with *top-down* prediction.

The same goes for engineering prototypes. Steam engine, motor car, radio transmission - each distinct, basic idea remains invariant while subordinate, dependent variations accumulate around; speciation clusters round an axis of original potential. This, again, spells mac- to mic-, theme-to-variation, major outlines followed by minor variations of detail. It is Darwinian anathema because it spells his theory's demise. In the same vein, concept seeds a forest not a single tree. This is what the fossil record shows. Macro-boundaries, representing intensely-integrated information packets, persist. Few variations at first proliferate, with time, into many species.

(2) They do.
(3) Such discontinuity is evident. We find no 'spectrum' of intergrading fossil transitions. Billions of intermediates, which Darwin himself recognised his theory required, should dominate the fossil record. In fact, they're practically absent. Because they're crucial 'missing links' are eagerly-sought and often 'found' - but so far every candidate has, on close scrutiny, been open to severe criticism, reinterpretation and thereby plausible rebuttal. There exist, moreover, numerous mechanical objections to the conceivable transformation of one operational form (say, a mouse) into another (say, a bat, whale or human). From molecule through organ to system and whole organism algorithmic flow-sheets of expected paths of metamorphosis are not supplied by intellectual let alone the fossil evidence.

Over 100,000 fossil species are known and the fact is that some, such as diatoms, micraster, graptolites or ammonites, show

---

[155] Charles Walcott discovered *Pikaia*, a chordate which seems similar to extant lancelets. More advanced than *Pikaia* is *Metaspriggina* found at Marble Canyon near the Burgess Shales. It has a backbone, well-developed eyes, nerves, nostrils, pharyngeal bars and strong fish-like muscles. It is, indeed, effectively an eel-like fish - from the Lower to Middle Cambrian! A similar couple of 'putative chordates' (*Haikoichthys* and *Myllokunmingia*) have been discovered on the other side of the world at a parallel site, Chengjiang in China. The features of *Metaspringgina* (many of which you possess) bring to mind Haldane's 'precambrian rabbit' - the discovery of which would have definitively persuaded him that the theory of evolution was false.

What, by the way, about jurassic rabbits? Although none are on public display over 400 mammal species from dinosaur fossil layers have been catalogued (Kielan Jaworowska *et al.*; Mammals from the Age of Dinosaurs: Columbia University Press 2004); also apparently represented are all major, extant animal phyla and plant divisions (Werner *et al.*: Living Fossils Vol 2).

[156] see Stephen Meyer: Darwin's Doubt for detail.

variation-on-theme (or micro-evolution). Such variations as occur are often trivial, reverse themselves and produce no new types. Indeed, while the complexity of ammonites fluctuated that of graptolites reduced towards simplicity. No partial forms leading to or from any of these types are found, nor even between genera within the type. An ammonite remained an ammonite and a graptolite a graptolite throughout their spans.

In short, such evidence of variation within type is continually found (both now and in the fossil past) but serial evidence of gradual, evolutionary transformation is lacking. Actual typological gaps rout theory's godless gaps that inferential missing-links should fill. The fossil record negates Darwinism.

(4) It does; and, as regards new types, it doesn't.
(5) Not destruction but origin of specific, complex bio-information is the issue.

Variations circulate within the perimeters of fixed archetypes. The record reveals that distinct biological phyla spring abruptly, fully-formed in typical bursts that proceed to variegate. As the late palaeontologist Stephen J. Gould long ago reflected,[157] 'the trade secret of palaeontology is out'.

The fossils state discontinuity but Darwinism, conversely, demands transforming continuity. Gould's own reconciliation, fact with theory, invented punctuated equilibrium. In this scenario long periods of stability are punctuated by erratic, missing-linkless rushes of evolutionary transformation. Now it's not seen, suddenly it is. Even if such unseen episodes have happened the basic question remains the same: can unaccountable mutations work the trick? Might they, as rapidly as sleight-of-hand, repeatedly supply the swathes of information macro-evolution mandates?

What response to fossil fact did you expect? Darwin's is a tinker's theory. Witless yet 'progressive' variations doodle innovation; and species somehow pioneer new phyla, not the other way about. *Thus irrational, bottom-up theory contributes nothing to an explanation of the <u>origin</u> of irreducibly complex biological mechanisms and their incorporation within every organism as a functional whole.* **Neo-Darwinism weights explanations. It habitually spins ill-fitting cloth to fit bare bio-facts. The cover story neither suits nor fits and, since its predictive vector's incorrect, it may be predicted that it never will.** None of this, we see, prevents an evolutionist hypothesizing to the mythic contrary; nor, on the other hand, need it hinder neutral parties from pragmatic, study of life's every brilliant, information-rich and present part.

---

[157] Natural History 86; May 1977.

Chapter 23: Neo-Darwinism Doesn't Work

An organism's book of life is made of *DNA*. This chemical, like ink and paper, carries information - letters, words, sentences and chapters integrated into a coherent book. Other chemicals cooperate to frame the way genomic information is expressed. You are, like every life-form, an expression of dynamic, coherent text. How is such text composed?

The Creator: Mutation

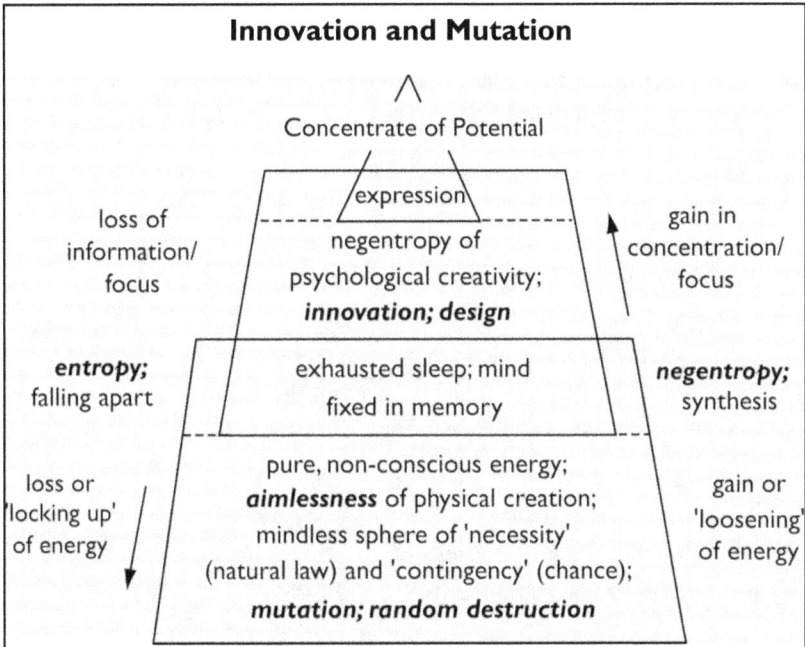

Creation and creations; active creativity of mind and passive, physical arrangements.

Meaning is gained by concentration of a mind, by focus of attention or the aim of intelligence. Intended stimulus generates coherent, purposive systems; it invents means more or less efficient to serve ends. The origin of codes and machines falls into this category; so does the origin of types of cell, tissue, organ, system and organism.

Meaningless movement (such as a sea of wind or waves) is derived from unintentional stimulus. Energy rings mindless changes; natural forces make and break up aimless aggregations. And the aimless, repetitive application of energy to informed constructions (like frost or rain weathering a bicycle) corrodes their sense;

random interruptions disrupt or jumble code. Engineers all know the elements will savage systems. Untamed energy destroys the purpose in designs and pulverises their coherence out of sight. For order it is terminal. Are biological mutations all disintegrators?

Matter and information are absolutely separate entities. *DNA* is mindless matter. **It needs therefore, be clearly understood that random variation in the combinations of its chains (in size or order) is <u>not</u> the equivalent of introducing fresh information.** Mindless rearrangement of original information actually scrambles it. Mutations scramble. Isn't each and every one an information loser with, for code and meaning, mostly deadly never innovative, systems-building tendency?

fig. 23.1

tam/raj	*Sat*
imbalance	*Balance*
system	*Governance*
variation-on-theme	*Archetype*
mutability	*Immutability*
informed	*Informant*
↓ *tam*	*raj* ↑
anti-life tendency	*life tendency*
incoherence/randomness	*systematic coherence*
accidental disintegration	*purposive integration*
cacophony	*harmony*
error	*accuracy*
babble/'noise'	*code*
meaninglessness	*meaning*
incorrect information	*correct information*
genetic/functional loss	*genetic/functional stability*
damage/disease	*health*
disorderly recombination	*orderly recombination*
mutation/misprint	*facsimile*

A scientific G.O.D, no less! A Generator of Diversity! Since natural selection fails entirely to innovate the second step in evolutionary lore was to propose a mechanism of creation. Therefore, the saying goes, a watery swirl of most unusual chemicals inscribed ancestral books of life. Next random rearrangements aimlessly 'constructed' more. Unpredictability is G.O.D's left hand at work. Mutation is the name for an unconscious twitch like this. *Only such random mutations as affect germ cells are heritable and, as such, the sole engine proposed for initial construction of a hierarchical language and all subsequent evolutionary innovation.* **Thus materialistic faith is vested in a core of unpredictability, a central lack of reason - mindless chance.** Is its vital proposition feasible or not? Did irrationality build you and me? We are about to see.

If you left ink and paper, would it in a billion years churn out a simple book? Or simply fade away? Don't hop behind a battered veil of age. Vast time, an essential ingredient in the evolutionary scenario, does *not* increase the probability for an improbable, complex and functional end-product. The reverse. Time's tide washes a sand-castle flat. An accidental molecular complexity would rapidly degrade into simplicity. The chance it would be randomised - antithesis of code and purposeful construction - swells all the time. *Indeed, in no reproducible conditions do highly complex yet different molecules such as DNA and polypeptides ever naturally form outside a cell - let alone work in perfect tandem.* **Time, however long or short, is not the issue. Informative intelligence, in the form of anti-temporal, negentropic coding, is.**

You can't agree? Might not mutations, if you found 'beneficial' ones, transform one type of body to another in the way accretive evolution needs? *'Beneficial mutations' (BMs) are to genetics as 'missing links' to evolutionary palaeontology - critical.* **Not just The Primary Corollary but The Primary Axiom of Materialism and its whole panoply of philosophical, political and sociological speculation, not to mention the paradigm of modern science, hangs on their slender thread.** *The whole of secular academy depends, for its verbose existence, on this evanescent gleam of hope, a key to unlock all of life's diversity that's called a BM ('positive' or 'beneficial mutation'). Can BMs rise to such occasion? Everything depends on this.*

This is the crux. The origin of bio-code's a problem *bottom-up* insistently refuses to confront. By an oxymoron its rationale is chance. Is information, that can anticipate, really the gift of Lady Luck?

In other words, what sort of accident or set of accidents is systematically 'positive'? To repeat: mutations, now known to be the result of copying errors or environmental damage to chromosomes by chemicals, radiation and so on, are random changes to cellular *DNA*. In the germ-line they are rare (one for between one and ten million copies) and, as the product of chance rather than coherent, systematic events, they are analogous to 'noise' in information theory. They show entropy of information; they degrade and not improve a body's reservoir of structural command.

Entropy of Information

Deleterious mutations (*DMs*) are not fit for purpose and so we need a special kind of information-adder, one that adds survival-value accidentally (*BM*).

Do *BM*s exist at all and, if they do, in what do they consist? Can they accumulate sufficiently to generate complex, systematic novelty in the awesomely concerted sorts of way we've glimpsed? Innovation and not minor variation, theme not margin-dipping, macro- and not micro-evolution have to be explained. Of billions of sounds our alphabet could make mind leaps to pick out only those that, in a sentence, express meaning and make working sense. Can *BM*s similarly leap vast combinatorial space and systematically, when they are clueless what a target-problem is, zap precise solutions that all bio-manuals engage? Could they invent a crucial protein fold, a metabolic pathway or an operational complex of domains? Every cell needs these in coded quantity. What pressure, in its engine room, can evolutionary steam build up; is it puff enough to drive life's train away?

Geneticists are sensitive to any rumble of a possible *BM*. If even one in a million mutations were constructively beneficial should not the literature be overflowing with reports? But, except for a few involving bacterial drug resistance, silence reigns.[158] A body is unaffected, maimed or killed by a mutation. Thus inherited *DM*s and 'neutrals' are, for sure, the vast majority. Are you incapacitated by a minor injury? 'Nearly-neutral' injuries, passing undetected under natural selection's radar, may accumulate. 'Bad' minors swamp the postulated 'good' but anyway, Ronald Fisher noted, random drifts would wipe out even 'good' propensity before it fixed or had a chance, with mates, to make a cumulative, multi-coordinated benefaction. Loss would, before emergent innovation could, occur. Genetic entropy of information is, contrary to what synthetic theory needs, always apt to increase. Entropy disintegrates; it makes noise of sense. Whence, therefore, such informative negentropy as, anti-randomising, compiles integrative codes? Do *BM*s embody it?

In fact, any *BM* would be invisible at the level of a whole organism and its small 'advance' overwhelmed by neutral or deleterious mutations long before the successive and exact chain of cooperative *BM*s needed for any novel biological system could 'evolve'.

Not knowing where you're going is a fundamental problem. It applies to 'junk', 'neutral' or any other kind of *DNA*. What should a first or following *BM* be? One can only be defined within a preconception. What is 'good' or what is 'bad' is only so when valued in anticipation of that end. **How can you even define a *BM* if you don't know where you're going?** Mindless evolution's natural selection doesn't know. And if there's no such thing as target how can 'progress' gradually mint a whole with integrated working parts? **With such lack of logic bang goes**

---

[158] see also this Chapter: Evolution in Action?

the irrational *PCM*[159] and therefore bang goes *PAM* as well!

Never mind! In the 1920's Ronald Fisher, Haldane, Sewall Wright and others, almost as unaware of a cell's molecular complexity as their Victorian predecessors, established the study of population genetics. However, the authors of neo-Darwinism's mathematical 'minder' incorrectly assumed that single mutations could, if each step in an incremental series conferred survival or reproductive advantage, generate beneficial variation; that they could achieve this end despite non-functional, intermediate positions; and that some mechanism for constructing novel genes/ proteins exists. Thus might the mathematics of genetic change assure the gradual, haphazard march of evolution.

Nowadays we're not so sure. Tight constraints of biological complexity should iron out cacophonic 'noise'. Why should integrated hierarchy suffer revolution at the hands of random 'renovation'?

Fact. A three-base codon specifies each acid in a protein.[160] Proteins comprise, on average, 150 of these. There are numberless ways, most meaningless, to combine a length of 450 letters composed from our own Latin alphabet; similarly, from all possible combinations of base-sequence vast numbers of proteins, most non-functional, could be constructed but life, like us with words, employs useful, non-redundant forms alone.

Just as 4^2 (four letters ranged in pairs) yields 8 combinations so 4^3 (four letters in codon-like groups of 3) gives 64. But, in the case of 450 bases coding for an average protein, 4^{450} astronomically inflates the number of possible combinations - far beyond particles in the universe ($\sim 10^{80}$), organisms that have ever lived ($\sim 10^{40}$) or seconds since the start of time ($\sim 10^{17}$)! Chance would, in such oceanic space, thresh aimlessly Yet a functional gene is far rarer than a meaningful, four-letter word. And the appeal's to steady streams of working winners in life's cosmic lottery - else chance is by impossibility eclipsed.

Our lonely protein, whose haphazard origin the calculations presuppose, needs stability to function. Such stability involves its 3-D structure, that is, the fold of its operational domain. Such fold is the lowest denominator for any strike by natural selection. In 'combinatorial space', however, bio-functional folding-up is very rare; thus, to transform from one functional 'island' to another demands crossing a 'sea' of non-functionality which may, at very best, not involve catastrophic loss of function. It may well, moreover, require multiple, serviceable mutations at different sites. Thus, to obtain a novel protein-fold (the smallest unit of structural innovation) necessitates multiply-coordinated mutations. The odds against a winner rocket up.

---

[159] see Glossary
[160] a codon may also, as a multiplex player, help bind transcription factors (TFs) or, which affects the interaction between TFs and target genes, affect histone-binding, the constitution of chromatin and the regulation of gene expression - see Glossary.

At Cambridge Douglas Axe experimentally discovered that the number of changes needed to produce a novel protein-fold exceeds those whose degradation will deliver it to natural selection's lethal arms. He estimated that the probability of a 'mutational trial' generating a single, specific, 150-residue operational protein is 1 in 10^{77}. This, even allowing for several working possibilities, is the measure by which, in vast 'combinatorial space', 'gibberish' outweighs 'operational sense'. You might therefore, most reasonably assert (since you'd need other changes simultaneously to build more novel proteins, simple metabolic pathways or the many other complex items even very simple cells involve) that sheer Mount Improbable cannot be mastered - not, at least, in gradual, evolutionary ways. **Darwinian theory is statistically dead.**

Is there an exit from this *impasse*? Accidental duplication of a gene can happen. Doubling up won't buy you innovation; but imagine that mutations, each one useless by itself yet (since it's in a duplicated section) neutral with respect to genome operations as a whole, gradually accrued. In this way, it's suggested, advantage might accumulate below selection's 'radar'. They'd add up (in theory, of course, correctly stored) in 'dead ground' apart from where mutation's general negativity could interfere with business-as-usual. With last clue solved a crossword's clinched; one lucky gene is left intact, the other varied until one day a novel, fresh-perfected gene salutes for active service. Evolution by gene duplication[161] might occur, innovation of a function might be found! Except a solitary protein is not innovative unassisted. Where would it fit? No innovative function or its organ can 'emerge' without a multiply-coordinated, freshly-coded co-emergence of coherent proteins, morphogenic agents, system and, perhaps, even body-plan - all while holding on to every part that might be useful in this formerly unknown, unplanned future! Why *should* useless, neutral, individual changes not be lost to drift or natural selection? Why *should* changes integrate? And what would happen to the 'junk' that made no bio-sense? High improbability piled higher approaches Peak Impossibility. **By such imagination evolution theory with collateral atheism stands or, on impossible confabulation, falls.**

Take a look at baker's yeast. Many of its genes seem duplicated so that perhaps, historically, some other fungus stuttered into wholesale duplication, doubling up its whole genome. More generous an allotment of spare copies you could hardly dream of; and analysis revealed that other species also seem to have descended from original 'double trouble'. Such yeasts show different mutations but, with millions of extra nucleotides and a hundred million years of evolutionary opportunity, what have they made of their career? Precisely no promotion!

A few proteins, co-opted or transformed to other functions, may perform like standard batteries for different appliances. But can re-

---

[161] suggested by Susumu Ohno.

arranging modular blocks of code create functional programs; or mobilising textual elements a fresh appliance make? Yeasts contain such elements a-plenty yet still no lucky strikes on novelty obtain. They change a little but stick yeasts. Is such permanence the best gene duplication and mobile genetic elements[162] can do? You could not, on present showing, take genetic st-st-stutter as a serious candidate of macro-evolution's mechanism or the author of a bio-logical appurtenance that's new!

Remember, though, the question's begged. You've smuggled information in. How did your copied gene originate? What's the source of integrated information? Shuffling speculations presuppose sufficient functional information is already there; and that, in combinatorial space, unobserved mutations blindly link up towards an unanticipated jackpot - all in gradual time! When all else fails nescience summons up '*de novo*' explanation - creative power of the imagination to 'long-jump' a gene from nowhere. But the source of information, it's agreed, naturally can't be mind!

The integrated webs of hierarchical biology impose tight 'flexibility-constraints'. Transgression is not tolerated so, at each undirected step, multiply-coordinated, disturbance-neutralising *BM*s would have to arise harmoniously within the system as a whole. And for evolutionary novelties to 'progress' these mutations would have to adjust correctly within the context of each hierarchical level of expression to ensure a healthy flow of information. No doubt, natural selection weeds out failures; the systems engineering question is, with such demanding standard, could blind accretion of so-called *BM*s ever build informative, transformative success?

The word is not 'evolve' but '*innovate*'. The question's critical. Can gene mutations randomly perform the trick - grand innovation in the form of life on earth? Ironically, if naturalism's G.O.D can't innovate, if its Creator can't create, then it's just another black box - an empty one because it's full of godless gaps.

Mutations are by-products of genetic operation over time. They embody entropy of information; the relentless, net effect of random mutation is degradation unto destruction of function. Entropy does not create; its direction is, for life's Titanic, pointing ($\downarrow$) down. **G.O.D but no creator - 'life's destroyer' were more apt a phrase**. Thence it might be argued that mutation, as an explanation for trans-speciation, is an unproven and, if reason is set dead against, mythic system of belief. **In short, mutations *devolve* not evolve**.

Evolution in Action?

So is there evolution, yes or no?

Variation-in-action certainly exists. All agree upon the limited plasticity of micro-evolution. Is, however, speciation evidence for transformist macro-

---

[162] see Glossary: transposon.

evolution? **If a new feature had arisen, yes; but if change occurs within parameters that describe existing structures then variation and not evolution has occurred.**

Thus if, as many do, you *equivocate* by calling variation evolution then evolution definitely occurs. But is it more than blur, a fallacy by common, verbal trickery to claim that evolution is just variation, variation does occur in organisms and therefore molecules evolved to man? That requires an extrapolation to unlimited plasticity. Such notion amounts, in fact, to the simple, pivotal and celebrated Darwinian theory of 'descent by gradual modification'. Is such never-actually-demonstrated transformation, macro-evolution, fact or fantasy? No equivocation. Let's resolve the pair apart. Random variation (so-called micro-evolution) doesn't aggregate to macro-evolutionary innovation of a program that's replete with integrated, modular routines. Here's why.

In science you experiment. Can you test evolution in the past? In his book 'The Edge of Evolution' Michael Behe demonstrates that nature has empirically tested Darwin. Do you want numbers showing how, by gradual mutations, life might step-by-step evolve a tree? **HIV, E. coli and malarial parasites satisfy the numbers game. Virus, bacterium and eukaryotic cell have reproduced, mutated and should have evolved through sufficient generations with sufficiently large populations to indicate whether neo-Darwinism's engine, random mutation, can bear the weight of a theory that would have it gradually innovate parts and body-plans by gradual but cumulative, useful steps - or, buckling, is crushed by numbers.** In nature and laboratory these and, to a lesser extent, fruit flies and other organisms illustrate Darwinian changes. Of what extent and quality are these? To what far edge can you observe extrapolation run? Is, when you evaluate the action, your conclusion that unlimited plasticity could burst all boundaries? If not the tree-of-life hypothesis would be, as Darwin feared, negated.

What, therefore, are the facts?

HIV, a nine-gene scrap of *RNA*, lacks editing formality and thus mutates about 10,000 times more wildly than non-viral cells. Each infected person is burdened with about a billion viral particles and over ten years incubates (since *HIV* generation-rate is about a day) 10^{13} viruses. Fifty million sufferers would have produced about 10^{20} copies in the last fifty years, that is, since the earliest version of *HIV* was found (1959), probably transferred to man from monkeys in the Congo. However, even with its prodigious evolutionary advantage no innovation whatsoever has, during its whole *blitzkrieg* of Darwinian variation, evolved! *HIV* **is still** *HIV*. One might reasonably conclude explosively mutating viral strains fatally infect the theory.

In 2008 Dr. Richard Lenski reported over thirty years' work using twelve separate (but initially identical) populations of *E. coli* bacteria.

After over 40,000 generations (equivalent to about a million years of human being) and billions of mutations (enough to affect every point along its *DNA*) one population incurred an advantageous mutation that allowed the bacteria to metabolise a chemical called citrate. This single significant change did not rely on citrate metabolism *per se* (code for which was already present but 'silent'/ 'switched off'). Instead a mutation reactivated the latent code by, in turn, either reactivating or perhaps co-opting the genetic material for a missing 'go-between'. This link-molecule, possibly reactivated after suffering a previous deactivating mutation, is crucial for switching the metabolism on. We ask, was new genetic information created? No.

A computer system may target specific combinations of digits on number-plates; an immune system or drug may likewise 'police' certain protein configurations. If, by accident, a number-plate sequence or, say, bacterial recognition site is damaged and thereby evades capture does such lucky break's survival amount to evolution-in-action? You can label such change 'micro-evolution' but does it presage any nascent transformation or a step in systematic innovation? No. Does *E-coli* 'macro-evolve' (or even start to 'macro-evolve') into another kind of organism? No.[163] On the contrary, **E. coli has, in the immunological case and in thousands of experiments, been greatly strained in every nucleotide of *DNA* but still remains *E. coli*** - a second macro-evolutionary bombshell!

All malarial parasites remain the ones they were. Interestingly, chloroquine resistance to malaria demands two specific, simultaneous point-mutations.[164] Such minimal multi-coordination involving a single protein in the system occurs at about 1 in 10^{20} (a hundred trillion) cells.[165] On abatement of dosage the resistant type soon secedes to wild-type again. Any less trivial 'advance' demands multiple simultaneous mutations on multiple occasions with consequent, coherent fine-tuning. The waiting-times inflate exponentially. How could any organism have evolved? Thus the sums appear to negate Darwinian and confirm non-Darwinian, holistic hypothesis.

What about the human, as opposed to parasitic, side of things? The job of haemoglobin in red blood cells is to carry, by means of some precise engineering, oxygen from your lungs to organs. Perhaps the best-known mutation to the highly engineered shape of haemoglobin occurs in a form called HbS. Sickle-cell anaemia is a textbook icon because it is supposed not only to illustrate evolution-in-action (which, if you mean variation, it does)

---

[163] At this rate turning a mouse-like mammal (possessing greater *DNA* editorial control than a bacterium) accidentally into a cow, platypus or human in 200 extra units of a million years each would seem, despite skeletal interpretation to the contrary, as absurd as impossible.

[164] Proceedings of the National Academy of Sciences USA, Vol. 111: 29-4-14.

[165] Behe M: The Edge of Evolution; The Mathematical Limits of Darwinism.

but also thereby underwrite large-scale transformation (which it doesn't). This disorder is, like cystic fibrosis, painful and debilitating. The mutant allele for cystic fibrosis is supposed to survive because it may confer some resistance to cholera; so sickle-cell haemoglobin relates to malaria. Thus serious negative effects with serious secondary malfunctions and suffering are played down in order to label a mutation 'positive' or 'beneficial'. *There is no doubt that sickle-cell anaemia demonstrates natural selection at work; it demonstrates a trivial, Darwinian sense of variation but no step towards the origin of biological novelties.* The trait is a defect; it shows neither increase in genetic information, physiological/ functional complexity nor structural complexity. *In other words, it adds nothing to an 'upward', 'progressive' evolutionary argument. The sickle-cell mutation is negative. It represents loss, bug or breakage in the system.* This example, commonly touted in textbooks as evidence for evolution, is no such thing. <u>On the contrary, it illustrates a fact of genetic science - unequivocally beneficial mutations are unknown and even those that raise tangential benefit are very rare. And, like all mutations of a gene, it represents not gain but entropy of information</u>. From such quality of 'proof' you might well infer disproof.

In short, then, micro-evolution is a prejudicial, biased word because it implies the existence of an extended process for which no hard evidence exists **The reverse. Macro-evolution is a theoretical phantom**. In this view, **evolution-in-action is simply variation-on-predesignated-theme**; this variation is always constrained by working systems already in place; and it is either coherent, due to in-built genetic potential or incoherent by Darwinian mutation. **Variation proliferates; the special theory (STE) is right. But the general theory of grand macro-evolution (GTE) is not.**

Non-Protein-Coding *DNA*

About 2% of the human genome codes for protein. If you assume that the only function of the genome (a cell's *DNA*) and transcriptome (its *RNA*) is to produce protein then you might judge, *bottom-up*, the other 98% accumulated ^jnu~&n*k.

Top-down, however, you'd expect a program (including a biological program chemically inscribed as a genome) to be written and initially 'go live' bug-free. You might expect, after a period of time in the manner of 'bumps and scratches' on a vehicle, a number of accidental but non-lethal bugs could accrue.

Applications are supported; teams fix bugs. The primary function of a biological genome is, however, to serve as an operational database for protein required to make an organism; and a database is useless if its data's not accessible. A large data bank, like disc storage, a library or even a single book needs be accurately indexed, accessed and regulated in its provision of correct information at the right time in the right quantity of copy. You would, therefore, expect all code except some small accumulation of 'nearly neutral' or 'non-lethal' bugs to be specifically involved in structural, indexing/ address-related, switching and other regulatory functions. **In fact, you might predict that 'junk' (already more humbly and precisely called 'non-protein-coding *DNA*') was a term born more in the darkness of ignorance than light of knowledge; and that any minimal, mutant fraction played a negative or at best neutral part in the computational expression of genomic information.**

In fact, organisms transcribe most of their DNA, including so-called junk, into RNA. A great majority of bases from small, sample sections were, by 2007, found to transcribe for either for protein *or* functional *RNA*. While perhaps 25000 human genes that code for protein have been revealed there may be up to 450000 '*RNA*-genes'. What's the point of them? *Why waste energy on transcription if they amount to useless and thus interfering junk?* Non-protein-coding regions of *DNA* engage multiple functions. They seem to be involved in, for example, the genetic operating system and thus regulate the addressing, transcription and (during translation and post-translation) editorial systems of modification by which genes are expressed as protein. Hundreds of 'genetic corpses' (so-called pseudogenes) have also been resurrected. Many such erstwhile 'dead-weights' are highly conserved which, in common parlance, means they are likely to be important; and, perhaps metabolic participants in the process of producing correct proteins in the right quantities at the right times in the right cells, they may actively transcribe. Thus 'pseudogene' might be construed as a regular misnomer and the description 'relic' a misinterpretation. Lastly, ex-*RNA* (extra-cellular *RNA*) has been found, travelling in body fluids, to transmit specific information between cells.

Contemporary molecular biology is, in effect, involved in a process of complex cryptanalysis. The ENCODE project (**ENC**yclopedia **O**f **DNA E**lements) is a public research consortium launched by the US National Human Genome Research Institute after completion of the Human Genome Project. Its remit is to winkle out all functional elements in the human genome. In 2012 a Cambridge co-ordinator, Ewan Birney, was quoted as saying that the genome is 'alive with switches, millions of places that determine whether a gene is switched on or off'. Already 'unexpected complexity' of regulatory networks has also been discovered in flies and sea anemones. Such genetic complexity, reasonably extrapolated to all animals, would indicate the evolutionary necessity for a Marvellously Complex Hypothetical Precambrian Ancestor.[166]

Bill Gates has commented on the computer-like nature of digitally coded *DNA*. Of all persons he might recognise machine code. Such is the language of systems programmers, *top-down* Natural Dialectic and the computer analogy applied to genomes. Since at least 80% of genomic performance has been identified as functionally significant, ENCODE's Birney told the BBC, "The term junk *DNA* must now be junked". *If the work of 'non-*

---

[166] Chapter 25: Evo-Devo.

protein-coding' segments is critical then 'junk' is absolutely 'bunk'.

In short, recent discoveries confirm that a modular, computational picture of the genome is correct; they thus as much support the informative predictions of a Theory of Intelligence as damn the 'flotsam and jetsam' notion of an evolutionary Theory of No Intelligence.

Hierarchical Language

To *inform* is to instruct. Informative instruction involves symbols, signals, messages and therefore purpose. Its purpose is communication of a meaning.

Mind is a communication centre. It receives impressions and informs response. It also communicates with body through, primarily, a medium called brain; brain is, therefore, also an information exchange. *Top-down*, the single, critical distinction is that mind is active, semantic and metaphysical but brain, perhaps life's most intricate form of matter, is passive, guided and physical.

Information is transmitted according to rules understood by sender and receiver. Linguistic rules aren't just chaotic sounds. Language engages layers of complexity - letters, words, syntax and grammar integrated into swathes of meaning; these it commits to physical arrangement in the form of sound, print or other medium. Nor is life a pretty, well-pressed pile of molecules; it is, layered with a book's complexity, an assemblage of instructions. **The problem for materialism is, always and unrelentingly, the origin of biological information.**

Even a unicellular organism is as irreducibly complex[167] as an instruction manual combined with its mechanical end-product. This manual's alphabet is grouped hierarchically in higher order units (codons, genes, chromosomes and the genome). These symbolic sequences are specifically related to operational purpose. Their digital component is enhanced by syntax that includes reading frames, start/ stop punctuation and multiplex expressions of control by which genetic intent is translated into proteinaceous phrases. And, at chromosomal level, structural syntax also helps determine semantic expression; for example, nucleosomes, chromatin-binding,[168] the relative arrangement of genes involved in anabolic and catabolic metabolism and control of interactions between transcription factors and specific target genes. It turns out, in such integrated circuitry, that *DNA* is not conductor but the score. Its melody's evoked by subtle layers of epigeny. Such instruments combine to orchestrate the music called an organism.

The beauty of this hierarchical involvement is astounding. Can engines run as sweet as music? This one courses on prefabricated tracks; its locomotion's programmed into every cell; at all levels and at every moment

---

[167] Chapter 21.
[168] see also Entropy of Information (codon footnote) and Glossary.

sophisticated feedback communicates between genome and its physical context. The integrity of such a micro-hierarchical library (the genome) is itself a fraction of layer upon layer of integrated chemical complexity that constitutes, by the transformation of genetic story, into a body's physical reality. Thus, since language is a way of organising any complex system, genetic information is not simply language-like. It is not 'as if' but, involving the characteristics of real linguistic symbolism, actually *is* devised. Such language neither accidentally nor even accidentally-on-purpose just emerges out of chaos. It does not accumulate changes without reason. It emerges out of mind *in order to* accomplish purposes.

What's happening outside the nucleus (circumstance called 'context') also trammels morphogenesis. Its molecular feedback both affects specific access to the *DNA* but also impacts the epigenetic markers that punctuate an organism's text.[169] Such punctuation adds specific nuance that involves at least six kinds of mechanism: spooling tags, histone-positioning codes, histone methylation and acetylation, cytosine methylation, use of larger protein tags, cortical/ cytoskeletal configurations and the precise disposition of materials in an egg at point of fertilisation. The latter - key priming - amounts to a launch trigger that has been called 'the zygote code'.

Firstly, therefore, spooling tags. Your *DNA* is superbly, functionally folded (2 metres of it folded into a nucleus approaching, in diameter, a million times as small). It is spooled onto proteins called histones; there are eight coordinated histones per spool and millions of spools per cell (indeed, it has been established that chromosomes contain twice as much protein as *DNA*). Two coils are wound round each spool or so-called nucleosome; using these spools *DNA* is further reeled, that is, super-coiled into exact topologies that are known collectively as chromatin. Histone octamers are important factors controlling bio-information.

Such superbly functional folding of *DNA* facilitates not only the reproductive dances called mitosis and meiosis but also (by blanking, locking or leaving accessible) helps implement the important 'third dimension' of nuclear super-computing.[170] In this case a positioning-code, repeated every ten nucleotides, specifically dictates nucleosome binding-points with *DNA*; it leaves open linker regions between the histones upon which easily accessible initiation sites for polymerase attachment and thus gene transcription are found. Thus *histone-positioning code* is another super-code that, working in concert, overlays the genes.

A third program involves a *histone acetylation/ methylation code*. Put simply, gene expression depends, as well as labelling, on the differential burial or exposure of *DNA* in chromatin. An acetyl device

---

[169] Also next section: Supercodes and Adaptive Potential.
[170] Chapter 19: A Nuclear Super-computer.

is, as required, positioned by one specialised protein and taken off by another. Acetylated spools are 'unlocked', 'open' or 'accessible' to facilitate transcription. There are many thousands of such tags and attendant proteins for each chromosome. Methyl tags, on the other hand, used singly, in pairs or triplets, 'close' or 'lock' a section of *DNA*. They silence expression; but lift the signal, change a switch-point and transcription's way is clear. It's not yet clear how tagging is decided. How is one cytosine correctly methylated and others, that would generate mistakes, ignored? However, it is certain that a *code of methylation* pegs the road through *DNA*. Information systems regulate each body like a vastly complex signal box.

Each main type is used in conjunction with phosphate and larger markers - sometimes in different permutations. Furthermore, these histone tags work as signals and switches in concert with nearby chromatin regulators and transcription factors. If the way a gene is wrapped changes, that is, if its chromatin combination is changed then its expression will change as well. And, in case of division, the cell generates hundreds of millions of new spools and markers. In this amazing way works the *first* facet of super-coded potential for genetic modification without disturbing the bonded sequence of bases. *Code on code, could mutations 'think up' such highly integrated, signal information?*

Direction-giving railway lines are called permanent way; but signals and switch-points are as vital as the lines of track. In fact they render flexible the path of 'trains of information' over 'static', chromosomal tracks of *DNA*. They grant such adaptive potential for the dynamic running of life's messenger as to explain much of the variation that a single coded system can express. The fact is, to reiterate, that the genetic 'permanent way' does *not* wholly control life's operations. Just as you need an operating system to relate an application program to a database so an arsenal of ancillary devices act as signals and points that regulate the various tracks along which different permutations of genes are expressed in different cells. The technological equipment includes methylating enzymes, the actual signal of a 'fifth base' (methylated cytosine), promoter segments, transcription factors, protein elements with 'tails' incorporating complex code, various other types of movable flag (for example, acetyl and phosphate groups), repressors and so on.

DNA outside a cell is as useless as track alone. A so-called 'self-replicating molecule' will not perform without cytoplasmic circumstance.[171] The gene-centric Central Dogma of Molecular Biology (proposed by Francis Crick) states that information flows *only* from genes to protein. **This dogma, although protein doesn't act as a template for *DNA* conformation, is radically incorrect.** Circumstances, such as dietary regime, intra-cellular or wider

---

[171] Chapters 20 and 21.

environmental cues (e.g. chemical, climatic or pressure-related), can vary genetic interaction and consequent expression. This happens in many, perhaps all, organisms. Check out tardigrade for fun! This *second*, circumstantial facet of super-coded potential for genetic modification can throw epigenetic switches that may (reviving, in a lesser way, the old Lamarckian idea) even be heritable. A special protein (called methyl transferase) makes sure that the requisite flags are raised on the newly-replicated strands. In this way gametes are 'primed' and, on life's journey, markers for fresh cells defined. If it were not so there'd be havoc down the line. Furthermore, if a re-set permutation were heritable then subtler, outside influence as well as random, track-buckling mutation could affect the future. *But nothing new would come of it - just points set for different diagrams of usage down the full set of genetic rails.*

If you compare a railway system's permanent way to information stored permanently in a cell's genome, then the possible routes that trains can take over the tracks are controlled by signals and switches (called in England points). The configuration of levers in a cell's nuclear signal box is called its 'epigenome'. For example, there are over 200 cell types in you, each with epigenetic super-controls appropriately set on the *same DNA*. In each case circumstantial messages have been transmitted and different permutations of 'levers' switched 'on' and 'off'. It may even be that, in this dynamic system, the same cell type/ tissue in different organs of the body has its tags set slightly differently; and that the switching algorithm in the same tissue is responsively re-set at different stages in life and/or according to changing constraints. Just as switching algorithms are central to the operation of information technology so epigenetic hierarchies of control are clearly central to the study of genetics and gene expression. Combined in this centrality is a *third* facet of super-coded potential for genetic modification. We've already met the huge swarm of *RNA* and other genetic communicants codified by what was once dubbed (by in-the-dark Darwinians) 'junk' *DNA*. In short, such informative hierarchy reflects the cell's dynamic set-up; it reflects permissible routes, that is, specific restriction of the full potential in that unit at a given time. Orders within orders; clever, very clever ARM or INTEL stuff!

Are points ever switched at random? Mistaken pathways soon disintegrate; scheduled traffic runs to dead-ends. Both random mutations and epimutations (accidental alteration of super-coding markers) are both dangerous - except in some cases, as we'll see and as *top-down* predicts, what ignorance calls 'mutation' may amount to appropriate, systematic buffering against changing circumstance.

This brings us to extra-nuclear circumstance - differential conditions in the cytoplasm and membrane such as micro-skeletal arrays (roads for molecular transport) and the so-called 'zygote code'. [172] This constitutes a *fourth*,

---

[172] see Glossary and Index: epigeny; also *A&E* Chapter 8: A Book Needs an Index.

cytoplasmic facet of super-coded potential for genetic modification. When it comes to eggs, sperm and development a whole fresh, brilliant world of super-coding heaves in view! New epigenetic configurations must and do accord with the growth of new cell types scheduled along the train of development. What, in the midst of this heaviest of biological traffic, causes tags to be re-arranged correctly and appropriately? Most pertinently, why should and how can the markers change because a cell concerned becomes an egg or sperm? They need to be because mother sex-cell tags are adult in pattern but gametes-as-zygote will need to pass through various developmental sub-routines. A different setting will be needed so the slate's wiped clean; and, after fertilisation, a second purge takes place but some imprints (especially regarding growth) are restored. Thus epigenetic patterns can be inherited (perhaps more often in plants and fungi) across generations. Super-coding structures must be correctly maintained or, as for example in partial deactivation of the second X chromosome in females, pathologies may develop. How, though, can the system cope with the apparently contradictory tasks of leaving a number of specific tag permutations for inheritance while at the same time clearly re-setting the switches to accommodate a dynamic, flexible regime of growth and development?

Moreover, if *only* genome counted you might implant this 'organ' into any egg-cell and, gene-centrically, expect its own species of organism to develop. In fact, gene-egg incompatibility kills. Cross-species cloning mostly doesn't work; and where it does (as with carp and goldfish) one's genome yields a cross reflecting the enucleated but still informative egg of the other.

How? Particular membrane patterns imprinted from the maternal egg serve as targets for specific protein logistics. Such patterns involve the specific construction, distribution and interaction of *RNA/* protein messengers, ion channels establishing electrical and electromagnetic morphogenetic fields and the extrusion of 'sugar codes' as cell-cell and protein recognition sites. Such wondrously informative process is not born of muddled inadvertencies, that's for sure. Ask manufacturers of super-chips if complex coding comes for free. Can natural forces fabricate such goods or is prime charge intelligence - intelligence allied to purpose that anticipates and thereby specifies a future shape? In this case the attractor is an adult form. Before you start the 'image' of an adult form must be in place; this is the 'anchor point' that pulls the guided process forward. **Embedded in an egg is concept of maturity. Development is a conceptual not an evolutionary business. It marks you the construction of a great idea! If you argue, then spell out how egg or sperm evolved.** No ersatz, arm-waving, naturalistic-only explanations, please. No plausibilities that, like the ones from Haeckel, the next two chapters are about to comprehensively dismiss. Evolution theory is full of promissory hypotheses that will, forever, lack precision since they're based on unpredictability; on top of which there is no guarantee these ever-varying guesses, even if they bask in plausibility, are in fact the least correct. This time, moreover, it is absolutely obvious that chancy explanations will not do. Certain concept is involved; informative potential takes up centre stage. Thus flow-charts - not an evolutionary non-concept - might explain the origin of eggs (with all surrounding apparatus that includes a male with sperm). But in the last 150 years no chance-bound flow-chart or other explanation whatsoever has been slapped upon the table - not even if selection's there to trim mutations that don't work. Do any? Name what you would need. And then you'd have to add the tranche of super-codes. In Chapter 25 the working of genetic switches, regulatory proteins and the extraordinary computational bio-logic that creates an adult form will be elaborated. **Core code and super-coding occupy the limelight of biology.**

It's as well to start to understand this battery of signals that a cell and, thereby, body needs to make it work. Explain how multi-layered specificity is conjured up by chance in time when this pair couldn't, in earth's lifetime, brew a jug of beer. If information really were a by-product that emerges out of 'noise' then you'd expect to find that low-grade, meaningless 'static' gradually evolved to hi-fi, complex, 'noise-less' directives. Fantasy! You'd also expect that quantity of *DNA* would correlate with this build-up. It does not. Check the jumble for yourself. **Research has signally failed to demonstrate that amount of *DNA*, number of genes or chromosome count demonstrates evolution. On the contrary, the quantity of *DNA* in genomes throughout the living world is dotted in a mosaic way, called the C-paradox, that is systematically anomalous and therefore puzzling for Darwin's theory.** This paradox notes that extensive differences in the quantity of *DNA* in a cell appear to bear little relation to an organism's complexity, size or type.

Super-codes and Adaptive Potential

Darwinism assumes an adaptational paradigm. It seeks to explain how, step by adaptive step, mutation and natural selection combine to 'contrive' winning functional novelties - even if, inexplicably according to such explanation, non-adaptive specialities (such as the patterns and colorations of butterflies, flowers and leaves) survive and prosper in abundance.

But are mutations that innovate fresh purpose and evolve fresh, basic body-plan or physiology ever observed? Trivial variations-on-type due to sexual recombination or mutation are widely found but never has an emergent system or nascent organ been caught graduating to full or, come to that, any degree of novel, integrated functionality - *although origination in either other-functional or pre-functional form is central to the theory of evolution.* Indeed, whether or not you propose (like Darwin) that nascent organs must be useful at all stages, nascence is difficult to spot and evolutionism very shy of quoting cases of such crucial missing links. Surely 'pioneers' should be visible today, burgeoning in organisms at various stages up to the integration of an efficiently functional new system? We certainly see circumstantial adaptations to what's already there but none of novelty. *Things have to work; but neither observation spots rudimentary intermediates nor has controlled experiment demonstrated a key Darwinian requirement - natural selection acting on random mutations to originate fresh function with its organ.*

It needs be noted here that two entirely different kinds of change to any construction manual or its product are possible - accidental or deliberate. *From both top-down and bottom-up perspectives random and non-random variations can occur.* The random mutations and unguided variations of **negative, natural selection**[173] tend, as do mindless bugs in a computer program, to degrade performance. And sexual meiosis is a lottery that shuffles but not changes a genome; lotteries exist *in order to* produce a limited bank of recombinations and evolve no new function, organ or type of organism.

Is there, on the other hand, any *top-down* **positive form of natural selection?** The last section likened a genetic program (or genome) to a railway system wherein myriad switches enable trains to achieve their purposes flexibly. Similarly, *AI* programs are written to buffer (or reflexively adapt) to unplanned challenges and accidents that circumstance throws up. **And multi-level code exists in nature. It might be seen to encapsulate adaptive potential.**

Options A, B, C, D and so on! Resilient bio-logic! A genetic 'nervous system'! Broad, inlaid pre-adaptive design would endow organisms with great reflex capacity to cope, over time, with environmental stresses and strains. A multitude of switches and their agents are known to inhabit *DNA*. Regulatory switching between subroutines using splicing, non-coding proteins, epigeny, transposable (jumping) genes, micro-*RNA* factors and so on may, as well as routine operations, allow for the *in-built control of adaptive potential*. **Super-codes are switching systems that amount to innate, responsive, adaptive potential.** Potential involves possibility and therefore dynamic but ordered flexibility. In this case reflex adaptation induced by environments is *naturally, positively selected.*

White fur in winter, brown in summer; down and contour feathers; stickleback fin and finch beak variations. There are many cases wherein a conceptual model allows, as in a machine, cybernetic 'choice' between specific switching systems and fail-safes. In computers, for example, the essence of whose informative flexibility resides in switching between modules and lines of modules, alternatives are predefined; and in all biological organisms switches are likewise designed to anticipate various adaptations for survival. Super-code tags genetic information so that appropriate buffering response to changing circumstance occurs. Tracks don't change but signalmen switch points for different paths: the bio-manual's *DNA* remains unchanged but epigenetic tags and splicing codes adjust the sections that are read and those passed by. It may be a century or more before the schedules, pathways, signals and their contextual operations are worked out but we know that orderly coordination is the

---

[173] see Chap. 22: The Editor; also artificial/ unnatural selection, Chap. 23: Galapagos and All That.

basis of biology. Densely-packed code when unpacked decodes to yield multi-level answers. The same linear inscription of information can vary in its local 'meanings'; its potential can adapt the product it encodes.

Wouldn't it be smart if, as you read an instruction manual for information, you could pick out what you needed following the 'go to' signs? It is as if the pages of a book were so intricately packed with information that, according to the way you read it, different messages appeared. Obviously, any disturbance to such a close-knit context would upset its whole integrity. Conversely, chance does not create the complex integrity of reasonable systems. Write a program or a book yourself incorporating various layers of cross-reference, feedback and adaptability for readers. Do you attribute any of your thought-through interactions to the mindless hand of chance? The biological point is the same. Codified genes and proteins are as house materiel. They don't organise location, large-scale pattern or logistics but are guided by contextual cue. This 'higher plan' or 'shape of house' is preordained. Epigenetics - codes governing codes - amounts to an extra informative dimension. Hierarchically arranged, its mechanisms build a towering, high-grade data structure. *A responsive genome incorporates multiple, overlapping and interactive code in a mode the IT profession calls data compression.* **How can first-class data compression arise without a prior informant?** Ask any software engineer. *Why should DNA and its controllers, packing information just as tight as any hard drive, be different?*

In short, accidents happen but for Natural Dialectic adaptive potential is not an accident. **It is a sophisticated buffering feature of every bio-system; and it starts where every practice starts - *top-down* from informative principle transmitted through a code.** Invariant code with interchangeable routines - this basis of biological flexibility is one well known to *IT*. It is the construction of hierarchical code. Sub-routines are nested within a master routine and switched on or off by means of operator genes or super-codes. What else but 'computer programs' are logical 'genetic strategies'? Strategies aren't accidents. Epigenetic switches are an efficient necessity programmed into genetic operations. *These super-codes constitute, far from a matter of evolutionary chance, an additional level of sophistication, a switching refinement that can signal routes through 'memory-routines', that is, through code-locked metabolic application programs stored on DNA.*

In the case of life this means information in the form of options for response to changing circumstance. It means *adaptive potential.* **We also call it <u>positive natural selection</u>.** Such induced selection is pre-programmed in the way that modern *AI*, guidance programs and, indeed, our own nervous systems are pre-programmed to buffer changing external conditions. Its predetermination is positive as opposed to the **<u>negative, unguided kind of natural selection</u>** that underpins the Darwinian version of events. No doubt, negative selection can occur. But it is random, not switched on. Its form of adaptation is due to chance and, as such, an effective bug in program.

Natural Genetic Engineering

The shrug of 'unaccountable chance' is insufficient answer. 20th century neo-Darwinism's cracked and gone. Dr. Reason must convert his stories to the 21st century's information era. In this case immaterial information's presence is inferred from the arrangement of material bio-molecules. Ever more sharply research brings the anti-evolutionary reality of life's molecular complexity, program and adaptive anticipation into focus. So how, far beyond the scope of chance, has such bio-physical sophistication been sourced; how, where 'science = materialism', can 'psycho-biological' be scientifically explained?

'Natural genetic engineering' is a fashionable 'systems' metaphor. Recall that bio-logic's bio-informatics works more enterprisingly than any manual's rationale or even a computer program's steps. Its powerful routines and lines of code are chemically inscribed, *top-down*, to engineer its symbolism into an embodied fact - one such as you or me. If an emphasis of randomness is minimised and programmed buffering advanced, two questions arise. **How do functionally effective, pre-programmed (that is, anticipatory) codes originate; and how might chunks of them change function?** Random mutation constitutes a bankrupt information-source so richer input is required. Subtle super-coding is one thing but 'combinatorial thinking' also suggests that shuffling modular routines and accreting extra chunks of *DNA* might accidentally obtain, by leaps, changed track-layouts and signal clusters whence new lines of function might evolve.

How? How, despite quality-control in the form of proof-reading and repair, might genomic operators be meaningfully tweaked? How might such as 'go-tos', 'dos', do-nots', 'turn-ons', 'turn-offs', cut-and-pasting and a host of other kinds of regulator be sufficiently but randomly re-spliced and innovative proteins engineered afresh? Various mechanisms for bulk data reorganisation are invoked. These include 'neutral selection within gene duplication',[174] inter-specific hybridisation, gene-splicing adjustment, horizontal gene transfer and, above all, transposable (mobile) genetic elements.[175] Transposable elements, whose rapid activity in all of us appears controlled by cellular programs, seem to act as regulators. *Indeed, if genome is considered a sensitive 'organ' of the cell, don't they precisely restructure its parts to buffer specific circumstantial exigencies?*

Heritable epigenetic flexibility operates to counter such exigencies; proteins (such as heat shock proteins) can respond with a burst of adaptive potential, that is, cascades of pre-codified variation; and a longfin squid (*D. pealeii*) adapts to environmental changes by editing up to 60% of its *RNA* transcripts. Genetic buffers may be very common. And naturalism might suggest that 'extra-level' circuitry of the genome, while mainly used for

---

[174] refer to this Chapter: Entropy of Information.
[175] see Glossary: transposon.

the maintenance and correct expression of bio-information, could, in its adaptive mode, service macro-evolution. Rather than simply play alternative cards couldn't bio-chips be upgraded seamlessly, mindlessly yet radically? Thus haphazard computation and not competition would drive innovation. All circuitry, including developmental, must, it's argued, have been naturally engineered. You'd just need retention of appropriate mutation, shuffling and reuse of critical domains, that is, of high-level architectural controls that specify an organism's form.

Deliberate programs involve 'alternative cards' called sub-routines. These are, it could well be countered, what we actually see. Such 'engineering' systematically supplies pre-codified and accurate answers. Its adaptive potential, hierarchically organised, helps subtly guide an organism through the different trials of life. And since such flexible guidance directs in a step-wise, conceptual, algorithmic way it is certainly not the product of haphazardly evolved 'engineering'. Indeed, it specifically acts to buffer the exigencies of chance.

'Knock-out' is a genetic technique whereby molecular biologists, in order to understand a gene's function, deactivate it by intentional, intelligent mutation. In species such as the fruit fly and nematode, *C. elegans*, all genes have by turns been systematically silenced. Such unnatural engineering is destructive not creative. But in 80% of cases the 'knock-out' itself is found silent, that is, has no effect. The effect of mutations is buffered by pre-programmed physiological networks that appear to compensate. If mechanism A fails, go to B or C. Such fail-safes predictably occur in all intelligent systems; forestalling accidents is the presumptive norm.

Therefore, can bio-logical buffers of such quality be obtained by lucky shuffling of a chunk of code? Are new body-plans just rehashed old? Does such thesis work? If you successfully rewired a dedicated micro-chip you might, in this developmental case, rewrite morphogenesis from top to toe. Mix 'n' match of core modules might possibly create not just new species but, as evolution needs, whole new families, orders, classes, phyla and domains. Disjunction yields abruption - Goldschmidt's hopeful monster theoretically lives again! In monstrous numbers since you must be one of them! Disparate invariance is well and liberally punctuated. Variation's run, at least in mind, amok. But mind amok is not, all engineers agree, good practice for design; and should 'seems designed' be an illusion then 'designed' were true.

In body's fact does *NGE* succeed? *Evolution is not observed.* Flies stay flies, roses roses and so on. And, fundamentally, the question isn't answered. ***Whence arise modules of information, programs with an end-goal, in the first place?*** Shuffling pre-programmed routines around, even if it worked, cannot explain their origin. **Tautologically, the question's begged, the answer is assumed.**

Chapter 24: An Impossible Dream

Anti-Parallel Interpretations of Biology

Inversion[176] describes the reflective asymmetry whereby, upon a gradient, one pole is converted to its opposite. Various flows, one way and to-fro, are fundamental to a cosmos that is dialectically explained. Caduceus, double helix and antagonistic spins suggest an intertwined and polar geometry of how the world's informed.

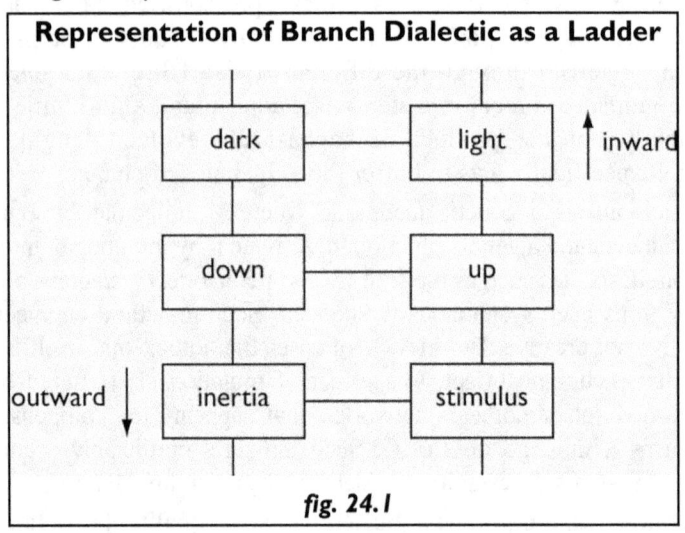

fig. 24.1

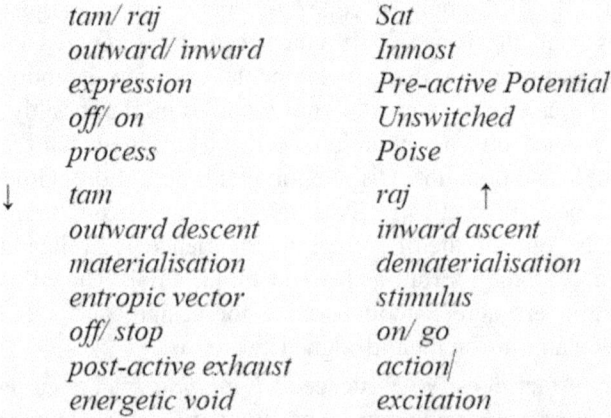

Opposites attract, repel but also complement; they bind, separate but still in quality are two-in-one; such staples nail the cosmic plan to nothingness. According to this plan the motion of materialisation flows,

---

[176] Chapter 2 and Glossary.

stream-like, 'down and out'; or from centre towards periphery, source to sink or, in dialectical terms, from right to left and downwards.

This is not, of course, a materialistic perspective. However, a hierarchical creation devolved from 'inside information' out to physical construction shows a mind-matter or informative-energetic twist not only reflected in Natural Dialectic's stacks but in biology - not least, we'll soon see, in you. Another fundamental inversion, which mirrors the whole scope of creation, is that offered by anti-parallel *top-down* and *bottom-up* perspectives. Holism (and with it Natural Dialectic) inverts materialism and *vice versa*. Neither pillar of faith disputes the fact of material existence but interpretations of how mankind began and, therefore, all that follows from their viewpoints is diametrically opposed. No union or resolution there.

Representation of the Ladder as a Double Helix

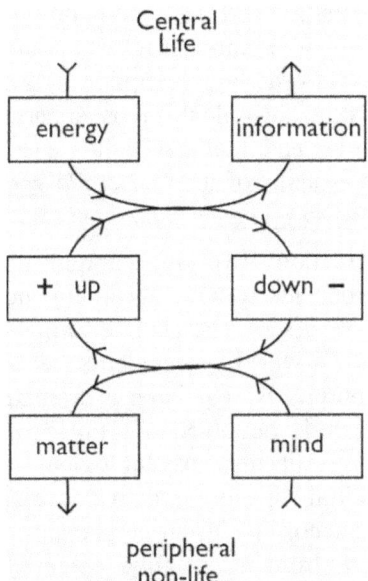

From pole to pole and side to side intrinsic to Natural Dialectic is polarity. It's binary, digital and switches, straight or gradually dimmer-like, on and off. <u>Note that the polar strands of this dialectical helix (information and energy) are arranged to run anti-parallel; they are also plaited into a double helix.</u> As regards **information** check this spiral against Chapter 14: Quality of Information: stack. As regards **energy** creation is a downward, outward 'muscular' process; information-gathering (or knowledge-seeking) is an upward, inward 'perceptive' process leading, at culmination, to the source of creativity - the natural potential that is consciousness. Your own *top-down* construction confirms a polar, spiral geometry of information. Your caducean template (Chapter 17) assumes this

> pattern; so does the physical expression of your informative *DNA*; nervous and hormonal systems work antagonistically.
>
> *fig. 24.2*

Evolution involves contrary twist. It involves, *bottom-up*, release from 'locked' conditions. The name of matter's stream, falling into fixity, is entropy; but, endowed with clever mechanisms, life pumps up against the flood. It treads water homeostatically so, for a while, its bodies are not swept away. Such dynamic equilibrium needs stably input energy yet, as we've learnt, raw energy destroys. It drops into another name for death - inertial equilibrium. Death can't liberate a form nor can entropic fall-out modify one life-type to another. No doubt, stimulus fires fixity to flux but fluids never by themselves create; they can't generate a bio-operational system. Such a system is, crucially, initiated not by energy but *information*. Biological eggs (and every other kind of cell) are, by prior prescription, operationally codified; and only mind can generate informative negentropy. Yet, tail-to-head, Darwinism omits informative an input. It presumes an orderly prescription by prescribing chance and natural selection in its stead. Could this twisted narrative be right? In fact, the greater understanding and intelligence the easier and more powerfully any purpose evolves systems of its choice. <u>Operational systems always rise from mind; they never chance from matter</u>.

Two ideas of creation: two anti-parallel ideas. They need be emphasized apart - one starts with absolutely nothing and the other nothing physical - that's Archetypal Information. One illogically 'evolves' the chaos of a world by accident; the other 'devolves' order of coherent cosmos purposely. The latter is dramatically reflected in the way a baby grows. It is devolved not only from primary physical information (an egg) but also psychological blueprint (archetypal memory and, from initial conception, active development of the 'human idea'). It 'emerges', according to strict plan, from 'inside outwards', from zygote to adult, from informative potential of an egg through developmental phase to reach a final climax, that is, end-result in the extension of maturity. Inside information is devolved to outward structure in the form of mortal coil. If human being is a cosmic metaphor might not spiral geometry by nature (and, therefore, dialectically) express informative potential?

By the same twist in reverse, how might material oblivion invert its own non-consciousness? It is without ideas; its passivity is utterly incapable of active thought. *Somehow, though, natural particles and forces are imagined to <u>evolve</u> purposive complexity - the state of biological bodies, brains and thereby minds with their ideas.* **Such impudence (*A Theory of No Intelligence*) represents the height of unreason, an axe to any logic in creation and at root, therefore, to human thought.**

A helix is a single thread. Cosmos is a polar pair. Its fundamental components are information and energy - a couple intimately entwined within a duplex called the 'conscio-material gradient'. You can spiral either way. Oscillations include up *and* down. And when you entwine two coils you have a double helix, a polar twine. You can scramble either way, that is, reversibly on Jacob's ladder too. In short, cosmos is informed, at heart, by polar opposition and all motions in between. Its strands are vectors Natural Dialectic knows as cosmic fundamentals.[177] **Thus a double helix with its anti-parallels well represents the roots of truth.**

Twists that Entwine

A psychological part of twists-in-tale involves the opposite directions of focus[178] - inward and outward. Forever torn, centaur-like, in the torsion between two opposite poles, awareness and material embodiment, which way do you turn? No doubt you want emotional entanglement; you seek intense, abiding happiness. How do you find it? Does not ecstasy, for which exists a universal craving, involve at root a *loss* of self? Is bliss not loss of self-centred identity at the same time as finding fullness in a beautiful experience, wholeness in a world of love and, especially, communion with a loved one? So which way will you go? Where is the music most complete?

The *materialistic twist* tends outward. You conclude that happiness resides in objects, physical events and the sensation of them. A popular pursuit of pleasure concentrates excitement of both physical and emotional involvements - interests and associations as opposed to isolation, absorption as opposed to boredom. Enthusiastic actions, togetherness - it is easy to see how energetic inversely mirrors informative ecstasy and how, for example, physical reproduction mirrors a psychological act of creation. Perhaps the most powerful physical drive to ecstasy is in the consensual, climactic, sexual union of male and female.

The externalising effect of sex or any other act of physical creation is diametrically opposite such contemplative spiral inwards. But whereas sexual reproduction is an act of creation that involves physical eggs the contemplative act involves a metaphysical egg. The germ of an idea is developed into a work of genius. This psychologically creative act cuts both ways. On the one hand, materialisation of an idea begets 'brain children'; development of potential involves the *creation* of corresponding arrangements.

On the other hand, hierarchical dematerialisation of a particular object translates it into symbol, lifts into general category and leads to *understanding* of its principles. Such *inward twist* turns hedonism on its

---

[177] Chapter 1.
[178] Chapter 0: Opposite Directions of Mind.

head. The reverse of a creation is[179] to understand it. It is to retrace the steps of a development, to comprehend the principles and evolve an understanding of its meaning, its purpose. To comprehend a seed-idea in its entirety is to commune with its creator. Time and distance disappear. By working back to its creator you can understand an artefact. What more natural an artefact than cosmos? What more natural than mind from which it issued and, therefore, than mind which grasps the threads again? Thus, *top-down*, the highest informative ecstasy is to retrieve the Cosmic Egg! This is the mythic, mystic quest. It is Communion with Transcendent Other and a comprehension of the cosmic patterns through which The Big Idea (including you and me) is realised.

Information is, we've seen, the potential for orderly behaviour and, thence, construction. Inside, immaterial information drops down to material, outside form; it falls through a triplex, complementary combination of informative, energetic and structural factors. The hierarchy externalises from codified potential into the form of physical behaviours and, biologically, the book of *DNA*. Every step in order. Programs (and, therein, signals, switches and messages realised by the chemistry of nerves, hormones and other factors) govern biological expression. Their outward adjuncts are biochemical metabolism and physiological function; and hardware, called organs and skeletal anatomy, within whose appropriate infrastructure software can express the pattern of its purposes. 'Dimensional eversion', informative to structural, involves both symbolic and reproductive elements; and, as regards its order of events, cell replication with development.

First, then, brain! It was noted[180] that psychological nuance was reflected neuro-physiologically. *Couples dance; the cortex of a human brain is split into two distinct hemispheres by a central junction called the corpus callosum.* Despite this there are no cortical centres with exclusive function nor, as might be expected if consciousness was simply a by-product of synaptic networking, loss of portions of memory or thought process if parts of either hemisphere are excised. Holism involves two-within-one. *As well as bilateral operation psychologists note evidence of a 'chiral' symmetry or bias of function that is apparently related to sexuality.* With different emphasis right and left-hand characteristics exist in either sex. As such, intertwined humanity is mirrored in the brain.

The bi-polarity of brain's 'morphogenetic symmetry' involves two characters. *The left hemisphere that governs the right-hand side of the body tends towards serial processing, stepwise analysis, logico-*

---

[179] *fig.* 13.3.
[180] Chapter 13 *passim*.

mathematical calculation and linguistic tasks. Reason focuses on local detail and arrangement of its circumstance. It analyses, classifies and cuts clear line. It thus relates to body-self's identity; and is related to the 'masculine' domain where energy (in the determined form of physical action) holds sway.

On the right-hand feeling's part communicates, 'entangles' and emotes. Such character tends, in computer-speak, to process information from its circumstance in parallel; and instead of plodding on methodically it absorbs 'globally'.

Hold out your hands. Place the left one over right. It will not fit unless you switch it upside down. Opposites employ a switch of mirror-like inversion. So with information and energy, mind and body; each part complements and exerts a range of influence over the other but where they meet a switchover, 'chiral' polarity or reflective asymmetry occurs.

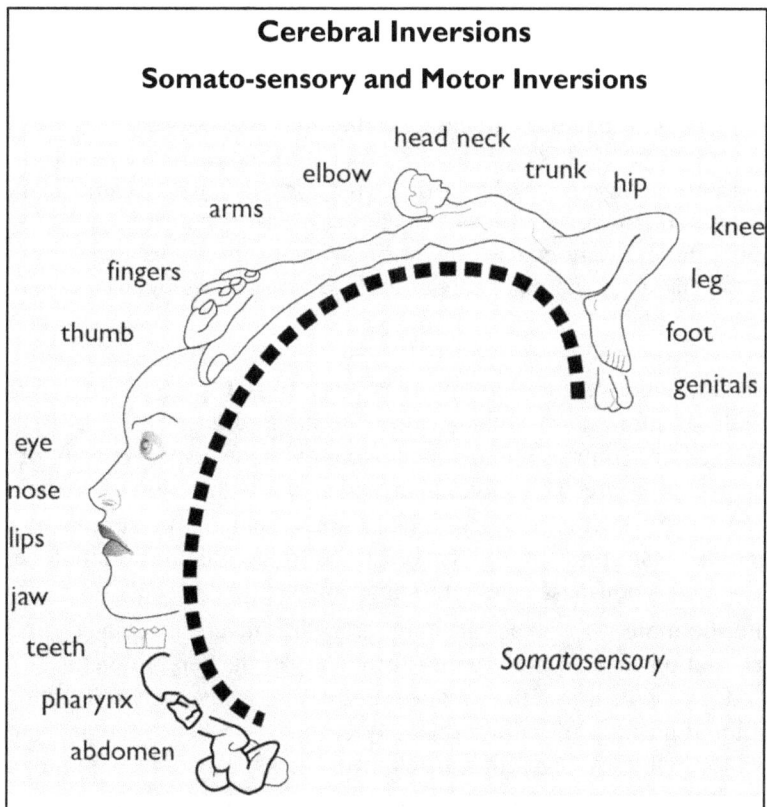

The arrangement of somatosensory and motor 'panels' is orderly. Right hemisphere governs left side of the body and *vice versa*. This inversion is compounded by a twist wherein base-of-body controls are located at the top, leading gradually down (but up the body) to facial and thence inward to abdominal controls.

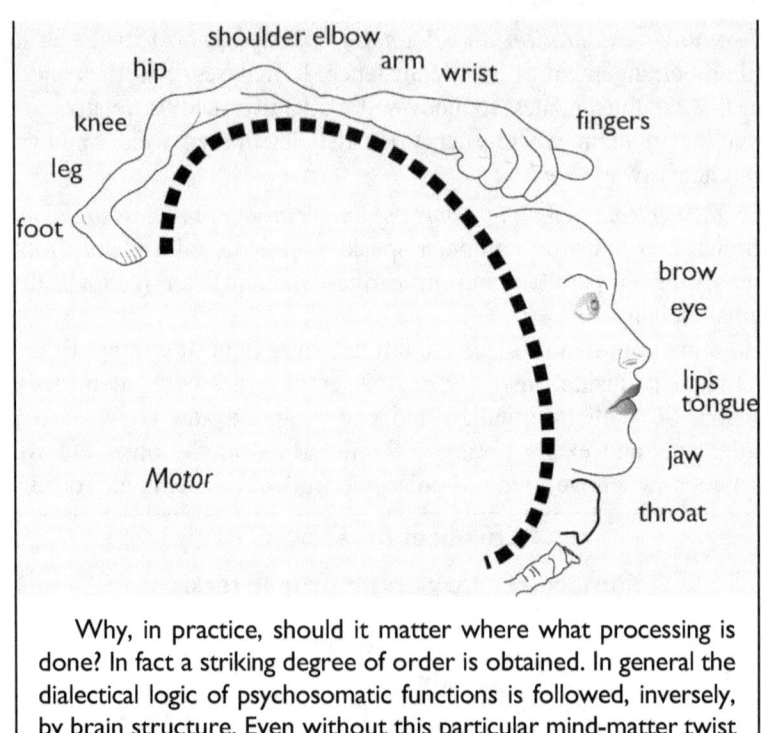

Why, in practice, should it matter where what processing is done? In fact a striking degree of order is obtained. In general the dialectical logic of psychosomatic functions is followed, inversely, by brain structure. Even without this particular mind-matter twist the three-dimensional geometry of cerebral architecture appears too efficient, coherent and reasonable to have evolved by chance.

fig. 24.3

Examine the cerebral cortex. It is the top-front part of our physical medium of information exchange, the internal dashboard of your nervous system. Motor controls appear, in an arc around the centre, in front of the somato-sensory arc. This physical arrangement, with action to the fore, is copied lower down. 'Sensory dorsal, motor ventral' means that motor output exits the spine through nerves at the front while information being raised to consciousness enters the spine at the rear.

Furthermore the areas of both arcs are inversely related to the functional position of body parts. For example, the cerebral genital area, embedded top-centre of the somato-sensory arc, is inversely related to the functional position of its base-of-body correlate - whence the arc 'curls' down through toes, legs, hips, trunk, head, arms, hands, face, mouth and inwards towards the throat and abdomen. A similar, inversion occurs on the forward motor arc and, of course, each integrates with the brain/body inversion, left lobe governing right side of body and *vice versa*. Well done, Lady Luck, if you evolved the measure of such symmetry!

Can't you see? Reflective asymmetry occurs with precious sight as well. Informative < > energetic; mind < > body; brain < > body and brain's own left < > right-hand decussation illustrates creation's

Polar, Hemispheric Inversion of Visual Faculty

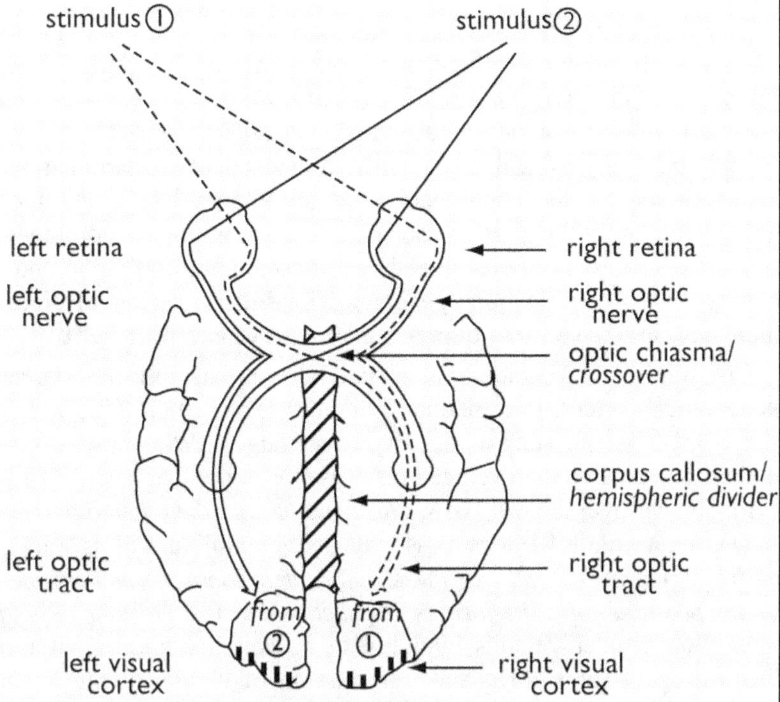

stimulus ①   stimulus ②

left retina — right retina
left optic nerve — right optic nerve
— optic chiasma/ crossover
— corpus callosum/ hemispheric divider
left optic tract — right optic tract
from ②   from ①
left visual cortex — right visual cortex

Just as signals to the left eye are for the most part processed by the right visual cortex, so the right hemisphere of the brain controls the left side of the body and *vice versa*. Such division is clear-cut. Although complemented by anatomical asymmetries (e.g. heart, pancreas, liver) the bilateral symmetry of human form is functionally balanced and aesthetically superb.

Surgical 'commissurotomy' is an operation by which connection between the two sides of the brain is severed. Cleaving the complementarity of left-right function gives rise to behaviour whence emerges a 'split-brain model' of hemispheric tendencies. Major surgery seems to have divided the faculties of mind and body according to the handedness of the brain. With its *corpus callosum* intact, however, the vast majority of humans use their undivided brain as a generally asexual unit laced with sexuality; they are 'sexlessly' proficient in all kinds of skills.

fig. 24.4

character.[181] Equally, we'll shortly see, such inversion informs sexuality.[182] Archetypal sex involves complementary asymmetries that range within whole but, at the same time, polarised human being.

Informative *DNA*, sexuality and nervous system are, in turn, reflections of the wider principle of complementary polarities. *This is symbolised in columns whose pair can, in a final twist, be dynamically expressed as a double helix.* Any behaviour, object or event, psychological or physical, is actually composed of a combination of characteristics - a dialectical stack. A stack allows, like genes upon a chromosome, multitudes of characters, that is, of principles combined. It therefore seems to symbolise the shape of complex information and the polar world its combinations generate. **The full flowering of Natural Dialectic is expressed as a binary spiral, that is, a double helix.**

The spiral is an ubiquitous natural shape that follows a precise mathematical pattern. In a Fibonacci number sequence (1, 2, 3, 5, 8 etc.) each number is the sum of the two preceding it. When two adjacent numbers are divided the averaged result is (for larger over smaller) 1.618 or (for smaller over larger) 0.618. *This ratio (φ) is called the golden mean* . A *DNA* molecule is 21 angstroms wide and, in a full turn of its spiral, 34 angstroms long. *The ratio of these adjacent Fibonacci numbers means that the potential for a life-form is, in practice, borne on a stack of golden rectangles.* 'Golden' chains bind life to its cells. Nuclear *DNA* is the chemical agent for storage and transmission of biological information. The very shape with which it fixes life to earth is expressed as multiples of the golden mean. *The shape of each gene, chromosome and genome is constructed according to the best principles of dynamic symmetry; life's book is printed with geometrical beauty.* **This suggests that, both with respect to components and structural engineering, its capacity for the fax carriage of vital information is optimal; and that *DNA* is[183] the best possible material to expedite, in a durable yet precise, efficient way, the critical use and replication of the book of life.**

The consideration of such *functional logic* provokes a question. Was an aesthetically and functionally meaningless stroke of chance the creator of such an illuminated manuscript? If so, how did a previously sterile, primeval marsh haphazardly scrawl its proto-nonsense in such an informatively relevant, rich style? Why on random earth should unsupervised puddle-chemistry fall involuntarily into the arrangement of a high-tech, code-loaded database - whose operation must have been a part of even an Ancestral Blob? The structural and functional elegance of the *DNA* twine suggests that, in fact, it was the combined product of high-tech

---

[181] see *fig.* 2.4.
[182] this Chapter: Archetypal Sex.
[183] Chapter 20: *DNA* Supremities.

engineering and information technology. If the Textbook about each type of organism was conceived, transcribed and transferred by Natural Science into the business of a biological shape, it would amount to quite a final twist-in-the-information! Consider the creation of purposive mechanisms by chance or by mind. Which idea involves a sleight-of-hand? Which evolves illusion?

Protein synthesis is yet another mini-metaphor, a reflection of the universal creative process. From a whole *DNA* genome specific chemical symbols of archetype (called genes) are transcribed; outside the nucleus mirror-images called m-*RNA* are, in turn, mirrored against translation (by t-*RNA*) into the building blocks of bodywork (amino acids and protein); a chain of amino acids is extruded from its ribosome and, with help from protein quality controllers called chaperones, guided to assume precise three-dimensional shape; finally, according to an archetypal plan, the bulk material of a classical, biological body is developed. **From initiator through conversion into target form - this fundamental order rolls life's patterns out.**[184]

The process is vital. Genetic information, introverted and thereby protected at the inside-centre of its anti-parallel strands in a tube-like scroll, is in the reading opened out. Introversion is unfolded; the message is specifically revealed, transcribed and passed upon a ribosome from code to extroverted shape, from symbol into its material expression. Like thought to deed, the meaning of an idea is translated to the body of its actuality. This is, *top-down*, devolution. Do plans happen randomly? The reverse; the dialectical process is and always was in order. Of this you are living proof!

Replication is for now. Extended survival involves inheritance by reproduction - the business of copying physical structures. *Reproduction is never, even in bacteria, a chemically simple, un-choreographed operation.* **Equally, it is conceptual because it is anticipatory**. *It is pre-programmed. It foresees death. It looks into the future when the present organism will cease to survive.* **It must, from the first or at very most within the first days, weeks or months of the first cell, have 'appeared' to meet this eventuality and, with it, the 'desire' for a species to transcend an individual's life.** Materialism's individual is a bunch of atoms. Can gangs of atoms really congregate by fluke - including fluked-up reproductive information to resist by their own configurative demise? If so, molecular felicity has invented, chance-wise, death-defying species. *If not, evolution's central mechanism, inheritance, is non-evolutionary.*

In biology there are thousands of obvious examples of forethought in design. **Informative and reproductive systems are just two of them; but each one surely tolls the knell for evolution.**

---

[184] Chapter 1: potential, action, exhaustion.

The Reproductive Archetype

Law not flaw is fundamental to biology. Firstly, The Law of Biogenesis.[185] This law is to be confused neither with Haeckel's false theory of recapitulation (his self-styled 'biogenetic law') nor infringement by the Hope of Abiogenesis.[186] No exception, except by mind's imagination, has been found to this First Law - although not for want of promissory trying.

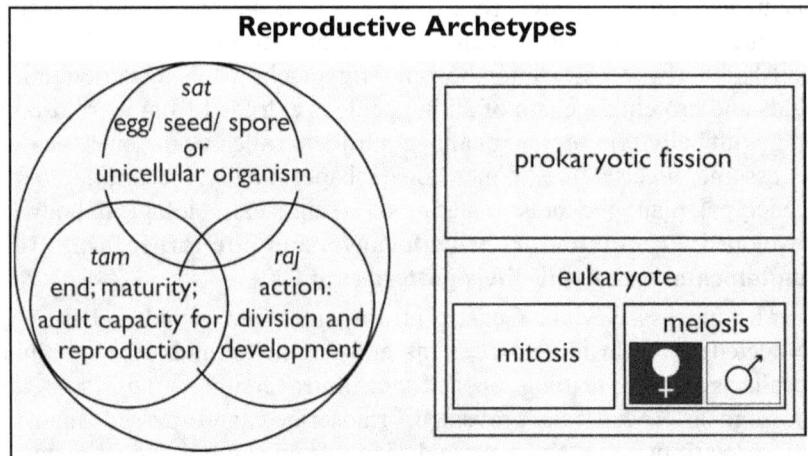

Reproductive Archetypes

- sat: egg/ seed/ spore or unicellular organism
- tam: end; maturity; adult capacity for reproduction
- raj: action: division and development

prokaryotic fission

eukaryote: mitosis, meiosis

In dialectical terms there exist a vector from prior, informative potential through cell division and development to target end-result, an adult form capable of further reproduction; and three basic but distinct modules/ sub-routines of the reproductive archetype. These are prokaryotic fission, asexual mitosis and the rarer but still very common process of sexual meiosis. Around the triplet of core archetypes many different reproductive organs and procedures are constructed.

fig. 24.5

Second, evolutionary fables break the second Law of Heredity. Organisms, without any known exception, reproduce their own types. Spore and seed are well-designed to conserve *DNA*; egg faithfully transmits its program too. Variation (by sexual recombination, mutation, environmental pressures and adaptive super-coding) happens all the time but, by this law, with limited plasticity. This is what, without exception, observation shows. Unlimited plasticity has never ever been observed and yet the raconteur, who claims authority from science and in whose mind it is conceived, protests that such plasticity has made us, every one.

Fact is that reproduction, operating flexibly around the norm of type,

---

[185] Chapter 20 *passim*.
[186] Chapters 20 and 21; also *A&E* Chapter 8: Growing Out of the Past.

but a form of homeostasis? It is, within this limited plasticity, life's mode of stable, generational survival. Thus its mechanism involves, as an orderly biological if not also psychological process, both *anticipation* (of an organism's death) and *purpose* (that its species shall persist). It occurs in sexual and asexual forms. The latter is sometimes considered more 'primitive' than the former but, in terms of functional efficiency no form of reproduction is more advanced or fitter than another. *Conservation and variation* - each has its clear, purposeful place.

Moreover, reproduction depends on chromosomes. Never mind 'survival of the fittest', without *the maintenance of chromosomal integrity*, necessary from day one, there can be no generational survival at all. Was such intricate, important maintenance quickly, accidentally obtained?

The Origin of Asex

The point of asexual reproduction is invariance. It is the conservative retention of complex information by accurate replication. One doubles. It becomes two of the same genetic circumstance, called clones. A clone is, therefore, a cell or organism with the same genetic constitution as its parent. Bacterial fission, budding, spore formation, vegetative reproduction and parthenogenesis are ways clones can occur.

Given this strong rationale you have to ask how various aqueous solutions gradually worked it out. How, for example, did asexual reproduction happily, immediately appear? And if it didn't do so straightaway how did the first bacterium's look-alike appear?

All cells make a copy of their database before division. Accurate reprint, that is, cloning leads to the reproduction of identical cells or, as in the case of your own development from a single cell, the zygote, myriad different kinds of cell. Of these, bacterial cells are conservative survivors *par excellence*. Their replication mechanism is one thing. Non-bacterial cell division is entirely another. After replication any eukaryotic cell not destined to become a gamete enters a profoundly efficient, conceptual postal system called mitosis. *Mitosis is a prize-winner.* Its essential purpose is to save intact written information of the genome every time a cell divides; the chance that chance could do this are statistically nil. Conceptual elegance combines with strict economy of working parts. A choreographed algorithm constitutes the ubiquitous foundation of eukaryotic cell division. Its irreducible components comprise a system of multi-chromosomal postage entirely different from the bacterial one; the mechanism employs a logically impeccable order of steps and complex kit (centromeres, centrioles, spindle and so forth) to expedite its accurate deliveries. Incomplete or faulty mitosis is useless. No organism sports half-way mitosis; faulty mitosis is, equally, a dance of death because information is scrambled or lost. **The evolution of mitosis is another Darwinian black box.**

The Origin of Sex

Asexual reproduction deals in clones. However, is such sameness all you want - a bland and sexless lack of any variation? How dull if there was no such thing as sex.

The point of sexual reproduction is variation on theme. It is neither to add nor subtract but to shuffle an organism's cards into new permutations and thereby deal new hands in an old game. In this case two become a different one. In principle (or type) each and every offspring is the same as its parents but in practice (or detail) different. A refreshing mix. This sex-driven difference also protects it from genetic damage that may exist in the parental genes - a mutant gene has only a 50/50 chance of transfer or, if transferred, is very likely to be masked by a 'healthy wild type'. And as well as promoting fitness it preserves adaptive potential that can, under certain circumstances, buffer changes in the environment.

So carry on! With sex the postal system differs. A double dance is at the heart of life's sexual gamble. *Meiosis, which underwrites the sexual reproduction of some single-celled and nearly all multicellular organisms, bears the hallmarks of a routine designed to extract maximum variation-on-theme at minimum cost in labour and materials.* Its precise, initial dance extends the mitotic algorithm so that, after replication, *two* reductions are made to produce a gamete (egg or sperm) with half the normal number of chromosomes - a state called haploid. The reason, an excellent and critical one, is to avoid exponentiation (i.e. doubling each generation) of the chromosomal number. The count is stabilised. If, for example, it is 46 normally then gametes with 23 each will, when recombined, recreate the diploid 46. The meiotic routine also involves two mechanisms that are not, because they are at the heart of sexual reason (i.e. variation) and its implementation (meiotic lottery), at all haphazard. Count its parts and processes. Analysis reveals that Darwinism's got a clutch of lethal problems here. **The fact is, hundreds of protein 'machines' maintain the integrity of each chromosome**. *Synapsis and crossing-over cannot occur without such maintenance; nor, without further gameto-genetic algorithms, can sperm or egg be made. Sex could not start to work.* These two, however, aren't enough. *After shuffling genes the chromosomes are 're-assorted'. Finally, the genomic level is addressed. Now, since a partner's been provided, half of two whole shuffled genomes are combined.* An informative 'love-letter', sperm, is posted; its delivery, as *USB* plugged into drive, into the female envelope catalyzes the developmental application. Three kinds of recombination. Three specific steps - at genetic, chromosomal and genomic levels (the latter including operational male and female body-parts) - without all three of which variation and survival could not flourish. <u>A well-maintained and deliberate lottery of meiosis coupled up with sexual intercourse (using the correctly prefabricated body-forms) generates</u>

variation;[187] but, equally, strict limits imposed by the maintenance of genetic integrity limit the degree of variation by common descent.

Remember, no meiosis, no fertility so no survival. If, on the other hand, everything's in order then the deal changes but the pack remains the same. Constrained flexibility obtains. Limited plasticity of type restricts lines of common descent.

So how, if it did, did sex evolve from asex? So many different times?[188] According to the theory of evolution, life on earth endured no 'adolescence'. Sexual maturity, in algae, jellyfish and other forms, was present in pre-Cambrian seas. Today we observe unicellular conjugation (in some bacteria and *Paramecia*) and an alternation of asexual and sexual life cycles (for example, in algae such as *Chlamydomonas*). Did sex evolve in each case separately, as second string to an asexual bow? Yeast, malarial *Plasmodia* and, perhaps, a parasite called *Giardia* can also sexually reproduce. Progressive, evolutionary thinkers speculate sex may have started in some single cell but, from a *top-down* perspective, their comparisons, probability analyses and computer simulations are unpersuasive to the point of feeble. Why? Because every case exhibits striking, many-stepped, cooperative and codified metabolic algorithms. These algorithms must produce the correct quantities of specific chemicals (e.g. pheromones) and induce the correct structures (e.g. meiotic choreography or micronuclei). If the code is not fully 'validated' or any step fails the whole process fails. Purposeless evolution must either produce the entire program at once or gradually by accidental mutations; the latter have no idea of target and, as individually useless occurrences, each would soon be selected out. Programmers, on the other hand, aver that exact programs involving multiple sequences of switches and signals set up to anticipate and thereby achieve a precisely targeted event do not appear by cosmic forces acting chancily.

Program is, therefore, of course the rational choice. *Top-down*, dialectic order runs[189] from potential (information) through the business of expression to end-product; it runs from egg through development to final, adult, reproductive form. *Purpose, anticipating then achieving such a form, is an attractor; metaphysical concept draws the present through a program to its future goal.* **It is reasonably predicted that a degree of stricture that**

---

[187] as does a second kind of deliberate lottery, the defining feature of an adaptive immune system. In this case a programmed process of hypermutation acts only on specific, binding site regions of immunoglobulin (but not germ-line) genes. In principle, such SHM 'spins' a repertoire of variation sufficiently diverse to counter any new-to-the-system antigen.

[188] see *A&E* Chapter 7: Seductive Trends for a fuller account. How, for example, could such complementary sexual structures as male stamen and female pistil on a flower or female clitoris and its male homologue, the penis separately and gradually evolve?

[189] *fig.* 24.5.

confirms the Law of Heredity but is entirely incompatible with evolution by common descent will, in due course, be measurable by the close study of nuclear chips, genomics. In other words, the Darwinian tree of life is dead. Meiotic complexity, incompatible with gradual evolution, kills it. You could say that sex kills evolution off.

Reproduction, sexual or asexual, would require prior archetype.

Archetypal Sex

Bottom-up, evolution must have sex. Sex alone yields flexible survival and accelerates the metamorphic 'rate of progress'. Thus, since theory is doubtless fact, sex evolved. Such epiphany cranks up a rumour mill of guesses how. Is flexible survival of the theory's fact revealed by red

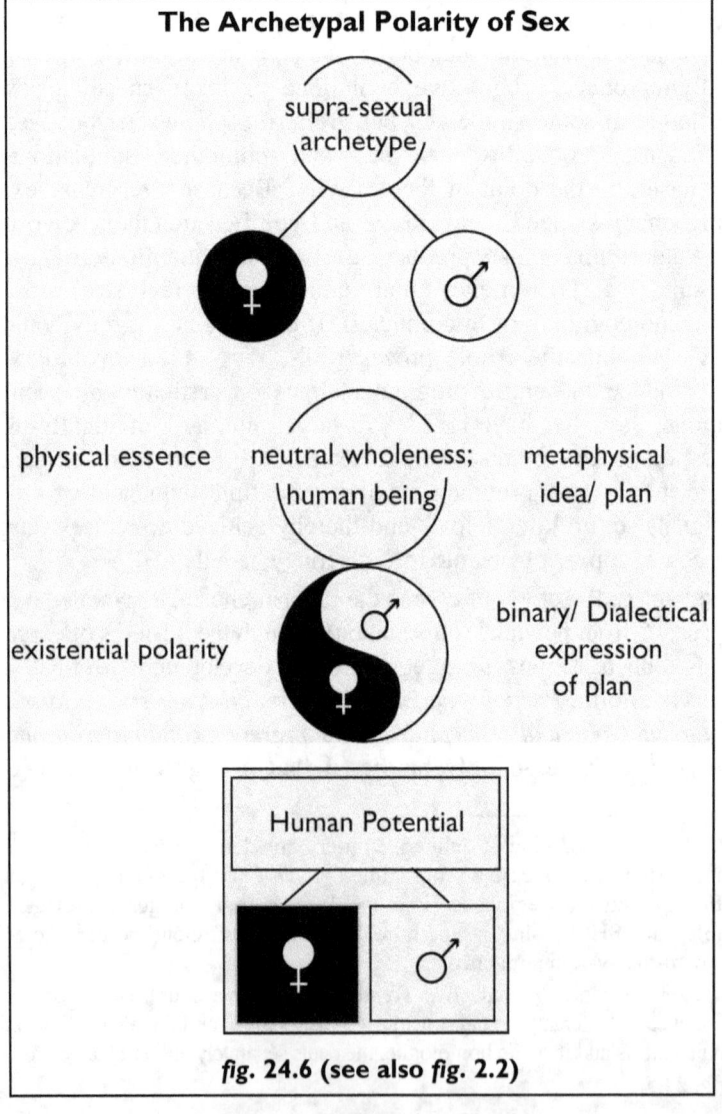

fig. 24.6 (see also fig. 2.2)

(eukaryotic) algae? Do same-size groups of worm-like, fossilised *Funisia* foreshadow coral reproduction? Or did both sexes in a single body (called hermaphrodite) perhaps split later into two? How on earth, the speculation perseveres, did life stumble on such cute 'investment strategy' as sex?

Gradually yet planlessly perchance? Sex is all about *provision* for the future but empty-headed Lady Luck can't even see a second in advance! Foresight to realise complex targets isn't in her character. So neo-Darwinism, lacking concept, proffers only feeblest hints as to the origin of sex. Past and present bodies might be sexual or not but description of them can't explain an origin derived from swathes of (coded) information. Repeatedly, we find, the secret of life's start-up and its carry-on is immaterial information. Even if you found the primal, sexual eukaryote you wouldn't have *explained* its sex.

Of course, feasibility's a friend to play with but, say, Darwin's speculations of 'fit copulation' and 'selective sex' are only gloss veneer. Fitness hones the quality of wild-type offspring - but what has this to do with *innovative origin* of instinct and associated body-parts? Moreover, how, when it can't see a single second in advance, can natural selection 'sort' for future functions? How should an asexual molecular configuration somehow 'conceive' that, after myriad functionally integrated intermediate steps, sex might improve its differently-constructed lot and therefore risk the gambit? **Yet, in principle, without foresight neo-Darwinism is logically selected for elimination.**

'Sex', our subject, is derived from (Latin *seco*) 'cut'; a being, such as human being, is divided into male and female parts. There's reproductively rational difference in body and why not, appropriate to role, in psychological experience as well? Of course, although of different average size, each sex displays the same intelligence, capacities and basic tendencies; and each harmoniously employs both complementary hemispheres of a single brain. Thus, can't the switched stack shown below be naturally right? Can't psycho-physical crossover *generalise* our human sexuality? Is not such nuanced ambiguity, such 'pink and blue' polarity, a fact? But many generalities involve exceptions too. Individual citations challenge stereotype. Of course, hormonal variations and switch malfunctions can occur; nurture or circumstance may also force exceptions to the natural, salient norm.

	yin/ yang	*Tao*
	two	*One*
	polar	*Neutral*
	sex	*Whole*
	female/ male	*Human Being*
↓	*yin*	*yang* ↑
	recipient	*initiator*
	female	*male*

But can the switched stack shown below be right?

tam/ raj	Sat
inclination	Symmetry
asymmetric imbalance	Balance
sexual duality	Single Whole
female/ male	Human Being
↓ tam	raj ↑
left hemisphere	right hemisphere
objective analysis	subjective synthesis
parts/ detail	whole picture
fact-oriented intellect	emotive connection
harder	softer
male bias	female bias
	psychological
↓ ✕	↑
physical	
yielding/ conforming	dominant/ leading
reactress	actor
reciprocator	donor
follower	inventor
below	above
softer	harder/ muscular
feminine	masculine

The brain is a plastic, dynamic organ affected by external events as well as by innate genetic potential. The latter *pro-actively* informs construction; but the former, in the way of so-called micro-evolution, merely generate *post-active* micro-variations and create neither grey matter nor the infrastructure of sex. Blind forces never innovate. How, therefore, could they create a complex, specific program of complementary polarity? In fact, research and dialectical principle indicate that within the range of sexuality there exist deep-seated, polar inclinations. In other words, there exist statistical 'spikes' that register male and female bias called 'sexual norm'. These two clusters suggest tendency, probability and definition to the point of reproductive reasons for their 'opposition'.

Sex is 'spectral'; each organism illustrates a range of partialities - even unto intersexual indistinction and, typically in plants and worms, hermaphrodites. Proportions of the cut (not least hormonal) are blended into all of us but, as regards our individual forms, distinction rules. Polar tendencies take sides. Sexuality is stereotypic by degree. Adam and Eve are dialectically bound so that, within the human archetype, each programmed gender is attractive to the other yet divinely different. In accordance with universal principles, such duality involves the paradox of two-in-oneness, reflective asymmetry and the interaction of opposites. Complementary bodies illustrate, in the energetic field, clear polar discontinuity; while at the psychological, informant end[190] the distinction is implicit, nuanced and fluid. It is subtle but obvious; and powerfully reasonable. **You could easily deny that sex is just an accident; instead, with hormones and its reproductive architecture well disposed, that it is a deliberate construction.**

Stereotypical dimorphic (male-female) variations in brain anatomy are first evoked soon after the chromosomal switch for genital differentiation has been applied. Left-hand analytical/ logical 'male' as opposed to right-hand (or perhaps whole-brain) empathetic 'female' emphasis of hemispheric use has already been referenced.[191] And at adolescence, such stereotypical emphasis is confirmed by hard-wired nervous connectivity established front-to-back in each male hemisphere (perhaps promoting better spatial, analytical and decision-making skills) and side-to-side across the corpus callosum in females (perhaps promoting improved imaginative and empathetic, social skills). You might guess holistic, multi-tasking female thinking complemented the somewhat exclusive, one-track focus of a man. And at the base, energetic end, with human procreative parts most dialectically and physically removed from brain, the distinction between male and female organs is obvious and practically fixed. In other words there exist two systematic, stereotypical norms, male and female, around which the majority in each sex are clustered. Male and female 'spikes' both show the whole range of non-sexual characteristics but also, in different individuals of either sex, a spectrum of 'more-or-less' masculine or feminine traits. Whether it is fashionable to emphasize difference or similarity between the sexes depends not on nature but on individuals and cultures.

Production and sexual reproduction show clear evidence of complementary design in the polar stereotype. There is, however, yet a further, critical aspect of inversion, one closely involved with the physical storage and transmission of information for use in biological reproduction, which it is now time to consider.

We use the word 'offspring'. We think of fresh life springing, as water from the earth, from womb. *Life is elastic, vital, sprung.* A single helix is a coil or spring. A spring is compressed prior to the release of its power. A double helix is, in effect, a double spring. We turn to ponder

---

[190] *fig.* 17.4.
[191] Chapter 13: To Build a Brain and this Chapter: Twists that Entwine.

the chemical structure of biological information storage, a molecule called *DNA*. Whether or not *DNA* acts as an archetypal aerial[192] both its structure and function are dialectical. At the start of every life cycle chromosomal *DNA* is compressed, in the process of mitosis or meiosis, prior to its release as the new life is sprung into operation. In a fresh cell *DNA* uncoils; cell division may continue through vibrant generations of re-coil and release. And the whole process of development is 'sprung'. At first life's body uncoils (develops) at great speed but the rate slows until it stops and, after a pause of poise on the cusp of adult youthfulness, enters a long and gradual decline towards death.

In other words, polar relativity is set within a framework that includes its point of origin - in this case nucleus and its carrier, egg. This point-source is variously described as *axis*, centre or point of governance. It is known as the potential or prior information from which an orderly, regulated behaviour of things derives. **Double helices, whether cosmic, nervous or genetic, are 'inside information bearers'.**

Thus, *top-down,* the facts are evolution's same but line of view opposing. Darwinian natural history, which asserts that the appearance of design is a 'powerful illusion', is itself seen as an upside-down, diminished version of the whole truth. Its interpretation of the facts amounts, overall, to a widespread illusion that has ensnared otherwise intelligent adherents.

In archetypal terms sex is polar. It is the product of precise, purposive pre-programming according to dialectical principles. <u>Each type of life form that exhibits sex is conceived of as a neutral whole divided into polar male and female sexual halves.</u> Unifying information and polarised energy combine to express a duality of complements within a unity. Biologically, therefore, archetype is neutral or, rather, hermaphroditic; and so, effectively, is *DNA* although it may incorporate a polarising switch (in humans, whose default form is female, the Y chromosome is part of mechanism). Neutral potential is expressed, as one of two preordained polarities. The information encoded in each sub-routine is passive but purposive. Sex is, therefore, *conceptually* hermaphrodite - which doubles up initial complexity. Not only switching code but asymmetrically-mirrored and yet complementary body parts to follow have to be conceived - not to mention the idea of gametes, fertile union and consequent development. All must straightaway be in working order so that natural selection cannot kill less off. Such spot-on precision is, from a *top-down* perspective, not an accident at all.

Two from one; duality from unity - this is the order of creation. *Top-down,* single informative potential yields polarity; the principle of 'sexual whole' is split into duality.

---

[192] Chapter 16: Synchromesh 2 and How Does The Connection Work?

Male and female thereby represent the natural order. They reflect the dialectical two-in-one nature of polarity, the cosmic play of opposites. Polarity[193] involves complementary (but anti-parallel) aspects of the same thing, in our case human being; and the dynamic expression of its symmetry involves bias, asymmetry or the differentiation we call gender.

In Indian or Chinese words, opposite vectors (tam/ yin and raj/ yang) are expressed within the unity of a (sat/ tao) conceptual archetype. The archetype is metaphysical; it is the potential from which (*raj/ yang*) male and (*tam/ yin*) female factors are physically expressed. The *yin/ yang* symbol (*fig. 2.2*) represents this polarised, covalently bonded whole. It embodies degree, relativity or emphasis towards one side or the other of the bond. The holistic, archetypal bio-logic is, therefore, that organisms exhibiting sex are conceived as a (*sat*) neutral, genderless whole in which the sexual halves constitute polar (*raj/ tam*) modules; their being transcends but also includes complementary opposites. **The sexual archetype (<u>not</u> a common ancestor) and its *DNA* coding are fundamentally hermaphrodite and incorporate both male and female potential.** Implicit polarity is, in different organisms, explicitly normalised in either a bisexual, hermaphroditic or a single-sexed body. In the former (e.g. a tulip or an earthworm) both modules are expressed. In the latter (e.g. a human) the organism is switched, during the development of heterosexual form, into one channel or the other. *This shows, from the outset, a masterful efficiency of design.*

Archetypal Sex Expressed

Did male and female form evolve in parallel together? If not, which came first - not only when it comes to sex but sexual maturity?

Bodies don't erupt from sperm; and, since an egg cannot precede its gender, one reasonably presumes that females blazed the sexual trail. Hold on! Eggs need sperm - both members need to have evolved smack-dab in tandem. Evolutionary musings ramble ineffectively. Let's cut to the chase.

The potential for sex resides, as for everything, in its central, informative aspect.[194]

Nobody understands how genes, which simply code for protein, act as moulds for the precise, purposive, large-scale shape of organs such as hands, eyes or sexual parts. Perhaps they don't. It is, no doubt, increasingly apparent[195] that a versatile, powerful bio-logic composed of epigenetic prompts and genetic switches delivers the appropriate molecular building-blocks. Equally, construction conforms to a critical, preordained overall plan in order to translate these chemicals into cells and phenotypic shapes as unrelated to their component

---

[193] Chapters 0 and 1
[194] Chapters 16 and 17.
[195] Chapters 23 and 25.

protein as is protein to nucleic acid. The shape of a heart is not caused by muscle protein; male and female sexual apparatus derive from the same tissues but are not of the same functional shape. What informs anatomy and physiology?

Scientific materialism correctly perceives that information rules the roost. A single *top-down* system runs each cell. Information loops from *DNA* to protein, from nucleus to cytoplasm and, cybernetically, returns from outside circumstance. Such information must, by philosophical decree, be physical and therefore what else will you nominate life's chemic royalty but genes - the passive information held on scrolls called *DNA*? How can such oblivious scrolls have sourced their own good sense except by chance? There is no other naturalistic way.

Top-down, however, this gene-centric emphasis both masks royal truth and crowns pretenders. It usurps original potential (*immaterial information*); it substitutes, for governor, the chemical inscription of decrees (*material genes*). Yet chromosomes, like books, mean nothing chemically; their value is the immaterial information they symbolically represent.

Both quantity surveyor and an architect submit their plans; quality of shape defines the secondary rules for quantity, that is, the timely delivery of materiel. As regards development and maintenance of the multicellular architecture of a whole organism hierarchical systems ramify *top-down* to regulate. For example, in many animals two of these branch literally from head-end down. They are concerned to integrate and coordinate the psychological and physical aspects of life and are identified as nervous and hormonal hierarchies. A third, whose task is to raise a multicellular body from the materials of earth, is an order of development that ramifies until it has, in adult form, fully realised original potential. Curiously, this developmental hierarchy has its top-level headquarters vested in a single cell that, adult-wise, inhabits the basement area. It is called a fertilised egg or zygote.

Thence each step in the development of a tiny worm, *C. elegans*, has been meticulously logged. But the origin of its dependency on the orderly sequence and operation of its 100 million *DNA* base pairs has not. After all, isn't logical control a signal of design?

So how would *you* have done it? If *you* needed to achieve the imperative of survival down the generations, how would you have copies made? Especially if you wanted to include some sober flexibility, that is, continually varied reproductions on a standard theme? *We can ask how an engineer might design a self-reproducing machine and ask if nature has preceded him in his intelligent logic. How might the least demand be made on the tissue or strength of a parent while at the same time encapsulating its potential?* <u>*A brilliant, optimally economical idea would be to reduce the parent to a single cell and then, from the symbolism of this cell, build up a new adult.*</u>

This is exactly what happens in nature. Each individual adult is 're-potentiated'. Its body is reduced almost entirely to a symbol, a directory, a coded book of what might be. Its spring is re-compressed into the top-level potential of an egg or sperm; these fuse and the offspring, a single-celled zygote, unfolds in a complicated but very precise program, a hierarchical developmental archetype, which eventually realises its goal - re-creation of the next adult generation.

And sex? If you wanted, theoretically, to design complementary sexes you might opt to specify each module using thousands of genes; *alternatively, and far smarter, you might conceive a main routine which included a 'gender switch'*. Such hermaphroditic concept's switch would trip a male or female line; it would call gender-informed sub-routines. In principle, therefore, at a mere flick the balance would be tipped. A hierarchical cascade of emphasis would ensure dimorphic forms occur. **This skilful concept, this consummately programmed switch is precisely what we find in bio-practice.** Indeed, dimorphic *and* hermaphroditic (but never multimorphic) sexual algorithms are found in plants and animals. From hermaphroditic potential derive uni-morphic hermaphrodite, di-morphic male/ female and other forms of expression such as, by successively alternating sexual *and* asexual forms, even manage to polarize the reproductive archetype itself. You know for certain, therefore, the same program can deliver, for example, sperm (including the set of over 600 proteins that compose its eukaryotic flagellum) and egg (with her own specialities). Of course, genetic switching *aberrations* (e.g. intersex, abnormal hermaphrodites or trans-sexual tendencies) may occur as well but, normally, sex is implicit in a neutral archetype's potentiality.

Nipples are worth a momentary diversion. If both male and female forms were originally rudimentary or, as nascent intermediates, pre-operative, what selective advantage was there for a milkless milk-production system to improve to the point, a few million years later, of working? Are male nipples, positioned like a female's at the centre of an areolar target, the vestigial remnant of a time before men had evolved from women; are they from an ancestral dream-time in which aboriginal, Amazonian parthenogens asexually reproduced; or, as Darwin wrongly suggested, may men have suckled in the deep past? It would have given the baby a double dose of survival so why, if they ever started, did they stop?

Nipples and breasts are certainly the product of a genetic database. What about their shape and techno-logical position? What about allied milk production and the untaught instincts of mother to bring child to her heart and suckle or of the child to suck? Is any instinct simply a product of chemical bonds, of a sequence of non-psychological molecules or of a network of electrical stimulations? In short, can anyone detail the gradual steps whereby a 'purpose' of lactation might have evolved its mechanisms? Not simply to style of outlet (nipple) but the whole production and delivery system synchronised to develop and operate in step and in harmonious association with all other evolving co-factors. **In fact, sensibly perceived, male breastless and female breasted nipple are, as penis and clitoris, examples of the differential expression of the human archetype.**

Immaterial information is explicitly expressed in various switched bio-permutations; variation on the sex-theme/ bisected-archetype runs through many kinds of organisms; and thus (it is the point of sex) within each kind's fertile intercourse individual variation-on-generic theme occurs. Nor, being metaphysical, can archetypal programs ever physically mutate.

In short, powerful economy of reproductive plan implies a clear purpose; highly intelligent strategy anticipates and targets its conceptual climax - a reproductive singularity or, two as one, a couple.

Talk of routines, switches, menus, *top-down* programming and latent sources of information is never the language of chance. **Top-down programs never happen accidentally.** *Since, however, matter can't anticipate it has no goal.* So goal-less evolution of a bio-program is, for every systems analyst, irrational. *It's therefore no surprise that the evolution of sex is another deep black box.* <u>Every information technologist knows utterly and absolutely that you do not construct sub-programs and their large-scale integration, systems, for no reason using just a bucketful of bugs</u>! **Yet this, an impossible dream that scathingly dismisses archetype, is the neo-Darwinian Theory of No Intelligence.**

Which Came First?

<u>*Did an egg precede its adult*</u>? You can't see a memory but have you ever seen a physical idea? Ideas are potential; from potential outcomes are evolved; an egg, packed with symbolic information, is as near to 'natural idea' as you will get.

E cellula omnis cellula. 'Only from a parent cell does daughter come'. Nevertheless evolution contradicts this primal axiom. It sees cells derived from the disordered 'egg' of a biochemical puddle; then it sees different types of cell evolve from precursors and from these, in serial order, transformations unto multicellular complexity. *As a simple egg 'evolves', by <u>planned development</u>, into an adult, so it's easy to presume such 'evolutionary algorithm' represents an original order of construction. The theory of evolution is based on the seductively similar yet opposite concept of <u>unplanned development.</u>* But where informed development is full of reason how can mindlessness anticipate a thing? How could teleological development evolve unplanned? Such evolution never did evolve! Evolution's *not* development; conflation of the terms leads/ misleads into equivocation. **Uninformed, deficient evolution makes 'developmental' nonsense.**

An egg precedes its chicken. Is the chicken simply an egg's way of making another egg; or an egg simply the fowl's reproductive *modus operandi*?

Did an adult precede its egg? Did fruit precede its branch? We noted that reproduction looks to the future; and thus sex *anticipates*. It is a conceptual process engaging machinery of irreducible complexity to achieve a target - generation after generation. Survival of the kind.

Which, after all, comes first - principle or practice? In what order did the homeostatic triplex (information processing, the flux of energy metabolism and structural permanence that includes reproduction) 'happen'? Does phenotype precede genotype, hardware software, adult zygote, chicken egg? *Is life bred from chance like chance itself completely uninspired - or, in design, inspired and awe-inspiring?*

Perhaps life, if it evolved, cannot be compared to well-conceived informative systems. *This is because no such system can arise by accident.* A computer system, especially, has to be complete and bug-free before it works. **It cannot be too clearly stated that the primary, basic component of any informed system is its plan.** A plan serves a purpose, the purpose is the goal and the goal is the end product. There is in life, as with any code or machine, an element of 'chicken-before-the-egg'. **In this view the adult is *in mind* before the egg.** How, therefore, can offspring precede parent? Adult, either conceptually or physically, must have preceded egg.

Egg and Adult Together. What has been noted bears repetition. **Wherever an apparent 'chicken-and-egg' situation crops up the puzzle is resolved by the introduction of purpose, design, information and mind rolled into one - teleology.**

Natural Dialectic takes a top-down, teleological perspective. It supports a systems approach to biology. A system's order, like that used by all human designers, starts with concept and finishes in material detail. Indeed, professional engineers and information technologists such as systems analysts are instructed by international design standards to program *top-down* starting with fully functional concepts called plans! The fact that technology insists on such a design process demonstrates that its physical systems nearly always contain irreducible (i.e. chicken-and-egg-*together*) mechanisms. How absurd, not to mention unprofessional, it would be considered if you were to gather any old nuts, bolts, sheets of metal and wires and 'worked' blindfold to try and finalise a purposeless non-machine with high level functions!

As far as chicken and egg are concerned, therefore, the simple answer of an information technologist is that neither came first. One did not precede the other. You make them together with the same object in mind - in this case a self-reproducible biological information system or, in other words, a living being. **You are one half of a metaphysical idea called 'human being'.**

Chapter 25: Extra-Grand Mythology

Creation myths! Aren't they atavistic nonsense from beyond the bounds of naturalistic science - like the world itself? Immaterial fairy-tales, non-scientific fables that, quite rightly, rational folk ignore. How, therefore, were the world and life-on-earth shaped?

What is thought but a projection on mind's screen? What is such image fixed but memory? What, therefore, is archetypal memory but cosmic image fixed?

Cosmos is, in scientific terms, a transcendentally-projected fine-geared miracle whose cause nobody understands.[196] Why, from metaphysic, shouldn't bio-physic have been awesomely projected too? How illogical is such informative mythology?

There's certainly no logic in the answer 'chance'. If you're made of faceless chance what logic is in you? Yet, we're seeing, reason, logic and intention make you up. You are one of bio-information's incarnations. Thus, the chance-borne theory of evolution would seem illogical as mythical - life's extra-grand materialist mythology!

The Origin of Growth and Development

	surrounding field-of-action	*Axis*
	physical expression	*Plan*
	material outcome	*Immaterial Concept*
↓	*detail*	*outline* ↑
	lower-level expressions	*top-level guidance*
	informed system	*informant system*
	micro-individuality	*macro-generality*
	minor instability	*major stability*
	variation	*invariance*
	target parts	*symbolic representatives*
	codified product	*generative code*

To nail the latter point home you require, on top of sex but immediately with it, development. Biological development involves the precise, deliberate and highly anticipatory process of transforming a single cell into a multicellular, multifunctional form of life. Of course, the *origin* of sexual reproduction has two lines of view, *top-down* and *bottom-up*; by informative potential or by chance. In this case, which line is the dominant wild-type, which the recessive product of mutation, error and defect? What is the natural reality? And what, in this same vein, is the reality of process following hard on sexual climax? Is not the

---

[196] *fig.* 12.1.

zygote's subsequent development towards a second sort of climax - adulthood? How came, to turn life's wheel unbroken, adulthood?

Bottom-up, materialism needs ascribe the double-authorship of these two very complex, integrated algorithms to the jiggling wiles of Lady Luck and, IQ-zero *inter pares*, time. How did this dumb pair unwittingly proceed? **Surely the <u>origins</u> of sex and development are not, like so much else, yet more Darwinian black boxes?**

In fact, as we've seen, neo-Darwinism has incorrectly assumed that gene/ protein mutation was the basic cause of both innovation and transformation; and that life's parts, not least its reproductive cycles, can, although involving hierarchical, highly articulated and minimally functional programs, nevertheless accrue gradually and randomly! These uninformative, irrational ideas are dead. What might replace them?

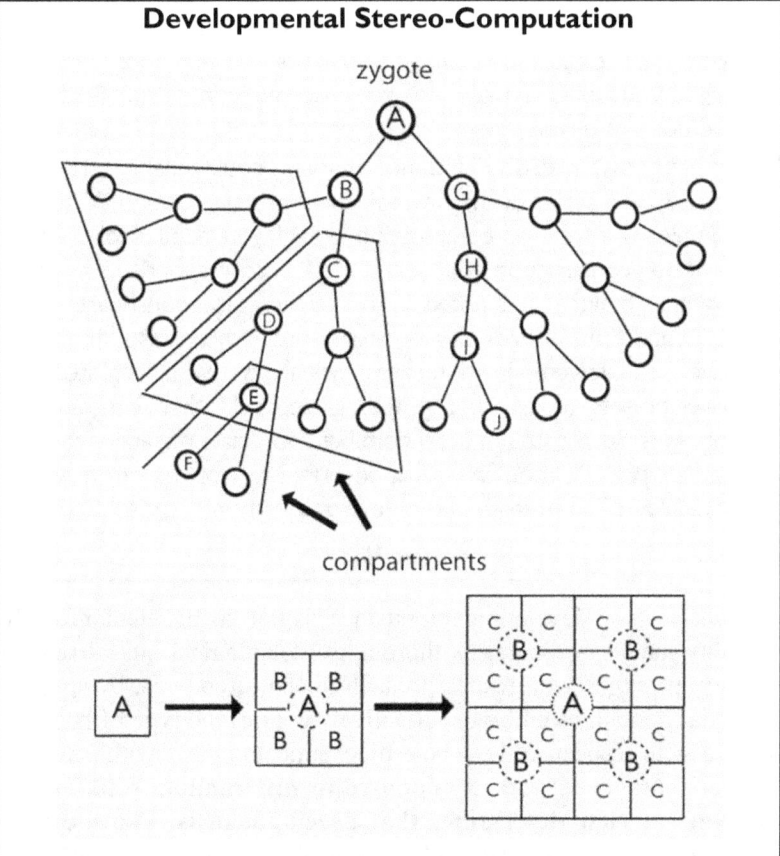

These schematic 'decision-trees' are data structures that represent the development of organism from zygotic **source to sink, that is, from top potential of a seed downstream** (past A, B, C etc.) **to adult form**. Nodes of influence appear and, in turn, give rise to others while they disappear. Routine A gives rise to B's and B's to differentiated C's etc. An increasingly determined, distinct picture emerges. Through stepping stones of stages sub-routines generate various compartments (or segments) that will become the integrated and yet self-contained tissues, organs and systems of a whole. In short, a preordained cascade, an anticipatory stereo-computation of breath-taking elegance controls, perhaps from conserved and universal basic principles, the development of all specific downstream details of a form of life on earth.

Evo-devo (the evolution of development) consists, as we'll shortly see, in the notion that mutations at a high level in the cascade (say, B, C or D) might cause large-scale effect 'downstream' (say, at F, G or lower stages) as subsequent details emerge; and that they might thus generate large-scale transformations in body-plan, create fresh coherent systems with their organs or simply adapt organic functions. Thus might you solve a major, as-yet-unsolved conundrum - the evolution of development.

Yet how, precisely and not vaguely, could changes to a protein or a series of them wreak such macro-evolution? What complexity of stereo-computation would you need evolve to navigate construction of, with all its correspondents, for example, brain?[197] It is noted, in this vein, that locations/ parts B, C, D and E in a developing organism must synchronise, that is, have co-evolved in intimate synchrony with collateral developments at G, H, I, J and, indeed, all other integrated segments. Such a stream of random but coordinate mutations that, in their jack-pot thousands, built an organism would seem quasi-miraculous! The strength of evo-devo plausibility will be discussed.

fig. 25.1

Top-down, the developmental business is not 'as if' teleological. It is, in reality and not analogy, thoroughly teleological and irreducibly complex. It is thus with respect to such nuclei of code as, by specifying the spatial domain of embryos, effectively control the type of body plan; and also with respect to the whole biochemical context within which its magisterial programs work. **As opposed to any 'random walk', the *top-down* line of view determines that development is, like the rest of reproductive process, <u>anticipatory</u>; it looks to the <u>future</u> with a goal in mind and is thus conceptual. Such exquisite, futuristic programming, useless without its end-product, cannot involve non-design; it has, therefore, to involve informative design.**

---

[197] see Chapter 13: Build a Brain.

To this end contextual epigenetic factors 'guide' genetic operations. A hierarchical, strictly coordinated program generates - from egg through general outline to increasingly 'determined' detail - an expanding network of decisions. Thus, from principle to practice, 'up-stream' conceptual calls establish, firstly, infrastructure. Outline. Clusters of master genes involved in the regulation of developmental patterns control at least the flow of materiel required in the early stages of that brilliant stereo-computation[198] otherwise known as morphogenesis. They amount, with egg as program's first-line, to high-level sets of switching routines branching to lower hierarchies of sub-routine that progressively express the detailed, 'down-stream' formulations of an organism's phenotype. Some of these routines have been conserved to the extent that, for example, genes from a mouse can be inserted into a fly and trigger the same desired result. You can interpret such inducing routines as so powerful that, although evolved early on, they are conserved simply because their loss or deformation is, because you lose the organ involved, catastrophic. This, however, by no means explains their 'must-have' origin within a stepped, *top-down* scheme of construction.

'Saturation mutagenesis' of bacteria and fruit flies has, by turns, mutated every location in their genomes. It has disturbed early genetic decision-making by the master genes but never produced viable offspring let alone implied a Darwinian necessity, the capacity to transform one sort of organism into another. Indeed, the integrated circuitry of regulatory gene networks that controls stepwise development tightly constrains the possible extent of *any* change let alone accidental ones. The effect of more than subliminal disturbance would, without systematic compensation, ripple through the process and maim or destroy it. This is exactly what we find. Miscalculation of parameter early in the construction of a machine may lethally affect it later; similarly with the 'outline' or foundational stages of biologically-engineered development. Later 'downstream' errors yielding low-level variation (say, 'wrong' colour or minor morphological aberration) are often tolerated.

How, therefore, does evolution explain the construction of body-plans with correlated systems, organs, tissues and cell-types? *Top-down*, hierarchical integrity strongly resists mutational interference. In fact, as far as optimal biological function is concerned, tight-knit, top-level circuits never vary; lower levels may, as variation on top-level themes, proliferate. Similarly, each organism's body-program, articulated macro-theme to its specific variations, is unique but also phylum-wide. No crack in coherence gapes and thus, notwithstanding the hidden invariant that immaterial information represents, macro-to-micro is proclaimed, in both *top-down* theory and experimental practice, as the rational way that

---

[198] see Glossary.

bodies work. From principle to practice, from conceptual generality to individually specific outcome life's particles accrete around a magnetic central axis, information. This axis is coherently represented by packets of chemical principle - epigenetic harmonised with genetic regulatory circuits. Such chips constitute the antithesis of *bottom-up*, unplanned aggregation. How, logically and rationally, could mindless, repetitive matter allied with chance have, even given endless time, aggregated life? Random, Darwinian-style evolution, micro-to-macro, is, by virtue of every organism's thoroughly-informed integrity, rendered as irrational as impossible. Natural forces couldn't wipe sweat from a forehead nor even, *ever*, serve a glass of beer!

The informative interpretation is, therefore, that intelligence develops plans; and, in each case, a *top-down* program from first principles (or main routine) issues regulations through a suite of sub-routines.

It's now clear why novelties do not evolve but, like the Cambrian 'explosion', suddenly appear. Life-burst! Information revolution! 20 of 27 animal phyla in the fossil record appear in a time-scale measured (at about 10 million years with the main pulse 6 million) as the wink of a geological eye. Theoretical impossibility has driven the production of a series of hypothetical Precambrian scenarios none of which[199] irrefutably assuage humanist anxiety let alone address life's core initiation problem - the source of its 24-carat information chips.

Other pulses left, according to interpretation of the fossil record, plants and fungi in their wake. Editions of the lower classes (order, family, genus, species) radiated later; as a *top-down* model predicts, variation (micro-evolutionary or relatively trivial in scope) proliferated. Relatively few species are found early on, abundant later. Firstly there's a basic model (say, The Rocket); then ramifications of variation-on-its-theme. Steam locomotive classes are evolved; improvements issue from the minds of engineers. What sets a basic bio-model up? As techno-logic whence does bio-logic flow? At least, later adaptations flow from either *pre-programmed* potential[200] or minor accidents (say, if a lopped-off chimney missed a low bridge or marginally increased top speed). But you don't build corporate networks let alone life's logic-circuits using lucky strikes in circumstantially haphazard ways.

Bio-logic

It is easy to organise a series of phenotypic forms in a way that suggests one could morph into the next, for example, that amphibian forms represent an extant 'island' intermediate in a chain of extinct, 'submerged' intermediates between fish and reptiles. *It is more difficult*

---

[199] see Darwin's Doubt by Stephen Meyer for a detailed account.
[200] Chapter 23: Super-codes and Adaptive Potential.

when you have to explain intermediate metabolic or developmental steps which are critical but useless outside the specific context of their purposive pathway towards target molecule or adult, reproductive form. On the one hand, either form of development involves numerous intermediate stages, each in its own right indispensable as intermediate configurations in a particular, progressive process of construction - one which dovetails into all the other relevant aspects of an organism's structure and function. On the other hand, no step has any selective value either *per alium* or *per se* except in the company of the full suite necessary to achieve the purpose of the pathway. How, thence, could a developmental sequence, from zygote through impotent, pre-hatched 'larval' or 'baby' forms evolve serially over even a few thousand let alone hundreds of millions of years? How did the first blastula let alone baby reproduce? In other words, the coded appearance of an intermediate can have no selective value unless it occurs in functional concord with the rest. On the contrary, it will suffer negative selection and disappear before any other of the series of links haphazardly appears. *Simple logistics eliminate gradualism*.

From one you grew into perhaps 100 trillion cells. From a zygote (of perhaps 100 trillion atoms) there developed, precise in time and shape, every facet of your adult form. Who bluffs that any multi-celled incorporation could evolve its own development through aeons to maturity? No-one suggests a spore or gamete was once, in itself, an independent organism that gradually transformed piecemeal into an embryo; or then, after much 'blind-watch-making', turned into a foetal form which then, by haphazard starts and stops, became an infant; so that eventually reproductively capable infants evolved by a series of chances into the reproductive climax of an adult. Such negation of orderly control would be as unreasonable as suggesting there was no intelligence behind stereo-computation.

Sex, metabolism, metamorphosis and morphogenesis are four anticipatory and therefore conceptual processes. **The fact is that the evolution of development is as much a black box as the evolution of any of them. Their bio-logic hammers nails squarely into at least four corners of Darwinism's and thereby materialism's coffin.** It is a fact that in the production of any machine both 'egg and chicken' almost concur - except that the 'egg' of principle and design inextricably *precedes* the physical realisation of equipment. *The root of archetype is conceptual.* **Indeed, in this view the architecture of an adult by which any multicellular species is normally judged is only the last, special frame in the series of a dynamic, hierarchical, developmental archetype.**

Thus, when it comes to reason, simple, systematic substitutions can be made. *Instead of 'time and chance' employ 'intelligence'. For 'conserved' or 'ancient' read 'high-level or upstream instruction' or 'efficiently devised'. For 'natural selection and mutation' employ,* respectively, *'editor and author of ideas'*. Thus finally, having transformed *'macro-evolution'* into *'body-plan stability'*, supplant *'phylogeny'* by *'bio-logic'*.

Logically Expressed

The attainment of target results from a completely controlled, hierarchical cascade of events might well be called a logical expression. Nowhere are reason, code and logical design as clearly displayed as in the exquisitely complex, simultaneously serial and parallel expressions of bio-logical construction and operations. *Thus the IT element that governs life is absolutely clear.* **Information is potential; metaphysic leads to physical expression.** Without plans life is nowhere. Plans grow trees but trees can't grow them. Trees grow reproductive flowers and fruits *in order* to secure their future. But, fact is, neither anticipatory codes nor, therefore, anticipated, complex, specified results ever emerge, unsupervised and accidentally, from mindless chemical events - though they can certainly derive from mind.

It should be obvious by now that life's databases are, in cybernetic practice, governed by an operating system composed of application programs. For example, extra-nuclear hormones, messengers and other circumstantial factors can reset epigenetic factors or call genetic routines; and internal genomic factors include an army of switches, locks and keys called variously initiators, promoters, repressors, selectors, inducers etc. Their agency is composed of *DNA* sequences (sometimes repetitious, sometimes not), strips of *RNA*, proteins, binding motifs, epigenetic methyl groups, chromatin, hormones, cyclic *AMP*, chemical gradients and other signalling (or messaging) molecules. As a whole they constitute a complex information system whose key function is, like the musician's, timely selection according to a pre-programmed score.

It is equally obvious that you access a library as you need specific books, not all at once. Differential gene expression is the way every cell requisitions its supplies for maintenance, repair, reproduction and often, with the latter, development. Control of access constitutes a critical part of the way construction processes are managed - *top-down* from coded message into physical actuality. Cell differentiation and stem-cell research depend on it. *Top-down networks run the show. They are always the product of purpose, design and its intelligent information.* Once it's accepted how life's physically expressed experiments can search accordingly for the expected cogs.

Bugs disrupt the smooth-running of logical expression. For machine and coded text alike what intelligent solution might a human best employ to rub them out? Repair shops and detective editors? Hey presto! **Life's real editor is not crude natural selection**. In fact, cells invest heavily to protect *against* the theoretical generators of evolutionary diversity. **Thus, far more sophisticated, the bio-logical saviour is a battery of genetic**

repair systems found to suppress the appearance of illogicality (random genetic mutation) in every single unit of life. An operating system of at least 130 genetic factors, including enzymes, cooperates to reduce error in copying the human genome to about one in three billion parts. Extensive correction mechanisms exist to cover all kinds of mutation. These include 'proof-reading', cuts and pastes that, for example, detect wrongly paired bases and repair them correctly. Electronic scans identify a broken strand. Together such factors may reduce mutations per nucleotide in bacteria to between 0.1 and 10% per billion transcriptions and ten times less for every other kind of organism. **Such ingenious investment emphatically resists the theory of evolution; it literally militates against.** After all, why should any random process care if what it writes is 'wrong'? Indeed, evolutionism welcomes bugs as, supposedly, the mechanistic 'spring of progress' leading to more intricate complexity of text, greater compass of abilities and quality of the genetic literature! So why, by carelessly evolving careful editors, should the process self-degrade? *On the other hand, debugging systems certainly suggest the text that DNA is carrying is 'right' and must not be degraded into nonsense. It suggests conceptual care incorporated into protein and the genes.*

So much rests, in universities, on evolution as an *a priori* mind-set. This world-view denies a *top-down* frame and so the problem's stuck in mud - how do you source such mass of proof-read information in an irrational, chance-blown way? Who writes intricately cross-referenced books like that? Illogical expression waves its arms in academic air.

Evo-devo

Evo-devo, a sub-set of aforementioned 'natural engineering', is another ploy.

As a first step presuppose the presence of the object of your explanation, the source of bio-information! After all, whence elsewhere could arise a special proto-cell - one that incorporates informative potential for development of all life's forms? Development's a *top-down* process; anticipation needs a strategy in place.

Next off, how could that top-level initiator, a single-celled 'egg-of-life', have been accidentally transformed by printing errors from the entirely different, basic genomic potential of a bacterium or even an asexual protozoan? Actual eggs are yet another game. Sexual apparatus[201] in completeness needs surround them. Although containing the same *DNA* as every cell in its developed, adult form, banks of switches need be integrated for complete control. Differentiation's not haphazard. It is automated. Precisely primed *in anticipation*. With what 'knowledge' such primeval proto-cell must have been first endowed! Did the primer (or its priming) start with a stream of must-have-happened accidents?

---

[201] Chapter 24.

Developmental biologists uncover switch-arrays and associated, complex, hierarchical cascades of regulation. They reveal top-level 'master genes' or 'tool-kit factors' that determine which parts of a body will develop out of others. These blocks of levers clustered as a signal box that rules developmental lines remind you of a railway running from a start to terminus. They remind you also of computer software. **For example, a cluster of control genes would represent a top-level or main routine from which genetic sub-routines were switched; *Hox*, *Pax* (for nervous system inc. eye genes), *Otx* gene family (for kidney, guts, brain, gonads) and other 'informative centralities' would, with the brilliant Boolean logic of a computer program, mastermind the developmental output of bodies ranging from a jellyfish or fly to you - with a somewhat different kind of logic to govern plants and fungi!**

Regulatory logic-clusters are found, it will predictably transpire, in all multicellular animals. For example, in four distinct blocks in vertebrates (called *Hox* gene clusters) each gene is responsible for triggering a cascade of sub-routines that will supply the right materials in the right place at the right time to construct a given segment of the animal in question. Obviously, therefore, a *Hox* gene for a worm will define (or, in computing terms, call) different or deeply differentiated organs (sub-routines) to ones for a fly, a horse or a human. In other words, the same genes can be responsible for initiating the development, in terms of systems or organs, of different outcomes. In one instance a gene may specify for a tail, a coccyx or the rear part of a fly, frog or grasshopper; and the gene that triggers development of your eye would, if transferred to a fly that lacked it, cause its blindness to be eyed - with a fly eye not a human one, of course!

Nevertheless you would expect any changes at the high level of main routine to initiate major changes 'downstream'. Indeed, you earlier expected that specific cases of incorrect high-level inductions, either natural or experimental in origin, could cause bizarre effects. The right potential would be realised, if the determining locks were 'sprung', at the wrong place and/or the wrong time. Recorded *Hox* (homeotic) mutations lead - with rare, minor exceptions - at least to extreme and crippling abnormality but, in most cases, spontaneous abortion. Neither outcome is a powerful harbinger of useful novelty.[202] In fact, therefore, the critical, computational algorithms of these *Hox* regimes are intensely conservative. Therefore how, logic from no logic in the form of chance mutations, could the whole tight-knit caboodle have gradually evolved? Why should such systematic elaboration have developed piecemeal in an accidental series? How, as in a metabolic pathway, could each fractional stage survive until the working whole clicked into place? *In the top-down view Hox would*

---

[202] Chapter 22: Origin of Type and Chapter 23.

represent a high-level, maybe main, routine and thus, allied with all connected sub-routines, be part of a line-by-line solution, a stepwise algorithm in the chemical expression of an archetypal program!

Pass back in time to when such blocks of logic must have been installed. Whence originated *Hox* genes or epigeny? Speculation provides, deep in time before segments, eyes, mitosis or anything else existed, Mysterious Ancestors. Bacteria presumably. **The evo-devo idea is considerably constrained by its fact that such a marvellous Ancestor must have been equipped with sufficiently networked circuitry and, perhaps, unemployed, spare genes for building complex bodies. Most importantly, it must have carried, in latent or embryonic form, hierarchical genetic logic; also corresponding pre-genes marked for bearing the extensive developmental themes for eukaryotic animals, plants and fungi!**

Why? Because of the rate these body-forms appeared in the Cambrian and other revolutions. And because genes controlling sexual reproduction and the development of multicellular bodies must have existed *before* any such embryo or adult body existed. But atoms never dreamed with formidable 'forethought'; such a concentrated information packet isn't something natural forces thought out - only human speculation at the other end of time! Such speculators are persuaded single-celled Ancestral Colonies provided regulatory logic that would, with tinkering to *top-down* systems somehow even then in place, forge proto-organs for eukaryotes. An avalanche of just-so errors (we're unsure which but so the evolutionary 'logic' goes) *must have* thrown together complex, cross-referencing structures such as permit motion, sense, nutrition, reproduction and so on to work cooperatively. In fact, it seems more likely that these programmed co-functionalities sprang from latent potential, that is, archetypal pre-formation.

No proof but materialism's animistic theory drives its wild, theory-necessitated apprehensions of the facts. No doubt, as has been discussed, trivial changes commonly occur. For example, the size of a stickleback's fin might be altered epigenetically or by a random mutation - but the fin *per se* is correctly developed and reasonably placed already! Why should accidental variation presage evolution of a system or a body-plan? What about the *origin* of fins or eyes or any other organ, not just minor changes to them? What about invertebrate or vertebrate developmental cascades; where did whole bodies come from? Biology has now begun to understand the preternatural mechanisms of development; but evolution can't explain the stunning forethought that's embedded in developmental bio-logic.

A Clap of Fragile Wings

How does chance evolve a path that targets goal? How, entirely ignorant of end-game, did as-yet-useless intermediates survive?

Missing links die of incompetence thus how did any metabolic pathway prophesy its own construction or feedback control; how (as, analogously, in the logical, consequential proof of a mathematical theorem) did thousands of correct steps on the path to reproductive adulthood arise by accident? Yet miracles make possible impossibility and what is probable soon morphs to certainty. Evolutionary 'must-have-somehow', 'just-so' stunts are pulled continually except, with metamorphosis, their bluff is called spectacularly. Butterflies have always thrown Darwinism, in an arm-lock, in a flap and on its back.

Back-to-front. The transformation from egg to beetle, bee, fly or, more strikingly, a butterfly illustrates a pattern of development that defies a gradual, practice-to-principle evolutionary explanation. Entirely different-looking phases, each perfectly formed for function, serially erupt. First egg; then, for a 'childish' caterpillar to become an adult moth or butterfly, it eats and moults. It keeps moulting exoskeletons (which are flexible like cat-suits and yet give it shape) for larger ones that form folded underneath the smaller outer sheath. At the last and largest size skin is shed by delicate manoeuvres revealing a cocoon, a chrysalis in which the future hangs. Then caterpillar body parts dissolve and build again into a butterfly called the imago that, emerging after several days, inflates its lovely wings by pumping blood into their veins.

Now check the wondrous microstructure of coloration that, using multiple pigments and reflections from precise arrays of mirror-like scales, is revealed as iridescent wings unfold. An epitome of engineering - watch the fluttering ascension of a butterfly. Could humbly waiting as a caterpillar eventually evolve a program (saved as what are called 'imaginal discs') for pupal development? In what nascent cocoon could enzymes 'know' how far they should dissolve a caterpillar's parts before re-building to transfiguration called a healthy butterfly. How long did some ancestral pulp hang round inside a chrysalis (that came from who knows where) until a suite of chance mutations magically (how else?) re-programmed 'mush' into the concept of winged flight? How, in fact, did pupal death rise straightway (not even in a generation) to a form by which type-butterfly appears? One must ask also how, without the benefit of plan to reach anticipated adult form, serial immature phases took hundreds of millions of years to accrete. *It is noted that before any organism could 'create' any new stage of development it would have first to reach its present stage's adult limit then add that new stage.* This is because it has to reproduce to create the fresh, 'advanced' offspring. A reproducing caterpillar? Such back-to-front order is patently absurd.

From principle to practice, conceptual information to biology - imaginal discs, homeobox genes and other features simply demonstrate the *top-down* venture that, at climax, claps on stage a butterfly.

Butterflies are symbols of nemesis. At the clap of silent, fragile

wings Darwinism logically dies; a giant is slain and at the same time flutter flags of life's innate, original intelligence.

Signs of obvious anticipation always flatten evolution. They squeeze the theory's time to death and thus compress it to impossibility. How can forethought be by chance? When is a plan not a plan? Is concept the same as lack of it? Biological evolution entirely *lacks* concept or target yet its theory equally lacks any serious explanation how, through many specified and integrated stages serial *and* parallel, the evolution of development '*must have*' occurred. Is this 'rationalism's' finest hour? **The fact is that all instances of metamorphosis (of which development in general is one) are another Darwinian black box.**

A Mutant Ape?

Where does such complexity leave you?

Welcome to a Hall of Smoke and Mirrors - palaeoanthropology! Convoluted, febrile, labyrinthine are the speculations that are woven round contentious bones. In this abbreviated volume there is not time to fairly, reasonably disentangle detailed fact from fiction and one line of vision from another. You'd need quite a lengthy monograph (see accompanying book *'A Mutant Ape? The Origin of Man's Descent'*). At this point suffice to say where Natural Dialectic's bio-logic leads. *Man is man and ape is ape.* **As in the present so the past all that is found is one or other.** The distinction's archetypal; archetype is metaphysical; and the inclusion of a hidden invariant, immaterial information, with material energy as fundamentals of our cosmos must attract a different view. This, since materialism *knows* that evolution of material cosmos, life and therefore humans *must* have happened, stands contentious in extreme! *Such assertion is, as bottom-up sirocco-witheringly insists, not only theoretically improbable, it is IMPOSSIBLE...*

As You Like It: Scientific Animism

If volume of print or dint of repetition were any criterion of truth it's certain once-upon-an-ancient-time that reproductive, energy-metabolising and other bio-miscellaneous chemicals precipitated out of sterile pools or barren earth. Mutually attracted, they 'self-organized'. Thence, locked into naturalistic commentary, wild-life narrators punctuate fine documentaries with animistic faith. Genes 'work as components of machines', cells are 'programmed', 'ingenious natural engineering solves the problem' and things always seem to 'find a way'. You'd think that, if a chemical, metabolism, organelle, cell, tissue, organ, system or a body needed any feature, animistic evolution 'willed' it to appear. But matter's mindless, aimless, pushed along. Such errant phraseology no more explains than it explains away the tacit *reasons* every bio-form is pulled by.

Reason is the gift of mind. How irrational, therefore, at the sharp end of informative development, to deny the source of information and assign the cause of life to energetic chance. Uneducated folk, whose rural circumstance might seem to harbour elements alive, animate their place with spirits, sprites and deities. Educated fellows, on the other hand, plump for chance. Since such cause is reason's absence 'reason-without-reason' is the frequent cry. Now a Great Sprite, evolution, chooses, solves, discovers and creates the mindless rationale of life. Since unpredictable's the motion of this fey, who can gainsay any guess at how it leaps from A to B? But this is scientific animism's spirit - eyeless progress in the image of a human mind.

Hail, therefore, Sprite Fortuna! More energy required? Add organelles, fresh coding, head, mouth, stomach and the rest disposed coherently in an appropriate body plan. And fossils, plausibly arranged, can prove what you believe. **Yet, unfortunately, the fact is there exists no natural Law of Innovation such as just-so plausibility requires. Nor jot of evidence, besides interpretation, that the progress claimed occurs.** Matter's aimless. Darwinism's world is mindlessly devoid of prescience - the foresight only mind supplies. Its irrational maelstrom can't, step by entirely groggy step, progress to build up integrated systems - not least since natural selection would cull starters that, like any uncompleted mechanism, didn't yet perform. The previous six chapters have shown clearly why.

Worse still, blind process does not generate 'advantages' informed by prescient code. The basis of biology is information and this information's always codified. Signals stored or carried various ways (including *DNA*) inform specific, purposive complexity. From faulty metabolic algorithms through to incoherent parts of systems, natural selection kills whatever doesn't work. Whence, therefore, in a way that's never been observed, did the initial information for each *innovation* (not just adaptation) formulate itself? *Raconteurs cry 'chance' but is chance, in the form of mutant code, sufficiently creative unto functions, targets and the interactive modes of complex life*?

Design, for Darwinism, lives or dies upon 'as if'; if this analogy be false as teleology is true the theory dies an instant death. No facts are changed. You might, quite naturally and without loss, select against irrational suggestion and thereby bin a bug that has semantically excluded and thus failed to develop modular typology. **Such imperious exclusion occurs in spite of the fact that everything able to be interpreted according to *bottom-up* phylogeny can equally and systematically be approached by such *top-down*, modular typology.**

Has Darwin Had His Day?

No and yes.

No. Darwin will remain remembered for his observations and research. He identified correctly and in part explained the fact of variation/ micro-evolution by notions that he learnt from breeders (artificial selection) and from Edward Blyth (natural selection). Variation, genetic mobility, mutation, natural selection and heredity are facts. And some protean form of evolutionary theory is essential for selected breeds of thinking to survive. Their naturalistic principles must not, on any count, be breached. Thus, since 19th and 20th century versions taught as practically law are nonetheless demonstrably deficient, the 21st must, by re-invention, sedulously plaster up the cracks. Thus, already, buttresses of 'natural genetic engineering' and 'evo-devo' have appeared to shore up faith. *Evolution is materialism's oxygen.* Lack snuffs; thus Darwinism's the doctrinal air it breathes.

Yes, however, if his grander speculations are found fiction. *A materialist will try convincing you that the aforesaid facts add up to life on earth. This is not the case.* They amount to the 'lesser half' of what is, combined with highly dubious hypothesis of chemical evolution and an unlimited plasticity of macro-evolution, presented as a whole truth - The Primary Corollary of Materialism, *PCM* or, if you like, the scientific *AVB*.

In previous pages we have hinted at a lack. **The fact of the matter is that nowhere in the bio-scientific literature does there exist a full, step-by-step causal account of the *origin*, that is, the gradual emergence of any specific, complex code, integrated pathway, structure or behaviour!** We are not discussing the persistence, spread or variation of an already-present bio-system but definitive description of its informed *origin.* Because such origin is historical it is invisible; and because it is presumed randomly caused the steps to its construction can't be accurately traced. Science is reduced to guesswork, inference and reasonable feasibility.

Thus feasibility replaces factual precision; a talent for inventive story-telling ubiquitously takes the lead. But what is feasible need not be true - especially if there's been a fundamental 'category mistake'. For example, Gilbert Ryle mistakenly presumed that no categorically distinct element of immaterial mind exists; and thus that, *a priori*, all schisms of materialism exclude an immaterial element of information. Thus, non-conscious matter is predicated the power to innovate. Bio-logic is extracted from biology. Macro-evolutionary feasibility is what rules Darwin's day.

Victorians didn't know molecular biology; nor very much about cells, metabolism or development. Access to the fossil record wasn't worldwide as today. Darwin's knowledge was, in these critical respects, very superficial. It is even possible that, if he knew what's known now, he would have retracted. He might have realised that bio-logic is what logic actually is, reasonably recanted and with grace conceded that the grander aspect of his theory of origins was wrong.

But, since most of his followers have not, we've reached a flashpoint. Awesome or awful - what is the abductive inference on origins to be? So great is the religious and intellectual investment in Darwinian/ secular mythology that mention of an alternative - informatively-projected bio-logic - provokes at mildest emotional indignation. Is, in the face of facts, such response a rational or a scientific one?

Not only lack of evidence impedes the macro-evolutionary case. The category of immaterial information strikes between its eyes. **It is well argued Darwin's whole hypothesis is built on the *false* idea that micro- leads to macro-evolution. And, where informative code is the product of mind, its attribution to natural forces is a prime category error that infects the sweep of secular academy.**

Once this infection's cleansed, yes, Darwin will have had his day.

Theories of Accommodation

Damage to a vehicle will not kill its engineer. And stabs in the wrong direction are generally called a miss. So, if Darwin didn't 'murder God', hang on! In as much as scientism flies from science why do clerics hanging on its lab coat fly behind? Why should administrators of an organised religion, bowed and bamboozled by 'progressive' huff and puff, preach what neither science nor their prophets teach? The framework of this book (several times reduced from fuller version's argument) highlights how ill-conceived the theory of evolution and thus accommodation of it is. Theistic and deistic theories want it, as their names suggest, both ways - 'divinity' *and* 'evolution', theology *and AVB*. Is it possible (unless you lose them in a cloud of waffle) to yoke vehicles of contradictory logic in this way? 'Easy!', senior theologians cry. 'Upon suspension of reality!' is, with atheism's, Natural Dialectic's counterthrust.

There are two pillars of faith - atheistic and theistic - but, honestly, an agnostic can't decide between them. He is, in all humility, not sure of the 'inferential proofs' with which the *PAM* and *PAND*[203] assail him. He cannot make his mind up.

All parties agree the obvious - physical forces and objects exist. We dwell among their mindless behaviours. We are part of their 'event'. After this smooth union immediate cracks appear. Does material constitute the whole or part of cosmos?

A *pure atheist* adheres to 'The Theory of No Intelligence'. Only oblivious energies exist. Thus, if Darwinism is correct, his 'grand unification of purposeless causation and the world of meaning' is correct. If not, he's wholly wrong.

Impure atheism summons 'natural teleology'. This means you add speculative laws of innovation, progress and self-organising goal-orientation to normal physical determination. Such abnormalities might then skew cosmic destiny to 'naturally engineer' the kind of 'progress' that produced terrestrial life - including atheistic academics! Chemistry

---

[203] see Glossary.

and physics (thus reductionist biology) entirely lack such vision of 'non-purposed teleology'. Such oxymoron, countering entropy, is in reality no more than promissory fantasy designed to counteract theism's pure form.

On the other hand a theist, who used to be called a 'natural theologian', currently adheres to one of three versions of a 'Theory of Intelligence'.

Firstly, a *pure theist* ascribes creation to the projection of preconceived ideas. In this case intelligence both precedes and co-exists with its informative projection, cosmos.

In the *second* version of The Theory of Intelligence *deistic evolutionists* (like Erasmus Darwin and his grandson Charles) allow an impersonal, 'externalised' kind of creator, the miraculous creation of souls and lawfully-behaving energy. Post-creation such creator plays no interactive part.

Thirdly, theistic evolutionists don't believe (correctly) that undirected chemistry can single-handedly create specific information, generate a functional system or, at length, write up humanity. Yet such theist is mind-locked to Darwinian 'progression'; he is fused with the general idea of evolution. *Thus, accepting a deistic start, he effectively embraces impure atheism except that speculative laws are replaced by intangibly divine tinkering.* Thus apparently random mutations are assigned to interference by an Intentional Tinker!

Deistic and theistic evolution theories both represent fairy-tale confusions. Materialism's 'useful idiots' assert the existence of laws or processes for which there is no evidence. In essence, they confuse study of polar aspects of the basic existential dipole - energy by material and information by immaterial science. *Holism includes both parties but, if 'science' is monopolistically equated with materialism, the only true compatibility between science and religion must rest in atheistic faith.* And atheism's central prop is accidental evolution as opposed to deliberate design. Chalk is certainly not cheese.

Is not the macro-evolutionary myth effectively a smokescreen diabolically obscuring truth? But many theologians and their circles fearfully and wishfully conflate. Such widespread intellectual dishonesty dwells in ignorance and schizophrenic blur. Science is the modern devotee's Authority. "In fact, what we mean by evolution is the world as created by God," claimed Papal Primate Gianfranco Ravasi. Does he mean God works through evolution and, just recently, an infidel enlightened His believers this is so? Perhaps, systematically substituting 'random chance' with an 'inscrutable Divinity', he imaginatively accommodates the *AVB*. Or thinks some law of innovation-by-oblivion exists - though would he dare insult technology with the assertion its inventions could have come about by chance? No doubt, evolutionism's infidelity nails the church to an uncomfortable crux. Be crystal clear - from Natural Dialectic's point of view accommodating churchmen hammer it to death as well.

Book 4: Community
Chapter 26: Community/Society

Is Natural Dialectic simply an intellectual exercise restricted to its relationship with academic science and philosophy? Or has it practical, real-life application? *The system of Natural Dialectic is, as much as being an abstract reflection of the way things are, an application program.* This program generates a template for both involuntary, instinctive and voluntary, chosen behaviours. It organises the pattern not only of 'hard science' and biology but politics, law and religion; it guides the aspirations of education; it is hard-wired into the humanities and, as such, is expressed in the very fabric of individual and social life. Its consequences, especially moral and psychological consequences, involve everyone. How?

Towards a Unified Theory of Community

From a *bottom-up* perspective humans are animals and mind evolved as a strange function of brains. Consciousness is a will o' the wisp, a phantom wafted methane-like from life's wet chemistry; it is an elusive, illusive by-product of atomically constructed, non-conscious neurological circuitry. The study of brain-chips is called psychology.

Top-down, the first compass of community is universal and absolute. It reflects a structure of creation whereby all things, psychological and physical, descend from an Absolute Source. In this sense alone is everything connected or, as some aver, 'is one'.

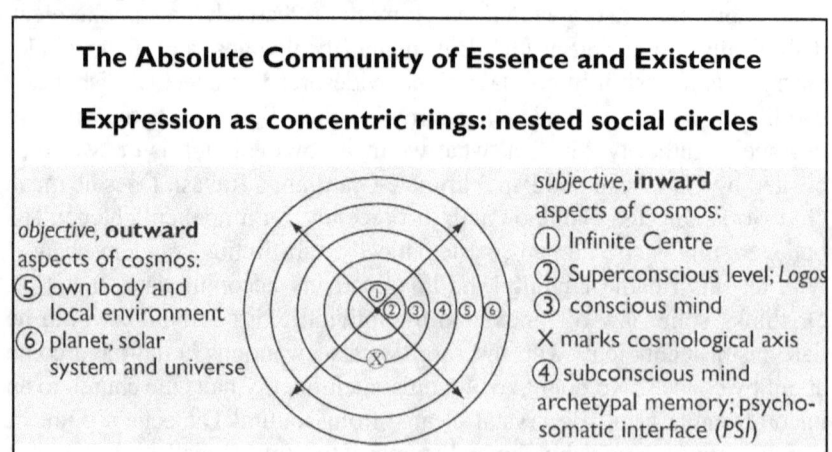

The Absolute Community of Essence and Existence

Expression as concentric rings: nested social circles

objective, **outward** aspects of cosmos:
⑤ own body and local environment
⑥ planet, solar system and universe

subjective, **inward** aspects of cosmos:
① Infinite Centre
② Superconscious level; *Logos*
③ conscious mind
X marks cosmological axis
④ subconscious mind archetypal memory; psycho-somatic interface (*PSI*)

The Relative Community of Existence from a Standpoint of X

Circles inward and outward from X

④ ecological relationship with other organisms; natural community of life on earth (*biota*)
⑤ interaction with *abiotic* factors i.e. the gaseous, liquid and solid environments of a biosphere

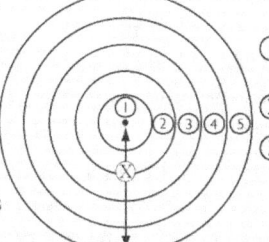

① Superconscious association, Communion, Reunion
② own mind, thoughts
③ friends, other persons, socio-political community of human population

(1) Super-Conscious association, Communion.
(2) Your normal experience/ mind/ inward perceptions.

This couple are internal from the cosmological axis, X.
The *ascending* vector
marks the possibility of voluntary return to the Centre.
The *descending* vector marks materialisation;
this is an outward, externalising process.

(3) Friends, other persons, the socio-political community of human population.
(4) Ecological relationship with other organisms; natural community of life on earth.
(5) Interaction with abiotic factors, that is, gaseous, liquid and solid zones of a biosphere.

(3), (4) and (5) mark a descent 'fanning out' from biotic towards an involuntary, abiotic periphery.

fig. 26.1 (see also figs. 3.5 and 14.3)

The second form of community is relative and, in this relativity, we need to identify a stable vantage point, a criterion. X marks the spot.[204] Cosmos and its ways are relative to you; you, fixed as far as you're concerned, perceive its pain and glory from your third eye's axial point of consciousness. Every individual eye of a beholder charts relationships with body and surrounding cosmos from this axis. This is its own but not the cosmic centre.

From this point X, however, social circles radiate in *both* directions.

Internally there exist the symbolic yet most real relationships of mind. They involve other persons, bio-forms, events, objects and the paraphernalia of a life spent learning. This is the experience of your mind-

---

[204] see Chapter 3: Cosmological Axis.

world. No doubt 'suction' of the senses towards external circumstance is overwhelming; it tends to overwhelm this inner world as sun outshines a universe of stars. But if, during the physical anaesthesia of profound, unbroken meditation, capacity to escape the bustling gravity of motion *and* motivation is realised and if peace is obtained then you will have ascended into inner space. You will have earned a voluntary discharge from existence and arrived at what is called *moksha,* *nirvana,* enlightenment etc. You would have shot to the tranquillity of Void, communion with Origin, association with Axial Truth.

Externally you descend from X through physical interaction with your environment. The rings spread first through your nervous system (which is the closest physical associate of mind) to the rest of your skin and bones and their neighbourhood. This is your space-time location. It includes other biological bodies to which you physically and, more or less easily according to type, psychologically relate. If you restrict these to the human type then you involve the dynamic of family, friends, neighbours, interest groups, nations or the international scene. You take a social part, which may involve institutionalised politics, law and religion, in your community.

At this point a *top-down* stack can be proposed:

	lesser truths	Truth
	relative	Absolute
	issue	Source
	duality	Union
	hierarchy	Top
	cycle	Axis
↓	*tam*	*raj* ↑
	down	*up*
	outward/ exit	*return/ inward*
	external circumstance	*internal core*
	division/ isolation	*communication/ unification*
	passive respondent	*active controller*
	involuntary	*voluntary*
	body	*mind*

More broadly, you can think of life-forms (biotic factors) in terms of individuals, populations or communities of different kinds of organism. This positive, inclusive perspective is called ecological. Thus ecology is about your 'wider body'. Such body includes a community of organisms (*biota*) that inhabit the surface region of our planet. This region, called an ecosphere, comprises living and non-living parts. The latter descend, as the rings convey, to include non-living elements (*abiota*) such as sources of energy, climate, ocean, water, mineral cycles and the soils of earth. The compound amounts to a stage on which, in different scenes, various players act. An ecological play is dynamic. In such a network the health and behaviour of each part affects its whole.

Firstly let's survey non-living factors.

Physical Part

House-hunters know the value of location. How beneficial is earth's real estate in space?

Heavy traffic isn't safe. Black holes, deadly radiation and exploding stars frequent the centre of the Milky Way. And the conurbation's borders house too few elements for planetary life. Large areas of 'suburban' spiral arms are also inhospitable but earth's globe plies business in the safest zone of town.

Not only our galactic but also our planetary zone is habitable. Life needs an endless supply of liquid water and thus strict, natural thermoregulation. Our thermal generator is a 'dwarf main sequence star'. Large, hot stars flood planets with bio-destructive ultra-violet radiation; if you want a low dose you'd be far enough away to freeze. If, on the other hand, our ordinary nucleus of power were smaller (like 90% of its galactic neighbours) then you'd have to orbit closer - but at the same time your rotation might get locked in a moon-like synchrony of orbit so that one side always frazzled and the other darkly froze. 5% nearer and things would have boiled long ago, 5% further and runaway glaciation would have frozen life out. Most unfriendly! Not a *'CHZ'* (continuously habitable zone). But earth's distance from our golden ball is fine. Moreover, critically, its magnetic fields generate a well-constructed shield (the Van Allen belts). They flip periodically but their strength is subject to exponential decay. Such a protective 'bubble', which any habitable planet would require, staves off lethal solar outbursts but allows life-friendly radiation through. How strong was it initially? How long, unless decay's reversed, before our vital magnetism slips away?

Our lives are hung upon our lucky star. Precise strength and character of the four basic physical forces keep it burning radiantly. Not only earth's distance from the sun but also its unusually circular orbit slung on a fixed radius and a specific kind of rotation round its own axis is each exactly felicitous. They're appropriate by distance because temperature on the globe's surface is, except at the poles, life-friendly; by circular orbit because this temperature is stabilised; and by speed of spin that avoids life-destructive boiling, refrigeration and violent, protracted wind-storms. Critically, for perhaps billions of years, maintenance of constant temperature at precise degree has produced that earthly *CHZ* - a liquid water-bath in whose fluids carbon-based molecules vital for life are stable so that bio-forms can thrive.

If life's solar lord clocks fine statistics what about our closer lady of the night? Four hundred times smaller but four hundred times closer, at the moment of eclipse she exactly covers him. With us she dances near enough to lock in motion so she never shows a dark side, just a single, silver, sunlit face. Our planet's tilt, which wobbles through the seasons and affects sea levels, has been finely damped and stabilised by moon's gravitational effect. Did, you surmise, a deluge of huge rocks crash long ago into creation of a tilting norm, a vital oscillation 22 through 24 degrees and at the same time generate its stabilising moon? If so then, oh, what planetary serendipity! This way lunar influence delivers benign climate, her caress dispenses temperate seasons to the brow of life on earth. And her rhythmic sway stirs up the seas. She lifts oceans then exactly drops them down again. The effect is not so large that continents are flooded and eroded each high tide and not so small that surface waters can stagnate. Marine environments are flushed and with diurnal freshness the communities of ocean bloom and thrive. Lunar periods influence other biological events and our consort (in company with vaster Jupiter) also shields mother earth from a barrage of comets and asteroids.

Not only moon but earth is of the right size in relation. Thus gravitation can sustain a viable atmosphere, oceanic swells and a reasonable proportion of tectonically active landmass. Each factor, like a right note, helps to swing along the couple's 'bio-centric' waltz. Of course, each could be a coincidence. From an accumulation of how many coincidences might one argue that there's more than chance at work? How many steps would make you think a form of motion was an orchestrated dance?

Pure energy (sunlight with harmful frequencies deflected or deleted) falls on gas. Earth's primary dynamic, sky, consists of a concentric suite of atmospheric shells. These transparently envelop the surface of our 'nuclear' globe (what kind of life survives an opaque atmosphere?). They drop from the ionosphere through mesosphere, stratosphere and ozone layer to climatic troposphere. Each, like a membrane, offers its particular aegis to the life within. The living planet floats, egg-like in a white of air, within the warm, deep womb of solar influence. Outside the influence of its star there stretches endless, barren, ultra-freezing space. You might construe the stratosphere and ionosphere as buffers, membranes, even subtle skin.

Fresh breath of life! Earth might be mostly made of iron, magnesium, silicon and bonded oxygen but clouds of water, carbon, free oxygen and nitrogen compose our friendly sky and, for the most part, bodies living under it. Atmospheric oxygen blocks dangerous radiation out. Its less reactive, diatomic form absorbs no heat and yet (at a safe 20% of the air) allows combustion of most hydrocarbons to yield energy.

Transparent, neutral, odourless! Whence, with its array of exceptional and beneficent properties, did all the water spring? Nested lower down the suite of air-light shells you find fluids of fertility - mists, rain, rivers and the oceans. These, lower atmosphere and ground water, assume the blood-like role of convectors, radiators and conductors. The media churn;

like an engine nature's universal solvent drives the earth's wheels round. Healthy currents help to circulate life's nutrients and refresh the tissues and anatomy of earth. For example, the hydrological cycle helps to moderate the climate, cleanse sky, flush earth and, geared to the tectonic cycle, weather rock formations into soils and sandstones. It fosters, mostly gently, life's fecundity.

The amount of water on the blue planet has remained stable. So when rivers pour megatons of salts into the oceans how do these avoid acceleration into dead seas? Above the norm (another specific stability maintained throughout life's lifetime) salinity is fatal. Is the random, geological formation of salt-pans enough to have held the narrow line or might life, such as cyanobacterial stromatolites, algae and corals, make and break lagoons (and thereby salt-pans) in a precise biological response to conditions in the water round them? Whatever its cause, the improbable is robustly upheld. Accident and illness are overcome.

The Mother's bones, nails and hair are soils and solid, crustal rocks that, washed by storm and stream, yield minerals. These minerals life's producers, plants, absorb. In fact eleven lighter, non-radioactive and so-called 'biogenic' elements (carbon, hydrogen, oxygen, nitrogen, phosphorous etc.) constitute 99.9% of body by weight and occur in the same proportions in all organisms. Fourteen more (including iron and iodine) are trace elements. Such surface appearance is complemented by heavy, subterranean elements. The radioactive ones are, while lethal to life, in another sense essential. The heat of their decay, particularly of the heaviest 'stable' element, core uranium, may have caused the processes by which molten iron fell to earth's central core and a crust, while 'floating' up towards the surface, differentiated into layers or zones of lighter materials with lower boiling points. While extra-terrestrial energy drives the world's superficial fluxes internal fires of tremendous heat power her geological circulation. Mobile areas of a molten iron core generate magnetic fields that deflect harmful solar winds, protect against them ripping off the atmosphere and interact with the electrical conditions of life. The heat, derived from radioactive decay, also drives volcanism. Volcanoes throw up irregular formations like mountain chains whose various habitats permit an abundance of ecological niches. If life's sac is sky, the earth's skin is a crust of islands (or tectonic plates) that float on seas of magma. These plates are, comparative to the globe's diameter, wafer thin - between 3 and 60 kilometres thick. This allows them to move, crumple, recycle rock and shape the continents; also to buffer the surface from its molten core. Partial melting of the mantle rock, peridotite, creates basalts, granites and other mineral-rich rocks. Basalt is volcanically extruded at constructive plate margins and all three at destructive margins. These are zones where rock is recycled down into the mantle. As well as spitting up volcanoes their contortions sometimes trigger, like a roaring bull tossing in the black depths, earthquake, *tsunami* and, for humans, brutal devastation. Thus, without the presence of uranium, rocks of the crust would not have surfaced nor water gassed into the sky and then condensed to seas. Neither is the radioactive heat that drives tectonic cycles too little otherwise 'no wheels would turn'; nor too much otherwise rampant volcanism would suffocate, poison, crush and bury life. The massive geological cycle is supplemented by the stress, deformation and metamorphic reformation of rocks by movements of large blocks of earth that crack or fault the crust. However, none of this life-friendly action would occur if our globe were larger; pressure and viscosity would induce a stagnant outer layer that inhibited tectonics; and such magnetic shield as deflects harmful radiation would be weakened.

Animate is coupled with inanimate. A system is a network of ideas or objects linked by common purpose. *By this definition life on earth is called an ecosystem.* An ecosystem includes living and non-living factors combined into a single self-regulating system. For James Lovelock's Gaia theory earth is a 'super-organism' made of all lives tightly coupled with air, oceans and surface rocks. Such a 'super-organism' maintains dynamic equilibrium. It comprises a totality that *seeks*, in a cybernetic manner, optimal conditions for life. Its variables include temperature, pH, salinity, electrical potentials etc. Biological controls include metabolic, hormonal and nervous systems of adjustment. In this case the *reason* for such coherent operation is not, in the last analysis, physical but metaphysical. **'Cybernetic homeostasis', like 'program', is a conceptual phenomenon. Its presence in any machine indicates an underlying purpose.** *Numerous biological sub-systems are coordinated under the overall purpose of the continuation of life, that is, of survival.*

For example, under normal circumstances the chemical reactions which happen in fluid media run to completion. Contrarily, in case of Gaia's homeostasis, "the chemical composition of the atmosphere bears no relation to expected steady-state chemical equilibrium… disequilibrium on this scale suggest that the atmosphere is not merely a biological product but more probably a biological construction...the atmospheric concentration of gases such as oxygen and ammonia is found to be kept at an optimum value from which even small departures could have disastrous consequences for life…. The climate and chemical properties of the Earth now and throughout its history seem always to have been optimal for life. For this to have happened by chance is as unlikely as to survive unscathed a drive blindfold through rush-hour traffic."[205]

However, Lovelock's naturalistic thesis roots earth's 'eco-physiology' in chance and natural law. Tightly coupled features of the biosphere *must* gradually *co-evolve*. Thus genetic inconstancy (in the

---

[205] James Lovelock: Gaia (OUP 1995 ps. 9/10).

standard form of random, mutative 'innovation') is invoked to explain the world's ecological order. Is such invocation powerful enough?[206]

Top-down, purpose informs the structure and function of every mechanism; mind is in machine; metaphysical drives physical.[207] Life builds its house with survival in mind. *The mind of Gaia includes this purpose. Its informant is life. Neither rock, sea nor air but life is its mind.* The 'super-organism' involves, as well as astronomical good fortune, a whole range of minds and archetypal memories.

Biological Part

Individuals, populations or communities of different kinds of organism are involved. Life's geo-physiological health, the poise on which all ecosystems and their multicellular inhabitants depend, is in great part the gift of bacteria. Whatever their mode of origin, microbes of the kinds that exist today always existed. In the sun and fresh, oxygenated air of the 'over-world' flourish aerobic types; in the dark, anoxic soils, sediments and mud of the 'under-world' slave multitudes of anaerobes. Tough and reliable, they toil relentlessly. They 'plough' the earth and continuously 'farm' organic substrates on which other organisms thrive. Bacteria might even, as the foundation of life's ecological pyramid, be construed as its primary, substantial, most important form of life; yet, working at the interface with inorganic matter, most 'robotic' too.

From their microbial foundation rises a dynamic construction through which sweeps the energy of sunlight. Through life's spectrum lower, less conscious forms serve higher. Through bacteria and plants to animals each hierarchical level tends to involve more complex, often larger and rarer bodies; and the levels are bound by networks of essential interaction. Steps up the vital pyramid combine producers (nitrogen-fixing bacteria and photosynthetic organisms) with chains of animal consumers and the department of refuse collection. Recyclers (teams of scavengers, detritivores and decomposers) mop up exhaust; they clean the dirt of expulsion and decay; they finish off the end. Provision, consumption, recycled waste. *You need all three.* Input, process, output. *Homeostasis needs all three.* Ecology is irreducibly, biochemically homeostatic and cyclical. Together every community and every level of life in each community cycles around each co-factor. Indeed, each organism plays one or more of the roles. You need three-in-one to peg the balance happily.

Yet now one actor has claimed centre stage. Happy balance is upset. Monocultures easily destabilise; they weaken linkages that make life's safety net so strong. Thus we're led directly to man's salient percussion on earth's natural society. Just as brain function is often studied through the effect of injuries so ecology discovers, more and more unto astonishing

---

[206] Chapter 23 *passim*.
[207] see Chapter 6.

extent, interconnections between components of healthy ecosystems that are being ruptured by mankind. *Indeed, at this point an acute paradox rears its head.* **Man - whose intelligence, forethought and powerful creativity should serve in the role of steward and conservator of life on earth - seems to have lost his better mind.** His innate curiosity, insatiable thirst for information and greed to efficiently extract rich living from the environment has led him to empirical, material science. Physical science wants to help; it wishes to improve our lives, increase our knowledge and, in a material sense, our rationality. Its heavily invested focus well serves the bodily division of a 'centaur' man. Organised, coordinated study of all physical behaviours has yielded dazzling mechanical and medical rewards. Science, while it ignores the metaphysical, has vastly contributed to anthropocentric wealth, health and comfort. Of course, research works towards benefit but at what eventual cost? Could such supposedly amoral but 'right-minded' enterprise become a wrecker? For good or evil, human will-power cuts a swathe across the world. Could its immaterial influence, in the hands of greed or lust or other vicious form of immorality, transmute material advance into a monster that might lead the passengers of an overpopulated, technologically-pillaged planet to catastrophe?

Nature's Negativity

An ill wind, evil cold, cruel sea and other natural challenges (not least, inescapably, the body's own calamities) may threaten life with suffering. They may spell fearful pain and death. Such 'evil', as we understand, does not involve intent or animosity. It is, as matter is, oblivious: not immoral but amoral, called *involuntary negativity*. Thus 'ill wind' does not blow with ill intent. It blows according to the fashion of inanimate design. In short, nature's so-called 'negativity' is really an insensible neutrality.

It is said that when his beloved daughter Annie died Charles Darwin lost faith in 'The Good Lord'. The vexed question was how, from his unhappy point of view, 'a loving Father' could have allowed such imperfections, bugs and pain. The deist thought his 'deity apart' was too apart and from aloofness should have intervened.

What, sir, were you, like many of us, really saying? That suffering and God are incompatible or cannot logically co-exist? Are you saying that the world of duality is, according to your way of seeing it, imperfect? What, therefore, is the nature of the perfect imperfection that you seek? That nothing ever hurts? That aging, death and its decay be banished? No power be negative? Neither rain if we want dryness nor ice if we seek warmth; or *vice versa*? Must there be a plentiful supply of everything we think, at any given moment, that we want or need? Should volcanoes not exist because Pompeii suffered? Or seas disappear because their stormy waves have killed? Should neither accident nor sickness blight? Would

Amoral 'Evil'

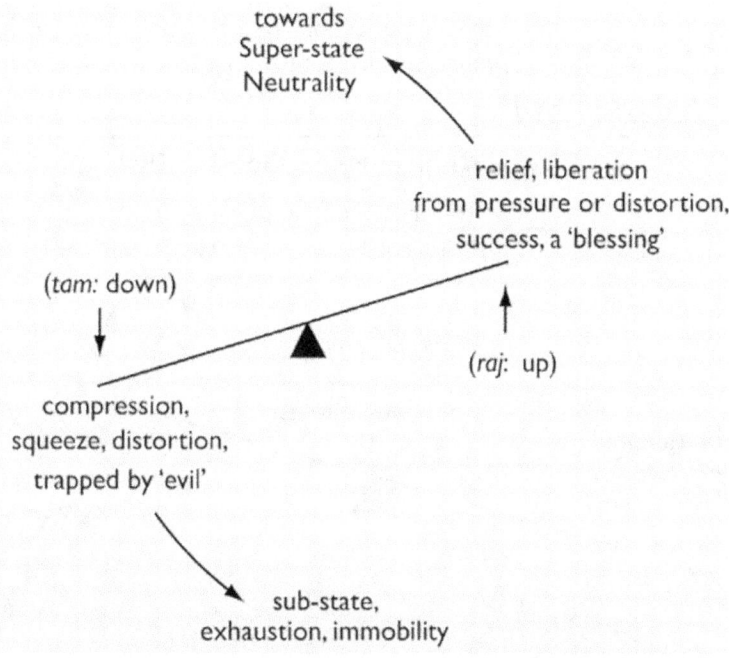

You *choose* (↑) moral or (↓) immoral action (see next section). Matter doesn't choose a thing. Energetic reactions always, due to automatic mindlessness, exhibit neutral amorality. This sub-state is the diametric opposite of Alive and Super-state Neutrality (below).

Polarity involves the notion of swing between two poles whose axis is a single pivot, Unity. This Axis is an original super-state whence at last issues its opposite, extreme down-swing. Thus obtains the natural anti-pole of sub-state darkness, pain, exhaustion and their discharge on the cosmic floor. Physically, amorality derives from the purposeless oblivion in which material operations act.

fig. 26.2 (see also figs. 2.1 and 2.2)

one, how could one create such a utopian but unnatural world? Is it not ironic? The one whose theory of evolution banished any 'Special Interference' in 'unbroken natural law' was the very one who, when it suited him, rejected a Creator for a lack of 'Special Dispensation'!

Darwin's prayers for Annie 'didn't work'. What does 'didn't work' mean? Does it mean we stamp our foot because we don't get all we want the moment we demand it; or make demands that counter natural law; or curse because we can't conjure suspension of its operation? Shall we rationally blame an engineer whose locomotive knocks us down? Isn't this, which many prayers adopt, a childish attitude? In demanding Darwin's sort of perfect world what is really called for is a cosmos rearranged to suit our whims and wishes. No natural 'calamity'. And yet if every one of nature's children kept on wanting different things, then how confounded would that nature be! There would be no end to muddle. Confusion, chaos and confliction of the elements! Nature can impact 'imperfectly'. Yet humans, when we suffer, pray such 'imperfection' be perfected while cursing the imperfect part. We cry, unrealistically, for paradise on earth.

The Nature of Evil

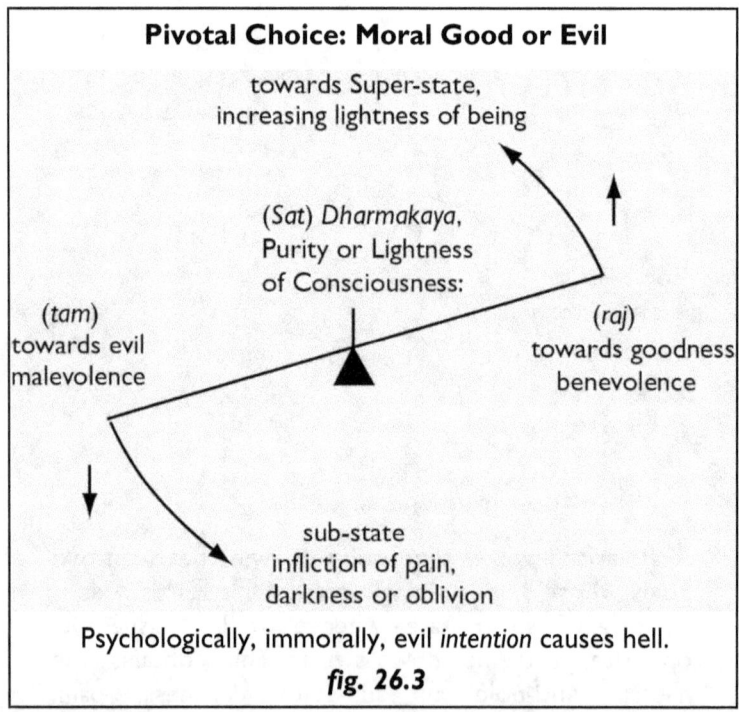

fig. 26.3

This stack makes clear which way the compass of life's moral currents swings. Evil deals, intentionally, in lies and pain. Malevolence is known as *voluntary negativity.* Such condemnation isolates, its burden weighs you down; conscious negativity, inflicted psychologically or physically, inspires a state of hell.

all below	*Transcendence*
lesser sorts of being	*Supreme Being*
vectored deeds	*Potential/ Poise*
balancing acts/ reaction	*Balance/ Pivot*
range	*Super-State*
range of shadows	*Essential Light*
moral spectrum	*Good*
↓ *negative act*	*positive act* ↑
division/ demonization	*unification*
from Truth	*towards Truth*
from Peace	*towards Peace*
body/ self-centred	*soul-centred*
passion	*detachment*
contraction	*expansion*
malevolence	*benevolence*
hate/ abuse	*love/ care*
crooked/ perverted	*straight/ open*
criminal/ demonic	*saintly*
a curse	*a blessing*
immorality	*morality*
descent/ darker	*ascent/ lighter*
negative wish	*positive wish*
tendency to create chaos/ cause pain	*tendency to order/ pain relief*
decline/ fall	*lift/ helping hand*
depriving	*sharing*
malefactor/ enemy	*benefactor*
evil/ sin	*goodness/ virtue*
darkness	*light*

Decline and fall; corruption; slide (or even race) towards moral death - you can't negotiate with vile's (↓) direction any more than dispute gravity. Constriction, isolation, pain - the compass is both individual and a social one. As energy so information suffers entropy; noise blurs morals most where strong desires combine with lack of love. 'Survival of the selfish, hardest fittest' drums its din.

God and suffering can, in a polar, vectored cosmos, logically co-exist. Why not? How could the world be always right from everybody's point of view? But if there is a God of love why should He create an entity of malice, cunning and perversity? Is, therefore, God father of the devil and, thereby, the root of evil? Who loosed the wolf of persecution on a world of suffering innocents? I cannot, cries apostasy, trust two-facedness or diabolical betrayal, Truth that lies, love that dies, light that sets in shadow - how can pain pervade a paradise? How, if there's a God, can evil thrive? Thus there isn't!

Light in darkness, darkness in the light - how else, except by contrast with its opposite, is any nature understood? How might Unitary Essence know itself except in the reflection of duality? Resistance, opposition and exhaustion - the world exists through shadows set against its light. Beneath transcendence truth is broken by polarity into opposing vectors; plus and minus are unbreakable a pair. But as free agents and not unconscious robots humans *choose* which way to act - and sometimes voluntarily create another's pain. Holistically-speaking, don't blame God but human choice for evil. Thus, if evil's sourced in men's own heads, moral struggle isn't cosmic but just local. So is the devil universal or anti-Christ the figment of an individual's nervous brain? Brain's an instrument of selfish chemicals called genes. Could a mindless 'demon gene' engender devilry? Materialistically, it's no-one's fault; morality's a dream; the devil's answer's in molecular biology.

A godly gene would have to be recessive. Weak. Not the wild, fit type. From the unknowing angle of a 'selfish gene' the problem is not evil; it is selflessness. Altruism. Who denies that in a fallen world the bestial side of life is struggle for survival? 'Survivorship' and 'reproductive fitness' win a day that's governed by genetic products called your limbic system and its master glands. To rape, cheat, pillage and exploit is thus implicit in your naturally selected genes.

What, therefore, might a well-evolved and cunning despot's genes inflict upon 'the enemies of scientific reason'? Deceive, purge, pulverise? Eradicate all threats to his genetic egotism's article of faith - any-cost survival? Ruthless, forceful and, the best of all, efficient tyrants soon create such 'excellence' as Dante's hell.

Racism. Humans are genetically 99.9% or more identical. If they descended from an original pair they would obviously all be, in this sense, blood relatives. Brothers and sisters for whom character not colour counts.. Darwin's title is, however 'The Origin of Species by Means of Natural Selection or The Preservation of Favoured Races in the Struggle for Life'. Clearly, in this context 'race' approximates to 'species' but, where 'race' is nuanced with an innate sense of superiority-and-inferiority, Darwin was, with the great majority of his contemporaries, of a supremacist mind-set. Did his avidly-supported cousin Francis Galton's pseudoscientific study of eugenics not imply such rationalistic moral code as 'fittest races' might inflict on cavemen, primitives and those in poor or technologically backward, 'third world' states? *Within the twentieth century materialistic shades of social Darwinism as diverse as totalitarian fascism and communism, socialism, colonialism, racial supremacy, apartheid, dog-eat-dog capitalism, scientific atheism, hedonism, humanism, anarchism and sheer nihilism have all, in some cases at the cost of vast human depression, inhuman suppression and cruel suffering, drawn doctrinal strength from the material heart and pumping blood of evolution theory's 'unofficial bible', The Origin of Species.*

Theocratic hypocrites may pervert religion to support cruel bestiality; but Darwinism, centrepiece of liberal, secular philosophy, *demands* progress by way of superior competitors. Lower varieties represented part of a ladder of 'inferiority' reaching up to 'us'! Such concept is embodied in Clark Howell's outrageously deceptive but iconic, textbook evolutionary sequence of chimp through stooping ape-men to Neanderthal and, at the top, upright *Homo sapiens*. Us! Colonialism's Darwinists saw their humanistic rationality, technology and education as far above the state of their subjects. Such evolutionary attitudes did not *cause* exploitation and slavery (which long preceded Darwin and succeed him still) but serve to justify. **How, if nature lacks morality, can such behaviour be 'wrong'?** It's natural!

Fascism. 'Higher race subjects to itself a lower... a right which we see in nature.' Hitler argues[208] that Darwinism is the only basis for a successful Germany; and in December 1941 he revealed to his Secretary, Martin Bormann, that his life's final task would be to solve the religious problem - the organised lie must be smashed. Franco and Mussolini also sprang from Nietzsche's *Übermensch*, apostolic Haeckel's febrile evolutionism and, therefore, once more Darwin's and his cousin's pseudoscientific corm. Indeed, world records for mass-murder germinated like black flowers of modern twentieth century evil from that bulb.

Communism. Stalin followed Lenin who followed Marx - for whom a signed copy of Darwin's book contained 'the basis in natural history for our views'. Mao Zedong, Pol Pot, Kim Il Sung and others descended from Marx by way of Stalin. Dialectical materialism is central to the thrust of soviet ideology. Its oppositional stance treats opposition to its chosen 'positive' as an excluded 'negative' with which it is at war and wishes to eliminate. Taking scientific atheism as 'positive' this antagonistic form of logic thereby demonizes metaphysic as prime enemy. Heaven is brought to earth; and earth's where paradise is lost. The 'system' cultivates materialism's fundamental lie. In Central Committee there presides A Totalitarian Lord; bureaucratic angels sing perforce in tune. Not communion but warfare, death and prisons rule. No atheistic culture has historically existed except its ideologists have red-flagged or brutally suppressed all 'backward' opposition to its canons and its whimsical moralities.

However, atheistic nihilism does not necessarily make for depression and evil - though it certainly helps. There are many atheists, humanists and agnostics of excellent moral character - thoroughly good people.

---

[208] Mein Kampf: Chapter 4.

Whether or not the principles that mark this goodness were transferred from long Christian tradition is a point for interested academics to debate because, in fact, it is the 'now' of behaviour that counts.

What is really the difference between a theist and an atheist? Because one party, doing what he has decided is 'God's will', is historically capable of as much evil as the other. One side believes in a Conscious Creator, transhuman, trans-physical and the other believes only in non-conscious physical energy (which we call matter). But where does the energy come from?

Genghis Khan was a Tengrist with interest in contemporary faiths. About 40 million died at his hands; nearly that number met death caused by Mao Zedong and 20+ million by vice of another professor of 'logical rationalism' and the communistic faith of scientific atheism, Stalin. In the world of truly mass murder Moslems, Christians and religion-hating Hitler take a wooden spoon. What can be said?

War. Some humans love to hunt and kill. War's the climax of such 'sport'. The devil's game. It inflames in-built passions to an internationally murderous degree until victory bestrides the spoilt and spoils. Evolution has, for the last 150 years, been pinned to martial and imperial masts. It may not directly instigate devilry but its theory, survival of the fittest, cannot logically condemn a fight.

You can understand that nature is not evil but, still, it doesn't take a genius to understand the nature of an evil act. What's, therefore, the source of evil's stream? Is the devil an unnatural fiend or a wretched feature of man's natural mind? Anger, greed, lust, pride, wish to dominate and laziness are swollen forms embodied egotism easily assumes. Sins are inflammations, devilry a cancer that inflicts a mind controlled by body but not body by its higher mind. Thus demons are, in dialectic script, the offspring of desire writ large, passions overrunning and anger's body of revenge. Evil is an opposite that's turned from goodness, spurned the central truth and turned its Luciferian back on light. **God did not but His risk, humans, do create damnation for themselves**. Devils spring in mind; they place themselves, by choice, in individual and universal quarantine; they are the body's egotistic choice when magnified into the fires of vice.

Towards a Unified Theory of Community: Social Part

How best, philosophers enquire, live life? What solvent best dissolves the harshness of our problems on hard earth, what solution flows towards universal happiness? Does cosmic law (polarity) provide an answer even if it's not what *ego* might desire?

Down-to-earth solutions to the problems of disorder, ignorance and suffering involve two species of utopia - one objective, outward social and the other inward, personal and subjective. *We'll deal with* *the outward, large-scale solution first and then, as was the plan, work inwards towards a microcosmic, personal Nuclear Solution.*

> **Religion, Politics and Law:**
> **A Dialectical Plan**
> **of the Way Communal Life Works**
>
>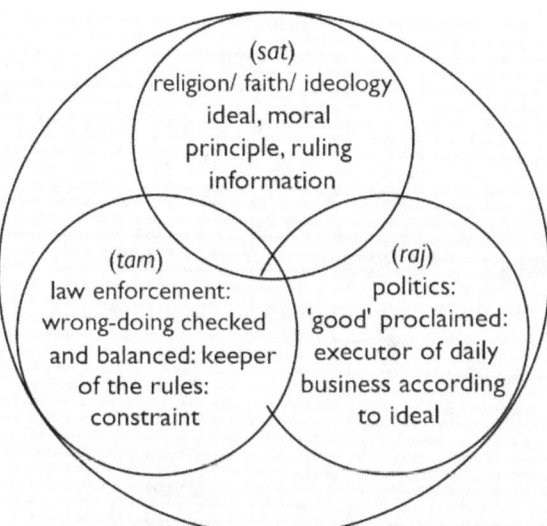
>
> The cosmic fundamentals indicate that, round a central axis, equilibrating oscillations swing; practices diverge, in more or less degree, from principle. Thus they indicate, as well as natural, social order. From prior ideal the practicalities of business (politics) and trammels (how behaviour's physically constrained) are each derived.
>
> Natural Dialectic further indicates that, in perennial search for peace, stability and happiness, fashions cycle more or less between the opposites. Cultural modes and mores don't progressively evolve; they oscillate between polarities. Sociologically behaviours - legal, political, commercial, sexual and so on - swing periodically between the (↓) left- and (↑) right-hand columns of a secondary stack.
>
> *fig. 26.4 (see also fig. 19.1)*

Therefore, at this point - the consideration of a **Unified Theory of Religion, Politics and Law** - we speak about *'external association'*. Such association comprises relationships between an individual and his circumstance. It includes both positive and negative interaction with non-living objects, any living organism and, closest, other humans. At the heart of human community are religious, political, legal, educational, economic and personal communications with each other.

The strongest, most important relationships are often with those in close physical proximity. This includes one's own body (habitually thought of as oneself), family, neighbours, colleagues and workmates. Of this group the closest are those 'on the same wavelength', that is, perhaps family but certainly friends. And of friendships potentially but arguably the closest kind ideally fulfils the criteria of both physical and metaphysical union - man and wife. Meeting in mind and body is a symbol full existential unification, that is, love. It represents positive, creative polarity, homeostatic stability and dynamic equilibrium. In the context of family these cosmic principles translate into security, continuity and healthy relationships; also, since a family is the basic unit of society, into social cohesion and stability. <u>What, therefore, should the instruments of social cohesion - law, politics and religion - most attentively and sensibly nurture but family and, within the family, its individuals. Is this your society's case?</u>

subsequent derivations	Ideal/ Norm
material/ immaterial	Immaterial
motion	Peace
variant/ swing	Pivot
relativity	Absolution
relative imbalance	Balance
eccentricity	Central Axis
↓ lower nature	higher nature ↑
negative act	positive act
wrong	right
deformity/ deformation	integrity/ integration
tendency to violence/ disorder	tendency to stability/ order
less Normal	more Normal
towards insanity	towards sanity
ego/ selfish self	disinterest/ selfless self
imposition	self-control
fall	self-correction
punishment	reward
confinement/ constraint	release
guilt/ debt/ burden	credit/ innocence
shame	merit
failure	success

Religion, politics and law - this homeostatic triplex guides the individual and, by extension, his society. Is not the foremost regulator of the three, from which ideals derive, world-view? Secular or, if immaterial precedes and governs the material state of man, non-secular religion sets ideals. Ideals, as far as realised, compose desired utopia.

Do you remember, yet again, the schizophrenic state-of-centaur paradox? These are creatures struggling with the tension in between their loving upper and body-absorbed, lower parts - between their metaphysic and its earth. They are us. What is a swollen ego but a case of psychological oncology? As moral harmony (care, manners and respect) wears thin so harsh correctors have to take its place. If voluntary morality is lost involuntary restraints must hold the balance, harsh reminders hold the line. As health is lost the medicine is dosed, the hardness of the surgeon's knife grows real. There have, in this case, always swung to action scales of religious law, law-making politics and justice by such formal, legalised morality.

In this case from Natural Dialectic's point of view best principle derives from Top Ideal. *Society's best formal answer is one that addresses evil issues from the State of Immateriality; our material solution is resolved, most definitely without recourse to science and without a test-tube anywhere in sight, by institutions charged with primary exercises in morality.* Morality steps centre stage. It is not scientific but it is for sure the bigger player. Material knowledge waxes grand, technologies progress but does morality improve as well? The purpose of the agencies of social order - religion, education, law and politics - is to combat imperfections, minimise all criminal behaviour and promote relationships. *The whole point of morality is maintenance of individual and thence social balance and, wherever necessary, restoration of dynamic peacefulness.*

Peripheral Religion

existence	*Essence*
spectrum	*Source*
hierarchy of riders	*Ideal*
lower principles	*First Principle*
consequent development	*Nucleus*
belief	*Knowledge*
religious faith	*Super-religious Fact*
world-tree	*Archetypal Seed*
↓ *darkness*	*light* ↑
peripheral formalities	*approach to nuclear core*
bondage	*freedom*
lack of intelligence	*intelligence*
rigid adherence/ dogma	*thoughtfulness/ debate*
hypocrisy	*sincerity*
rule by fear	*rule by wisdom*
arrogance/ intolerance	*toleration*
wrong	*right*

Of course, society needs structure and bureaucracies more rigid and extensive than an individual's paperwork. **The structure that encrusts around his nuclear ideals is called a man's religion.**

How men deliver principle in practice is therefore dependent, above all, on what a group believes is true about itself, its purpose and relationship with natural law.

Religion means 'a system of belief' and, whether vague or clearly framed, is therefore unavoidable. Whether atheistic or theistic, oriental, occidental or plain secularly 'liberal' it frames relationship with cosmos. It defines, informing us about our origins and heritage, our standing in the world to which we have been born. It is intrinsic to the basis of our thinking. Even a majority who do not think about it still wave with the sway of social norms. Relationship with both the physical and metaphysical domains of nature influences the fundamental form and thereby patterns of behaviour found in every community. <u>It's down to origins again</u>. We spent chapters in discussing them. *How you orientate yourself is central to the type of world you build. And if the place of origin is myth, then a society's myths are crucial. Truth is even more so.* Myth and truth are at the heart of a religion but each kind, atheistic or theistic, propels the other far apart.

Nuclear Religion

For materialism there is only one way. There is no direction but its own wherein all life is evanescent and thought dances, like will o' the wisp, inexplicably upon the waters of oblivion. If this is hopeless and a counsel of despair grow up, the creed exclaims, grow strong because that's just the way this harsh, unfeeling, hopeless cosmos is.

If, on the other hand, you call Essence, in the terminology of physics, Symmetry, then forms of different religions break into localized expressions of its Law. Law is Symmetrical. Its Archetype is from all angles at all times and places just the same. The Principle of Nature's Nothingness, the (N)One, emerges with a real and central meaning. If religion naturally exhausts in scattering from its sacred core into profanities; if like ivy repetition sucks and dries vitality; and if tradition suffocates its shoot then of what nature is the quintessential symmetry? What is the sort of nucleus that seeds varieties of devotee; whence develop creeds and cultures as religious progeny?

Religion's core is its Ideal, Transcendence and Top Teleology. Its Great White Truth is, the reverse of flight from matter's 'cold, steel-hard reality', an eminently rational undertaking to commune with a Subjective Friend and thus discover nature's Natural Heart. This Heart is nowhere outside to be found. Despite their imperfections, quirks and quaintness world faiths all promote Your Inward, Immaterial Centre that is Life.

Do you remember the dialectical models (Chapter 0) - concentric spheres, Mount Universe and fundamental harmony? Humanity across the earth, historically and now, is crying out the theme. What is a pyramid if not an elevator to the stars; and ziggurat a ladder earth to heaven? **Aren't you, fabulous in beauty and complexity as Universal, Pyramidal Man,[209] equally a temple of the living God? Whose access to the holiest of holies is the portal of your single, central eye? Mind's eye looks down upon a world of soul-less things but, swivelled upward towards the Apex, seeks illumination's jewel. You step up; and, in communion with Essential Soul, are transformed into supreme and living paradox.**

Personal enlightenment rests at the apex of Mount Universe.

Politics

Below the third-eye's balcony observe life's market-place.

Survey the action, business, economy; seething crowds, cross-currents of humanity; turbulence of social ocean, individuals jostling with each other in a way that keeps (or fails to keep) the peace. What is a crowd but a community, communities a town and towns writ large across a wedge of continent? And all, down to the finely-tuned chemical and biological conditions of our planet needed to exploit Promethean fire, technologically civilised. Individuals in society, men in a collective state, states dealing with each other in a way that keeps (or fails to keep) the peace - this is the cauldron, human politics, into which birth throws us where we boil.

At the heart of any business, even cosmos and the dialectic that describes it, rests a point of balance round whose comfort zone there swings dynamic equilibrium. The logic indicates that bodies normalize by the avoidance of extremes, that is, in the relaxed pursuit of light, poised and middle ways. As with biological so with legal and political dynamics. **Resilient balance minimizes stress so that equilibration is, for body politic, the basis of its politics**. The purpose of a government is steering towards its stated goals. Are these not, in a human nutshell, education and, in providing for a healthy body, keeping healthy peace? *Its purpose is to protect its subjects' sense of equilibrium by maintenance of law and order in its realm.* Employing metaphysical ideal how best, philosophers enquire, to legislate?

Law

Governing or governed. Your perspective changes with the hat you wear. Do you remember, yet again, the centaur? Animal and rational parts combined? Part of me wants action as I please; the other fraction - conscience, reason or the fear of punishment - says 'no'. In this case

---

[209] *fig.* 17.4.

what about internal politics? Self-government. Are conscience, self-control and an internal law to be preferred? Or don't they count and you invite a visit from the officers of law? Rules leash animals with reason; they frame your practices with principle. You are constrained within a mental box of regulation. If you flout the ideals that it represents then an external leash of conscience must restrain. Those officers detain and teach you with the pain of punishment. They might even lock you in a hard box called a cell. Did home and school not teach you this? As far as they did not they failed.

Towards a Unified Theory of Community: Individual Part

The hard box is outside. Now it's time to turn within. From community we turn towards communion - but with what or whom? **Turn from X^{210} towards the Real Deal, Nature's Inward Positivity.**

Self-Government

No crime! No prisons, courts or punishments! Do such societies, outside monastic, still exist in which self-government eliminates the reckless, disrespectful element? Where, really, does self-government exist but mind? So crimelessness has great potential. Nature's Inward Positivity exists in every human but its exercise needs choice.

This choice is simple but its impact huge. You'll get exterior law with police and CCTV. You choose big brother or interior self-government. Governed, governing. Law is an immaterial structure imposed from outside on yourself; equally, invisibly, you also use the information principles endow to regulate your own response to outside factors from within. Internal politics. Is your self-regulation coded like a cell's or, like instinct, set in automatic reflex? Are genes the real, inmost self or, although your body uses preordained responses, do you also play a game of psycho-politics involving active choice?

You might claim it's genes for everything! You can't see them but can blame them. Such absolutely passive entities aren't conscious of their 'motive'. Like computers they have no intention but, 'fork-tongued', they do! So powerful is a programmed hormone, some believe, that it leads you by nose. If good morality and altruism help survival that's what genes allow. If not then cut them out - that's fine as well. Evolution claims an explanation for all things.

What, though, about the other rationale? What about the immaterial and subjective one - devotion, honour, bravery, self-sacrifice or mother love? Whose folly is the giving of your heart in love? What if, far from deluded, all the saints were nearer than logicians or 'genetic motives' to the Truth?

---

[210] *fig.* 26.1.

This is the nub. From which nucleus, cellular or psychological, is self-government derived? Is morality a function of your *DNA* or quintessential mind? If the latter how might nuclear association be contrived?

Individual Association

Further in. Take nuclear religion to your own (and everyone's) extreme. *Internal, individual association* involves just one relationship. How well do you live within yourself? How to realise The Real Deal - Your Self?

Return. As *fig.* 26.1 shows, this chapter is about (↑) return from the periphery of creation to its Natural Centre. Isn't the Aspiration of a human life, naturally inlaid but much neglected and distracted, to distil the mind's pollution to a pure distillate and thus, as every mystic always told you, reunite your life with Life? Essential Psychological Unification: Arrival Home. No doubt science cannot, though a scientist being human can, understand the theory and practice of a science of the soul. But can you now grasp that, at the Heart of Individual Association, is Reunion - Reunion of the self with Quintessential Self?

	body/ mind	*Soul*
	lesser selves	*Self*
	personae	*Psyche*
	concentric rings	*Nucleus*
	relativity	*Truth*
	relatively sane	*Sane*
↓	*external*	*internal* ↑
	descent	*ascent*
	exit	*return*
	lower self	*higher self*
	from Creator	*to Creator*
	from Centre	*towards Centre*
	from Goodness	*towards Goodness*
	criminal/ evil	*respectful/ good*
	insane	*sane*

If you are secular in outlook you cannot because it can't exist, can't happen and is neurologically insane. How can nerves (that do your thinking) meet in Oneness with the Truth? The idea is immaterial, irrelevant and primitive in fantasy. But haven't you, thinking thus and locked into material partiality, in turn wholly locked the bigger picture out? **If you allow a natural element of information, immaterial**

consciousness, then the strivings of mankind since history began are understandable. They are materially germane to mankind's greatest movement - a return towards individual association with the 'place' where all began. Such singular relationship is immaterial, of core relevance and the highest of realities.

Inward shifting; trans-religious movement, a possibility that's always been incorporated in the human frame, is mankind's future evolution. It frees from physical and psychological confinements; it transcends materialism's body and dogmatic creeds. Contemplative association is internal, outwardly invisible and independent of material circumstance. Metaphysic's distillation purifies awareness. Its full flowering is not only super-intellectual but, where conscious is your present state, super-conscious. Much more wide awake!

What first and final paradox! *Now, as much as in the past, the affairs of men have need of such a healthy frame of reference as Natural Dialectic's. Now, as ever, most prefer equivocation.* The remedy is hard to swallow; its regime is hard to follow. Its logic neither replaces nor competes with any 'orthodox' expressions of Absolute Morality, rather underlines their Centralising Mode. **This, it clearly indicates, is the right-hand and ascending path of Peace. Thus one would hope that comprehension acted as a universal pain relief.**

Chapter 27: Up and Away

'When I first wrote my treatise about our system, I had an eye upon such principles as might work with considering men, for the belief of a deity, and nothing can rejoice me more than to find it useful for that purpose.'

Sir Isaac Newton, *Principia Mathematica* 1687.

Where Does the Data Actually Lead?

Materialism may accuse - *top-down* holistic Natural Dialectic only reaches an 'unnatural', 'pseudoscientific' conclusion due to its first premise. Sauce for goose, however, is for gander too. **Materialism takes its primary assertion, I take mine. Both are philosophical; neither is a scientific one.** *Logic derived from materialistic imperative thoroughly inverts my own - not with respect to current, naturalistic operations but to tiered structure and historical origins. And vice versa.* That is to say, each assertion yields a world-view that is self-consistently derived from a chosen starting-point; it yields a belief dependent on one of two axiomatic origins. One of the couple (a cosmos immaterially informed or not) must be the final case; and by Socratic principle good science always follows, without prejudice but with impartiality, wherever data leads. We have come to our conclusions. Now for the summary of summaries.

If you choose to leave the immaterial element of information out of your equation then chance and 'natural selection', operating in the frame of physics, must rule; if you include it then reason (whose inventions perform within the same frame of physics) becomes the 'bottom line, top conclusion'.

This book's narrative has added nothing to material fact because informative reason and fresh perspectives are both immaterial. **Nor is knowledgeable consciousness (with which you are most intimately familiar and which, indeed, you *are*) some arcane, novel or bizarre addition to the known world. The reverse. It is The Prime Datum.** Without it there could be no *known* world; nor humanly constructed orders. And, in the universal case, simple addition of an immaterial, informative dimension changes nothing physical but at the same time changes everything. This prime datum is the reservoir of information by which the order of projected, psycho-centric cosmos is defined.[211] Prescient reason and not chancy action dominates creation's day.

---

[211] Chapter 6: Top Teleology.

Bottom Line, Top Conclusion

relativity	*Absolution*
appearance	*Reality*
lesser truth	*Truth*
relative order	*Order*
↓ *objective*	*subjective* ↑
energetic	*informative*
non-conscious matter	*conscious mind*
ignorance	*knowledge*
involuntary/ reflex	*choosing/ voluntary*
randomness	*purpose*
chance/ accident	*intention/ design*

We've got many questions and want answers. It is an exhilarating experience to explore our world using the gifts of a powerful mind and healthy body. Observing present facts and speculating on the past or future we use, as Chapter 0 at the outset clarified, opposite directions of mind and we choose one of two pillars of faith. It is natural we concur upon material facts. It is also important to realise that the interpretations of anti-parallel world-views will reach radically different conclusions with respect to the original source of humanity, cosmos and the fabled 'meaning of life'.

Bottom-up, materialism's 'grand mythology' intimately involves eternal matter (or a universe from absolutely nothing); the principle of evolutionary accident; consciousness as an unexplained excrescence of brain's electrical activity; and the logically subsequent religion of life vested in body with its 'soul' relegated to a figment of physical imagination (whatever that is). Death is, thus, the absolute terminus of your existential theatre.

In the light of such world-view the culmination of exploratory endeavour might seem, perhaps ironically, to reach for the stars and thus astronautically expand forever to the outward heavens - to an endless vale of death, darkness and inhumanity!

Why ironically? Such outward aspiration is of course the polar opposite of *top-down* endeavour - to go within, 'know thyself' and reach the immaterial source of life. In this case information gleaned is not reliant on feasible fables (Chapter 25: As You Like It: Scientific Animism') involving life by accident. Nor does it rely on passing clouds of speculation involving the evolution of consciousness such as, for example, an implausible story of planetary orbits causing climate change that somehow drove the evolution of human brains in the Rift Valley of Africa many moons ago. On the contrary, by internal investigation (concentrated meditation) informative consciousness is discovered to be an immaterial cosmic element, an independent metaphysical component whose *derivative* is physical order. The presence of its source is in us all; if man's uniquely powerful psychology is correctly focussed inwardly then, no miles away, he can can travel to the source of mind, that is, the Cosmic Source; he can experience Communion and thereby know (see Chapter 5) Top Teleology.

Materialism vehemently disagrees. Holism notes, however, man's endowed with both capacities; he's equipped to travel both ways so that the two objectives, finding outer truth and inner truth, aren't mutually exclusive. Thus let's express the Top Conclusion in a second way.

The past 200 years of occidental philosophy represent a thrust, at first tentative but increasingly brazen, to subvert holism and systematically convert materialism's 'rational' mind-set into total explanation of our world. Why so stress the promotion of materialism, materialistic creeds and their central support, the theory of evolution? Firstly, to sustain the authority of physical science and its methodology as the paramount and, in the last analysis, only true mode of knowledge. And secondly, to eradicate any challenge to such authoritative claim by apparently conflicting, dogmatic and sometimes bellicose forms of metaphysical irrationality viz. ritualistic religions.

However, forswearing the bathwater of unscientific dogma, such determination throws out baby's innocence as well. Are immaterial experience, consciousness, abstract reason, symbolism, mathematics, information, sensitivity and conceptual subjectivity not, composing knowledge, valuable? Mind's metaphysic has a character that's, obviously, apart from physic's. Psychological behaviours involve interest, curiosity and purposeful activity; their forces are will and desire; and mind tends to seek underlying patterns, order and emotional fulfilment. No more are these atomic than morality. Immateriality is not unreal. Its intrinsic vector, unlike the ($\downarrow$) scattering of material entropy, is ($\uparrow$) negentropy of focus, curiosity, the gathering of intelligence and understanding of life's principles. Involuntary matter is exhaustion-bound but voluntary intelligence seeks liberation through the truths and understanding that it finds.

As physical life depends on physical light couldn't metaphysical life (of minds) depend on metaphysical radiance? Top conclusion is that energetic purity is matched by an informative purity (illumination). A Prime and Single Datum, uncreated consciousness, substantiates an orderly, hierarchical cosmos. Call it spirit if you like. And if you don't and think there's no informative ingredient then how did fixed and integrated, energetic patterns issue out of nothing? How did chance precede a thing? The source of reason, immaterial mind, can plan. Physics and biology have roots in reason; cosmos is, like living cells, logically informed. **Thus the bottom-line is, from the view of Science and the Soul, that materialism and its central prop, evolutionary theory, are at best but half-truths. Each lacks metaphysic and a primary source of information.**

The wheel goes round. In this sense Science and the Soul is revolutionary. **A systematic rethinking of nature, man and his institutions is demanded. It needs, in recognition of non-random creativity, informative light that clears two centuries of darkening, materialistic cloud. It needs turnover that resolves an astigmatic irrationality promoted, in main thrust, by Darwin (body), Marx (politic) and Freud (mind).**

Lux et Veritas

'*Lux et Veritas*' proclaims Phelps Gate at Yale. What kinds of light and truth are meant? Where modern man is secular 'God is my light' (Oxford's motto '*Dominus Illuminatio Mea*') makes certain nonsense for his fashionable philosophy. Conversely the motto may, as inspiration from which cloistered corridors of academe descend, be true. Might Phelps' light and truth be metaphysical as well? '*Magna opera Domini exquisita in omnes voluntatis eius*' is inscribed on the portal of the Cavendish Laboratories here in Cambridge. 'Great are the works of the Lord, carefully studied by all who so desire' is not a maxim that has deterred a galaxy of scientific luminaries making profound discoveries there since evangelical Presbyterian James Clerk Maxwell, the scientist *par excellence* of physical light, oversaw its construction.

Are information, creativity and mind distinct subjective lights? Or are they simply made of gravity, electromagnetism and the elements of atoms multiplied in certain but uncertain and mysterious ways? **If subjectivity is not composed of these last three but rather *immaterial* mind and matter interact, then Natural Dialectic stands.**

It has, therefore, been argued systematically that one-eyed materialism, aiming low, misses here, there and everywhere, a gesture of informative magnificence. On the other hand, holistic argument is clean, clear and natural as air. Its logic, Natural Dialectic, is systematic and, within simple premises, self-consistent. *On balance, I accept its force of truth.* **Anyone is free, without subjection to aspersion, so to do**. Experimental, operational science works entirely undisturbed. *Feel intellectually free, therefore, to doubt materialistic faith - because its atheistic tale, made plain, is no more than one of speculation. Fundamentally, it misses natural information wholly* *out. Thus its imperious narrative is, rudely stripped, no more than naked bluff, assiduously cultivated puff, a clutch of bets developed cleverly around aforesaid PAM and PCM.*[212]

The PCM is neo-Darwinism. Some want to believe in it. But the mill of variation round an archetypal axis does not grind progressive evolution out. *Thus, also feel free from a holistic and informative perspective to reject that theory.* **Nobody *need* believe it represents original truth.**

In the final analysis, therefore, it is with happiness I note that Sir Isaac, with whom my life has shared the same town, eloquently subscribes to the same non-materialistic purpose as mine. I also humbly offer the words of '**Science and the Soul**' as a less exalted pointer towards higher and, in our time, somewhat neglected Truth; I hope its concepts bring you to the Cosmic Upper Gate.

existence	*Essence*
relative truth	*Truth*
duality	*Unity*
↓ *division*	*unification* ↑
material focus	*immaterial focus*
towards illusion	*towards Truth*
no thanks/ ingratitude	*thanks/ gratefulness*

I hope also that you have enjoyed the exploration of a philosophical architecture whose *motif* is the binary pattern of Natural Dialectic. We have, considering both inanimate and animate, circled round the universe. *What is, from the evidence, the final line and highest of conclusions? What, at the most natural core of cosmos, is the nature of Truth? The choice, the faith is at the end between material chance and an Informative Creator.*

In a cosmos finely tuned for life, with Darwinian evolution logically counted out, the Dialectic shows the latter, a Creator, is a choice most open, rational and likely Real and True. In the Glossary check 'evolution' one last time. The Primary Axiom of Natural Dialectic rejects the second usage but, accepting all the others, glories in the fourth. *It thereby deduces that the character of cosmos is, beyond the dreams of physical cosmology, life-friendly; and that the Nature of Nothing is, paradoxically, the Natural Heart of everything.* **In short, the practical and most rational aim of philosophy can only be to lighten and enlighten; its arrow points you, right-hand upwards as in blessing, towards Essential Truth.**

---

[212] see Glossary.

Glossary

A

adaptive potential: involves pre-programmed, super-coded switches and recombinant (transposable) refinements intrinsic in the genomic program of any particular biological type (*SAS* Chapter 23).

ahamkar: pre-scientific term; conscio-material band/ grade - conscious band of mind; sometimes identified (only partially correctly) with *ego*; faculty of self; *ahamkar*, involving identity, frames thought; habitual identification with own physical body; also with friends, family, community, study, work, country or cosmos as a whole; intellectual analysis is a 'knife' that dissects according to the interpretations of this frame.

akash: pre-scientific term; possible conscio-material band/ grade - archetypal field; translated as sky, space, void within (and, some argue, from) which all events occur; also, as a higher (metaphysical) element from which the physical world derives, termed 'ether'; 'inner skies' are psychological; the 'outer sky', physical *akash*, is vacuum or *inter omnia* space.

allele: a sister gene; you have two copies of life's book, one from mother and the other from father, so that each gene from father has a correlate 'allele' from mother and *vice versa*.

anti-entropy: *see* **negentropy**

archaea: phyletic anaomaly; prokaryote differing significantly from bacteria regarding cell wall, membrane, genes and some metabolic pathways; unlike bacteria its replication and transcription is of eukaryotic type; chirally distinct codification for membrane lipid (L) glycerate-1-phosphate, as opposed to (R) glycerol-3-phosphate in all other organisms; t-*RNA* also unique; origin and evolutionary 'progress', if any, unknown.

archetype: basic plan, informative element; conceptual template; pattern in principle; instrument of fundamental 'note' or primordial shape; causative information in nature; 'law of form'; nature's script; Natural Dialectic's 'holographic' edge, omnipresent but invisible because it's metaphysical; the psychosomatic place where metaphysic and its physic meet; morphological attractor or field of influence in universal mind; the subconscious component of universal (natural) mind comprising archetypes; prototype-in-mind (maybe related to Platonic ideas, Aristotelian entelechies and/ or Jungian archetypes) whose potential matter is seen as hard a metaphysical reality as, say, particles are physical realities; program(s) naturally stored in cosmic memory - simple in terms of inanimate physical 'law' (of particles and forces), complex in terms of animate structure/ function/ behaviour; information stored in a typical mnemone; in biology, metaphysical correlate of biological type/ super-species that is physically expressed in code as *potential* form; abstract or metaphysical precursor; the collective unconscious of a type e.g. human type; as thought is father to the deed or plan is prior to ordered action, so archetypes precede physical phenomena; pre-physical initial condition of matter.

Archetype*:* Primary First Cause; *Logos* (*see* Glossary and Index).

ATP:	Adenosine TriPhosphate, life's standard bearer of chemical/ heat energy; a cell's agent of energy transmission; a biological 'match' or 'battery'; an active cell may discharge many thousand units of *ATP* per second to drive its metabolic machinery; these are recharged by respiration; *ATP* also plays a critical informative role in the transmission of nervous and possibly other signals.
AV:	Authorised Version e.g. *AVS* authorised version of science.

B

base:	significant component of a *DNA* nucleotide: a letter in 'the book of life': there are 4 bases in the genetic alphabet - A (adenine), G (guanine), C (cytosine) and T (thymine): in the case of *RNA* base T is replaced by U (uracil).
base pair:	the conservative accuracy of genetic inheritance and the elegant construction of *DNA* are both dependent on a base-pairing rule viz. G pairs only with C and A with T (or U).
big bang:	unconscious singularity *see* also **transcendent projection**
biomimetics:	also known as biomimcry; fast-expanding field of scientific study and imitation of the biological production of codified substances and processes; research in order to better inform the processes of design (engineering) and technology (production) for human purposes.
black box:	process or system whose workings are unknown.
boson:	*see* **elementary particle**.
buddhi:	pre-scientific term; conscio-material band/ grade - conscious band of mind; faculty of intellect; instrument, whether sharp or blunt in an individual case, of learning and discovery; analytical tool to educe physical and metaphysical patterns; pragmatic and hypothetical power of reason; crucial to gauge physical circumstance for survival and metaphysical principle for optimising state of mind.

C

caduceus:	staff of Hermes/ Mercury the communicator, intercessor and informant deity; the messenger of metaphysic, carrier of thought is a power mythologically trivialised; symbol, including double helix, used by Natural Dialectic to represent basic human infrastructure, that is, the archetypal form of man.
chakra:	pre-scientific term; conscio-material band/ grade - subconscious mind; metaphysical modulator; also called a node or plexus; device for the wireless transmission of *prana* to the electrical systems (e.g. nervous) of an organism; psychosomatic gate; trans-dimensional (metaphysical to physical and *vice versa*) transit-point for informative signals; a mechanism including an antenna (receiver), transformer (between 'voltages' of *pranic* energy, transducer (between electrical and *pranic* conveyance of charge), simple harmonic oscillator (a 'heart' controlling *pranic* flow) and distributor (of *prana* through a network of meridians); archetypal channel; specific mind-body broadcasting interface; morphogenetic informant especially important in the process of bio-development; lowest metaphysical component in the

hierarchical transmission of power throughout macrocosmic creation and its microcosmic reflection as mankind and other forms of life of earth; *chakras* are power hubs (and thus, like suns, 'controllers of regions') which exist on a cosmic as well as individual-body scale; conscious chakras (or focal concentrations) of higher mind are not included in this cosmology; because it cannot observe or experiment on metaphysical apparatus science has not developed any understanding beyond the physical expressions of informed energy; see also *prana*.

chaos: a confusing notion with three main but disparate implications - emptiness, disorder and randomness; Greek word meaning chasm, emptiness or space; structureless 'profundity' that pre-existed cosmos; *prima materia, prakriti* or primordial energy structured by regulation of divinity, archetype or natural law; anti-principle of cosmos i.e. disorder; any case of actually or apparently random distribution or unpredictable behaviour; also apparently random but deterministic behaviour of systems (e.g. weather, electrical circuits or fluid dynamics) sensitive to initial conditions.

chitta: pre-scientific term; conscio-material band/ grade - conscious mind; attention *per se*; pure, formless (or boundless) intelligence in which forms of thought are projected; source of ideas and creativity; psychological focus.

chloroplast: organelle in plant cells containing photosynthetic apparatus.

chromatin: a nuclear complex that, using histone proteins, helps package, reinforce and control the expression of genetic *DNA*.

chromosome: a 'book' in the 'encyclopaedia' of life; the human genome contains 46 chromosomes.

cladistics: method of classification using diagrams called cladograms; organisms are collected into groups on the basis of shared (homologous) features; homologies are tallied and numerical rather than speculative, evolutionary/ phylogenetic links drawn up between organisms; cladism is thus a powerful, neutral, objectively detached tool of analysis and for this reason the technique enjoys growing popularity among the world's taxonomists.

code: the systematic arrangement of symbols to communicate a meaning; code always involves agreed elements of morphology (the form its symbols take), syntax (rules of arrangement) and semantics (meaning/ significance); without exception such prior agreement between sender (creator/ transmitter) and recipient involves intelligence.

codon: a 'word' in the genetic language; stands for an amino acid or a full stop; since more than one codon may stand for a single amino acid the genetic code is sometimes ill-perceived as 'degenerate'.

conscio-material dipole: basic, binary structure of existence; slope of creation that, in dialectical description, extends from the immaterial pole (a concentrate of Pure Consciousness) to a material pole of pure non-consciousness - the physical plane; drop/ descent from

Essence through existence; drop from Conscious Singularity to unconscious singularity (black hole); a hierarchical description of polar creation simply modelled *passim* by the use of spectrum, concentric rings and, step-wise, ziggurats; immaterial, subjective element (information) tapers on a sliding scale with material, objective element (energy); 100% objective is, for example (*fig.* 0.14), 0% subjective - physical nature is a special case of consciousness, its total absence at creation's spectral base; *vice versa*, at Top, Transcendent Nature (Essence) there is 0% objective form; and ratios, from top to bottom, in between; from this perspective any 'sharp' division between mind and matter becomes illusory; it is no more real than exists between, say, UV and microwave radiation in the 'rainbow' continuum of their e-m frequencies; cosmos is a conscio-material spectrum, a taper of consciousness to, at base, its absence (fig. 0.14).

convergence: the tendency of unrelated organisms to evolve similar characteristics; in the case of *divergence* adaptation/ speciation from an original feature occurs (e.g. beaks of finches); *convergence*, involving the unrelated, mosaic occurrence of similar features (such as the camera eye, viviparity and thousands of other instances), runs counter to Darwinian expectation; it means that such codified features must have evolved independently many times over; evolutionary explanations of this profound yet ubiquitous puzzle may thus involve speculations such as appeal to non-random 'deep bio-structure', 'principles of evolution', 'morphological laws' or 'inevitablility' granted by imaginary natural laws of codification/ innovation; for a design theorist the bio-codification and engineering of 'convergent' forms derives from either an original use of modular programming or, in the case of so-called micro-evolutionary variation, from in-built adaptive potential flexibly but appropriately activated by genetic switches and epigenetic markers.

cosmic fundamentals: cosmic psychological and physical qualities (see Chapter 2); basic states or tendencies; universal ingredients whose mixture is variously expressed in every object and event.

cosmological axis: human pivot; the point at which subjective and objective perception meet; eye-centre; third eye; thought centre; *ajna chakra.*

cosmological principle: idea that, on a sufficiently large scale, the distribution of matter in the physical universe looks just about the same from any vantage point; it therefore has neither centre nor, being infinite, edge - unless of course, its space is somehow spherical.

cosmos: often applied to physical universe, universal body; from Greek word denoting orderly as opposed to chaotic process; involuntary pattern of nature; also equated, including metaphysical mind, with existence as a whole; seen, dialectically, as a projection through the template of metaphysical archetypes; **the umbrella title of the series and website of books, Cosmic Connections, could, with reference**

         to the Natural Dialectic which structures its *CUT* (see Glossary: unification), equally be called Orderly Linkages.

creation: origination; physical or psychological arrangement; mind creates with purpose, matter without; creation means active production but also passive result; a creation will have been informed by force of mind and/or matter.

D

dialectical stack: stack of opposites; columnar expression of polarity; there are two kinds of stack - primary or non-vectored and secondary, vectored; primary (essential) stacks set (*Sat*) Unity against (↓ *tam*/ *raj* ↑) duality (for elaboration see especially Chapter 2 and *figs*. 1.1, 24.1 and 24.2); secondary (existential) stacks represent the various kinds of polarity from which the changeful web of existence is composed; each pair of polar 'anchor-points' implies a scale or dynamic range that runs between 'paired opposition' or 'complementary covalency'; stacks do not necessarily list synonyms or make equations; *their perusal is intended to promote connections because consideration of connections tends to help unify/ collate/ organise one's working comprehension of any matter in hand.*

dialectic: a form of debate between positions of polar opposition (argument and counter-argument or thesis and antithesis); the motion of to-fro discussion that results in resolution (synthesis) whereby points of view are aligned; balance, compromise, neutral ground, golden mean and central truth are aspects of this synthetic (*Sat*) fundamental; paradoxically, two become one; union supersedes division; Natural Dialectic (see also *PGND*, *SPFP*, *RSP and WE*: Glossary) suggests that dialectical motion reflects the polar, to-fro or oscillatory dynamic of creation; the continual disequilibrium of nature (called motion) always seeks its various balances

dialectical stack: stack of opposites; columnar expression of polarity; there are two kinds of stack - primary or non-vectored and secondary, vectored; primary (essential) stacks set (*Sat*) Unity against (↓ *tam*/ *raj* ↑) duality (for elaboration see Chapter 2 and *figs*. 1.4 and 2.2); secondary (existential) stacks represent the various kinds of polarity from which the changeful web of existence is composed (see *figs* 1.4 and 24.1); each pair of polar 'anchor-points' implies a scale or dynamic range that runs between 'paired opposition' or 'complementary covalency'; stacks do not necessarily list synonyms or make equations; *their perusal is intended to promote connections because consideration of connections tends to help unify/ collate/ organise one's working comprehension of any matter in hand.*

diploid: having full genetic complement with one copy of chromosomes from each parent e.g. you have 46 chromosomes, 23 from mother and 23 from father.

DNA: a complex chemical; a large bio-molecule made of smaller units, nucleotides, strung together in a row; a polymer in the form of a double-stranded helix; a medium superbly suited to the storage

	and replication of 'the book of life'; 'paper and ink' on which the genetic code is inscribed; an organism's 'hard drive'.
dukkha:	imperfection, suffering.

E

electromagnetism: physics of the field that exerts an electromagnetic force on all charged particles and is in turn affected by such particles: light/ e-m radiation is an oscillatory disturbance (or wave) propagated through this field; light; light paradoxically involves a perfect, polar balance between contractive/ magnetic and radiant/ electric components.

elementary particles: science has discovered and, for the most part, experimentally verified, over fifty elementary particles; these are divided, in simple terms, into bosons (force carrying particles) and fermions (separate particles); bosons include photons (which mediate the electromagnetic force), gluons (which mediate the strong nuclear force), W and Z particles (which mediate the weak nuclear force), possibly gravitons (which mediate the gravitational force) and also possibly a Higgs boson (which may mediate a proposed mass-giving field); fermions include two main groups - six quarks and leptons (six electron/ neutrino types); derived from quarks are strongly interactive composites called hadrons; hadrons include baryons such as protons and neutrons and (perhaps a little confusingly) bosons such as short-lived mesons.

entropy: a measure of the amount of energy unavailable for work or degree of configurative disorder in a physical system (see second law of thermodynamics); inertial aspect of an energetic, material or conscious gradient; diffusion or concentration gradient outward from source to sink; drop towards 'most probable' outcome i.e. inertial slack; a measure of disintegration or randomness; expression of the (*tam* ↓) downward cosmic fundamental; a major property of matter, closely coupled with materialisation; in a closed system, which the universe may or may not be, this tends the eventual loss of all available energy, maximum disorder and the exhaustion of so-called 'heat death'.

enzyme: biochemical widget; protein catalyst without whose type metabolism (and therefore biological life) could not happen.

epigeny: genetic super-coding; contextual punctuation; chemical modification of *DNA*; also extra-nuclear factors that may cross-reference with genetic expression.

equilibrium: three modes of equilibrium are (*sat*) balance of poise or pre-active potential; (*raj*) dynamic balance occurring in all regular cycles, wave-forms andcybernetic homeostasis that is basic to the stability of life-forms; and (*tam*) inertial equilibrium that results from diffusion of information or energy; it equates with exhausted inaction or 'flat', impotent rest; such post-active inertia represents the most probable distribution of energy/ matter with the least energy available for work viz. the most random arrangement permitted by the constraints of a system; expressed in psychological terms as ignorance,

	unconsciousness or sleep; see also equilibration, *karma* and *fig.* 1.1 'Pivoted Existence'.
Essence:	(*Sat*) Supreme or Infinite Being; Substance (perhaps Spinoza's Substance) 'prior to' or 'above' existence; Pure Consciousness/ Life; Peace that transcends all psychological and physical action; the root of an essentially undivided universe; Conscious Singularity; Uncreated One within which and whence all differences have their being; Apex of Mount Universe; goal of saints/ 'philosopher kings'; the 'point' at which All-Is-One.
eukaryote:	non-prokaryote; any organism except bacteria and blue-green algae.
evolution:	there are today *four* main usages of this word; each 'loading' derives from the original Latin, 'evolvere', meaning to unroll, disentangle or disclose; the *first two*, physical and biological, are conceived as natural/ mindless processes; the *second, mindful pair* is of psychological/ teleological import; specious ambiguity may conflate or switch between the fundamentally separate pairs of meaning. **Firstly**, in the scientific context of physics and chemistry, the word is used to describe change occurring to physical systems; the laws of nature can't, it seems, evolve through time but stars, fires, rocks or gases can. **Secondly**, though also subject to the 'rules' of entropy, biological evolution is a theory of *random progression* from simple to complex form; it thereby implies increasing, codified complexity; while retaining the 'hard loading' of physical science it also, ambiguously, claims that codes, programs, mechanisms and coherent, purposive systems - normally the province of mental concept - self-organise by, essentially, chance; such confusion, the basis of naturalism, is compounded by failure to distinguish between, on the one hand, ubiquitously observed variation (called micro-evolution) and, on the other, Darwinian 'transformation' between different sets of body plan, physiological routines and associated types of organism - such 'black-box macro-evolution' as is never indisputably observed; to evoke a naturalistic ambience it is fashionable to use 'evolved' interchangeably with or to replace the words 'was created', 'was planned' or 'designed'; finally, it is noted that the coded, choreographed development of a zygote, packed with anticipatory information, through precise algorithms to adult form is the absolute antithesis of blind Darwinian evolution. **Thirdly**, man certainly evolves ideas; intellect can evolve 'purposive complexity'; we invent all kinds of codes, schemes and machines; we devise increasingly complex theories and technologies; and we evolve an understanding of natural principles; this, which all parties accept, is an informative, psychological sense of 'evolution'. The **fourth** sense of evolution, at least as near to the original Latin as the other three, is the spiritual usage; immaterial spiritual evolution, unacceptable to materialists and unknown to physical science, is at the very heart of holism; in this voluntary sense of evolution

practitioners cast off material attachment, evolve and merge into the *Logos*; evolution (or, perhaps better, centripetal involution) of the soul is their great business; their aspiration is to unite with The Heart of Nature.

evolution pre-Darwinian: minority/ anti-mainstream pre-Socratic snippets and sense-based Epicureanism lionized by interpretations of post-18th century materialists; virtually undetectable eccentricity in Chinese, Indian and Islamic literature; natural selection treated by creationists al-Jahiz and Edward Blyth; Buffon, a non-evolutionist, addressed 'evolutionary problems'; Lamarck (evolution by inheritance of acquired characteristics); hints in poem by Erasmus Darwin.

evolution Darwinian: mechanism - natural selection; major tenets - common descent (inheritance), homology and 'tree of life'.

evolution neo-Darwinian/ synthetic: as Darwinian, except synthetic theory adds random mutation as the mechanism for innovation; also adds a mathematical treatment of population genetics and various elements (e.g. geno-centric perspective) derived from molecular biology.

evolution post-synthetic phase: natural selection and random mutation acknowledged as mechanisms insufficient to source bio-information; post-Darwinian evolution invokes mechanisms from hypotheses such as *NGE* (natural genetic engineering) and 'evo-devo'; holistic possibilities also address the origin of complex, specified and functional bio-information.

existence: which 'stands out' from background 'nothingness'; the apparently divided universe; seemingly disparate, finite things; all motion/ change/ relativity; all psychological and physical events.

exon: specifies the amino acid sequence for a protein; m-*RNA* after protein editors have removed introns.

F

fermion: *see* **elementary particle**.

field: any extent wherein action either physical or metaphysical but of a certain kind occurs e.g. field of battle, influence of mind or magnetism; the scientific definition is limited to a collection of numbers varying from point to point - such as a scalar field of contours on a map - or numbers with direction - such as a vector field showing speeds and directions of wind.

first causes: check *figs*. 3.3, 5.1, 9.1, 11.1 or 11.3. First cause is first motion in a previously undisturbed, pre-conditional field.
First Cause Psychological is Archetype, Potential Informant or (see Chapter 5: Top Teleology) *Logos*; attributes of this Primary Source and Sustenance of Creation include omnipresence, omnipotence and omniscience; **first cause physical** is also called potential matter or archetypal memory; as the secondary source of creation it precedes physical phenomena; as such it is, transcending physical appearances, metaphysical; this 'physical nothingness' is therefore, paradoxically, the source of everything composing

astronomical cosmos; it consists of their being or essence as opposed to their becoming; its void, with respect to the presence of finite phenomena, appears infinite; attributes of immanent archetype, the primary informant of our non-conscious, energetic universe, include omnipresence and omnipotence.

G

gamete: sex cell with half of full genetic complement i.e. a single set of chromosomes.

gene: generally means a basic unit of material inheritance; section of chromosome coding for a protein; digital file; a reading frame that includes exons and introns; the old one gene-one protein hypothesis is incorrect; in fact, by gene splicing, a particular piece of *DNA* may be used to create multiple proteins.

genome: total genetic information found in a cell: think of the genome as an instruction manual for the construction and physical operation of a given organism.

genotype: the genetic constitution of an organism, often referring to a specific pair of alleles; the prior information, potential, plan or cause of an effect called phenotype.

gravity: in physics an attractive mass-to-mass force or warping of space-time; in Natural Dialectic the term is redefined more broadly - the agency of its (*tam* ↓) downward vector includes all psychological and physical factors of materialisation; such 'gravitational' factors and their properties are listed in the left-hand column of Secondary, Existential Dialectic; they include pain, pressure, confinement, strong nuclear force, mass, electromagnetic binding, inertia, entropy, 'standard' gravity and so on; gravity might be summarised as 'negative power' or 'the principle of death'.

H

haploid: having half the full genetic complement, as in the case of sex cell.

heterozygous: having different allelic forms of a particular gene.

holism: opposite of reductionism; the view that a whole is greater than the sum of its parts; the extra metaphysical (immaterial) ingredient is identified by Natural Dialectic as information; information implies the purposeful design, development and arrangement of contingent parts in a working system; may operate according to a Logical Norm.

hologram: a 3-d photograph made with the help of lasers. Unlike a normal photographic image each part of it contains the image held by the whole.

homeostasis: vibratory or periodic control of a system to obtain balance round a pre-set norm; the mechanism of its information loop involves sensor, processor and executor; the operative cycle works by negative feedback; psychological (nervous) and biological cybernetics; the informed basis of biological stability.

homeotic gene: gene (e.g. *Hox* gene) involved in developmental sequence and pattern; high-level co-determinant of the formation of body parts.

homozygous: having the same allelic forms of a particular gene.

I

illusion: is the cut between illusion and delusion an illusion? illusions, apparently outside the mind, appear real; a delusion, in it, we think real; neither, mind allows, is real or true.

information the immaterial, subjective element; information occurs in three distinct modes; informative *potential* is action's precedent; this potential is both the source and substrate of all psychological activity; we know this substrate of life and consciousness; *active* information inhabits its own centre, mind; mind knows, feels, purposes, creates, codifies and recognises meaning; it is also, by way of secondary and subconscious forms of archetype) a physical entrainer; thus *passive* forms of information, either in memory or physical objects and events, reflects active; in other words, *passive* information is stored as subconscious 'files' in memory; and it is fixed in the expressions of non-conscious matter according, universally, to the archetypal behaviours of natural bodies or, locally, to particular schemes of life; that is to say, both the constructions of life-forms and the inanimate cosmos are the physical product of stored concept.

informative entropy: loss of information due to degradation of its carrying medium; such a medium may be metaphysical (mind) or passive and physical (for example, computer files or genetic code); and its entropy may be metaphysical (loss of memory, focus or consciousness) or physical (for example, genetic mutation); the informative correlate of such degeneration is diminished organisational capacity, meaning or thrust of original purpose.

informative negentropy: gain of informative clarity; increasingly focused, purposive specificity; associated with knowledge, wisdom, grasp of principle and pristine construction; machines are a good example of informative negentropy.

intron: genetic control panel; n-p-c (non-protein-coding) segment(s) spliced from an m-*RNA* transcript prior to translation; introns include regulatory elements (to variably promote or inhibit gene expression) and addressing factors of the genetic operating system; gene-attached information lending specific flexibility to protein manufacture.

inversion: turning upside-down or inside-out; reversing an order, position or relationship; in a hierarchical sense inversion is allied with the reflective asymmetry of opposite poles; information outwardly expressed; pole-to-pole reversal integral to dialectical structure; inversion represents, of the two anti-parallel vectors on creation's conscio-material gradient, the (*tam*) centrifugal vector; as opposed to (*raj*) centripetal eversion; various kinds of inversion (cosmic and micro-cosmic (biological)) are discussed in these books.

K

karma: action; law of cause and effect, that is, balance between action and reaction; equilibration such as underlies all mathematical equation; a deed with implications of the reactions or 'payback' it provokes; fruit or result of previous thoughts, words and deeds; applies as rigidly to metaphysical (psychological) as physical events.

L

lepton: *see* **elementary particle**

levity: agency of the (*raj* ↑) upward vector; dialectical converse of gravity; psychological and physical 'levitatory' forces lift or stimulate; they are listed in the right-hand column of Secondary, Existential Dialectic and include light, heat, excitement, dematerialisation, release, negentropy, focus of interest, affection and so on; physically, levity includes anti-gravity or the intrinsic property of matter's absence, space; generally summarised as 'positive power' or 'a buoyant principle of liveliness'.

logic: analysis of a chain of reasoning; principles used in circuitry design and computer programming; 'normative reason' relates to the basic axiom(s) of a given standard e.g. *bottom-up* materialism or *top-down* holism; three main logical thrusts are: (1) inductive (premises/ observations supply evidence for a probable/ plausible conclusion) as in the case of experimental science working *bottom-up* from specific instances to general principle: (2) abductive (best inference concerning an historical event): and (3) deductive (conclusion in specific cases reached *top-down* from general principle): two pillars of logic are holism and materialism; holism employs mainly deductive/ abductive operations and a Logical Norm; materialism tends to inductive/ abductive operations whose axis is non-conscious force and chance.

Logos: First Cause; Prime Mover; Causal Motion that sustains creation's conscio-material gradient; labelled with many names at different times, places and languages; *Logos*, transcending mind, is Conscious; therefore, <u>Who</u> is *Logos*?

LUCA: last universal common ancestor.

M

macrocosm the physical universe of astronomy and cosmology; dialectically, the whole of existence (i.e. both universal mind and universal body) as opposed to individual, microcosmic objects and events - including the human body.

macro-evolution: large-scale, non-trivial evolution; process of common or phylogenetic descent alleged to occur between biological orders, classes, phyla and domains; includes the origin of body plans, coordinated systems, organs, tissues and cell types; unexplained by mutation, saltation, orthogenesis or any known biological mechanism; sometimes called 'general theory of evolution' (*GTE*); crucial but unseen, hypothetical extrapolation from micro-evolution; may or may not occur; an extrapolation from Darwinian micro-evolution vital to sustain a 'progressive' materialistic mind-set, is conjectural alone.

manas: pre-scientific term; conscio-material band/ grade - both conscious and subconscious bands of mind; 'mind-stuff'; 'clay'

	moulded by the the hand of thought and perception; metaphysical 'material' on which direct formative action of *chitta* occurs; 'film' on which the perceptions of mind are developed and, at the same time, 'screen' on which they are perceived; receptor for sense perceptions and storage silo of such impressions as 'seeds' or 'files' of subconcious memory; substance of archetypal field, in other words, of universal archetypes (cf. typical mnemone); mental form and energy.
mantra:	archetypal symbol; psychological transformer; authorised form of words repeated to exclude other thoughts; examples include the 'Hail Mary', '*Om Mane Padme Hum*' and, materialistically, 'evolution made…', 'in time nature designed…' or similar incantation.
maya:	partial truth; world of forms and forces; illusion that changeful cosmos is the ultimate reality; motions and perceptions composing *maya* are thus, set against Essential Truth, more or less unreal; becoming wise to the nature of *maya* yields liberation from its cosmic veil.
meditation:	Lat. *medius*, middle; coming to Centre; mind and body dropping away.
meiosis:	shuffling the information pack: variation-on-theme; mechanism for the production of haploid gametes; genetic postal system for sexual reproduction.
metabolism:	body chemistry.
metaphysic	= non-physical/ immaterial/ psychological/ unnaturalistic; physically expressed as specific/ intended arrangement/ behaviour of materials; physical behaviour reflects metaphysical blueprint; involves element of information; involves symbol/ code/ abstraction/ logic/ reason/ mathematics; also meaning/ message/ goal/ teleology; also consciousness/ mind/ life/ experience/ feeling; and also morality/ force psychological/ emotion; involves innovation/ creativity/ art/ invention/ aesthetics.
microcosm:	an entity that reflects the universe by containing all its basic constituents. Used especially of the human state where it may refer to both mind and body or, in a purely physical context, body alone.
micro-evolution:	misnomer; non-progressive, small-scale variation within a species or, more broadly, between strains, races, species and genera; variation/ adaptation within type; trivial Darwinian changes that may occur by natural selection/ ecological factors acting on genetic recombination, mutation or adaptive potential (q.v. this Glossary); sometimes called 'special theory of evolution' (*STE*), micro-evolution/ variation is a fact.
mitochondrion:	organelle in eukaryotic cells containing the apparatus for aerobic respiration.
mitosis:	conservative copying and delivery of genomes in cell division; genetic reprinting; genetic postal system for asexual reproduction.
mnemone:	a division of memory whether individual or universal: an individual's two divisions are *personal mnemone* (likened to a

working cache or data store) and *typical mnemone* (likened to a *ROM* or an operating system); typical mnemone is, in effect, a program consisting of three subroutines - *signal translation, instinct* and *morphogene* (for more information see Chapters 15 - 17); it is also a synonym for natural, archetypal memory in universal mind; in short, it is a body's **metaphysical DNA**. Further than the character of each bio-type of organism it also includes the 'instinct' of matter (i.e. cosmos). Natural Dialectic's definition involves no 'cultural' connotation whatsoever and is thus wholly distinct from evolutionary psychology's use of the word.

mobile genetic element: transposon, retrotransposon, insertion sequence, other non-protein-coding *DNA*, n-p-c *RNA* fragments and various protein regulators that together expedite the operating system of a genome.

morphogene: one of three sub-routines of typical mnemone or archetypal memory relating to physical construction; morphological attractor; the component of subconscious mind associated with electrochemical function and thereby body; just as you might not guess from the picture on your TV screen or object from a 3-d printer the nature of the electromagnetic messaging that creates it so you might not guess a body's shape from its *DNA* or the messaging agent that links archetypal mind with body; morphogene is the dominant, perhaps exclusive, aspect of mind in unconscious organisms such as plants or fungi..

morphogenesis: the development of biological structure; more generally, the production of physical form.

mosaic: the presence of permutations of codified sub-routines or similarities of form and/or function scattered in organisms unrelated by lineage.

mutation: accidental change to genetic code.

mysticism: quite different from objective, it is the subjective science; not philosophy, religion or opinion but practice to achieve communion with natural, inner, immaterial truth; esoteric as opposed to exoteric, materialistic discipline; 'science of the soul'; as gyms and physical action are to athletes so meditative exercise and psychological stillness are to mystics; involves psychological techniques to achieve a clear, rational goal - purity of consciousness and thereby understanding of the fundamental nature of the informative principle, mind; since life is lived in mind a mystic seeks consummate knowledge of life's source and sanctum, that is, communion with its deathless heart; adepts were, are and will be 'Olympian' meditative concentrators.

N

nano-biology: biology of structures/ physiologies involving a few atoms or molecules; 'extremely small biology'.

nanotechnology: technology at atomic and sub-atomic level as is, basically, life's.

naturalistic methodology: also known as 'methodological naturalism': is, strictly, not concerned with claims of what exists or might exist, simply with experimental methods of discovering physically measurable behaviours; thus only materialistic answers to any question (e.g. how biological forms arose) are deemed 'scientific' or 'scientifically respectable'.

negentropy: opposite of entropy; lowering of entropy; expression of the (*raj*) upward-pointing cosmic fundamental closely coupled with stimulus, dissolution and dematerialisation; a measure of input, cooperation or synthesis; motive/ fluidising aspect of an energetic, material or conscious gradient; gain of energy, configurative order, information or consciousness in a system; when used in terms of information negentropy involves gain in order or understanding of principle from which different actualities derive; a measure of the amount of concentrated/ conceptual information, specific, intentional complexity or conscious arrangement in a system; a natural and essential property of mind.

Nirvana state of enlightenment; 'non-condition'; *nirvana* is devoid of existential motion; extinction of existence (i.e. perpetual change) leaving Essence Alone; pure soul; psychological super-state; Buddhists call such transcendence non-self or the Formless Self.

non-existence: where creation = formful existence, non-existence is formless; the polar opposite of physical space and time is Transcendent Potential; such pre- or super-existential formlessness is non-existent; Absolute Non-Existence is Essential; however relative non-existences of two kinds also occur; the first kind is metaphysical/ subjective and therefore psychological; it involves the absence of a specific psychological form or event; unconscious oblivion is one such non-existence; the second kind involves the local absence of a possible physical event (an object is a 'slow event'); impossibilities are non-existences but imaginations of non-existence (including symbolic abstractions, hypothetical entities, physical absences, absolute emptiness and the number zero) exist; furthermore, the nothingness of space and time, the zero-point of calculus and zero's empty set together constitute the basis of physical science and mathematics.

non-protein-coding *DNA*: occupies probably 95% of eukaryote and 80% of bacterial genomes; associated with the genetic operating system; may include some genuinely redundant misprints or duplications but now thought for the most part critical to the flexibility, efficiency and even possibility of gene expression; once thought of as useless, degraded information and ignorantly called 'junk *DNA*'.

non-protein-coding *RNA*: n-p-c *RNA* is also called nc-*RNA* (non-coding), nm-*RNA* (non-messenger) or f-*RNA* (functional); functional *RNA* molecule not translated into protein; many 1000's of different specimens include classes of t-*RNA* (transfer *RNA*), r-*RNA* (ribosomal *RNA*) and, commonly involved in the regulation of gene expression and other intra-cellular tasks, micro-*RNA*, double-stranded si-*RNA*, pi-*RNA* and so on; also, for intercellular communication, ex-*RNA*.

nucleic acid: *see DNA* and *RNA*.

nucleosome: a 'reel' composed of histone proteins around which chromosomal *DNA* is precisely wrapped; repeated nucleosomes allow the *DNA* to form a bead-like structure that can coil and super-coil; *DNA*, nucleosomes and other factors compose chromatin.

nucleotide: basic, triplex unit of nucleic acid polymer; monomer composed of phosphate and sugar (the 'paper' part) and base (the 'ink letter'); letters' of the genetic alphabet are (G) guanine, (C) cytosine, (A) adenine and (T) thymine. In *RNA* thymine is replaced by (U) uracil.

O

***Om*:**   universal sound, fundamental reverberation, basic truth; initial motion of Potential Information; sometimes spelt *Aum*, a Sanskrit word whose Semitic transliterations are Am'n, Amin and Amen; see also First Cause, *Logos*, *Kalam*, *Shabda* etc.

order: regular, regulated or systematic arrangement; organisation according to the direction of physical law; passive information by which things are arranged naturally (with predictable but non-purposive complexity) or purposely (with innovative or specified complexity); mind, generating specified complexity in the order of its technologies and codes, actively informs; the orders of mind are meaningful, the orders of matter lack intent; see also cosmos.

organelle: cellular sub-station; discrete part of a cell; sub-cellular compartment having specific role such as informative (nucleus), energetic (mitochondrion, chloroplast), constructional (ribosome, Golgi body) or other.

P

***PAM*, *PAND*, *PCM* and *PCND*:**   philosophical gambits; see Primary Axioms and Corollaries.

phenotype: the effect of causal potential; result of the development of prior, informative 'egg'; outward expression of inner plan; sensible appearance of an organism as opposed to its genotypic scheme: the whole set of outward appearances of a cell, tissue, organ and organism are sometimes called a phenome (*cf.* genotype/ genome).

photosynthesis: process by which inorganic carbon is introduced to the biological zone and energetic sunlight fixed as a crystalline molecule of storage, a sugar called glucose.

phylogeny: evolutionary history; relationships based on common or evolutionary descent.

potential: poise; latent possibility; potent non-action that precedes any particular action or creation; in science potential energy is defined as the energy particles in a system (or field) possess by virtue of position/ arrangement; gravitational, electrical, electrochemical, thermo-dynamical and other kinds of potential are recognised; in dialectical terms mind precedes matter, information precedes the pattern of material behaviour; information is energy's pre-requisite potential; in this case *informative potential* involves two conditions; firstly, a pre-existential/ essential state of pure potential; secondly, a pre-material, metaphysical fact of potential matter, archetype or laws of nature; if potential's pre-active equilibrium is related to the voltage of a full battery then aspects of psychological 'voltage', whose currents drive intentional behaviour, are purpose, will and plan.

potential matter: see archetype.

***prakriti*:** pre-scientific term; conscio-material band/ grade - whole spectrum of existence; complements Essential *Purusha*; universal energy; generic term for nature; screen and light on which the show of creation is projected; 'clay' with which the potter of conscious experience works to produce form; thus also (*figs*. 6.2, 7.5 and 13.6) identified with the objective, energetic as opposed to subjective, informative side of conscio-material cosmos; fundamental substrate whose root exercise involves continual recombinations of three cosmic fundamentals (Chapter 2); interplay of these qualities, attributes or tendencies intrinsically inhabits every object and event; it generates all patterns, forms and forces, whether in psychological or physical regions of the universe; nested, *prakritic* layers are hierarchically arranged like a grid of stepped-down voltages from a power station; in a second comparison, as the waveband of visible light is part of a much larger electro-magnetic spectrum, so the bands of higher and lower conscious mind, subconscious memory and non-conscious physicality compose the spectrum of conscio-material *prakriti*; in this way, for example, *prana* is a low-level, 'infra-red' expression of *prakriti* operant at the *PSI* border where subconscious archetypes (Chapters 15 and 16) give rise, with location, to quantum phenomena; and its lowest, 'radio' level of expression, peripheral to the full creation, is gross physical energy whose various transformations are expressed as the operations of physics, chemistry and biology.

***prana*:** pre-scientific term; possible conscio-material band/ grade - subconscious and physical (lowest) levels of *prakriti*; lowest metaphysical bandwidth of *Logos*/ Om; Chinese *qi* or *ch'i*; associated, as in the yogic practice of *pranayama*, with breath and thereby life of material body; also with light (visible band electromagnetism) and oxygen (specifically, negative ionic charge); identified as the archetypal energy of subconsciousness and the operations of typical mnemone (Chapter 16); subliminal, psychosomatic or mnemonic energy of universal mind at archetypal grade; supports *PSI* (psychosomatic) traffic between universal memory and quantum agencies in the case of both biological and and physical formations; 'infra-mental' band called potential matter; vibration underlying perpetual atomic motion; five pranic bands are analogised with visible light's rainbow, each frequency correlated with an expression of elemental character in physical phenomena or with a level of biophysical expression (*fig.* 17.11); wireless *prana* is identified as the metaphysical life-force of the physical body; in living organisms (including you and me) it is processed by way of metaphysical apparatus called node or '*chakra*'; each of a hierarchical series of such nodes operates as an antenna, transformer and distributor; the system acts as a transducer of *pranic* frequencies to those of bio-electromagnetism and electrical charge; the highest bio-frequency resonates with our '*ajna chakra*' or third eye behind the forehead; this in turn, is subservient to '*sahasrara*', the 'thousand-petalled lotus' supplying 'voltage' (and thereby current) to sustain the physical universe; having entered the human system by resonance with the '*ajna*' antenna *prana* is passed through a grid of aforementioned nodes, well-known to yoga and arranged down the spine; each distribute frequencies appropriate to its body area by a network of *nadis* or meridians identified by medical acupuncture; being metaphysical the pranic system cannot be physically tested by empirical, scientific experiment but only by inference (e.g. a cure); for this reason some proponents of occidental medical science dismiss *pranic* mechanisms of the mind as 'pseudoscientific' and, having thus 'rationally' condemned, proceed to narrowly and unwisely dismiss the broader immaterial fraction of holistic order wherein such components play a crucial part.

Primary Axiom of Materialism (*PAM*): all objects and events, including an origin of the universe and the nature of mind, are material alone; cosmos issued out of nothing; life's an inconsequent coincidence, a fluky flicker in a lifeless, dark eternity.

Primary Axiom of Natural Dialectic (*PAND*): there exists a natural, universal immaterial element - information; immaterial informs material behaviour; a conscio-material dipole that issues from First Cause informs and substantiates both mental (metaphysical) and physical creations; there is eternal brilliance whose shadow-show is called creation.

Primary Corollary of Materialism (*PCM*): the neo-Darwinian theory of evolution, that is, life forms are the product, by common descent, of a random generator (mutation) acted on by a filter called natural selection; such evolution is an absolutely mindless, purposeless process; the *PCM* is a fundamental *mantra* of materialism.

Primary Corollary of Natural Dialectic (*PCND*): the origin of irreducible, biological complexity is not an accumulation of 'lucky' accidents constrained by natural law and death; forms of life are conceptual; they are, like any creation of mind, the product of purpose.

prokaryote: non-eukaryote; bacterial type with little or no compartmentalisation of cell functionaries.

promissory materialism: belief system sustained by faith that scientific discoveries will in the future justify/ vindicate exclusive materialism and, as a consequence, atheism; confidence that

protein: technology will solve (more often than create) the problems the world faces; may involve a call to progress towards the technological provision of its 'promised land'.

protein: factor made from a specific sequence of amino acids to perform a specific task; 'informative' protein includes some hormones; skin, hair, bone, muscle and other tissues are made of 'structural' protein; 'functional protein' called enzymes mediates all stages in cell metabolism, that is, it catalyses all biochemistry.

***PSI* (psychosomatic interface):** psychosomatic border; the level of mind-matter interaction; bridge between metaphysical and physical dimensions; potential matter; 'gap of Leibniz'; 'fit' of mind to matter; point of linkage between subconscious mind and non-conscious matter; gearing between instinct/ archetype and the behaviour of material objects and energies; as in the case of physical law, psychosomatic influence is both general in potential and local/ specific in engagement.

psychological entropy: a measure of loss of concentration, focus of attention or consciousness; loss of 'mental energy' or aptitude; the drop from waking to sleep; loss of knowledge, information or sensitivity; the gradient from intelligence through stupidity to oblivion; an expression of the (*tam*) downward cosmic fundamental in mind; a tendency predominant in lower, egotistical or selfish mind; increasing level of ignorance, anguish or immorality; loss of integrity, psychological disharmony or disintegration; see also *information entropy*.

psychological negentropy: a measure of gain in order; an increase in concentration, focus of attention or consciousness; gain in sense of purpose, 'mental energy' or aptitude; the rise from sleep to waking, 'dark to light' or unhappiness to happiness; gain in knowledge, information or sensitivity; the gradient of learning and spiritual evolution; an expression of the (*raj*) upward cosmic fundamental in mind; a tendency predominant in higher mind; increasing level of contentment, understanding and the natural morality of happiness; the ascent towards psychological radiance, harmony and integration. The converse of psychological negentropy involves *entropy of information*.

psychosomasis: operation across the psychosomatic border; mind/ body interaction; the one-way imposition of archetypal pattern on *physicalia*; the two-way exchange of information in sentient organisms through the agency/ medium of subconscious patterns; *see* also synchromesh 2 (Chapters 16 and 17).

***Purusha*:** pre-scientific term; conscio-material band/ grade - conscious; Pure Consciousness, Source of Life, Subjectivity and Creativity; Universal (*Sat*) Potential; boundlessly pre-active, in action Prime Mover, First Cause or Top Governor on the universal scale and order of creation; given many other names in many languages; ultimate subject of worship, praise and love.

Q

quantum: minimum discrete amount of some physical property such as energy, space or time that a system can possess; quantum theory

	states that energy exists in tiny, discontinuous packets each of which is called a quantum; an elementary discontinuity; an elementary particle e.g. photon or electron.
quantum level:	matter-in-principle; 'internal', 'causal' or 'subtle' matter; the vibrant or energetic phase of physical organisation; zone of sub-atomic particles and forces; step (on cosmic ziggurat) between potential and bulk matter whose aspect is sometimes extended to include atomic and molecular interactions; small-scale substance underlying large-scale, sensible appearances.

R

raj:	(↑) upward, levitatory or stimulatory cosmic vector.
reductionism:	opposite of holism; the materialistic view that an article can always be analysed, split up or 'reduced' to more fundamental parts; these parts can then be added back to reconstruct the whole; a whole is no more than the sum of its parts.
religion:	etymology debated between Latin *religare* (bind) and *relegere* (review); *religio* means dutiful and meticulous observance; currently religion means world-view, mind-set or basic faith; whether of materialistic or holistic belief, it involves the non-negotiable substance of an individual or community's truth - notably as regards origins; antagonism between holistic practice and the naturalistic methodology of science is, because the couple deal with separate but complementary physical and metaphysical dimensions, flawed; a materialist/ atheist 'binds meticulously' to an evolutionary mind-set, a holist to pantheism or a Living Creator; in the case that self-deception is crucial to successfully deceiving others which, holism or materialism, is the religion that is ultimately true?
resonance:	the tendency of a body or system to oscillate with a larger amplitude when subjected to disturbance by the same frequencies as its own natural ones; thus a resonator is a device that naturally oscillates at such (resonant) frequencies with greater amplitude than at others; resonance phenomena occur with all kinds of vibration, oscillation or wave; their sorts include mechanical, harmonic (acoustic), electrical (as with antennae), atomic and molecular.
respiration:	the controlled release of energy from food.
ribosome:	site of polypeptide (protein) synthesis.
RNA:	a single-stranded nucleic acid polymer employed in three different forms during the process of protein synthesis; in computer terms might be likened to a portable memory stick as opposed to *DNA*'s hard drive.
m-RNA:	is used to transcribe a base sequence from *DNA*. It 'photocopies' a gene and carries this information to a ribosome.
mi-RNA:	short micro-*RNA* molecules are important regulators of genetic expression.
r-RNA:	is part of the make-up of the protein-manufacturing station called a ribosome.

t-RNA:	critically translates genetic 'words' (see 'codon') into amino acids: 64 such operators form the link between code and the actuality of a functional protein.

S

sanskara:	character trait; groove, habit, obsession or repetitious mode of thought proportional in depth to the intensity of desire, force of impact or impression that created or sustains it.
samsara:	existence, phenomena, the place of cycles and, therefore, reincarnation; non-essence or, in Buddhism, what is not *Nirvana*.
sat:	'top' or essential cosmic fundamental; 'vector' of balance, neutrality.
science:	Latin *scire* (know); knowledge; commonly understood as the practical and mathematical study of material phenomena whose purpose is to produce useful models of the physical world's reality.
scientism:	a philosophical face of official, *de facto* commitment to materialism; today's majority consensus of what the creed of science is; an -ism born of *PAM*; a faith that all processes must be ultimately explicable in terms of physical processes alone; like communism, a one-party state of mind; a doctrine that physical science with its scientific method is ultimately, the sole authority and arbiter of truth; a set of concepts designed to produce exclusively material explanations for every aspect of existence, that is, to colonise each academic discipline and build its intellectual empire everywhere; 'scientific fundamentalism' closely allied, when expressed in social and political terms, with 'secular fundamentalism', sociological interpretation of behaviour and the fostering of a humanistic curriculum.
secular fundamentalism:	*PAM* as applied to the worlds of nature and of human society.
secularism:	concern with worldly business; lack of involvement in religion or faith; secularism is generally identified, as defined by the dictionary, with materialism; for a secularist the ultimate arbiter of truth is human reason - ideas are open to negotiation so that even morality is relative; however many liberal agnostics, atheists and humanists argue that their metaphysical, philosophical system also embraces so-called 'universal' moral values and, as opposed to zealotry or the logic of evolutionary faith, a liberal politic of 'philosophical live-and-let-live'.
siddhi:	marvellous, miraculous or 'super-natural' psychic ability that, at the point a practitioner actually masters it, becomes natural.
stereo-computation:	stereochemistry involves study of the relative spatial arrangement of atoms in molecules; in biology a 1-D line of informative code (whose 3-D constituents bear no figurative relationship with their informed product) give rise to relative 3-D spatial arrangements at all levels from molecular to systemic and whole-body. Such targeted generation may be termed bio-logical stereocomputation.
sub-state:	*opp.* super-state; impotence, discharge, exhaustion, final stage in the expression of potential; fixity; non-conscious base-state; state 'below/ subtendence; extreme negativity/ (*tam*) condition.

sufi: mystic, Islamic 'heretic' of whom the most influential is perhaps Jalal-ud-Din Rumi, a disciple of Shamas of Tabriz.

super-state: potential; source of possibility; causal metaphysic/ archetype; state 'before' or 'above' subsequent expression; immanence; transcendence; precondition; (*sat*) priority.

symmetry: closely allied with the (*sat*) characteristic of balance; aesthetically pleasing balance and proportion; geometrical balance or interactive process such that some feature of an action remains invariant, that is, conserved; the symmetry of an entity (such as a sphere, empty space or natural law) or feature (such as energy) that remains the same at all times everywhere from any local point of observation or through every transformation is called 'higher' or 'continuous'; if a feature is conserved only when an object or process is moved, turned or viewed at certain angles or under specific conditions its symmetry is called 'lower' or 'discrete'; the symmetrical properties of a system may be precisely related to corresponding conservation laws and *vice versa*; various kinds of symmetry independent of space-time coordinates are important to both quantum and classical physics; scale symmetry occurs when a reduced or expanded object keeps its shape but not its size (as with Mandelbrot fractals); dialectical symmetry also involves *informative potential*; its metaphysical archetypes inform principles, laws or determinant fields that exist prior to action and, from their possibilities, govern actual outcome; such 'configuration of the world' is absolute and, beyond entropy, stable; it is negentropically immune from decay; by contrast, the 'free' symmetry of potential energy is inherently unstable and (like a pencil balanced on its tip) liable to spontaneously 'topple' or 'break' into the least energetic of a range of circumstantial possibilities; such spontaneous symmetry-breaking, the basis of diversity, represents an expression of 'deep symmetry' or archetype under local conditions and is therefore called by physicists 'contingent'.

T

tam: (↓) downward, gravitational or inertialising cosmic vector.

tanmatra: pre-scientific term; possible conscio-material band/ grade - subconscious band of mind; *tanmatras* involve the least metaphysical, most nearly physical band of mind; they are traditionally thought of as mental ideas, psychological forms or the Platonic ideals of physical perceptions (e.g. notions of heat, light and motion in a flame or fragrance in a scent); as *chakras* are immaterial structures dealing with energy, so *tanmatras* deal with image, quality and form; they represent the qualitative aspect of matter and are the 'device' that allows mental grasp of physical effects; as such *tanmatras* are an instrument of potential matter, the archetypal processors of image; as particle to wave so *tanmatra* to *prana*; and as sound is plucked from a tuned string so each of five *tanmatras* is like a string creating, by resonant association, one of five *pranic* 'notes' (see Glossary: *prana*); *tanmatric* apparatus represents a stage in the hierarchical translation of incoming (matter to mind) or outgoing (mind to matter) signals across either individual or universal psychosomatic border (see *figs.* 15.8 and 15.9, also Chapters 16 *esp.* signal translation and 17); the two-way traffic across this border means they equally act, as prism or lens to light, as media for the orderly projection of *prana* into subtle (quantum) events and thence, lower down, gross aggregates of called bulk matter; as such they are, in conjunction with *prana* as power source, the pre-physical mechanism that expresses, in the physical vacuity of archetypal field, the fundamentals of material phenomena; whether simple, single or in 'complex opera' that codes for bio-symphony, *tanmatras* transmit 'sounds' (vibrations or frequencies) which translate into cymatic messages (check Index: Chladni) exciting the emergence or maintenance of physical form; in short, they were the final agents in the initial creation of physical form and force; the wireless physical expression of cooperant *tanmatras* and *pranas* is known to quantum physics; the wired, bonded or aggregate expression involves the chemistry, physics and biology of condensed or bulk matter; and, as electricity supports the running of a machine, they iteratively, correctly support our starry universe.

tattwa: pre-scientific term; possible conscio-material band/ grade - both subconscious and physical bands; means 'that-ness' or 'not-self'; by nature, using their intellect/ *buddhi*, philosophers seek to analyse, categorise and argue so that, in the case of *tattwa*, the number of items listed varies considerably according to tradition; basically, however, it amounts to a 'catch-all' description of human, animal and inanimate condition; five well-known *tattwas* correspond exactly to Greek, Latin and medieval European elements of ether, fire, air, water and earth (see Chapter 10: Old Vacuums); on the subconscious side of *PSI*, in archetypal memory of universal mind, these correspond to five *tanmatras*, five *pranic* 'notes' and five lower *chakras* (*fig.* 17.11); and on the physical side to five informative senses and five energetic organs of action (*fig.* 0.8 and Chaps. 14: Lower Physical Loop, 15: Psychosomatic Linkage and 16: Signal Translation); the five elements/ states of matter are variously defined; ether is space, upper air (home of the gods) or psychosomatic archetypal potential (which, being metaphysical, is physically unseen); it is related to the throat chroat *chakra*, seat of dormancy; there follow gas (air *tattwa* related to breath, oxygen and heart *chakra*), energy (fire *tattwa* giving stimulus for change, heat and light whose hub is the solar plexus), liquid (water and its osmoregulatory and waste expulsion systems) and solid (the earth *tattwa* of related to bio-mass and its sense of pressure/ touch whose *chakra* rests at the supportive base of the spine).

teleology: the doctrine that there is evidence of purpose in nature; doctrine of non-randomness in natural architecture; doctrine of reason ('for the sake of', 'in order to', 'so that' etc.) and intent behind biological and universal design.

third eye: place where you think; point of metaphysical focus between and behind the eyebrows, that is, just above the physical eyes; HQ/ seat of mind beyond the sensory world; cosmological eye-centre; gate through which meditative concentration can pass; single way that leads within.

transcendent projections: **psychological:** see Chapter 5 Top Teleology, Index: Archetype and *figs.* 2.6, 3.1, 5.1, 9.1 and/ or 11.1; **physical:** see Chapters 8, 9, 11, 12; Glossary: archetype; Index: transcendence, archetype, cosmo-logical language; *figs.* as above and 12.1; such projection involves an orderly, energetic expression from either metaphysical or physical nothingness, that is, unseen potential; an instantaneous 'miracle' that issues from 'within' non-conscious physicality; transcendently emergent, finely tuned expansion from 'inner' metaphysic into 'outer' material/ natural law; 0-dimensional singularity (paradoxically everywhere at once) from whose prior pointlessness all points perhaps began; cosmic seed whence, *ex nihilo*, the world developed; projection whose appearance, once physical, is visible and perhaps described but certainly not explained by big bang theory; transcendent projection of archetype is possibly, to the constrained sensory and intellectual states of human mind, ultimately incomprehensible; its invisible mechanism, the practice of materialisation, may remain a fact beyond material understanding. **biological:** if matter is developed memory (Chapter 9: How Does Nothing Physical Work?) then see Chapters 16: *passim* and 19: Conceptual Biology; see also Glossary: mnemone and archetype; Index mnemone, archetype; and *figs.* 2.6, 3.1, 16.1 and 19.1.

transposon: 'jumping gene'; ubiquitous genetic element found in all prokaryotes and eukaryotes so far investigated; *DNA* segment that can, by enzyme, be cut from a one site (the donor) and joined to another (the target); a retrotransposon is moved through the mediation of *RNA* and reverse transcription back from *RNA* to *DNA*; transposons and retrotransposons play a key, functional role in gene expression and regulation; a kind of retrotransposon, SINEs and LINEs are thought to compose 35+% of the human genome; such elements may be flanked by terminal repeats that allow specific, operational variation in different types of cell; from an evolutionary view they comprise functionless viral imports; from a *top-down* view it is predicted they form a dynamic, intrinsic element of the genome involved in gene regulation, genetic shuffling as (epigenetic) response to buffer circumstantial exigency and, just as important, structural agents able to reshape a chromosome to meet specific genetic demands.

transcription factor: a protein that, binding to a specific *DNA* sequence, regulates genetic transcription.

U

unification: simplification: details are unified by their working principles, themes or programs; better to perceive intrinsic principle is to simplify or unify an understanding; progressive unification of forces is the grail of physics: Clerk Maxwell unified electricity and magnetism; electroweak or *GSW* theory brought in the weak nuclear force; now the goal is to include the strong nuclear force (*GUT*), gravity in a super-force and show that, in essence, particles and forces are interchangeable (super-symmetry and *TOE*); Natural Dialectic, also working with the maxim 'All is One', includes what sums to a hierarchical *TOP* or Theory of Potential (*see* especially Chapters 5, 6, 7, 9, 16 and 19, also *fig.* 5.1); potential is the absolute from which variant orders of relativity derive; the equivalent of *TOP* is *CUT* (**Cosmic Unification Theory); Natural Dialectic is a vehicle of *CUT*, whose aim is to build cosmic connections, that is, orderly linkages towards a Holy Grail of Unification**; the Great Connector, that is, Unifier is consciousness; the subjective potential for mind is consciousness and the objective potential for matter is archetypal memory; such archetypal element unites psychology with the physics of natural science; it is the informative precondition of physical and biological form.

universal mind: cosmic grade; also called the 'mind of nature' or 'natural mind'; as a biological body is a specific though complex arrangement of universal chemicals so individual mind partakes of a particular, equally minuscule fraction of the metaphysical components of universal mind; *see* also archetype.

V

vector: existential dynamic; a vector has both direction and magnitude; it illustrates direction of travel with respect to a model or a secondary stack used in Natural Dialectic; fundamental vectors (↑ and ↓) denote relative gain or loss of information or energy; and, similarly, motion towards and from the axis, peak or source of a cosmic model; in this case, magnitude is inferred to occur on a scale between any pair of opposites, for example, the relative proportions of black and white in the grey-scale between these opposites; use of the word is general, metaphorical rather than specific; opposite members of a stack may involve metaphysical factors (e.g. love/ hate, beauty/ ugliness) as well as physical; thus its spectra do not necessarily concern physical motion or mathematical calculation; its 'field of relativity' extends beyond non-conscious elements; in this respect a Dialectical vector is similar in principle but not the same in practice as that defined by physics or biology.

virtuality: exotic component of quantum physics; para-physical feature of the quantum vacuum; immaterial substrate of material phenomena; inner (where solidity's the outer) edge of physical reality; ephemeral 'virtual particles' rise and sink back into a 'void' thought to teem with their 'fluctuations'; virtuality is identified as the agent of such important actualities as the strong nuclear force (resulting from interaction between virtual mesons and gluons), vacuum polarisation, the Coulomb force (between electric charges and mediated by the exchange flight of virtual

	photons) and so on; not used in the computer sense of a continuum between real and imaginary circumstance; see also *ZPE*.
Vitruvian man:	where art meets science Leonardo's 'universal man' demonstrates an architectural symmetry, excellence of composition and, microcosm unto macrocosm, a reflection of the universe; Da Vinci's connection, from his notebooks, is quite the opposite of Darwin's doodle (Chapter 5); if, with ratios and rationality, it demonstrates mathematical perfection then does not design of larger cosmos demonstrate it too? Natural Dialectic certainly concurs with Leonardo's logical submission.

Z

zero:	zero (the number) is a metaphysical entity, one critical to mathematics; zero (the fact) means, for Natural Dialectic, nothing in two senses; in the *negative sense* it means an absence of perception (psychological oblivion) or absolutely nothing physical (as naturalistically prescribed to precede, say, a big bang or as the nature of a theoretically perfect vacuum); negative sense may also be construed as (*tam*) an extreme sub-state, sink or emptiness; for materialism 'absolute nothingness' may involve natural law and its mathematical description; what, one may enquire, is the source of such 'eternal metaphysic', what is the nature zero-physical?: on the other hand, in a *positive sense* zero refers to source, pre-existent potential or (*sat*) higher cause-in-principle; for example, information (which is zero-physical) transcends/ precedes a course of action; information that passively governs the operation of cosmos derives from immaterial archetype.
ZPE:	zero-point energy; quantum vacuum; vacuum energy of all fields in space; residual energy of all oscillators at 0°K; concept first developed by Albert Einstein and Otto Stern; intrinsic energy of vacuum; the ground-state minimum that any quantum mechanical system, in particular the vacuum, can have; remainder, according to the uncertainty principle, when all particles and thermal radiation have been extracted from a volume of space; residual non-thermal radiation; irreducible 'background noise'; 'quantum foam'; the potent, microscopic side of quantum vacuum (as opposed to impotent, macroscopic vacuum left by the apparent lack of anything); subliminal 'rumblings' of immaterial weak, strong and electromagnetic fields (called *ZPF*s); seething, jostling ferment of subliminal waves and particles in emptiness; a flux of unobservable 'virtual' matter and anti-matter that may or may not appear as the basis of observable forces such as electromagnetism, charge and perhaps inertial mass and gravity; a subtle facet of levity; the anti-gravity of dark energy (or the cosmological constant) has been postulated as a component of *ZPE*; suggested 'mother-field' support for electron orbits, atomic structure and thus the phenomenal universe.
zygote:	fertilised egg.

SAS Icon

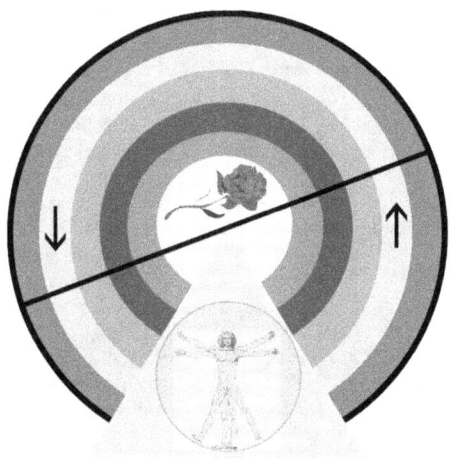

**What is your place in cosmos?
How does the system work?**

This icon (see also in colour on back cover) symbolizes central elements of Natural Dialectic. The concentric rainbow, running from white source out to black, peripheral sink, represents a conscio-material gradient of creation. From quiet, inner hub arise bands of informative mind and, outwardly, energetic matter.

The three cosmic fundamentals (balance, up and down) are represented by a pivot and the arms of scale.

In turn the pivot represents a triangle. Triangle (pyramid, ziggurat or mountain) is a second major model of creation. Within this triangle of truth is encapsulated, microcosm of the macrocosm, a human being such as you and me.

What might a rose upon transluscent
source convey?

For more information:
**www.scienceandphilosophy.co.uk
www.cosmicconnections.co.uk
www.scienceandthesoul.co.uk
www.michaelpitmanbooks.co.uk**

Index

A

abductive reasoning   *see* logic
abiogenesis *see* also Law of Biogenesis and Chapters 20-21 *passim*
     biochemical puddle 309, 314, 328, 385, 399
     chemical evolution   81, 290, 309, 311, 313, 339, 351, 414
     chemical evolution of specific code 323
     experiments   313, 324, 336
     life-by-matter   81, 91, 312, *see* also innovation/ imaginary law
     materialistic faith   324, 337
     scientific dogma   26, 336
     spontaneous generation   309
     vital moment   91, 351
absolute bio-logical
     biological minimum   334
absolute physical/ apparent absolute
     (non-)apparent space-time immateriality   41, *see* space-time
     0K (zero degrees Kelvin) 64, 136, 159, 188
     apparently innumerable continuity of space/ time/ space-time 158, 162, 167, 197, *see* also space/ time/ space-time/ (physical) nothingness
     archetype   *see* archetype/ potential matter/ first cause physical
     big bang   *see* big bang/ transcendent projection
     black hole *see* black hole/ subtendence
     boundless cosmos *see* cosmos/ infinity
     death   *see* death/ survival
     energy   *see* energy/ motion/ change
     infinity/ indefinity   *see* infinity
     physical non-existence/ metaphysic   134
     physical trinity   154, 156, 173
     Planck size/ virtual absolution   167
     potential matter *see* potential/ archetype
     present absence/ basic physic/ nothingness of space & time   *see* space/ time/ space-time/ nothingness/ void
     scientific cut-off point   *see* glass ceiling
     source of physical form   *see* archetype/ potential matter
     subtendence   *see* subtendence/ negative power
     velocity of light (c)   65, 136, 166, 167
Absolute Psychological   *see* also First Cause/ Potential/ Consciousness
     (N)One/ The Infinite   130, 133, 134, 154, 164
     Cosmic Source of Relativity   36, 64, 417
     Essence as *opp.***/ beyond relative existence**   44, 49, 51, 134
     Morality   73, 74, 75, 439
     Most Important Single 'Thing' 228
     Natural Centrality   283
     Subjectivity/ Objectivity   95
     Top Viewpoint/ Value/ Criterion   45, 46, 70, 71, 74
     Truth/ Reality   69, 70, 71
act of creativity   *opp.* act of perception; mind > matter/ materialisation of idea *see* also innovation/ creator
act of perception   *opp.* act of creation/ dematerialisation/ matter > mind *see* also understanding/ knowledge
active complexity *see* purposive complexity
active information   *see* information
adaptive potential   308, *see* also epigenetics
     beak sub-routine   345
     coloration   373
     editorial   317, 366, 408
     epigenetic versatility   88, 369, 373
     resilient bio-logic   373, 389
     transposon?   375
Adleman Leonard   304
adult form   *see* biological development/ reproduction
agnosticism   *see* faith
AI....   *see* intelligence/ computation
Ain Sof   66, *see* also Essence/ (N)One
alchemy
     chemical evolution   309, 337, 339
     information from matter   *see* creation stories/ innovation/ matter/ miracle/ Law of Non-Conscious/ Naturalistic Non-Innovation
Algazel (Al-Ghazzali)   48
allele   *see* gene/ Glossary
Alpha
     Moment   74, 95, 164, 165, 221, 381, 438, *see* Consciousness
     Source   155, 175, 180, 186, 191
alphabet *see* also code/ language/ cosmic language/ genetic grammar/ quantum agency
     cosmic elemental forces and particles   81, 100, 177
     genetic 5-letter   301, 319, 321
     protein 20-letter   319, 321
aminergic system   *see* neuro-hormonal *system*
amino acid   *see* protein
ammonite   354

amorality/ choiceless behaviour of objects
see morality
amygdala 208, 234
analogy *see* homology/ convergent evolution

anathema
    archetypal 305, 344, 354
    Christian/ reincarnation 282
    materialistic/ life-after-death 282
    secular 25, 117, 291
    teleological 124

anatomy
    passive/ structural *opp*. active/ functional physiology 297
    passively informed/ codified/ bio-cybernetic/ reflexive 209
    wired/ physical 239, 273, 297, 381
    wireless 239, 257, 260, 273

animistic language/ animism 301, 339, 410, 412
ante-natal psychology *see* reincarnation/ life/ death
anthropic principles 181

anticipation *see* **teleology/ reason**
    matter can't anticipate *see* code/ mind/ matter/ sex/ bio-logical development

anti-entropy *see* negentropy
anti-gravity *see* levity
anti-order/ disorder *see* informative entropy

anti-parallels
    (*raj/ tam*) cosmic fundamentals 29, 36, 201, 396
    cosmic/ dialectical stack vectors 29, 36, 378, 380, 396
    counter-intuitive proposals (dependent on perspective)
    from *bottom-up*/ physical
        cosmic hierarchy *see* conscio-material dipole/ cosmic models/ hierarchy
        immateriality *see* information/ metaphysic/ psychology, *see* information/ metaphysic/ psychology
        mind-over matter *see* mind
        quantum aspects *see* quantum aspects/ matter-in-principle/ archetype
    from *top-down/* metaphysical
        conscious experience < sub-atomic particles and four forces *see* consciousness/ information/ mind/ neuro-scientific reductionism/ reality
        life by chance/ evolutionary theory *see* chance/ randomness/ evolutionary theory/ intelligence/

information materialism *see* materialism/ atheism/ faith
mind = matter *see* neuro-scientific reductionism
dialectical opposites/ complementary antagonists 379, 385
directions of focus 21, 68, 224, 380, 418
DNA strands/ double helix 378, 386, *see DNA* molecule
informative/ energetic dipoles 21, 50, 52, 54, 59, 268, 306
interpretations
of biology 290, 306, 309, 377
of community 417
of physics 109, 128
of psychology 202, 215
inversion 36, 52, 53, 104, 137, 377, 378, 428, *see also* inversion/ reflective asymmetry
knowledge inward/ outward action 200, 201, 226, 268, 380, 418
negentropy/ entropy 49
sensory/ motor system 207
world-views *see* world-view/ faith
apoptosis/ programmed cell death *see* death
appearance *see* illusion/ reality
Aquifex aeolicus 333
archetype
    > cosmic grammar 93, 98, 101, 157
    anathema 305, 344
    archetypal memory 99, 118, 119, 129, 132, 135, 153, 154, 155, 157, 160, 198, 207, 233, 235, 237, 240, 244, 249, 252, 266, 272, 274, 294, 302, 379, 424
    archetypal program 157, 236, 249, 250, 265, 272, 305, 306, 308, 352, 353, 367, 386, 392, 398
    archetypal projection *see* creativity/ reason/ creation
    archetypal software 240, 246, 267
    archetype = potential matter *see* potential matter
bio-definitions *see* bio-classification
bio-energetic 328
bio-logical 249, 262, 274, 275, 305, 306, 308, 350, *see also* typical mnemone/ morphogene/ attractor/ psychosomasis
discrete template/ theme/ type/ invariant principle 107, 108, 109, 113, 137, 160, 260, 267, 306, 344, 350, 355
evolution of? 253
harmonic 177, 196, 259, 260, *see also*

harmonic oscillation/ vibration/ resonance/ music
hidden invariant 91, 109, 116, *see* also information
hologram/ holographic transmission 116, 117, 119, 250
immanence/ potential/ latency *see* potential
immediately transcends physical cosmos *see* first cause physical/ super-state/ transcendence
informative potential 266, 269, 305, 306
informative source/ metaphysical egg/ cosmic reason 108, 119, 132, 267
intrinsic/ space-time embedded quality 118
macro-boundary 345
mathematical abstraction 120, 157
memory *see* archetypal memory
metaphysical control 83, 246, 250
natural reason/ cause/ law 58, 95, 119, 136, 157
not constrained by physical law 114
physical no-time 166
pre-physical form 144, 149, 154
program 101
psycho-biological 239, 240, 246, 249, 251, 254, 257, 260, 262, 269, 271, 275, 306, 331, *see* also bio-logic/ potential/ polarity
psychological/ metaphysical entity 117, 138, 156, 267, 301, 353
subconscious form/ file in universal mind 235, 269, 272
typical mnemone *see* mnemone/ bio-logic/ potential
wireless transmitter/ receiver 267
Archetype 381
   Archetypal No-Time 167
   Cosmic First Cause 132, 435
   First Cause Psychological 50, 132, 160
   Logical 50, 93, 101, 128, 135, 177, *see Logos*
   Primal Motion 93, 132, 135, 204, *see* also *Logos*
   psychological *see Logos*
   Transcendent Projection 213
architecture of sub-consciousness *see* mnemone
Aristotle 47, 158, 246
Ars Moriendi *see* Book of the Dead
as if *see* metaphor
ASC *see* consciousness
asex *see* reproduction
Aspect Alain 116
association *see* resonant association/ unity/ community

astronomical bodies 167, 181, 194, 420
atheism (religious materialism)
   faith in lifelessness/ matter 192
atheism/ informal canon *see* faith
atmosphere 312, 313, 421
atom *see* also sub-atomic particle
   atoms-to-man? 24, 25, 84, 87, 181, 193, 200, 290, 309, 386, 443
   atoms-understanding-atoms? 59, 76, 115, 121, 136, 183, 197, 202, 204, 429
   collapse of wave function 116
   electron/ extra-atomic influence 141, *see* sub-atomic particle
   enduring structure 146, 159, 167, 178, 194, 279
   harmonic oscillator 142, 167, 259, 260, *see* also resonance/ music
   intra-atomic space 69, 146, 151
   non-conscious/ oblivious 15, 69, 262, 410, *see* oblivion/ non-consciousness
   nucleus/ intra-atomic influence 141, 146, 170
   perpetual motion 70, 146, 177, 180
   Planck's view 127
   quark-proton/ intra-atomic influence *see* sub-atomic particle
   symbol of physical order/ chord/ word 100, 137, 170, 177
ATP 204, 302, 328, 329, *see* also energy metabolism
attention *see* focus
attractor
   attractor = archetype 50
   flexible psychological > desire 222, 306, *see* also psycho-logic/ volitio-attractive force
   inflexible > first cause physical 117, 119, 129, 306
   inflexible/ bio-logical 266, *see* also archetype/ bio-logic
   instinct 251
   magnetic influence 254, 306
   morphological attractor/ morphogene morphogene/ typical mnemone
   pull-to-future/ behavioural constraint 49, 117, 306
   sub-conscious record 50, 242, 254, *see* archetypal memory/ potential matter/ mnemone
Attractor
   pull-to-Super-Consciousness *see* Archetype/ *Logos*
automaton *see* nature/ matter/ machine
AVB/ Authorised Version of Biology 197, 290, 299, 312, 338, 414, 416
AVP/ Authorised Version of Physics 105,

132, 140, 180, 290
AVS/ Authorised Version of Science   197, 290
Axe Douglas   361
axiom 15, 16, 22, 59, 68, 125, 200, 216, 399
Axis   *see* Central Essence
axis/ centre-line 59, *see* source/ pivot/ point of balance

B

bacterium
  archaeo-bacterium............ 331, 333
  attribute 218, 243, 251, 313, 319, 328, 333, 336, 345, 386, 388, 404, 408, 424
  E. coli   334, 344, 351, 363, 364
balance   *see also* neutrality/ symmetry/ equilibrium/ homeostasis/ health
  balanced imbalance/ conservation law 118, 168, 177, *see also* equilibration/ law
  dynamic equilibrium/ cyclical tendency/ cosmic homeostasis 88, 136
  equal-and-opposite effects cancel/ equation satisfied/ resolution 118, 120, 121
  harmonic counter-point 107, 138, 159
  imbalance > motion/ change 148, 159
  inertial equilibrium/ slack/ impotence   54
  informative equilibrium/ balanced orchestration/ homeostasis 30, 31, 35, 36, 38, 109, 120, 121, 166, 168, 213, 223, 225, 273, 275, 293, 302, 422
  karmic account/ natural equation/ cosmic governance 121, 168, 286, *see also* law/ equilibration
  peace/ resolution   51, *see also* equanimity
  pivotal   28, 31, 34, 35, *see also* axis/ centre-point/ zero-point/ dot
  post-dynamic impotence   138
  pre-dynamic potential/ poise 30, 33, 44, 51, 119, 138, 147, 204, 221, 424
  psychological well-being/ basis of mental health 213, 222, 273, *see also* homeostasis
  Sat cosmic fundamental/ quality/ axis/ hub/ core   34, 35, 120, *see also* essence/ centre/ source
  social equilibrium   72, 433, 436
Balance
  Cosmic Axis/ Hub/ Essence   165
Bani   *see* Shabda
base   *see* genetic linguistic hierarchy
Bathybius haeckelii   334
Behe Michael   320, 363
being   *see* essence

Bell John   115
Best Criterion   69, 71, 74, 75, 138, 205, 227, *see also* Truth/ Reality
Bible   276
big bang   *see also* transcendent projection/ miracle/ zero-point
  Big Accident/ Big Miracle   122, 187
  first cause physical?   91, 187, 188, 190
  illogically, from absolutely nothing 41, 113, 126, 186, 190, *see also* zero-point
  inflation   41, 185, 192
  naked singularity turned inside-out/ 'white hole'   184, 187, 192
  no-bang   126
  supreme initial order   *see* fine-tuning
binary opposition   *see* polarity
bio-classification   *see also* homology/ tree of life/ archetype/ species
  archetypal discontinuity 344, 345, 350, 354, 355, 407
  archetypal/ typological level 242, 248, 250, 266, 267, 269, 306, 342, 348, 349, 350, 351, 352, 364, 389, 392, 413, *see also* plasticity limited archetype
  ≈ genus/ family level   266, 351
  bio-logical definitions   351
  discrete template   250, 305, 346, 355
  macro-boundary   343, 344, 351, 355
  natural unit of bio-logical program 239, 248, 250, 254, 265, 266, 272, 351
    cladistics   346
    conscio-material mode   272, 294, 296, 298
    evolutionary continuity/ phylogeny   309, 344, 345, 346, 413
    holistic/ dialectical classification   237, 238, 248, 271
    Linnaeus Carl   343
    natural unit of bio-logical program 306
    standard hierarchy   343, 345, 352
    typology   349
  X. archetypalis 242, 248, 263, 269, *see also* caduceus/ archetype
Biogenetic Law (Pasteur)   *see* Law of Biogenesis
bio-illogical/ accidental   262
bio-logic   104, 405, 407, *see also DNA*/ genetic hierarchy & grammar/ information/ code/ circuitry/ computation/ archetype/ teleology/ design
  bio-logical explosion/ code-generating big-bang   311, 338, 353, 405
  low-level, local variation within high-level, general invariance *see* variation/

principle
modular flexibility   see modular programming/ code/ computation
purpose-packer 85, 213, 317, 372, 405, see also program/ anticipation/ teleology
reason-in-biology 275, 305, 396, 405, 407
reflective asymmetry   212, see also inversion/ reflective asymmetry

bio-logical development
adult form/ end-frame   see reproduction/ egg
anticipatory procedure/ codified algorithm > goal 263, 352, 399, 406, 412, see also anticipation
archetype > physical expression 209, 242, 255, 267, 306, 307, 410
codified/ teleological as *opp*.
uncodified/ aimless evolution 190, 212, 262, 371, 396, 397, 399, 403
conceptual/ immaterial/ metaphysical > factual/ material/ physical   255
evo-devo   408
evolution of?   403, 406
fundamental order of   54
hierarchical procedure 307, 403, 404, 405, 406, 407, 409, see also computation
hierarchically informed control 307, 407
Hox genes   409
informant program > informed body 104, 247, 262, 269, 306, 307, 390
informative potential bio-logically expressed 207, 267, 397, 404, 407
informative seed > adult form 104, 198, 247, 254, 262, 267, 293, 379, 390, 396, 397
informative/ brain-to-body inversion 212, 273, 382, 383
inversion/ inside information outwardly expressed 53, 54, 295, 378, 382, 403, 406
molecular/ metabolic pathway 247, 321, 327
morphogene   242, 255
morphogenesis   see morphogenesis/ evo-devo
program egg > adult   411
stereo-tree   402
symbolic code > physical expression 307, 398
the equivocation   399
top-down procedure   404

bio-logical language see alphabet/ genetic grammar & linguistic hierarchy/ code
biomimetics   253, 335, 350, see also Glossary
bio-rationality/ core bio-rationality see code/ language/ bio-logic/ informative potential/ genetic grammar & linguistic hierarchy/ super-code
bio-reason   see bio-logic
biosynthesis
intelligent bio-recreation   335
Birney Ewan   366, 367
Bischof Marc   256
bit/ binary digit   77

black box
artificial/ non-evolutionary   311
biological   311, 322, 412
evolutionary 204, 229, 230, 233, 234, 236, 246, 252, 253, 262, 311, 322, 332, 337, 362, 388, 399, 406, 412, see also black box biological
material   157, 191, 193
metaphysical/ psychological 229, 230, 234, 236, 246, 247, 252, 253, 305

black hole   see also apparent absolutes/ extreme subtendence/ death/ pain/sin/ immorality/ negative power
material oblivion   see non-consciousness
negative extreme   54
physical   64, 167, 184, 194, 420
psychological   see sleep/ sub-consciousness/ negative power

black smoker   314
Blyth Edward   86, 341, 414
Bohm David   115, 116, 160
Bohr Niels   45, 68, 71, 114, 116
book of life   see *DNA*/ genetic linguistic hierarchy
Book of the Dead   282
Born Max   115
Bose Satyendra   127
bottom-up   21

brain   see also neuro-hormonal *sys*tem
a brief architectural scan 205, 208, 268, 381, 383
cerebral hemispheres 213, 268, 381, 382
climactic ingenuity of construction 209
corpus callosum   381
dashboard for body's vehicle 206, 209, 211
evolution of?   209
faculties constraining mind ⟺ body interaction   211
information modulator   260, 298
mind = brain   see consciousness/ materialism/ neuro-hormonal system/ reality
mind-body decussation 212, 273, 382, see also switch

neural oscillation/ brainwaves 220, 225, 233, 279
neuro-imaging 202, 211, 225
no-life-in-it 213, 218, 367
psycho-somatic mediator 203, 213
brane 196
breasts 398
Buddha 69, 74, 95, 149
Buddhahood *see* Consciousness
Buddhist Sutras 276
Burgess (including Marble Canyon) shales *see* fossil
Burr Harold 255, 256, 267
butterfly 411

C

C. elegans 376, 397
caduceus *see* also archetype/ mnemone/ conscio-material dipole/ man
  archetypal memory 269
  conscio-material gradient/ *H. archetypalis* 271, 272, 296
  double helix 269, 377
  human archetype/ *H. archetypalis* 263, 269, 271, 273
  logical/ conceptual human program 248, 269
  psycho-somatic mediator *see* psychosomasis
  sub-conscious informant
  symbol of medical profession 269
  typical mnemone 269
  vis medicatrix naturae 269
Cambrian explosion *see* fossil/ bio-logic
Cantor Georg 133, 134
Carsonella rudii 335
Causal Information
  Logos 139
cause *see* also archetype/ reason
  archetypal causation *see* first cause physical
  can cause itself < boot-strap logic 48
  can't cause itself 47, 48, 61, 168, 401
  causal mind *see* (higher) Archetype/ *Logos*/ (lower) archetype/ mnemone
  cause = reason 90
  cause inwardly-to-outwardly devolves effect 301
  cause-and-effect *see* balance/ mathematics/ *karma*
  conceptual-teleologic pull <> energetic push 49
  First Cause Psychological *see* First Cause
  horizontal causation 56, 95
  informative template > energetic behaviour 59, 206, 252, 301
  lack of *see* chance/ randomness

quantum causation/ matter-in-principle > matter-in-practice 38, 57, 61, 81, 83, 100, 132, 137, 141, 154, 160, 177, 178, 250, 260, 407
  sufficient/ principle of causality 48
  teleological code > bio-logical effect 306, 406
  'transcends' effect 170, 203, 206, 401, 406
  vertical causation 56, 95
Cause *see* First Cause
cell *see* also molecular/ cellular machinery
  abiogenetic impossibility 123, 275, 311, 320, 328, *see* also abiogenesis
  basic (electromagnetic) unit of life 63, 198, 254, 255
  bio-electrics 160, 209, 239, 255, 256, 260, 267, 268
  cell machinery *see* molecular/ cellular machinery
  cellular language *see DNA* molecule/ language/ code
  chemical evolution 85, 192
  codified chemistry 81, 262, 294, 304, 315, 328, 336
  collection of very specific chemicals 15, 192, 198, 247, 315
  complex biotechnology 294, 336
  conscio-material gradient *see* hierarchy
  cybernetic/ algorithmic operation 423
  dormant/ sub-conscious mind or typical mnemone in every cell 240, 242, 243, 247, 249, 250
  information incarnate 198, 262, 304
  information primary/ material arrangement secondary 300
  intra- and inter-cellular information systems 294, 297, 298, 307, 391, 395, 397, 408, 435
  mind machine 305
  minimal genome/ functionality 333, 335
  no biopolymer outside cell 315
  organelle *see* molecular/ cellular machinery
  psycho-biological aspect 240, 242, 268, 295
  psychosomatic aerial 248, 249, 252, 255, 256, 260, 267, 394
  soft musical machine 63, 260
  spontaneous generation 339
  theoretical immortality 285
centaur *see* man
Central Dialectic *see* Primary Dialectic
centralisation
  negative/ de-centralising tendency > (*tam*) extremity/ blackness/ inc. ignorance/ oblivion *see* negative

power/ lifelessness/ mass
positive/ centralising tendency > (*raj*)
extremity/ radiance/ inc. understanding
    see positive power/ light/ life/
illumination
Centrality   *see* Essence/ Balance/ Source
centre/ centrality 35, *see* source/ pivot/ axis/ point of balance/ centralisation/ zero-point
cerebral hemisphere left   384
cerebral hemisphere right   384
cerebral inversion   384
CERN (European Organisation for Nuclear Research)   105, 171, 185
chance   *see also* randomness/ probability/ order
  absence of defined cause   122
  accident/ coincidence/ contingency   22, 77, 110, 262, 291
  chance-by-design/ lottery   198, 389
  code = anti-chance 96, 98, 99, 301, 305
  HUP (quantum uncertainty principle)   113, 114, 115, 116
  Lady Luck 95, 106, 109, 111, 117, 123, 124, 192, 196, 299, 309, 336, 392
  luck/ unexpected, unpredictable event   85, 110, 112, 121, 299
  mind = chance-reducer/ anti-chance   85, 99, 123, 125, 173, 325
  mindless/ passive/ unintelligent 'creator'   73, 76, 84, 87, 95, 104, 113, 119, 123, 149, 261, 291, 392
  Penrose computation > cosmos-not-by-chance   193
  unreasonable/ incalculable/ untestable   275, 291
change   *see* motion/ space/ time/ variation
chaos   *see also* chance/ randomness/ dysfunctional logic
  boundless space   157
  chaos theory   326
  pattern-less confusion   74, 111, 113, 222, 413
chemical evolution   *see* abiogenesis
  mechanism of abiogenesis   *see* chance/ randomness
  promissory faith   *see* faith
chemistry   55, 100, 102, 103, 137, 145, 146, 178
chemoautotroph   331
chiaroscuro   *see* conscio-material dipole
chicken-and-egg   102, 312, 315, 316, 317, 320, 322, 323, 327, 328, 331, 400, 406, *see also* black box
chirality   *see* mirror image/ reflective asymmetry/ bilateral symmetry
Chladni Ernst   257
Chlamydomonas   390
chloroplast   *see* molecular/ cellular machinery
chloroquine resistance   364
cholinergic system   *see* neuro-hormonal system
Christ   74, 95, 221, 282
Christ Pancrator   *see* Logos
chromatin   *see* DNA molecule/ protein/ molecular machinery
chromosome   *see* molecular/ cellular machinery/ genetic linguistic hierarchy
  maintenance of chromosomal integrity   388
circuitry
  computer chip/ integrated circuit   103, 241, 247, 304, 307, 329, 370, 375, 399, 404, 408, 410, *see* computation/ code/ mnemone/ DNA molecular operations
  cybernetic/ bio-systematic   79, 88, 166, 211, 293, 302, 305, 327, 407, 423, *see also* homeostasis
  electronic   72, 178, 240, 258, 340, 409
  genetic circuitry   240, 304, 307, 338, 369, 376, 404, 405, *see also* DNA/ computation
  nervous   209, 234, 417
cladism/ cladistics   *see* bio-classification/ homology
classification   *see* bio-classification
climax/ ladder up   196, 288, 379, 399, 406
CMBR   *see* cosmic background radiation
code   *see also* information/ computation/ purpose/ DNA molecule/ alphabet/ bio- & cosmo-logical languages
  anticipates   102, 211, 247, 301, 316, 359, 371, 375, 386, 398, 399, 406, 407
  bug   *see* mutation/ chance/ randomness/ entropy
  code = anti-chance   323, 325, *see* probability/ chance/ randomness/ order
  cosmic   *see* cosmic language
  cryptanalysis   366
  evolution of genetic code?   320, 358
  genetic/ bio-logical   86, 207, 212, 250, 254, 255, 262, 293, 301, 304, *see also* bio-logic/ alphabet/ genetic grammar & linguistic hierarchies
  hierarchical   *see* super-code/ hierarchy
  idea registry 84, 100, 244, 249, *see also* memory/ DNA/ computation/ language/ archetype
  informative potential   100, 198, 235, 240, 247, 298, 304, 398
  is metaphysical 81, 248, 249, 255, 263, 323, 415
  language < mind   *see* information/

innovation/ informative negentropy/ source
languages bio- and cosmo-logical *see* alphabet/ quantum aspects/ cosmological language/ vibration/ genetic grammar
meaningful/ grammatical structure 26, 79, 98, 99, 177, 262, 299, 300, 302, 315, 316, 321, 332
moral/ social/ legal 432
program/ teleological design/ algorithm 63, 79, 85, 98, 99, 103, 124, 128, 177, 208, 224, 226, 236, 241, 301, 304, 316, 318, 327, 349, 375, 398, 399, 407, 409, 423, *see* code/ bio-logic/ nucleic acid operations/ archetype
signal organiser 98, 177
source of coded information *see* mind
specific/ purposeful information 100, 275, 299, 301, 318, 356, 367
splicing code *see* spliceosome/ molecular & cellular machinery
super-code *see* epigenetics
symbolic representation *see* alphabet/ language
top-down programming > main/ master routine 108, 405
top-down programming > sub-routine 249, 251, 263, 345, 350
codon *see* genetic linguistic hierarchy
commissure 384
Communion *see* Consciousness/ Unity
communism 430
community *see* also unity/ relativity/ anti-parallels
external
ecological 419, 424, *see* ecology
religious (faith outwardly expressed) 418, 432, 435
society as *opp*. individual 218, 417
socio-political association 418, 433, 436
internal
association absolute 418, 435, 438, *see* Communion/ Consciousness/ Psychological Absolute
association relative 245, 418, *see* mind/ experience/ psycho-logic/ faith
complementary opposites *see* polarity/ reflective asymmetry/ dialectic stacks
complexity non-purposive 95, 326, *see* also nature/ law/ archetype/ passive information
complexity purposive 96, 275, *see* also order/ innovation/ creativity/ active information
computation *see* also code/ switch/ machine/ informative reflex/ archetype

bio-operating system 88, 245, 299, 304, 307, 321, 327, 335, 365, 409, *see* also *DNA* molecule/ genetic circuitry/ genetic grammar & linguistic hierarchy/ archetype
hierarchical control of expression 323, 324, 356, 367, 368, 370, 376, 386, 390, 398, 404, 407, 409
self-editorial capacity 304, 317, 407
bug-free 400
computer = mind-machine 79, 103, 240, 280, 304, 437
modular programming 307, 408
stereo-computation 402, 403, 404
top-down programming > main/ master routine 108, 304, 306, 353, 398, 405, 409
top-down programming > sub-routine 108, 128, 240, 244, 304, 306, 353, 371, 387, 398, 409
Turing machine 304
concentration *see* also focus/ source
biological/ concentrated information/ code/ high negentropy/ order 110, 293, 321
gradient upward/ sink > source *see* conscio-material gradient/ dematerialisation/ expression of (*raj*)
cosmic fundamental physical/ energetic-massive 33, 35
Concentration
Pure Concentrate of Consciousness 35, 46, 67, 69, 75, 93, 217, 221, *see* Consciousness
concentric/ radiant rings *see* cosmic models
conceptual development
act of creation 94
anticipatory procedure/ codified algorithm > goal 99, 102, 171, 247, 338, 381, 395, 399
archetype > physical expression 171, 247, 306, 404
conceptual approach to biology 292, 294, 306, 406, 407, *see* also bio-logic/ archetype
fundamental order of 81, 84, 95, 96, 171, 379, *see* also hierarchy
immaterial/ conceptual/ metaphysical > material/ factual/ physical 104, 159, 171, 177, 209, 211, 242, 306, 371, 407
inversion/ inside information outwardly expressed 53, 96, 104, 208
teleological exercise 254, 262, 295, 395, 399, 403, 411
condensation reaction 315
conscio-material (c-m) dipole *see* also Primary Axiom of Natural Dialectic

and Mount Universe/ cosmos
  anti-parallel vectors   36, 60, 380
  basic existential dipole   52, 56, 58
  caduceus/ human reflection of gradient   *see* caduceus/ mnemone/ hierarchy
  c-e conscio-energetic spectrum 29, 32, 56, 66, 237, 294
  cell   *see* also mnemone/ hierarchy
  c-m gradient/ spectrum/ slope/ scale/ chiaroscuro   30, 32, 36, 51, 56, 91, 131, 176, 177, 206, 210, 263, 271, 380
  cosmos/ creation as dipole   20, 22, 28, 35, 46, 51, 52, 53, 61, 69, 380, 440
  informative/ immaterial > energetic/ material coupling   52, 56, 58, 59, 80, 96, 186, 200, 204, 206, 210, 218, 263, 271, 272, 275, 294, 380
  informative/ subjective/ psychological > energetic/ objective/ physical coordinates 51, 52, 56, 200, 204, 206, 272
  sliding scale/ proportional representation   32, 59
  source-to-sink   22, 29, 32, 33, 51, 71, 176, 228, 263

consciousness   *see* also metaphysic/ active information/ active mind; 'con-scio' means, in Latin, 'I know together'
  altered states of (*ASC*)   96, 217
  basis of awareness/ experience 16, 34, 45, 57, 69, 91, 147, 174, 203, 204, 206, *see* also knowledge
  evolution of your *sine qua non*?   204
  human focal point   *see* eye-centre/ cosmo-logical axis/ focus/ concentration
  illusion suffering an illusion 24, 69, 203
  immaterial element/ metaphysical entity 15, 16, 22, 54, 66, 84, 126, 200
  informative potential   *see* information/ mind/ conceptual development
  intelligence   *see* intelligence
  materialistic view/ illusion suffering an illusion/ neuro-scientific reductionism   15, 23, 24, 68, 69, 84, 110, 126, 200, 202, 229, see also illusion/ reality/ matter/ neuro-hormonal system
  mind = ever-changing/ formful flux 96, 271, 273
  mind as prism/ filter of circumstance   148
  mind/ body exchange   *see* **psychosomasis**
  nervous excrescence   24, 121, 200, see also illusion/ reality

  qualities/ levels   60, 94, 211, 219
  seamless perception/ unitary coordinator   18, 127
  subjective dimension   *see* subjectivity
Consciousness   282, *see* also Essence/ Absolute
  Communion 66, 93, 95, 138, 228, 278, 381, 418, 419, 436
  Cosmic Axis   218
  Dharmakaya/ Nirvana/ Samadhi/ Moksha/ Radiant Communion/ Transcendent Bliss   70, 75, 93, 205, 221, 419
  Essence/ Absolute Truth   69, 91, 93, 137
  Experience of Pure Consciousness 66, 75, 93
  First Cause Psychological   135
  Highest Value/ Ultimate Criterion 72, 75, 205
  Illumination/ Enlightenment   45, 46, 48, 72, 74, 75, 174, 218, 221, 228, 278, 419
  informative potential/ latent mind 34, 46, 69, 91, 130, 137, 217, *see* also source/ potential/ information
  Natural Substance   93, 204
  Primary Component of Creation   127, 200, 204
  Psychological Principal/ Supreme Being/ Quintessence   204, 218
  Pure Concentrate of Consciousness/ Life/ Soul 46, 54, 66, 69, 82, 93, 130, 202, 204, 217, 221
  Source of Informative Mind   42, 103, 138, 204, 218
  Superintendent Pole opp. base-pole's oblivion   221
  Super-State 63, 82, 83, 204, 205, 217, 221, 228, 283, 439
  Uncreated/ Uncaused   48, 204, 442
consciousness-in-motion   *see* mind
conservation
  archetypal information   351, 388, 404, 406, *see* archetype/ information
  by natural selection   *see* Blythe/ Darwin/ natural selection
  ecological   424
  natural law   118, 177, *see* also invariance/ symmetry/ law/ equilibration/ archetype
  of bio-modular sub-routines   213, 308, 350, 353, 404, 409, *see* also homology/ code/ archetype
  typological (of wild-type) 267, 388, *see* also archetype/ mnemone/ bio-logic
constant   *see* physical constants
contemplation   *see* also meditation

inward concentration of attention 19, 20, 67, 69, 222, 225, 439
   metaphysical connection 132, 225
   opp. sensation 219
convergent evolution 88, 307, 308, 347, 348, 349, 350, *see* also Glossary/ computation/ code/ modular programming/ adaptive potential/ homology/ mosaic distribution
coral 422
corpus callosum 384
cosmic background radiation 185, 187
Cosmic Centre *see* Essence/ Supreme Being
cosmic egg *see* first cause physical
Cosmic Egg *see* First Cause/ Metaphysical Egg/ Archetype
cosmic fundamental
   fundamental order of creation 30
cosmic fundamentals *see* also dialectical operator
   biological expression 294, 295, 297, 383, 387, 404
   dialectical operator 34
   fundamental order of creation 27, 32, 34, 38, 58, 60, 80, 83, 84, 94, 129, 133, 154, 176, 200, 219, 229, 271, 273, 295, 298, 432, see also conscio-material dipole/ cosmic models/ hierarchy/ source > sink
   physical expression/ link to physics 32, 34, 117, 170, 294, *see* archetype/ potential (matter)/ quantum aspects/ matter
   psychological expression/ link to psychology 34, 94, 117, 210, 219
   raj upward/ active/ levitational vector 34
   Sat pivotal/ neutral/ equilibrator 34
   social expression 432
   tam downward/ passive/ gravitational vector 34
   trinity/ triplex nature 34, *see* also hierarchy/ Natural Dialectic
cosmic mind *see* universal mind/ mind/ archetype/ implicit universe
cosmic models
   (clockwork) machine 72, 102, 103, 113, 127, 259
   concentric rings/ vibrations/ waves 32, 33
   conscio-material (c-m) gradient 33, 51, 206, 163
   Cosmic Pyramid 33
   four-square pyramid/ ziggurat 29, 33, 57, 59, 60, 201, 283
   hologram 119
   Master Model/ Central Metaphor/

Music 101, 257, 259
mind-machine/ computer 103, 127
Mount Universe 33, 34, 57, 83, 130, 137, 151, 436
organic/ 'instinctive' cosmos 101
scale/ balance 30, 31, 36, 51
scientific *see AVP/ SMPP/ SMC*
spectrum 29, 32, 33, 51, 56, 237
cosmo-logic 94, 109, 118, 124
   fundamental order of creation *see* also order/ cosmic fundamentals/ conscio-material dipole
   fundamental order of physics 34, 132, 169, 171, *see* also nature/ order/ cosmic fundamentals/ trinity
   fundamental order of psychology 34, *see* also cosmic fundamentals/ hierarchy/ trinity
   inversion/ inside information outwardly expressed 53, 96, 109, 129, 378, *see* also conscio-material dipole/ anti-parallels/ order/ conceptual development
   low-level, local variation within high-level, general invariance *see* variation/ principle
cosmo-logical axis 66, 67, 75, 205, 224, 239, 418, 437, *see* also cosmos/ eye-centre/ point X
Cosmo-Logical Axis (Universal) 67, 72, *see* also Pivot/ Zero-Point/ Central Essence
cosmological constant 184, *see* also levity/ (dark) energy
cosmo-logical language/ score *see* also alphabet/ language/ quantum agency/ *Logos*
   atom/ chord/ word 100, 170
   forces/ textual (re-)arrangement 170
   grammar 38, *see* also alphabet/ cosmic language/ law/ Natural Dialectic of Polarity
   language of physics & chemistry 93, 98, 137, 170
   meaningful/ grammatical structure 81
   molecule/ phrase 170
   particle notes > atomic chords > molecular phrases 81, 100, 170, 177
   sub-atomic particle/ note/ letter 170, *see* also quantum aspects
cosmos *see* also creation/ conscio-material dipole/ universe/ polarity/ cosmic fundamentals
   archetypal pattern 48, 109, 118, 124, 135
   can't create itself 48, 186
   cosmic dipole *see* conscio-material dipole

cosmic program/ variation-on-archetypal-theme   see cosmic language/ quantum aspects/ fine-tuning
cosmogony/ transcendent projection 32, 110, 188, see also big bang/ source/ transcendence
cosmological extravaganza   183, 196
cosmos < vibratory sustenance   see vibration/ energy/ perpetual motion/ harmonic oscillataion/ appearance
dynamic/ vibratory text 66, 69, 99, 101, 127, 148, 177, 199, 257
grammar   38, see alphabet/ cosmo-logical language/ law/ Natural Dialectic of Polarity
hierarchical   22, 57, 59, 81, 129, 154, 263, see also hierarchy/ conscio-material dipole
information-carrier   58, 59, see also informative potential/ nature/ matter/ archetype/ law
macrocosm/ cosmos   30, 75, 129, 135, 177, 263, see also creation/ existence
microcosm/ man   30, 201, 213, 263, 272, see also man
orderly/ non-chaotic   27, 56, 83, 107, 108, 121, 122, 379, 442, see order/ law/ cosmic fundamentals
perpetually imbalanced balance/ perfect imbalance   see motion/ balance/ equilibration
source 22, 64, 95, 156, see also source/ transcendence/ creativity/ informative potential
trinity/ triplex nature 28, 34, 38, 53, 57, 59, 60, 130, 200, see also cosmic fundamentals/ hierarchy/ Natural Dialectic
universal body 121, 137, 168, 170, 180
counter-intuitive proposals see anti-parallels
C-paradox   372
creation/ act and fact   see also conscio-material dipole/ fundamental order of creation/ cosmos/ innovation/ creativity
alpha point   see source/ first cause/ archetype/ energy/ information
bio-logical/ archetypal projection   305, 413
cosmo-logical projection   see cosmo-logic/ cosmo-logical language
counter-creation/ destruction   94
created object/ form/ event   27, 35, 42, 51, 100, 102, see also sink
creation stories 24, 89, 109, 122, 171, 196, 197, 311, 337, 340, 415
creative source see also information/ energy/ cause/ source/ mind

evolutionary mechanism 309, 363, 413
ex nihilo   112, 122, 143, 188, 193, see also nothingness/ potential
expression of *tam* vector/ materialisation   44, 62, 137, see also impotence/ energy
field of relativity   29, 31, 37, 116, 196
hierarchical order of materialisation 27, 29, 33, 54, 57, 60, 84, 98, 128, 133, 169, 228, 263, 267, 292, 308, 379, 381, 392, 395, 400, see also order/ cosmic fundamentals/ conscio-material (c-m) dipole
inversion 53, see symmetry/ reflective asymmetry
macrocosmic/ universal   see also transcendent projection/ archetypal expression/ conscio-material dipole/ big bang
microcosmic/ individual instance
mindful   35, 95, 96, 207, 380, 386, see active information/ innovation/ creativity/ psycho-logic
mindless   156, 157, see passive information/ reflex cause/ nature/ non-consciousness
object of psychological or physical expression   67, 133, 177
opposite vector/ perception/ comprehension/ knowledge   94
orderly constraint/ reduction of freedom   116, 148
perpetual motion/ change   see motion/ energy
quantum see quantum aspects/ potential/ archetype/ cosmo-logical language
time-span impotent/ intelligence potent   173
creativity   see innovation/ intelligence
active information   see also information/ mind
hierarchical order of   80, 83, 91, 95, 96
idea/ source of invention 21, 34, 104, 222, 269, 306
matter is creatively impotent   80, 311, 335, 358
metaphysical/ super-natural/ immaterial/ psychological causation   78, 87, 120
subjective (re-)arrangement of information   35, 78, see also force psychological/ teleology
time is creatively impotent   25, 85, 107, 173, 311, 335, 336, 372
creator   see also causal information/ force psychological/ teleology
Creator immaterial/ not external   177
informative source   71, 121, see also

creativity/ source/ Source
inventor/ designer   25, 85, 89, 93, 104,
   see also innovation/ creativity/
   purpose/ teleology
   Logos/ Sound and Light   25, 66, 93
Crick Francis   126, 369
CUT (Cosmic Unification Theory)   143
cyclical equilibrium   *see* motion/ vibration/
   homeostasis/ *karma*/ metaphysical
   equilibrium/ reincarnation
cyclical time   *see* time - cycle
cymatics   257

D

dark energy   *see* energy
dark matter   *see* matter
darkness
   opp. light   *see* light/ negative power/
      evil/ unconsciousness
Darwin Annie   425
Darwin Charles   85, 117, 309, 338, 341,
   343, 344, 354, 363, 413, 414, 416, 425,
   429, 430
Darwin Erasmus   416
Darwinian puddle   309, *see also* abiogenesis
data item   100, 123
dating methods   172, *see also* time/ fossil/
   geology
DC time   169
de Broglie Louis   115
death   *see also* oblivion/ survival/ Chapter
   18 *passim*
   codified self-sacrifice/ apoptosis   303
   type-1 view of death   279
   type-2 view of death 279, 281, 286, 288
decentralisation   *see* diffusion/ entropy/
   centralisation
deep-sea volcano   *see* black smoker
delusion   *see also* illusion/ Glossary:
   illusion
   'only my way is right'   26
   code-specifying matter   103, 110, 307,
      321, 336, 338
   dogma assumed reality   26, 439
   irreducible/ codified complexity by
      accident?   332, 334, 372
   law of non-conscious innovation   91,
      110, 111, 325, 338, 413
   machine-making matter   102
   material anticipation or conceptual
      teleology   307, 323, 327, 338, 386,
      413, *see* also anticipation/ matter/
      teleology
   metaphor assumes reality   26, 144,
      196, 281, *see* also creation stories
   molecular self-animation 309, 312, 337
   scientific and religious   23
   special delusions   *see* reality

dematerialisation/ dissolution
   (*raj*) upward/ levitational vector (↑)82
   expression of (*raj* ↑) upward/
      levitational vector   36, 65, 82, 228,
      418
   inward/ raising process   82, 418
   psychological/ sense > symbol/ matter
      > mind   19, 380, *see* focus/ sensation
Denton Michael   337, 348, 352
depolarisation (two > one)   *see* neutrality/
   equilibrium/ unity/ love
design   mind-behind/ plan *opp.* chance/
   mindlessness *see* force psycho-
   logical/ cause/ reason/ metaphysic/
   teleology
destiny   283, 289, *see* also fate/ *karma*
determinism   113, 115, 148, 149
developmental potential   *see* informative
   potential/ bio-logical & conceptual
   developments/ idea/ archetype/ code/
   (metaphysical) egg
devolution *opp.* dissolution   44
dharma   *see* also *karma*
   natural law   168
dialectical operator   *see* also cosmic
   fundamental
   raj/ yang upward/ active/ levitational 34
   Sat/ Tao pivotal/ neutral/ equilibrator
      34
   tam/ yin downward/ passive/
      gravitational   34
dialectical stack   37, 38, 51, 378, 380, 385
diatom   354
differentiation   *see* variation/ (dis-)unity
diffusion   *see* entropy
   gradient downward/ source > sink   *see*
      conscio-material gradient/
      materialisation/ expression of (*tam*)
      cosmic fundamental
   opp. concentration/ centrifugal/ >
      peripheral sink   442
dinosaur   172
Dirac Paul   145, 147
discontinuity   *see* limited plasticity
dissipative structure   326
division   *see* duality/ negative power
Dixon Malcolm   323
DNA **molecule**   *see* non-protein-coding
   accidental authorship?   359
   as microchip/ rigid memory   98, 250,
      321, 373, *see* also circuitry
   chemical information carrier/ paper-
      and-ink   85, 98, 100, 283, 293, 318,
      321, 385, 397
   chromatin control   368, 369, 407
   coil   395, *see* also mitosis/ meiosis/
      epigenetics
   computation   *see* code/ computation

482

    core bio-rationality?   252
    database/ genome/ information store
        85, 98, 107, 250, 252, 254, 283, 301, 304, 321, 365, 370, 388, 407
    storage density   321
        deoxyribose sugar   319, 321
        double helix/ anti-parallels   319, 320, 350, 378, 380, 385, 395
        genetic engineering   375
        handedness   316
        hereditary material   285, 386
        hierarchical control   *see* computation
        informative improbability   318
        junk   365
        linguistic constituents   *see* genetic grammar/ genetic linguistic hierarchy
        linkage (3-5)   319
        local physical expression of general metaphysical archetype   81, 207, 249, 254, 262, 266, 293, 296, 342, 369
        major operations   *see also* molecular/ cellular machinery
    protein synthesis   322
    replication   322, 323, 388
        metabolic informant   327
        mitochondrial/ plastid *DNA*   329
        non-protein-coding 347, 365, 366, 367, 370, 407, *see also* super-code/ epigenetics/ *RNA*/ switch
        not self-replicating   59, 322, 323, 331, 369
        nuclear business   295, 322, 385, 408
        nucleotide/ base lack of chemical affinity/ neutrality   320, *see also* genetic linguistic hierarchy
        only found in cells   319, 358
        optimal chemical mechanism   385
        phosphate   319, 321, 369
        physical information storage   304, 350
        proportion of 'junk'   88, 359, 361, 366, 367, 372
        secret of life?   221
        solenoid/ aerial?   249, 256, 260, 395
        specific informative complexity   301, 318, 319, 335, 366
        super-code   *see* epigenetics
        symbolic bio-code/ vital text   85, 98, 249, 262, 293, 301, 304, 356, 381, 397, 408, *see also* language/ code/ gene/ genetic grammar/ genetic linguistic hierarchy
Dobzhansky Theodosius   293
dog   343, 345
dogma
    naturalistic & religious   *see* delusion/ reality/ knowledge
    dormant mind   *see* sub-consciousness

dot   29, 35, 167, 169, 190, *see also* source/ centre
double helix   *see DNA* molecule/ anti-parallels
dream   *see* sub-consciousness
duality   *see* polarity/ reflective asymmetry/ dialectic stacks
duck-billed platypus   348
dysfunctional logic
    all is matter (materialism = holism)   22, 73, 85, 88, 106, 121, 124, 246
    can randomness specify teleologically?   73, 84, 88, 124, 247, 323, 355, 359, 368
    chance substituted for design   87, 110, 209, 358, 405, 412
    for materialism = holism   22, 87, 121
    informative code from matter   209, 323, 324, 408
    irrationality 84, 85, 106, 121, 129, 323, 358, 443

E

E. coli   *see* bacterium
Eccles John   209
ecology
    abiotic component/ earth, water, air, space   419
    biosphere/ ecosphere/ ecosystem   418, 419, 423
    biotic part/ community of organisms   424
    CHZ - continuously habitable zone   420
    conservation   424
    ecological cycles   422
    Gaia/ eco-physiological homeostasis   421, 423, 424
Eddington Arthur   120
editorial capacity/ *DNA* and protein quality control/ adaptive potential   *see* genetic grammar/ epigenetics
education   *see* psychological development
egg   *see also* First Cause/ cosmic egg/ biological development
    bio-logical   110, 190, 207, 264, 267, 268, 371, 389, 398
    condensed-information-pack   395
    evolution of?   396
    metaphysical egg   *see* potential/ metaphysic/ first cause/ idea
    outworking from within/ inversion   177, 379
    physical idea   399
    potential/ capability   177, 250, 371
    source as *opp.* adult sink   207, 399
    source/ initiator   207, 264, 379, 399, 404, 408

ego
   inflamed > sin   431, 434
   intellectual executive   222
   my way/ me-me   90, 228, 429, 431
   self-awareness   148
Einstein Albert 27, 117, 127, 140, 141, 148, 187
electrical charge 18, 53, 118, 145, 147, 160, 178, 237, 239, 255, 256, 332, *see* sub-atomic particle
electromagnetic radiation   *see* light
electromagnetism   *see* force physical
electron   *see* sub-atomic particle/ polarity physical
embryological homology   348
embryonic polarity   267
Empedocles   51
ENCODE project   366
energy   *see* also cosmos/ conscio-material dipole/ motion/ polarity existential/ matter > existence   70
   archetypal behaviour *see* law/ potential/ archetype
   dark energy/ lambda force   105, 143, 182, 185, *see* also levity
   energetic coordinate/ secondary component of existence   29, 51, 55, 104, 204
   energetic input/ stimulus   *see* negentropy
   energetic order   *see* fundamental order of creation/ vibration
   energetic trinity   *see* cosmic fundamentals/ physic's primal trinity
   energy metabolism   *see* metabolism/ cell/ molecular-cellular machinery
   exhaustion   *see* sink/ impotence/ entropy
   kinesis   *see* motion
   material equivalence   *see* matter
   matter-in-principle   *see* matter-in-principle/ cosmo-logical language/ score/ quantum aspects
   non-conscious/ objective   *see* objectivity/ non-consciousness/ matter/ teleology
   not = information   55
   passive   *see* automaton/ passive information
   potential   *see* potential
   protean basis of physics   44, 55, 156, 173
   reflex informant/ 'inside' mass   *see* harmonic oscillation/ vibration/ resonance/ Chladni/ matter
   vectored   *see* cosmic fundamentals/ vector/ polarity/ anti-parallels
   whence came/ comes energy?   22, 41, 184, 186, 187, 191, *see* also source/ motion/ vibration/ harmonic oscillation/ *Logos*
engram   234
enlightenment *see* illumination/ knowledge/ understanding
Enlightenment   *see* Illumination/ Communion/ Consciousness
entropy *see* also energy/ information/ sink/ sub-state/ gravity/ Glossary
   archetypal immunity/ no impact on natural law   193
   energetic
2nd law of thermodynamics   77, 168, 169, 262, 325, 379
exhaust from system/ run-down/ drag 49, 226
expression of physically dominant (*tam* ↓) vector   49, 169, 262, 325
fall > sub-state sink/ gravitational vector of recession   169, 232, 262, 326
open systems/ biological bodies   325
   informative
biological evolution would need negentropic information gain   379
fall to sub-conscious sink   232
genetic/ chromosomal mutation   *see* mutation/ randomness/ information
increasing noise/ loss of meaning   78
informative loss/ > ignorance/ > unconsciousness   *see* chaos/ randomness/ diffusion/ (de-)centralisation
randomising/ diffusive tendency   49, 168, 169, 255, 325, 359, 362, 365, 379
steady-state theory   180
enzyme   *see* metabolism/ protein
epigenetics/ epigeny   *see* also adaptive potential/ switch/ hierarchy/ genetic regulation
   cytoplasmic/ cortical inheritance   368, 371
   epigenome   370
   fifth base *see* genetic grammar/ genetic linguistic hierarchy
   histone
acetylation/ methylation code   368
positioning code   368
tail   369
   super-code   368, 369, 370, 372, 373, 374
   zygote code   368, 371
equanimity   *see* peace/ balance/ pivot
equilibration   *see* balance/ *karma*/ equilibrium
equilibrium
   equilibrium dynamic

bio-stability   *see* homeostasis
chemical   315, 423
cyclic action/ wave/ vibration   36, 136, 146, 168, 226, 262, 283, 293, 325, 379, 423, 433, 436
    social   433, 436
    equilibrium inertial   *see* also sink/ exhaustion
field of no possibility/ exhausted end-product   36, 54, 232
    equilibrium potential   *see* source/ poise/ (*sat*) cosmic fundamental
field of possibilities   33, 35
point of balance   277
Equilibrium/ Equanimity   *see* Peace
essence
    lesser essence/ relative or existential being   *see* existence/ individuality/ (dis-)unity
Essence
    (N)One   56, 133, 154, 429
    Archetype   132, 271
    Complete Certainty   93
    Cosmic Pivot/ Centre/ Axis   31, 104
    Cosmic Potential   35, 39, 44, 45, 48, 130, 134, 149, 151
    Essential Equation   134
    opp. existence   38, 51, 52, 418
    paradoxically includes existence 37, 38, 44, 45, 46
    Source of Mind   91, 133
    Supreme Being   30, 37, 93, 133
    Truth Absolute   67, 69
    Ultimate Reality   67, 137
Essential Dialectic   *see* Primary Dialectic
Essential Equation   134
eternal matter *see* matter/ steady state theory
Euclid spacecraft   185
European Union   89
evil   227, 427, *see* also pain/ hell/ immorality/ de-centralisation
evo-devo   *see* biological development, *see* biological development
evolution
    meaning of the word   *see* Glossary
evolutionary theory *see* also Theory of No Intelligence
    chemical evolution   *see* abiogenesis
    convergent evolution *see* homology/ analogy/ convergence
    Darwinian/ chance and natural selection   341, 347, 379, 430
    error of category   111
    evolution of natural law?   110
    evolutionary continuity   *see* unlimited plasticity/ bio-classification
    gambler's game/ mindless/ unplanned development   301, 362, 399, 415, *see* also chance/ randomness/ irrationality
    implication of 'improvement'/ non-conscious innovation/ progress   *see* innovation/ Law of Non-Conscious/ Naturalistic Non-Innovation/ matter/ passive information
    materialism's *sine qua non*   15, 24, 300
    materialism's version of life's truth 300
    naturalistic inference/ faith   87, 345, *see* also faith
    neo-Darwinian/ synthetic   253, 293, 307, 309, 326, 399
    post-synthetic   375, 408
    pre-Darwinian   341
    progressive evolution of the theory of evolution   346
    progressive evolution of theory of evolution   352
    promissory faith   *see* faith
    punctuated equilibrium   *see* fossil
    replacement principles   406
    theistic evolution   *see* faith
ex nihilose nothingness/ potential/ creation
excarnation/ disembodiment   277, 282, *see* also incarnation
exclusion principle   146
exhaustion   *see* also energy/ impotence/ nothingness/ sink
    expression of (*tam*) cosmic fundamental   33
    opp. potential   33
existence   *see* also Glossary/ polarity/ motion/ conscio-material dipole/ relative illusions
    appearance < motion   70
    basic coordinates/ coefficients   54, 55, 69
    becoming   165
    compound of mind and matter   44, 48, 54, 55, 174, 290
    conscio-material field/ psycho-physical nature   51, 69, 91, 273
    creation-is-motion   37, 134, 177
    di-polar 28, 32, 38, 51, 52, 56, 174, 290
    Essence-caused/ Essence-dependent/ Essence-in-action 37, 39, 48, 62, 134, 151
    existence < vibratory sustenance   *see* vibration/ energy/ motion/ harmonic oscillataion/ appearance
    existential equation   134
    hierarchical order   130, 249, *see* hierarchy/ order/ archetype/ cosmo-logic
    informative/ energetic coupling   174
    metaphysical non-existence/ physical presence   69
    opp. Essence/ non-Essence/ non-

essential pole 37, 44, 52
perpetual motion/ perfect imbalance/ *samsara* 134, 177, *see* also motion/ cosmos/ creation
physical non-existence/ metaphysical abstraction 69
projection/ relative illusions 69
standing out and apart 52
variation-on-archetypal-motions *see* motion/ energy/ archetype/ cosmos/ appearance/ cosmo-logical language vectored 28, 30
zone of relativity 37, 48, 64
existential dialectic *see* secondary dialectic
expanding universe 187
experience
  anti-experience/ material oblivion 123, 149
  basis = consciousness 204, 206
  brain mediates physical experience 203, 211, 213, 225, *see* also mind/ brain/ neuro-hormonal system
  complete certainty = being/ lesser essence = 'I am here' 91
  graded/ sliding-scale/ hierarchical priorities 71, 72, 165
  living being 26, 93, 174, 202, 203, 211, 278
  made of molecules-in-nerves? 24, 59, 72, 121, 202, 209, 215
  out-of-the-body *see* OBE
  sole/ subjective form of knowledge 18, 20, 35, 45, 69, 91, 204, 226, 242, 418
  sole/ subjective form of understanding 20, 35, 74, 91, 128, 183, 222, 276
Experience *see* Supreme Being/ Consciousness
experiment
  metaphysical *see* thought/ contemplation/ meditation
  physical *see* naturalistic methodology
explicit universe 117
exteriorisation *see* polarity - active/ motor tendency
  entropic/ motor 225
external association 432
extinction *see* survival
eye-centre 20, 67, 75, 205, 208, 224, 239, 245, 273, 418, *see* also cosmological axis/ third eye/ concentrative focal point

F

faith *see* also world-view
  agnostic 14, 415
  animistic 339, *410*, 412
  atheistic 22, 25, 84, 124, 187, 196, 300, 361, 415, 416
  deistic 416
  faith in the unseen 150
  holistic 19, 20, 282, 416
  in miracles *see* holism/ materialism
  materialistic/ naturalistic 87, 172, 192, 197, 261, 285, 307, 311, 358, 414
  promissory 68, 143, 371, 387, 416
  religious 14, 20
  sceptic/ faith in doubt 125
  scientism 24, 71, 73, 110, 197, 202, 279, 334, 337, 415
  secular 74, 124, 261, 311, 358, 414, 430, 443
  theistic 22, 25, 416
  theistic evolution 416
  two pillars of 20, 26, 72, 76
family 433
Faraday Michael 158, 178, 255
fascism/ national socialism 430
fate 283, 289, *see* also destiny/ *karma*
female polarity *see* sex/ polarity/ reproduction
fertility 390, *see* also survival
Feynman Richard 125, 159
Fibonacci series 385
field theory 247
fine-tuning
  biological 307, 338, 350, 360, 364, 365, 368, 404
  computational 103
  cosmic 109, 190, 194, 197
  cosmic music/ soundless harmony 109, *see* also vibration/ harmonic oscillation/ cosmic models
  forces and particles articulate *see* cosmo-logical language/ resonance
  Goldilocks enigma 109
  six numbers 183
first cause physical
  archetypal *see* potential matter
  bio-logical 305, *see* creation/ archetypal projection
  harmonically informed energy 101, 177
  physical 48, 91, 110, 128, 132, 135, 138, 140, 153, 154, 176, 177, *see* also archetype/ potential matter / big-bang/ transcendent projection/ Glossary
  psycho-biological 246, 248, 262, 305
First Cause Psychological
  Archetypal *see* Transcendent Projection/ Logical Archetype/ *Logos*
  Cosmic 22, 48, 66, 71, 101, 135
  Cosmic Egg (is metaphysical) 35, 109, 154, 156, 177, 192

cosmos not self-caused 58
Dialectical 38, *see* also Potential
Natural Scientific Cause 48
Primal Motion 47, 48, 50, 135, 152, *see* also *Logos*/ Glossary
Psycho-logical 128, 132, 152, 154, 176, 200, *see* also *Logos*/ Logical Archetype
Uncaused 48
First Principle 72, 153
Fisher Ronald 360
flat-universe perspective 129
focus *see* also information/ concentration/ centralisation/ meditation
    anti-parallel directions 19, 21, 224
    concentration of attention 19, 20, 32, 69, 96, 97, 165, 204, 208, 225, 283, 356
    concentration of information 293, 321, *see* also principle/ informative potential/ law
    energetic 33
    informative/ point of attention *see* eye-centre/ interest
force
    physical
    cosmic operators 49, 52, 53, 63, 100, 106, 137, 140, 141, 142, 158, 162, 176, 177, 178, 184, 209, 247, 254, 258, 326, 420, *see* also vector/ cosmos/ cosmic language/ cosmic fundamentals
        whence? *see* also energy/ motion
    never codifies 110, 371, 379, 405
    whence? 107, 110, 144, 193
    psychological *see* psycho-logical volitio-attraction/ will-power/ desire/ reason/ ego/ cause
fossil
    abrupt appearance 353
    Burgess and Maotienshan shales 353
    Cambrian information-revolution 338, 351, 353, 405, 410
    chronology 172
    comparative analysis 353
    local variation within general invariance 353
    missing link 88, 347, 353, 354, 358, 405
    palaeontology's trade secret 355
    Precambrian stasis 405
    punctuated equilibrium 340, 355
    record 347, 351, 352, 353, 354, 405, 414
    systematic gaps/ discontinuity 353, *see* also *tree of life/ homology/plasticity limited/ typology*

free will 148, 149, 282, *see* also freedom/ will-power
freedom
    < freedom = expression of (*tam*) cosmic fundamental 148
    > freedom = expression of (*raj*) cosmic fundamental 148
    free will and determinism 148, 282
    conditioned free will/ degrees of choice 148, 222
    relative constraint 116, 148, *see* also probability
frequency *see* wave/ vibration/ energy
Fresnel Augustin 158
frozen time *see* sub-consciousness
fruit fly 344, 351, 376
functional complexity *see* purposive complexity
functional logic 41, 102, 103, 261, 262, 273, 334
fundamental order of creation *see* cosmic fundamentals/ cosmo-logic/ conscio-material (c-m) dipole/ nature/ order

G

Galapagos finch 344
Galileo Galilei 140
Galton Francis 429
Gamov George 187
Gates Bill 366
gecko 335
gene *see* also genetic code/ protein/ switch/ genetic linguistic hierarchy
    allele 341
    by chance? 318, 322, 323, 357, 361, 362
    dominant/ wild-type 266, 344, 345, 389
    duplication 340
    gene duplication 361, 375
    gene tree discordance 346, *see* also tree of life
    genetic buffer 376, *see* also epigeny/ super-code
    genetic mutation *see* informative entropy/ mutation
    genetic operating system *see* computation/ bio-operating system
    geno-centric myth 267
    jumping gene/ retrotransposon 305
    jumping gene/ transposon 305, 373, 375
    micro-hierarchy/ genetic regulation/ expression 267, 304, 307, 320, 366, 368, 369, 370, 372, 375, 404, 405, 409, *see* also computation/ circuitry/ epigenetics/ order/ switch/ hierarchical control

natural genetic engineering 340, 375, 414
non-protein-coding (n-p-c) *see* super-code/ epigenetics/ *DNA* molecule
   pool 341
   pseudogene 366
   recessive 266
   switch *see* switch, genetic and epigenetic
genetic code *see* code/ genetic grammar & linguistic hierarchy/ epigenetics/ bio-logical language
genetic drift 359
genetic grammar *see* also *DNA* molecule/ alphabet/ linguistic hierarchy/ code
   base pair complementary fit (A-T/ G-C) 320, 333
   binary/ digital code 85, 304, 321
   central dogma of molecular biology 369
   DNA alphabet/ A, G, C, T and U 319
   editorial cohort 408
   epigenetic markers/ punctuation 338, 367
   fifth base punctuation/ methylated cytosine 369
   protein alphabet/ 20 amino acids 319
   syntax 98
genetic linguistic hierarchy *see* also *DNA* molecule/ (bio-logical) language/ code/ genetic grammar/ hierarchy
   chromosome/ chapter 319, 367
   codon/ 3-letter word 318, 321, 322, 360, 367
   gene/ sentence 267, 301, 319, 367, 386
   genome/ encyclopaedia/ book of life 240, 249, 250, 262, 283, 285, 305, 319, 321, 322, 356, 365, 366, 367, 372, 375, 381, 397, 408
   nucleotide/ base/ letter 302, 314, 315, 318, 319, 320, 321, 360, 367
   start codon/ capital letter at start of sentence 319
   stop codon/ full stop 319
genome *see* genetic linguistic hierarchy
geology
   catastrophism 172
   uniformitarianism 172
ghost/ geist/ spirit 174
Giardia 390
glass ceiling 83, 108, 179, 188, 237
Gnostics 282
godless gap 23, 88, 110, 126, 185, 192, 290, 326, 334, 337, 338, 355, 362, 415, *see* also innovation/ evolutionary black box/ missing link
God-of-the-gaps 416
golden mean 181, 385
Goldschmidt Richard 376

Gould Stephen 355
governing template *see* archetype
Grant Peter and Rosemary 345
Granth Sahib 276
graptolite 355
gravity/ anti-levity *see* also polarity/ cosmic fundamentals/ anti-parallels/ force physical/ materialisation/ Glossary
   dialectical sense
cosmic influence of (*tam*) descendent vector *see* also cosmic fundamentals/ negative power
dark/ physically dominant force 36, 185, 217
decentralisation/ externalisation > inc. non-consciousness/ heaviness/ darkness 178, see also subtendence/ sub-state/ non-consciousness
decentralisation/ externalisation > increasing darkness/ heaviness/ non-consciousness 137
drag/ drain/ resistance/ sink/ inertial influence 36, 137, 154, 178, *see* sink/ matter/ inertial equilibrium
energetic loss > exhaustion/ finished state *see* entropy
gain in concentration > focus/ strengthening/ informative and/or energetic potential *see* also centralisation/ concentration/ unity/ order/ focus/ potential
Higgs mechanism *see* also mass
loss of concentration > dispersal/ weakening *see* diffusion/ variation/ randomness
strong nuclear force *see* also atom/ sub-atomic particle
subjective - informative loss > noise/ nonsense/ non-conscious state *see* informative entropy/ psycho-logic/ randomness/ oblivion
subjective > inc. suffering/ igorance/ unhappiness/ pain 82, 222, 226, 227, 282, 427, *see* also morality/ evil
tendency to compress/ contract/ bind/ aggregate 82, 137, 161, 178, 185, 192, 222, *see* also materialisation
   scientific (and dialectical) sense
Einsteinian space-time warper 140, 192
mutual attraction of physical bodies/ gravitation/ gravitation 49, 52, 141, 184
Newtonian action-at-a-distance 140, 142
quantum gravity 141, 142, 143
GTE (general theory of evolution) *see* macro-evolution/ transformism/

innovation/ type/ archetype
GUE (Grand Unified Experience)   see holy grails
GUT (Grand Unified Theory)   see holy grails

H

H. archetypalis   see also man/ archetype/ first cause physical/ typical mnemone
   caduceus   see caduceus
   human program   248
   memory man   248, 302
   psychosomatic form/ mind-side   249, 257, 263, 268
   subconscious/ psychological side   255, 257, 263
   typical mnemone   268
H. electromagneticus   see also man/ electrical charge
   electrodynamics   239, 249, 255, 257, 263
   non-conscious bioelectrical aspect   239, 254, 256, 266
   psychosomatic form/ body-side   254, 263
   quantum physique/ quantum man   255, see also matter-in-principle
   wireless/ radiant anatomy   238
H. sapiens   see also man/ bio-logic
   classical/ biological form   238, 248, 268, 296
   molecular and visible body   257
   non-conscious expression of bio-logic   254, 268, see also anatomy/ bio-logic/ bio-logical development
   sub-state shell   238
   wired/ physiological anatomy   238
habit   see sub-consciousness and personal mnemone
hadrosaur   see dinosaur
Haeckel Ernst   371, 430
Haldane John B. S.   312, 360
harmonic oscillation 136, 176, 257, 260, see also resonance/ vibration/ music/ atom
Hawking Stephen   126
health
   archetypal/ psychosomatic resonance   246, 255, 267, 274, 275
   ecological   420, 424
   harmony-in-action   101, 226
   homeostatic stability   223, 224, 303
   pre-set normality   226, 275
   psychological   228
   vis medicatrix naturae   246, 267, 269, 275
   wild-type   389
heaven   see positive power/ levity/ super-state

heaven-on-earth/ utopia   427, 431, see also utopia
Heisenberg Werner   113, 115
helix   see DNA molecule/ caduceus
hell   see negative power/ gravity/ conscious sub-state
hereditary law   see Law of Heredity
hermaphrodite   see sex/ archetype/ neutrality/ reproductive polarity
hidden invariant (cancels randomness)   see archetype/ information
hierarchy   see also creation/ cosmos/ cosmic ziggurat/ fundamental order of creation/ *top-down* perspective
   bio-logical hierarchies   291, 294, 301, 302, 303, 306, 407
   computational hierarchies   304, 307, 338, 374, 398, 404, 405, see also computation/ informative hierarchy/ code
   descending reduction of power   171, 274
   dialectical   27, 33, 34, 38, 108, 170, 318, see also fundamental order of creation/ cosmo-logic/ cosmic fundamentals/ Natural Dialectic of Polarity
   hierarchical genetic code   360, 367, 368, 370, 381, 403, 409, 410
   informative > energetic act of creation   63, 80, 83, 84, 94, 108, 169, 251, 255, 272, 273, 274, 304, 318, see also conceptual development/ creativity/ information
   models   see cosmic models
   neuro-hormonal   see neuro-hormonal system
   of truth/ reality   69
   psycho-biological   262, 268, 272, 273, 308, 403, see also caduceus/ morphogene/ archetype/ bio-logical development
   social   431
   symbolic   33, 59, 338, 358, 367
   taxonomic   see bio-classification
   triplex/ 3-tiered
general/ cosmic 22, 31, 38, 42, 56, 57, 59, 80, 130, 137, 169, 378, 442, see also conscio-material dipole
sub-division physical   34, 57, 61, 108, 137, 138, 140, 170, 179, see also matter (in-principle & in-practice)/ potential matter/ archetype
sub-division psychological 34, 60, 66, 94, 218, 219, 228, 229, see also mind/ experience/ subjectivity
Higgs mechanism   144, 179, see also sub-atomic particle/ Higgs boson

hippocampus 208, 234
histone coding   *see* epigenetics/ epigeny
Hitler Adolf 430
HIV 363
holism
   anti-holism/ materialism 23, 442
   both immaterial-material/ psychological-physical/ informative-energetic components of cosmos included 21, 27, 45, 69, 91, 280, 378, 381, *see* also conscio-material (c-m) dipole
   dialectical expression 30, 378, 440, 443, *see* also Natural Dialectic of Polarity/ polarity/ conscio-material (c-m) dipole
   expanded science 278
   miracles
of existence/ cosmos   *see* creation
   non-Darwinian 364
hologram   *see* archetype/ cosmos
holy grails
   biological/ seed of life *see* abiogenesis
   CUT holistic (physical and metaphysical) Cosmic Unification Theory 143
   GUE psychological/ metaphysical - Grand Unified Experience/ Highest Goal/ Holy Grail/ Communion 93, 95, 138, 139, 221
   GUT-TOE physical - Grand Unified Theory or Theory of Everything 140, 141, 143
   psychosomatic/ mind-body connective mechanism   *see* psychosomasis
   super-string theory 141, 143
   theory of quantum gravity 142
homeostasis   *see* also health/ balance
   bio-systematic 166, 226, 269, 293, 302, 303, 327, 342, *see* also circuitry/ bio-logic/ archetype
   dynamic/ developmental norm 303
   negative feedback 224, 303, 433
   norm/ keeping the balance/ cyclical equilibrium 166, 213, 224, 230, 231, 303, 325, 422
   psycho-biological 225, 268
   psychological 213, 224, 225
   social 433, 436
   triplex mechanism/ sensor/ regulator/ effector 213, 225, 268, 292, 303, 400
homology *see* also bio-classification/ tree of life/ computation/ code/ modular programming
   analogy 347
   archetypal routine/ bio-modular programming 306
   central Darwinian tenet 347
   evolutionary interpretation 346, *see* also phylogeny
   homological discordance 346, *see* also tree of life
   local variation within general invariance   *see* variation/ code/ modular programming
   molecular homology 346
   mosaic distribution 308
   typological interpretation 349, 352
hopeful monster 376
hormonal system
   passively informative/ reflex 209
Howell Clark 430
Hox gene cluster 409
Hoyle Fred 180
Hukm   *see Logos*/ Sound and Light
human   *see* man/ *H. sapiens*/ caduceus
Hutton James 172
Huxley Thomas 334
hyper-dimension 143, 144

I

IAC (inheritance of acquired characteristics) 341
Ibn Arabi 158
idea   *see fig.* 6.1 - The Order of an Act of Creation
   dawning/ realisation of possibility   *see* metaphysical egg/ innovation/ creativity/ source
   rationalisation of possibility   *see* reason/ principle/ teleology
   registration of possibility   *see* code/ memory
Idea   *see* First Cause/ Metaphysical Egg/ Top Teleology
ignorance *opp.* knowledge 232
illumination
   knowledge/ grasp of meaning/ wisdom 35, 48, 74, 174, 222, 225, 381, 435
Illumination   *see* Consciousness
illusion *see* also delusion/ reality/ Glossary
   Buddhist view 69
   deception/ error 90
   existential appearances = relative/ lesser realities 69, 71
   geno-centric vision.................... 71
   living matter/ chemicals alive? 325
   macro-evolutionary transformism *see* macro-evolution/ unlimited plasticity
   oblivious design 173, 308, 326, 327, 335, 375, 395
   partial or absent truth 68, 69
   principle of bio-physical evolution 111, 266, 286, 337, 340

immaterial matter   *see* archetype/ potential matter
immateriality *see* metaphysic/ information/ psychological component/ material immateriality
immorality *opp.* morality > darkness   *see* morality
immortality
   eternal matter?   81, 151, 180, 197, 198
   genetic   283, 285, 301
   logical/ mathematical/ metaphysical principles   120
   oblivious/ material   167
   three brands   283
   Transcendent Super-State   165, 221
   Uncaused Cause/ Pure Deathless Life   48
implicit universe   117, *see* also universal mind/ archetype/ potential matter
impotence   *opp.* potential *see* also exhaustion/ subtendence/ sub-state
   biochemicals outside a cell   255
   expression of (*tam*) cosmic fundamental   *see* also materialisation
   final cosmic phase/ non-conscious energetic matter   100, 167, 204
   no more possibility/ exhaustion/ end-point   36, 43, 53, 147
   teleological   *see* chance/ randomness/ Lady Luck
incarnation   *see* also excarnation/ reincarnation
   biological organism 198, 236, 262, 276, 277, 316
   typical mnemone/ archetype involved   289
index fossil   172
individuality   *see* (dis-)unity/ *ego*/ separation
infinity   *see* also absolute physical/ apparent absolute
   apparently innumerable continuities of time/ space   135, 197
   collapse of serial rationality   193
   existential/ apparent infinity   55, 135
   infinitesimal   144
   innumerable units/ indefinity   134, 138, 197, 326
   mathematical/ numerical infinity 134, 326
   physical transcendence/ potential matter/ archetype   135, 137
Infinity   *see* Absolute/ Natural Essence/ Trancendence
   Cosmic Potential   66, 137, 179
   Metaphysical/ Psychological Absolute   69, 135
   Source of Mobile/ Finite Forms   138

term of the Essential Equation   134
inflation   *see* big bang/ levity
Informant Rationale *see* *Logos*/ First (be-)Cause/ teleology
information   *see* also code/ teleology/ mind/ reason/ force psychological active/ passive 78, *see* also information - dialectical sense of mode
bio-information
bio-explosion/ code-generating big-bang   338, 353, 405
bio-information   121, 198, 262, 285, 293, 299, 301, 304, 318, 319, 346, 365, 385, 407, *see* also code/ bio-logic/ archetype/ bio-logical language/ *DNA*/ epigenetics
cell metaphysical/ seen as information   300, 327
information man   *see* memory man/ *H. archetypalis*/ caduceus
informative coordinate/ primary component of biology   *see* code/ bio-logical language/ genetic grammar/ genetic linguistic hierarchy
life-form = information incarnate   110, 262, 290, 293, 300, 401
specific, complex information   290, 319, 332, 335, 338, 367, 374
cosmic element
cosmic pole   32, 46, 51, 55, 69, 127, 263, 395
immaterial/ metaphysical/ psychological   19, 29, 32, 49, 58, 77, 79, 87, 110, 117, 137, 247, 308, 404
informative coordinate/ primary component of existence   55, 77, 91, 104
not = energy/ not from matter   58, 79, *see* also matter/ innovation/ creativity
dialectical sense of balance
'down'   *see* gravity/ entropy/ information loss
'rest'   *see* equilibrium/ centralisation/ understanding
'up'   *see* levity/ negentropy/ information gain
dialectical sense of direction
'inward/ negentropic'   156, 203, 207, 359, 379, *see* also perception/ knowledge/ concentration
'outward/ entropic'   49, 74, 80, 95, 203, 207, 325, 357, 358, 359, 365, 428, *see* also action/ creation/ diffusion
dialectical sense of mode
active/ conscious   55, 58, 72, 78, 80, 96, 97, 110, 129, 249, 280, 294, *see* also mind
active/ informant domain   273
passive/ archetypal medium/ first cause

physical 110, 132, 133, 237, 250, 267, 268, 306, 345, 350, *see* also archetype/ informative potential & psycho-biological exchange
passive/ informed domain 274
passive/ unconscious 55, 58, 63, 78, 80, 94, 97, 98, 100, 129, 235, 294
    focal point *see* also focus
central/ nuclear deployment 177, 247, 294, 395, 435, 438
informative nucleus 67, 239, 247, *see* also third eye/ eye-centre/ cosmological axis/ concentration/ centralisation/ meditation
    hidden variable 91, 404, *see* also archetype
    hierarchy
> order > logic 84, 100, 121, 177, 261, 395, 397, *see* also law/ hierarchy/ fundamental order of creation
conscio-material scale 22, 56, 59, 71, 80, 186, 206, 220, 263, 274
informative hierarchy 59, 83, 94, 140, 206, 208, 219, 268, 291, 301, 302, 305, 308, 338, 362, 365, 367, 368, 374, 383
    information technology 103, 304, 370, 399
    informative density
DNA 321, 338, 354, 355
of principle 129, 140, 169, 249, 374, 437, *see* also law/ principle/ morality
    informative entropy *see* mutation/ noise/ randomness
    informative transmission/ exchange
(bio-)physical 320, 385, 394
command & control/ central processor 19, 34, 77, 121, 200, 206, 227, 242, 251, *see* also mind
psycho-biological 48, 54, 63, 101, 156, 203, 210, 213, 225, 237, 239, 251, 256, 257, 260, 264, 269, 383, 442, *see* also harmonic oscillation/ vibration/ resonance/ psychosomasis
psychological 225
    informative trinity *see* Super-Consciousness/ consciousness/ unconsciousness
    informative/ metaphysical reflex *see* computation/ code/ archetype/ sub-conscious instinct
    material/ scientific view - information as an aspect of automatic energy 77, 183, 229
    neuro-scientific reductionism 91
    neutral archetype expressed in polar form 395
    passive *see* information - dialectical sense of mode
    Shannon information 77, 78
    source > outcome (sink)
causal information/ precedent trigger/ behavioural guide 22, 24, 35, 44, 62, 66, 72, 78, 93, 97, 177, 262, 291, 306, 397, 409, see also active mind/ idea/ ideal/ perception/ psychological force/ innovation/ code/ archetype/ natural law
informative potential/ precedent latency 33, 44, 58, 91, 99, 106, 108, 132, 154, 176, 268, 292, 298, 306, 323, 371, 372, 379, 403, *see* also egg/ anticipation/ potential/ archetype/ program
physical expression/ bodily behaviours *see* physical phenomena/ bio-logical development/ bio-logical forms
physical storage *see DNA* molecule/ database/ other physical storage media
source of information 87, 91, 99, 102, 135, 155, 183, 294, 340, 355, 367, 410, *see* also mind/ archetype/ psychological force/ *Logos*
storage metaphysical 63, 107, 119, 235, 273, 394, *see* also memory/ archetypal memory
symbolic information - use physical & metaphysical - *see* language/ cosmo- & bio-logical languages/ (bio-)information/ alphabet/ code
Information - Cosmic Source of *see* Essence/ First Cause/ Cosmic Egg/ *Logos* etc.
Information Centre *see* Essence/ Consciousness/ Super-State/ Potential
informative potential *see* information
infra-conscious = sub-conscious 66, 218
initial conditions *see* First Cause
Initial Projection *see Logos*
innovation 96, *see* also creativity/ intelligence
    creativity 21, 66, 78, 84, 86, 87, 93, 96, 125, 154, 269, 300, 336, 338, 350, 362, 425, 443
    imaginary law of non-conscious innovation 86, 102, 103, 110, 111, 172, 326, 349, 413, 415, 416
    large-scale 359
    matter can't create code 86, 88, 99, 102, 110, 308, 335, 338, 361, 362, 363, 424
    molecular? 340, 357, 358, 359, 361, 363, 375, 402
    rapid 338
    rational creation 120
instinct

archetypal behaviour 63, 98, 129, 138, 155, 157, 218, 252, 272
behavioural reflex 94, 215, 224, 437
dormant sub-structure of mind 242
genetic basis? 252
hormonal connection/ morphogenic link 126, 225, 255
mnemonic file/ sub-routine 63, 215, 240, 241, 242, 244, 249, 252, 256, 293, 306
natural (bio-)law 72, 75, 90, 100, 208, 252
nervous connection/ signal translation 235
sub-conscious/ mnemonic mechanism 50, 63, 205, 225, 240, 256
sub-rational/ involuntary/ thoughtless 222, 417, 437
in-swing 277
intelligence
    ability to design/ intelligent design 85, 86, 87, 104, 110, 124, *see* also purpose/ teleology/ innovation/ creativity
    AI/ artificial intelligence 103, 280
    constructions stored as memory/ program/ device *see* also program/ memory/ machine
    grasp of principle 33, 44, 58, 86, 125, 227
    increasing intelligence > increasing negentropy/ order/ understanding 358, *see* also negentropy
    intelligent/ purposeful selection or experiment 345
    logically/ rationally adept 125
    message/ relevant data 58
    metaphysical power/ acumen 81, 124, 227
    Natural Intelligence 129
    nowhere but brain? *see* neuro-scientific reductionism/ neuro-hormonal system/ dysfunctional logic/ brain/ reality
    Theory of Intelligence 125, 367, 416
    Theory of No Intelligence 91, 353, 367, 379, 399, 415
    unconscious intelligence = passive information 299, *see* also instinct/ program/ computation/ machine/ creation (created thing)/ physical phenomenon
intention *see* reason/ informative potential
interest/ enthusiasm *see* focus/ love/ centralisation
interiorisation *see* polarity - sensory/ contemplative tendency
    negentropic/ perceptive 224

internal association 438
invariance *see* variation/ conservation/ natural law
inversion *see* creation/ conscio-material dipole/ symmetry - reflective asymmetry/ anti-parallels/ biological development/ switch
irrationality *see* dysfunctional logic/ reason
irreducible complexity = irreducible organisation 88, 102, 104, 109, 303, 318, 323, 326, 327, 328, 332, 334, 400, *see* also machine/ minimal functionality/ purposive complexity/ chicken-and-egg

J

Jacob's ladder *see* double helix/ anti-parallels/ *DNA*
Jeans James 127
jellyfish 285, 307, 349, 390
Jenny Hans 258
Johnson Phillip 89, 123
junk *DNA* *see* also *DNA* molecule/ super-code/ epigenetics/ hierarchy
Justinian 282

K

Kabbalah 276
Kalam-i-Illahi/ Kalam/ Kalima *see* Logos/ Sound and Light
karma *see* also equilibration/ balance
    action/ reaction 31, 49, 121, 168, 283
    action-in-the-cosmic-balance/ equal-and-opposite effect 121, 168
    action-in-the-cosmic-balance/ equal-and-opposite-effect 49
    cosmic resolution 168
    cosmic tendency (psychological and physical) to balance the equation/ equilibrate 121, 168
    dharma 168
    fate (you receive)/ destiny (you make)
    fields of action (psychological and physical) 31
    give-and-take 168
    sanskara/ buried memory 288
Khan
    Genghis 431
Kim Il Sung 430
knowledge *see* also understanding/ illumination
    active/ seeking answers 70, 77, 109, 127, 442
    passive/ rote 26, 77, 125, 222, 282, 336, 344, 353, 439, 442
    understanding explicit physical cause > cleverness 17, 35, 77, 202, 318
    understanding implicit psychological cause > wisdom 35, 48, 55, 74, 276, 435

Koran 276

L

Lack David 345
Lady Luck/ Fortuna/ Capricia *see* chance
Lamarck Jean-Baptiste 341, 370
laminin 270, *see also* caduceus
language *see* also code/ *DNA* molecule/ bio- & cosmo-logical languages
  chemical   *see DNA* molecule
  descriptive/ instructive 98
  genetic/ bio-informative   78, 85, 98, 301
  grammatical/ syntactical structure   98, 262, 300, 304, 319, 367
  symbolic information   26, 78, 79, 85, 88, 98, 262, 300, 301, 304, *see* also symbol/ code
  universal (music and mathematics)   257
  vehicle of communication/ meaning   55, 78, 98, 255, 304, 367, *see* also information
Laplace Pierre 113
law   *see* cosmos/ nature/ order/ archetype
  anthropic principles 181
  balance/ find the norm   142, 168, 282, *see* balance/ equilibration/ homeostasis
  bio-logical   *see* **bio-logic/ archetype/ mnemone**
  cosmo-logical   93, *see also* nature/ cosmo-logic/ cosmos/ *Logos*
  evolution of physical law? 193
  external/ ethical canon & governmental law 72, 74, 93, 122, 437
  internal/ moral code   72, 437, *see* also morality
  invariance and probability 113, *see* also quantum aspects/ variation/ conservation/ symmetry
  latent instruction/ informative potential   58, 83, 107, 117, 125
  machines accede to physical law   97, 102, 335
  metaphysical/ archetypal   77, 120, 129, 249, 259, *see* code/ logic/ mathematics/ grammar
  natural law = eternal 'necessity'   83, 96, 112, 121
  natural law/ natural behaviour/ physical reflex   16, 36, 58, 63, 78, 79, 99, 112, 113, 119, 121, 122, 125, 157, 192, 431, *see* also archetype/ cosmo-logic/ bio-logic/ instinct
  natural psycho-reflex/ instinct 100, 208, 252
  relative moralities/ rules/ ideologies 75, 436
  resonant enforcement   101, 177, 252, 258, *see* also harmonic oscillation/ vibration/ resonance
  rule and regulation   108, 436
  symbolic/ grammatical   98, 157, 177, *see* also cosmo-logical language/ score
Law
  Highest Court   138, 438
  Natural/ Essential/ Absolute   *see* Communion
  Law of Biogenesis   309, 351, 387
  Law of Heredity   350, 387, 390, 399
  Law of Motion   168
  Law of Non-Conscious/ Naturalistic Non-Innovation   86, 110, 311, 326, 413, *see* conversely innovation/ imaginary law
  le Chatelier's principle   *see* equilibrium chemical
Leggett Anthony 116
Lenin Vladimir 430
Lenski Richard 364
lesser truth/ reality   *see* appearance/ truth/ reality/ conscio-material grading from Reality
levity/ anti-gravity   *see also* polarity/ cosmic fundamentals/ anti-parallels/ force physical/ dematerialisation/ Glossary
  dialectical sense
  centralisation/ internalisation > increasing lightness/ consciousness 185
  cosmic influence of (*raj* ↑) ascendant vector   36, 160, *see* also cosmic fundamentals/ positive power
  lift/ dynamic influence   66, 74, 160
  light/ heat/ electrical charge   156, *see* also light/ energy/ electrical charge/ negentropy
  light/ psychologically dominant force   217, 438
  subjective - informative gain > accurate message/ sense/ conscious state   *see* informative negentropy/ psycho-logic/ order/ teleology
  subjective - spiritual ascent   221, 228, *see* also concentration/ centralisation/ Transcendence
  subjective > inc. understanding/ happiness/ well-being/ love   74, 82, 222, 225, 226, 228, 282, 432, 435
  subjective-informative/ objective-energetic stimulus   48, 136, 156, 160, 177, 178, 192
  tendency to decompress/ expand/ unbind/ release   36, 74, 82, 160, 185, 192, 222, *see* also dematerialisation/ dissolution

scientific (and dialectical) sense
vacuum energy   *see* space/ quantum aspects/ *ZPE*
vacuum potential/ mass-cancellation 154, 185
Λ force/ 'dark energy'   185
liberation   *see* freedom
Liberation   *see* Transcendence/ Super-State
life   *see* also consciousness/ experience/ essence
   holistic inclusion
   biological form = embodied mind   55, 67, 107, 247, 254, 279, 305, 315, 325, 353
   holistic definition   48, 54, 75, 130, 174, 200, 221, *see* also Essence/ Soul/ Consciousness
   informatively negentropic/ purposeful 123, 227, 325, *see* also reason/ order/ negentropy/ intelligence/ teleology
   life-form = information incarnate   100, 173, 174, 239, 251, 262, 277, 293, 300, 304, 336
   metaphysical informant   15, 123, 200, 280, 334
   mind/ consciousness/ immaterial element   20, 96, 200, 204
   opp. death/ matter/ non-life   15, 24, 64, 69, 84, 202, 323
   psychological purification/ evolution of the soul   *see* centralisation/ focus/ psychological concentration/ meditation
   vectored (ascent/ balance/ descent)   19, 73, 228, 444, *see* also cosmic fundamentals/ opposite directions of mind (Chapters 0 and 14)/ morality
     life's golden chains   385, *see DNA* molecule
     life's highly specified/ codified bio-complexity 85, 96, 100, 107, 275, 304, 319, 327, 333, 336, 340, 372, 407, *see* also bio-logical language/ information/ code
     life-fit/ friendly universe 109, 183, 188, 420
     life-style   72, 75, 90, 138, 433
   materialistic reductionism
   a puddle evolved to think about itself?   24, 25, *see* also faith/ evolutionary theory/ naturalistic methodology
   accidental atomic configuration?   84, 87, 110, 122, 181, 193, 203, 290, 309, 367
   chanced from a pennyworth of chemicals?   15, 117, 197, 286, 301, 309, 311, 360, *see* chemical evolution/ abiogenesis
   electronic after-thought/ an experience of nervous atoms?   203, 209, 280, *see* neuroscience/ reality
   materialism's miracle   *see* materialism/ reality - special delusions
   reductionist definition   84, 181, 202, 221, 290, 311, 325
     wheel of life   282, 285
light
   information-carrier   18, 101, 136, 255, 274, 302
   life-giver   101, 255, 421
   physical
   apparent absolutes   65, 142, 166
   balanced electro-magnetism/ subtle polarity   65, 141, 176
   fossil light   *see CMBR*
   harmonic electro-magnetic oscillation   176
   iced light/ *ATP*   298, *see* also *ATP*
   least material levitational entity   61, 65, 136, 156
   matter's holy ghost   136, 176
   particle-antiparticle annihilation   28, 160
   photosynthesis   174, 302, 329, *see* also polarity bio-energetic
   pure radiant energy   61, 63, 136, 174
   quasi-immateriality   176
   spectral-wave continuity-photonic discontinuity   32
   wave-particle   28, 136
     psychological   *see* also Illumination/ Enlightenment
     guilt-less state   79, 286
     knowledge-understanding   174
     Light of the World/ Logical Illumination   66, 93
     volitio-attractive push-pull current   96, 149
   psychosomatic/ psycho-biological
   'body of light'   *see H. electromagneticus*
   medium/ connector   136, 174, 239, 257
   speed of (c)   *see* absolute physical/ apparent absolute - velocity of light
Light   *see* Essence/ *Logos*/ *Shabda*
Lightfoot John   171
Linnaeus Carl   *see* bio-classification
Lister Joseph   338
Little Miracle   *see* abiogenesis
logic   *see* reason/ cause/ cosmos/ bio-logic/ Glossary
   abductive   91, 300, 340, 414
   algorithmic   129
   deductive   120, 129
   inductive   129
   tautology   172
Logical Norm   *see Logos*
Logos   *see* also Causal Information/

Psychological Potential/ Creator
Creative/ Cosmic Current 135
First Cause/ Primal Motion/ Initial
Projection 71, 84, 93, 128, 149, 174,
see also Creator
Primary Logic 93, 98
Sound and Light 135
Source 33
Source of Mind/ Psychological
Substance 149, 204
Vibratory Communicant 47, 66, 93, 135
Lord Deliberate 106, 111, 112, 123, 124,
125, *see* also creativity/ anti-chance/
teleology
love *see* also positive power/ interest/
enthusiasm
communicant/ gives meaning 47, 93,
101, 135, 148, 177, 433, 439, *see*
unity/ depolarisation (two > one)/
positive psychological force
compassion/ radiant beneficence/
blessing 70, 74, 91, 149, 439, 444, *see*
also saint/ holy grails/ resonance
critical quality 73, 74, 226, 444
erotic 380
milk of life/ life-giver/ psychological
nutrient 437
of wisdom = philosophy 17, 71, 222
opp. selfish passion 228, 428
positive time-stopper 165
resonant association *see* music/
positive power/ harmonic
oscillation
selfless empathy/ unification 149, 380,
433, 437, *see* also (self-
)transcendence
warmth of interest 165, 380, *see* also
interest/ enthusiasm/ focus/
concentrated (light of)
consciousness
Love
Brilliance/ *GUE*/ Unity/ Communion
149, 221, 428, 444
Lovelock James 423
LUCA (Last Universal Common
Ancestor) 290, 333, 335
luciferin 307

M

machine/ mechanism *see* also irreducible
complexity/ minimal functionality/
teleology
biological organ/ system/ body 15, 63,
97
cellular mechanisms/ machinery *see*
molecular/ cellular machinery
computer/ mind-machine 103, 240, *see*
computation

conceptualised construction 101, 102,
211, 304, 322, 334, *see* also
conceptual development/ innovation/
creativity
functional design 85, 97, 102, 211, 317,
323, 335, *see* functional logic/ bio-
logic/ minimal functionality
irreducible to physics and chemistry
102, 299
metaphysical/ Natural Dialectical
machine 40
passively informed 103, 113, 300
robot/ automaton 101, 104, 317, 423
teleologically-intelligent mechanism
102, 111, 294, 302, 323, 334, 397
whole greater than sum of parts 254,
299, 318, 335
macro-evolution/ transformism 307, 308,
343, 344, 345, 351, 353, 363, 414, *see*
plasticity unlimited
macro-evolution and innovation 307,
344, 353, 363, 414, *see* also
innovation-imaginary law of
no hard evidence 344, 345, 363, 365
Mahabharata 276
Main Dialectic *see* Primary Dialectic
main routine *see* computation/ code; *top-
down* programming
malarial parasite 363, 364
male polarity *see* sex/ polarity/ reproduction
man
as microcosm *see* cosmos
centaur 19, 222, 227, 425, 434, 436
centaur paradox 434
H. archetypalis/ memory man *see H.
archetypalis*
H. electromagneticus *see H.
electromagneticus*
H. sapiens *see H. sapiens*
metaphysical man/ memory man 248,
273, *see* archetype/ caduceus/
mnemone
mutant ape? 412
mandala 258
Mandelbrot Benoit 136, 326
Mao Zedong 430
Maotienshan shales *see* fossil
marriage 433
Marx Karl 430
mass *see* matter/ energy/ Higgs mechanism
Master Analogy of Natural Dialectic *see*
music
master gene *see* homeotic genes
master routine *see* code/ computation; *top-
down* programming
material psychology *see* neuroscience
materialisation *see* also creation
decentralising/ centrifugal/

496

externalising process 148, 378, 418
expression of (*tam* ↓) downward/ gravitational vector 29, 32, 36, 44, 54, 60, 62, 82, 95, 128, 133, 137, 148, 228, 263, 267, 292, 379, 418
    hierarchical order   see fundamental order of creation/ creation/ creativity/ cosmic fundamentals/ hierarchy
    informative/ energetic loss *see* entropy/ gravity
    of idea/ purpose/ design   see act of creativity/ psychological force/ teleology
    One > many apparently isolated ones/ units   134
    orderly fall-out/ concretion/ fixation see exhaustion/ impotence/ subtendence
    psychological motor output/ symbol > muscle/ mind > matter 19, 226, 378, 380, see also sensation/ creation/ objectivity

materialism   *see* also faith
    force and particles alone   59, 67
    fundamental error   247
    fundamental problems 86, 91, 102, 103, 110, 172, 326, 367, 413, 415, *see* also innovation/ source of information/ evolutionary theory/ black box
    illusions of   *see* illusion/ reality/ matter/ consciousness
    informal cults of matter - humanism/ scientific atheism/ scientism 72, 84, 202
    materialism's style of death/ oblivion 'regained'   276, 279, 435
    materialism's godless gaps *see* godless gaps
    miracles
energy/ matter *see* big-bang (from nothing)/ transcendent projection
    of life   *see* abiogenesis
    specifically informant matter *see* imaginary law of innovation/ matter/ code/ information
        philosophy/ world-view 15, 16, 19, 21, 23, 111, 229
    scientific self-definition excludes metaphysic   22, 375, *see* also naturalistic methodology/ atheism
materialism's immateriality   *see* immateriality/ nothingness/ space-time

mathematics
    archetypal geometry   157, 167, 181, 259, *see* also vibration/ dot/ cosmic models
    cannot describe psycho-logical actions/attributes 28, *see* glass ceiling

equation/ balance 31, 35, 118, 168, *see* also balance/ equilibration/ *karma*
Fibonacci numbers   181
fractal geometry   326
geometry of illumination   136
is a rational/ psychological/ metaphysical exercise   *see* also symbol/ logic/ reason/ computation/ code
models nature 33, 121, 181, *see* also cosmic models
numerical metaphor 120, 180, 215, 326
of population genetics   360
physical world in mind   121
space-time/ relativistic geometry   167
strange equations   31, 126, 134, 138, 150, 151, 181, 215, 318, see also the Essential Equation
string/ quantum   115, 141
symmetrical/ polar geometry 377, 379, 382, *see* also symmetry/ mirror image/ reflective asymmetry/ anti-parallels/ inversion
symmetry 143, *see* also symmetry/ conservation

matter   *see* also energy
    absolutely passive/ reflexively informed   100, 123, 166, *see* archetype/ first cause physical
    aimless/ creatively impotent 80, 85, 95, 107, 172, 173, 311, 335, 358, 372
    can't anticipate   385, 399, *see* also anticipation
    can't create code   325, 338, 385, *see* innovation
    can't reason   247
    can't create mind/ information   *see* mind/ information/ creativity
    condensed/ bulk matter = cosmic sink 29, 33, 36, 62, 64, 83, 147, 153, 156, 161, 208, 300, *see* also oblivion/ omega/ expression of (*tam*) cosmic fundamental
    cult of matter   *see* materialism
    dark matter   105, 182, 185
    dark matter/ lambda force   143
    energetic entropy > sink   *see* sink/ sub-state/ entropy/ exhausted phase
    eternal 151, 180, 197, 198, *see* also steady-state/ multiverse
    glass ceiling   83
    illusory?   70
    made of energetic patterns   55, 68
    matter = informant void   79, 276
    matter = specifically constrained energy   140
    matter-in-practice/ classical matter/

bulk form 57, 61
matter-in-principle/ quantum forms 57, 61, 83, 127, 132, 154, 260
  mind = matter/ brain see (neuro-)scientific reductionism/ neuro-hormonal system/ consciousness/ materialism
  no-life-in-it 25, 46, 57, 64, 85, 156, 204, 280, 312, 325
  no-mind-in-it 15, 16, 19, 22, 23, 24, 25, 44, 51, 53, 55, 57, 64, 68, 104, 111, 200, 203, 216, 249, 262, 280, 379, 412
  non-conscious see non-consciousness
  objective form-generator 125, see also passive information/ objectivity
  physical essence see archetype/ potential matter/ potential (energetic-physical)
  potential matter see potential matter/ archetype/ typical mnemone
  subjective annihilation/ sink 58, 155, 156, 300
matter-in-practice see matter
matter-in-principle see matter/ energy/ force physical/ sub-atomic particle/ quantum aspects
Maxwell James Clerk 141, 158, 159, 178, 255
M-conjecture 143, 196
mechanism see machine
meditation
  specific discipline of contemplation 20, 75, 217, 419
meiosis see reproduction/ sex
memory see also sub-consciousness/ mnemone/ archetype
  physical
  DNA/ books/ databanks/ any data storage medium see code/ computation/ language/ cosmo- & bio-languages/ DNA molecule
  evolution of memory? 234, 253
  recording/ playback heads see brain/ engram/ hippocampus/ amygdala etc.
  psychological/ metaphysical
  idea registry 84, 100, 236
  personal see mnemone
  recording/ playback (recollection) 235
  psychological/ physical
  psychological ◇ physical/ psychosomatic codification see psychosomasis/ signal translation
Mendeleyev Dmitri 146
metabolism see also irreducible complexity/ computation/ cell
  abiogenetic origin? 315, 328
  algorithmic
  codified pathway as an application program 321, 327, 328
  enzyme as sub-routine 327, 328
  informant/ anticipatory chemistry 315, 328
  programmed/ cybernetic chemistry 315, 327
  specified complexity of code/ computation 333
  energetic operations
  photosynthesis 293, 302, 313, 315, 328, 329
  photosynthesis ◇ respiration/ bio-energetic cycle 302, 329, see also molecular/ cellular machinery
  respiration 274, 293, 302, 328, 329, 332
  enzyme see molecular/ cellular machinery/ protein/ machine
  evolutionary problem 320, 328, 332
  informative operations see information/ informative transmission/ hierarchy/ genetic operating system/ neuro-hormonal system/ teleology
metamorphosis 411, see also morphogenesis/ biological development
metaphor see also cosmic models
  comparative descriptor/ simile 26
  evolutionary 'as if' 75, 84, 118, 126, 304, 403, 413
  genetic 'immortality' 301
  image/ metaphysical form/ symbol of the invisible 31
  mathematical equation see equilibration/ symbolic information/ mathematics
  philosophical metaphor see holistic symbolism/ metaphysical argument/ teleological construction
  religious metaphor/ analogy 26, 286, 425
  scientific metaphor 29, see also mathematics/ big-bang/ atomic/ evolutionary/ brain-as-computer/ scientific inc. cosmic models
metaphysic see Glossary; also information/ informative component of creation/ immateriality/ cosmic pole (of dipole)/ code/ consciousness/ mind/ psycho-logic/ archetype/ language/ cosmo-logical language/ bio-logical language/ reality/ subjectivity/ nature/ life
  metaphysical egg 26, 27, 34, 87, 95, 100, 123, 129, 195, 262, 266, 306, 350, 379, 380, 386, 397, 399, 400, 406, see also idea/ archetype/ first cause/ innovation/ conceptual development
  metaphysical exercise 121, see reason/ logic/ focus/ meditation/ mind
  not constrained by physical law 48,

114, 118
Metaphysical Egg   see First Cause/ Archetype/ *Logos*
Metaphysical Tri-Unity   see (N)One/ Nothingness/ Unity/ Infinity/ Trinity/ Essential Equation
Metaspriggina   354
methylated cytosine   see *DNA*/ epigenetics
methylotroph   331
Meyer Stephen   325, 354
Michelson Albert   158
micraster   354
micro-evolution/ variation-on-theme see variation/ species/ plasticity
Miescher Friedrich   318
Miller Stanley   313
mind see also information/ consciousness/ psychological force
   = metaphysical entity   49, 53, 55, 62, 66, 104, 205
   anti-chance/ negentropic anti-randomiser 125, 334, see also chance/ randomness/ negentropy/ order/ reason/ creativity/ active information/ psychological force/ teleology
   consciousness
   = active information 29, 55, 58, 125, 221, 417
   animal perception   222, see also centaur
   consciousness-in-motion/ waking mind   47, 125, 204, 294
   human normality is triplex
      egotistical mind/ knowledge-to-fulfil-desire/ body-rationality 222, see also animal perception
      lower mind/ ignorance/ depression/ confusion/ negative intent/ evil-rationality   222
   informative element/ informant/ instructor   63, 67, 125
   loops
      lower physical/ mind-in-practice 222, 225
      upper metaphysical/ mind-in-principle   222, 224
   purposeful/ teleological force   96, 123
   subjective form/ image-generator/ formful consciousness   66, 125
   thought/ feeling/ emotion are metaphysical see metaphysic/ reality/ subjectivity/ neuroscietific reductionism
   unit of thought/ thought-particle   87, 213, see psychological force
   voluntary/ flexible   417, 437
   equilibrator   213, 273, see also homeostasis/ balance/ *karma*
   informative command & control

centre   19, 55, 138, 206
mind machine   see computation/ cell
mind over matter   58, 60, 61, 71, 275, see also hierarchy/ cosmos/ conscio-material dipole
mind-in-practice   219, 222, see mind/ third state of consciousness/ lower physical loop/ ego
mind-in-principle   219, 222, see mind/ first and second states of consciousness/ upper metaphysical loop/ principle/ *Logos*
non-consciousness
special case/ no-mind/ subjective absence/ impotence   see non-consciousness/ impotence/ matter/ energy/ subtendence/ objectivity
unit of thought/ thought-particle   see also neuro-hormonal system/ neuro-scientific reductionism
sub-consciousness   see also sub-consciousness/ mnemone/ archetype/ first cause physical
= dormant condition/ memory   246
= passive information   229
involuntary/ reflex   417
matrix of matter   127
mind <> body transmitter/ receiver   see psycho-biologic/ psychosomasis
universal mind 19, 42, 46, 50, 61, 106, 115, 118, 119, 121, 135, 137, 138, 144, 157, 160, 166, 198, 235, 242, 247, 252, 272, see also implicit universe/ archetype
minimal functionality 88, 109, 332, 333, see also irreducible complexity
miracle   see materialism's/ holism's definition   192
mirror image   see also inversion/ symmetry/ polarity
bilateral symmetry 205, 213, 346, 381
chirality   316
matter/ antimatter   145
reflective asymmetry   145, 273, 377, 381, 383, 394, see also symmetry
missing link   see fossil/ godless gaps/ black box
atheistic evolution's godless gaps 416
developmental/ anticipated target 211, 212, 247, 254, 267, 338, 371, 387, 399, 406, 410, 411
genetic   358, 359, 362
metamorphic   411
molecular/ origin of sequential nuclear & metabolic operants   320, 321, 327, 328, 406
nascent form   308, 342, 372
theistic evolution's God-of-the-gaps

mitochondrion   416 *see* molecular/ cellular machinery
mitosis   *see* reproduction
mnemone *see* also archetype/ memory/ sub-consciousness/ resonance/ psychosomasis/ Glossary
    bio-software/ bio-psychological operating system   240, 242, 246, 263
    memory/ psychological storage 50, 58, 80, 118, 218, 233, 234, 235, 302, 306
    mnemonic control/ first cause
      physical   248, 257, 263
      personal   218, 240, 241, 242, 244, 245, 248, 283
    long-term memory   235, 245
    psycho-active 'bio-chron'   166
    short-term working memory   235, 245
      psychological storage formula   245
    sub-conscious
    function = informant   218, 273
    psychosomatic mechanism 218, 240, 246, 252, 260
    structure   242, 246, 249, 260
    typical
    archetypal   218, 226, 235, 236, 237, 240, 241, 242, 246, 247, 248, 249, 256, 257, 269, 271, 293, 306, *see* also archetype/ first cause physical
    instinct   104, 215, 251, 283, *see* also instinct
    morphogene   254, 256, 263, *see* also morphogene
    psycho-biological 'bio-chron'/ cyclic bio-equilibration   166
    signal translator   251, *see* also signal translation
    tripartite structure   249, 256
    typical with every cell 250, *see* also cell/ dormant in every cell
modular programming   *see* code/ top-down programming/ computation/ typology/ homology/ sub-routine/ analogy/ mosaic distribution
    bio-module   304, 306, 331, 349, 350, 353, 376, 396, 408
moksha   *see* Consciousness
molecular/ cellular machinery   *see* also cell/ bio-logic/ irreducible complexity/ minimal functionality
      at moment of abiogenesis?   324
      chaperone   386
      chromatin   368, 407
      chromosome   264, 301, 322, 358, 367, 368, 371, 385, 389, 395
      energetic complements/ input-output
    chloroplast   297, 329
    mitochondrion   256, 297, 329

enzymes/ metabolic pathways as bio-machines   247, 256, 275, 301, 308, 315, 316, 319, 321, 327, 328, 408, 409, *see* also metabolism/ protein
extra-cellular matrix (*ECM*)   256
informative complements
brain-body/
    nucleus-cytoplasm/ informed field of action   299
    informative complements/ 'brain-body'
nucleus/ control centre   294, 368
    nucleosome   368
    polymerase components   319, 322
    replisome   323
    ribosome   317, 322, 324, 386
    splicing/ spliceosome   317
structural complements/ 'flesh & bones'
containers (e.g. membranes/ cytoplasmic gels)   255, 260, 293, 299, 316, 370
organisers (e.g. centrosome/ centriole/ cytoskeleton/ mitiotic spindle)   256, 260, 299, 388
monophysite schism   282
Monopole   *see* Unity/ Essence
moon   *see* astronomical bodies/ time
morality
    absolute/ relative   73, 75, 430, 439
    amorality
    instinctive/ animal behaviour/ innate incapacity to discriminate   *see* also neuro-hormonal system/ instinct/ mnemone
    non-conscious behaviour of matter   425
    chemical/ morality < neuro-hormones?67 chemical?
    morality < genetic *DNA*?   67, 72, 429, 437, 438
    morality < molecular amorality? 72, 429
    immorality
    anti-social behaviour   428
    evolutionary survival-of-the-fittest 'morality'   429, 430, 437
    harmful intention/ crime/ evil/ curse/ *see* sin
    moral entropy > sin   428
    metaphysical quality 26, 59, 121, 442 morality
    Dialectical expression   417, 444
    externally/ culturally imposed - formal/ canonic/ ethical regulation   *see* religion/ law/ politics/ community
    human equivocation   439, *see* also centaur
    internally self-imposed - informal self-government/ discipline   *see* principle/ ideal/ Reason

no scientific remit 90, 429, 434
socially critical 122, 125, 227, 228, 433, 434
   quality of intention 72, 228, 428, 434
Morley Edward 158
morphogene *see also typical mnemone/ archetype/ psychosomasis*
   archetypal memory 242, 255
   first cause physical/ formative influence *see* potential matter/ archetype
   generic form 254, 263, 266, 269
   immaterial/ wireless broadcast 264
   influences every cell 250, 254, 263
   logical/ conceptual bio-form 262, 263, 267, 306
   metaphysical blueprint 250
   metaphysical software/ program 254, 306
   psycho-biological mechanism 256, 263
   psycho-biological/ archetypal 'intelligence' 242, 250, 254, 266
   psychosomatic mediator/ connector 243, 255, 264, 306
   sub-conscious structure 250
   sub-routine of typical mnemone 241, 242, 244, 254, 256, 260, 264, 293
   template for biological hardware 254
morphogenesis
   archetypal *see* morphogene
   biological development 262, 268, 307, 338, 368, 371, 404
   evolution of? 262, 406
   resonant *see also* resonance/ harmonic oscillation/ Chladni
   resonant modality 260, 262
mosaic distribution 87, 88, 306, 308, 349, 350, 351, 372, *see also* Glossary/ convergence/ adaptive potential/ homology/ variation-on-theme
motion/ change *see also* space/ time/ relativity/ energy/ mind (consciousness-in-motion)/ existence (perpetually-in-motion)
   < imbalance 148, 159
   > relativity 31
   action-reaction *see karma*/ Law of Motion
   cyclical equilibration *see* vibration
   dialectical sense
   'down' *see* gravity
   'rest' *see* equilibrium/ balance
   'up' *see* levity
   fundamental nature of existence 31, 70, 107, 119, 133, 134, 142, 156, 168, 377
   information-in-orderly-motion/ music/ language 101, 259
kinesis
   anti-expression by (*tam* ↓) inertial cosmic vector *see* gravity
      deceleration 36
      resistant inertia/ drag/ > static fixity/ finish/ stop *see* exhaustion/ negative power/ sink
      special case/ no-motion/ rest *see* impotence/ negative equilibrium
   expression of (*raj* ↑) kinetic cosmic vector *see* levity
      acceleration & non-linear motion 36
      fluid aspect of energy 52, 177
      stimulus/ energetic input *see* levity/ causal agent/ force psychological or physical
      unchanging velocity 36
   expression of (*sat*) vectorless balance
      special case/ no-motion/ rest 140, *see* source/ potential/ positive equilibrium
   mobile thought *see also* consciousness-in-motion/ active information/ mind
   motion's second nature *see* time
   motivation/ emotion *see* force psychological
perpetual motion
cosmic/ classical 31, 70, 177
kinetic theory of matter 70, 177
sub-microscopic/ quantum 52, 70, 146
whence? 180
   primal *see* Primal Motion
Mount Universe *see* cosmic model
mouse 344, 404
MRI (magnetic resonance imaging) 211
multiverse *see* creation stories
music *see also* harmonic oscillation/ resonance
   Master Analogy of Natural Dialectic 101, 177, 258
   natural music/ alphabet of notes 100, 170, 258, *see also* alphabet/ cosmological language/ score/ Orpheus
mutation *see also* informative entropy/ *DNA* molecule/ gene/ innovation
   're-codifies' code? 359, 362, 369
   BM (beneficial mutation) 358, 359, 365
   bug/ randomiser/ noise-maker/ information-loser 266, 358, 362
   directed/ saturation mutagenesis 336, 351, 376, 404
   DM (deleterious mutation) 358
   epimutation 370

high-level/ 'upstream'/ homeotic
mutation 409
mutation > innovation? 307, 320, 340, 344, 357, 358, 360, 361, 404, 410
NM (neutral mutation) 359, 366
random mutation/ Darwinism's engine 71, 197, 262, 266, 299, 301, 309, 320, 338, 345, 357, 358, 362, 363, 365, 373, 375, 389
    sickle-cell anaemia 365
myth/ mythology   *see* creation stories

N

Nanak 74, 221
narcolepsy 231
nascent form   *see* missing link
Natural Dialectic of Polarity
    application program 417
    columnar structure 21, 39, 41, *see* dialectical stack
    Cosmic Dipole 49, 62, 271, 378
    Cosmic Infrastructure 27, 28, 131
    Cosmic Models   *see* cosmic models
    Essential Monism and existential dualism 46
    Essential Principal 44
    existential principal 44
    holistic 30, 378, 400
    Master Analogy/ music 101, 257, 259
    Perennial Philosophy 28
    Primary and Secondary Polarities 51
    Primary Axiom 44, *see* also Glossary
    Reconciliation 27
    Trinity/ Triplex Nature 33, 34, 35, 38, 51, 212, *see* also cosmic fundamentals/ cosmos/ hierarchy
natural genetic engineering (*NGE*) *see* gene
natural law   *see* nature/ law/ order/ invariance/ conservation
Natural Nucleus *see* First Cause/ Essence
natural selection/ Darwinism's regulator
    86, 308, 309, 311, 321, 337, 338, 341, 342, 343, 345, 359, 361, 373, 407, 413
    negative selection 342, 373
    positive selection 373
    unnatural/ mind-led/ breeder selection 373, *see* species; artificial/ breeder selection
Natural Teleology *see Logos*/ First Cause
naturalistic methodology
    limitation 87, 247, 281
    materialistic assumption/ scientific method 21, 173, 195, 263, 290, 300
nature
    holism
physical cosmos = base-end of creation 48
    holism (including energetic physic)
informative metaphysic is natural 86, 230, 308, 368
natural logic = reflex behaviour *see* also law/ cosmos/ cosmo-logic/ archetype/ bio-logic
natural order/ fundamental order of creation 28, 42, 50, 74, 118, *see* also cosmic fundamentals
naturalistic order/ laws of nature by chance? 83, *see* also chance/ randomness
no law of physical innovation *see* innovation/ evolutionary theory/ illusion/ Law of Naturalistic Non-Innovation
psycho-physical nature 23, 24, 48, 72, 84, 413
universal mind and matter 19, *see* also conscio-material (c-m) dipole/ informative-energetic coordinates/ cosmos
unnatural/ intentional interference 314
unnaturalism = purposeful design/ teleological engineering 20, 74, 84, 86, 102, 118, 192, 318, 403, 413, 427
naturalism (excluding informative metaphysic)
immaterial = metaphysical = unnatural = super-natural 230, 308, 413
natural sciences/ scientific naturalism < naturalistic methodology 23, 48, 106, 195, 247, 290
naturalism = materialism 19, 22, 23, 55, 136
nature = physical cosmos 48, 137
NDE (near-death experience) 281
negative power *see* also gravity/ morality/ (*tam*) cosmic fundamental
    > isolation/ division 51
    > materialisation/ confinement 51
    > social sickness 428, 432, 436
    amoral expression of (*tam*) vector 82, 222, 282
    evil/ devilry 427, 436
    hell 282, 283, 426, 427
    intentional infliction of pain/ death *see* evil
    involuntary negativity 425
    involves informative/ energetic loss *see* entropy
    negative extremity 54, *see* also subtendence
    selfish passion *opp.* love 222
    unintentional infliction of pain/ death 72, 425, *see* also morality/ amorality
    violent psychological/ physical locality 148
    voluntary negativity 427
negentropy *see* also energy/ information/ order/ levity/ Glossary

energetic expression of (*raj* ↑) fundamental 36
highly ordered cosmic start 169
stimulus/ energetic input/ gain/ gain 225
  informative
anti-randomising character of mind 80, 359
stimulus/ informative input/ gain *see* information/ creativity/ force
psychological teleological/ anti-randomising tendency 74, 80, 86, 358
  levitational vector of rise 326
  motion/ change towards source 225, 228, 326, 443
  problem for evolutionary theory 111, 326
  psychologically/ metaphysically/ informatively dominant vector 49
neo-Darwinism *see* index STE/ GTE/ also glossary PCM and evolution
neo-Platonists 282
neoteny 262
neuro-hormonal system *see also* brain/ anti-parallels/ polarity psycho-biological/ *DNA* molecule/ consciousness
  alive as biochemicals 203
  aminergic/ cholinergic systems 217
  chemical messenger/ informative hormone 250, 381, 407
  CNS (central nervous system)/ trunk route/ spine 209, 213, 275
  equilibrator 226
  evolution of? 213
  hierarchical 206, 208, 213, 274, 383, 397, *see also* hierarchy/ informative hierarchy
  homeostasis/ sensor/ regulator (brain)/ motor 226
  involuntary/ autonomic 275
  made of atoms/ molecules 59, 121, 136, 197, 200, 203, 204, 381, 429
  neuro-modulation 217
  neuro-scientific imperative *see also* neuro-scientific reductionism
Buddhist doctrine as genetic figment 429
Christian quintessence as a neural structure 429
devil as synaptic pattern 429
genes > nerves/ hormones > morality? 73, 429
Islamic Allah as ionic flickering 429
nerves = thought/ memory? 71, 73, 203, 213, 225, 234
systematic excretion of consciousness? 23, 121, 197, 202, 204

Truth isn't metaphysical 438
  non-conscious/ not alive 18, 59, 136, 200, 204, 209, 429
  passively informative/ reflex 63, 209, 381
  Pax6 master control/ main routine 409
  physical reflector of psychological events 201
  polar/ antagonistic hormones 226, 230
  psycho-neuro-immunological system 275
  psychosomatic mediator/ information exchanger 18, 201, 242, 245, 269, 292, 383, 397
  somato-sensory arc 383
  specifically codified 18, 100
  top-down order 205, 207, 209
  voluntary/ peripheral 275
neuroscience *see* neuro-hormonal system
neuroscientific imperative
  soul as figment of imagination 103
neuro-scientific reductionism *see* neuro-hormonal system/ faith/ dysfunctional logic/ information/ life/ consciousness/ intelligence/ mind/ reality/ materialism/ illusion/ special delusion
neutral evolution *see* gene duplication
neutrality *see also* balance/ equilibrium
  archetypal 395, *see also* Polarity Essential/ potential/ unity/ sexual archetype
  atmospheric 312
  balance of equation/ resolution 121
  creation-field 145
  detachment/ disinterest 34, 245
  genderless 396
  negative
  inertial equilibrium 147
  post-dynamic impotence 43, 147
  post-reactive phase/ slack 147
  void 147
    non-polarised 176
    pH of water 327
  positive
  fullness 34, 145
  potential equilibrium 44, 52, 53, 145
  pre-active potency/ poise 44, 52, 145, 147, 395
  Sat cosmic fundamental/ quality 34, 37
Neutrality *see also* Essence/ Polarity Essential/ Super-State
Newton Isaac 49, 103, 140, 141, 142, 158, 167, 171, 440, 444
nipple 398
Nirvana 93, 221, 419, *see* Consciousness and Glossary, *see also* Glossary
nitrogenous base *see DNA* molecule/ genetic linguistic hierarchy

noise *see* randomness
non-consciousness 137
  absolute ignorance/ perpetual oblivion 21, 32, 54, 59, 66, 149, 152, 166, 204, 218, 236, 271, 379
  biological/ biochemicals are lifeless 222, 276
  body-state inc. *Homo sapiens* 63, 97, 239
  cosmic base-level/ sub-state/ sink 22, 48, 53, 61, 167, 204, 208, *see also* impotence/ sub-state/ sink/ subtendence
  energetic nuclei (suns/ stars/ atoms) 247
  inanimate condition 69, 204, 276
  matter's sole commonality 121
  natural law = reflex behaviour 155, 194
  objective being 51, *see also* energy/ matter
  passive information 97
  physical phenomena/ forms 22, 54, 81, 110, 204, 218, 237, 239, 243, 269, 276
  psychological nadir *opp.* Psychological Zenith = Consciousness 204, 237
  psychology of physic 61, 128, 137, *see also* potential matter/ archetype
  reflex/ automatic dimension 55, 63, 104, 110, 154, 209, 210, 229
  sixth state/ special case of consciousness (none) 234, 237
  subjective exhaustion/ void 54, 137, 204, 276
  subjective timelessness 154
  subtendent pole 63, 204
  three grades 167
  uncreative condition of matter/ material energy 16, 18, 22, 25, 51, 55, 82, 301
non-existence *see* zero-point/ nothingness/ Glossary
Non-Existence *see* Essence/ Super-State/ Glossary
non-protein-coding *DNA* *see DNA* molecule; non-protein-coding
noology 201
(N)One 42, 134, 150, 435, *see also* Essence/ Potential/ Nothing
norm .... *see* homeostasis
Northrop Filmer 255, 267
nothingness/ void
  absence/ anti-presence 41, 48, 110, *see also* non-existence
  absolutely nothing 41, 48, 110, 112, 113, 126, 156, 179, 191, 341, 379
  ex nihilo - from non-existence existence 41, 112, 122, 126, 188, *see also* big bang/ zero-point

infinitely thinner than thin air 125
information weighs nothing 79
informative *and* energetic void/ void of voids/ impotent space 161
informative potential 108, 118, 137, 153, 156, 186, 190
materialism's immateriality/ basic physical coordinates *see* absolute
physical/ materialism's immateriality/ impotence/ exhaustion/ space/ time/ space-time/ apparent infinities
materialism's void (physical/ never metaphysical) *see* space
matter = informant void 78, 153, 217, 276
negative/ post-active exhaustion 43, 54, 151, 154, 157, 161
nothing archetypally constrained *see* archetype
nothing-physical/ something-metaphysical 15, 25, 42, 48, 107, 156, 186, 379, *see also* potential matter/ archetype
positive/ pre-active potential 43, 44, 107, 122, 129, 132, 146, 151, 153, 156, 160, 186, 190
potential/ latent field of possibilities 44, 91
space *see* space/ physical vacuum
time *see* time/ no-time
unconscious oblivion 55, 66
Nothingness/ Void
  Essence/ Potential 16, 44, 134, 153, 444
  Non-Existence/ Supreme Being 44
  Source 37, 41, 42, 110, 134, 152, 153, 154, 177, 258
  term of the Essential Equation 134, 151
  Void/ (N)One 42, 130, 177
NREM sleep 233
nucleotide *see* genetic linguistic hierarchy/ *DNA* molecule
nucleus *see* atom/ force physical/ polarity/ cell machinery
Nucleus *see* Natural Nucleus/ Super-Nature

O

OBE (out-of-body experience) 104, 281
objectivity *see also* naturalistic methodology/ non-consciousness
  attempt to eliminate subjective context/ prejudice concerning observation 18, 20, 24, 71, 74, 88, 126, 200, 279
  detachment/ pure perception 18, 20, 21, 24, 126, 200, 237

instrumental/ mathematical treatments of material phenomena 72, 120, 202
material condition/ objective being 27, 48, 55, 61, 100, 104, 115, 233, 276, see also oblivion/ matter
materialism's work-horse 104, 200, 279
matter's special case of subjectivity (zero) 26, 137, 233
outward sensational as *opp*. inward contemplative focus 20, 127, 237, 279
scientific ideal 20, 24, 88, 126, 200, 237
Objectivity   *see* Essence/ Absolute/ Consciousness
oblivion   *see* non-consciousness
opposite pole from Super-Consciousness 62, 80, 221, 229, 286
Ockham William of 114, 143, 159, 197
Om 66, 93, 258, *see also Logos*/ Sound and Light
omega   *see* sink/ sub-state/ energetic exhaustion
Omega   *see* Alpha/ Consciousness/ Cosmic Source
Oparin Alexander 187, 312
opposites   *see* polarity/ dialectic stacks
order *see also* cosmic fundamentals/ code/ hierarchy/ reason
    active order of mind 79, 96
    archetypal   *see* archetype
    codified order   *see* code
    codified self-assembly by chance? 267, 275, 306, 312, 326, 410, *see also* matter/ code
    codified/ bio-logical self-organisation 110, 254, 315, 317, 318, 327, 334, 368, 388, 395, 396, 406, 411
    dialectical   *see* Natural Dialectic
    e principio/ from-principle-to-practice
    fundamental order of creation 22, 28, 29, 34, 40, 44, 58, 129, 133, 154, 386, *see also* cosmic fundamentals/ cosmo-logic/ conscio-material (c-m) dipole/ nature
    informative communication ◇ order 83, 93, 98, 257, 381, 396, 440
    natural law/ reflex behaviour/ automatic order 50, 98, 118, 160, *see also* law/ archetype/ cosmo-logic/ nature
    non-purposive complexity 78, 95, 98, 107, 399
    opp. chance 107, 112, 121, *see also* chance/ randomness
    order in chance? 115
    passive order of matter 79, 88, 95, 98, 103, 311
    psychosomatic order 240, 248
    purposive/ specified complexity 78, 96, 97, 173, 334, 340, 350, 365, 368, 379, 399, *see also* teleology
    schematic order 315, 317, 321, 327, 382, 388, 399, 400, 402, *see also* conceptual development/ psychological development
    social/ rules and regulations 72, 433, 434
    top-down order 42, 60, 62, 83, 84, 100, 119, 169, 203, 207, 308
    uncodified/ physico-chemical self-organisation 108, 119, 140, 315, 326
    vibratory   *see* resonance/ harmonic oscillation/
orexin 231
organ of teleology   *see* mind
organelle   *see* cell machinery
organic development   *see* biological development
organisational-with-operational integrity *see* machine/ mechanism/ irreducible complexity
Origen 282
Orpheus 101
oscillation   *see* wave/ harmonic oscillation/ vibration
out-swing 277, 288
ovoid/ egg-like *see* egg/ metaphysical egg/ code/ codified source

---

P

pain *see also* negative power/ gravity/ black hole
    < immorality 72, 428
    expression of constrictive (*tam*) vector 222, *see also* negative power
    intentional infliction/ self-defence/ moral redress 28, 35, 41, 44, 47, 50, 427, 72, 74,
    intentional/ sadistic infliction   *see* evil
    involuntary/ ignoble quality 226
    isolator 428
    negative time-stopper 165
    unintentional/ reflexive infliction 72, 425
    violent psychological locality 245, 282, 428
palette of creation 117
pangenesis 341
panspermia 336
Paramecium 390
Paranada   *see Logos*/ Sound and Light
Pascal Blaise 95
passive complexity   *see* non-purposive complexity
passive information   *see* information
passive mind   *see* sub-consciousness

Pasteur Louis 309
Pauli Wolfgang 115, 146, 147, 159
Peace *see* Supreme Being/ Super-State
peace/ rest 28, 54, 72, 91, 221, 419, 434, 436, *see* also balance/ equilibrium
pendulum *see* vibration/ wave
Penfield Wilder 209
Penrose Roger 116, 120, 121, 193
pentadactyl limb 347, *see* also homology
Penzias Arno 187
perception *see* experience/ sensation/ idea/ active information
perfect mystic *see* saint
perpetual motion *see* motion/ existence (mind-and-matter)/ atom
personal memory *see* mnemone/ personal mnemone
photon *see* light/ sub-atomic particle
photosynthesis *see* light/ metabolism/ polarity bio-energetic
phylogeny *see* bio-classification/ tree of life/ Glossary
physic's primal trinity 133, 154, *see* also cosmic fundamentals/ absolutes
  physical/ energy
    first apparent absolute *see* space
    second apparent absolute *see* time
    third apparent absolute *see* energy
physical anaesthesia/ metaphysical aesthesia *see* meditation/ contemplation
physical constants 16, 107, 120, 142, 144, 183, 184, 185, 193, 198, 259, *see* also law/ order/ archetype/ invariance
physical objects/ events/ phenomena *see* energy/ matter/ polarity
physics 15, 34, 55, 77, 100, 102, 103, 105, 109, 137, 185
physiology
  active/ functional as *opp.* passive/ structural anatomy 297
  electro-physiological 268, 274
  glandular 273, *see* also neuro-hormonal system
  passively informed/ codified/ bio-cybernetic/ reflexive 209
Pikaia 354
Pippard Brian 101
pivot *see* point of balance/ centre-point
Pivot *see* Central Essence/ Axis/ Zero-Point
pixel/ quantum 117, 145, 167, 250, *see* also cell
Planck Max 115, 120, 127, 141, 144, 145, 146
Plasmodia 390
plasticity *see* also variation/ species
  limited > variation/ micro-evolution 342, 344, 363, 387, 390

  unlimited > transformation/ macro-evolution 342, 343, 344, 363, 387, 414
Plato 115, 120, 246
Plum Brook vacuum chamber 159
point X *see* cosmological axis
poise *see* equilibrium/ potential
Pol Pot 430
polar coordinate *see* energy
polar dialectic *see* secondary dialectic
Polarity - Essential *involves Primary Dialectic*:
  Absolution ◇ relativity 50
  Consciousness ◇ conscio-material scale (> base non-consciousness) 210, 213
  Essence ◇ existence 38, 52, 93, 429
  Neutrality
  Balance ◇ levity/ gravity 36, 51, 178, 429
  Neutrality ◇ + / - 52, 53, 145, 147, 160, 429
  Potential ◇ polar expression 44, 91, 145, 294, 298, 306
  Sat ◇ *raj/ tam* 34, 36, 47, 378, 429
  Soul ◇ psychological ◇ physical bodies 91
  Unity ◇ polarity/ duality 16, 35, 45, 47, 51, 53, 93, 145, 206, 305, 382, 395, 396, 429
    Nothing ◇ something 43, 50, 145, 155, 160
    One Truth/ anti-parallel world-views 21, 379
    Primary ◇ Secondary Dialectic 21, 108, 378
polarity - existential *involves secondary dialectic*
  bio-logical/ informed
  bio-energetic 297, 298, 328, 329, *see* also molecular/ cellular machinery
  neuro-hormonal 206, 207, 208, 211, 212, 213, 224, 226, 235, 382, 383, 397
  reproductive 265, 266, 371, 379, 387, 388, 389, 394, 395, 396, 397, *see* also reproduction/ sex/ bio-logical development
  physical/ informed
  large-scale/ matter-in-practice 36, 52, 55, 141, 145, 178
  quantum/ matter-in-principle 38, 136, 141, 145, 176, 178
  psycho-biological
  active ◇ passive information 81, 96, 273
  archetypal message/ anti-parallel *DNA* duplex 297
  archetype ◇ physical expression 235, 237, 246, 266, 269, 271, 272, 273, 293, 306, 395
  immaterial ◇ material structure 259

informant ◇ informed   52, 55, 77, 208, 273, 297, 395
life ◇ non-life   54, 192, 198, 269, 276
mind ◇ body 54, 96, 205, 212, 213, 225, 246, 269, 302
perception ◇ action   38, 206
psychosomasis   237, 268, 272
universal mind ◇ universal body   96
  psycho-logical/ informative
  contemplation/ sensation   219, 224
  mobile experience/ immobile memory   235
  oscillations
    asleep ◇ awake   218, 221
    one > two/ polarisation/ negative emotion   *see* negative power/ division/ resistance/ entropy
    reflexive ◇ voluntary   218
    two > one/ depolarisation/ positive emotion   *see* positive power/ interest/ unity/ love/ communion/ negentropy
politics external
  ideological forum   122, 124, 419, 432, 433
  market-place   72, 73, 436
  social construction   434, 436
  strategic survival   72, 436
politics internal
  self-government   437, *see* morality
Polkinghorne John   59
Polyani Michael   102
polymerase   *see* molecular/ cellular machinery
Popp Fritz-Albert   255
Popper Karl   132
positive power
  > dematerialisation/ release   51
  > immateriality   51
  > social health   432, 436
  expression of (*sat*)/ (*raj*) cosmic fundamentals   82, 222, 282, 444
  heaven/ highly positive state of mind   282, 283, 436
  informative/ energetic gain   *see* negentropy
  love *opp.* selfish passion 51, 66, 69, 70, 74, 222, 226, 431, 433, 437
post-mortem psychology   282
potential   *see* also neutrality/ source/ first cause physical/ potential matter/ archetype
  energetic-physical   *see* also egg/ cosmic egg/ archetype/ bio-logical development
  adaptive bio-potential   *see* adaptive potential/ epigenetics
  basic properties derived < field-in-space 29, 41, 117, 140

bio-logical 154, 246, 250, 261, 266, 267, 272, *see* code/ egg/ archetype/ bio-logical development
bio-nuclear potential/ *DNA*   100, 318, 341, 344, 350, 408
electrochemical   255
general/ typical > specific/ individual 344, 370
material order/ archetype > matter-in-principle > matter-in-practice 61, 100, 106, 145, 160, 169, 250, 405, 407
nuclear/ quantum expression of code/ blueprint   100, 266, 381
physico-chemical potentials *see* Glossary
quantum agency/ matter-in-principle   44, 52, 81, 100, 177
sexual   *see* sex/ neutrality/ polarity
source of typical action 41, 102, 106, 252
  informative + energetic   *see* also first cause/ source/ archetype/ nothingness
  (latent) field of possibilities   33, 43, 44, 52, 71, 145, 177
  apparently formless/ unseen source 29, 33, 37, 44, 134, 186
  archetypal/ potential time/ apparent eternity   171
  cosmic order/ potential > action > exhaustion   32, 33, 34, 44, 52, 53, 54, 71, 106, 298, 433
  general/ typical > specific/ individual   409
  harmonic oscillation/ resonance/ music   101, 258
  hierarchically superior   58, 132, 170
  informant (mind) > informed (physical pattern of behaviour) 58, 63, 65, 71, 91, 129
  inmost/ 'nuclear'/ essential precursor of expression see information/ informative potential
  opp. exhaustion   151
  point of balance/ poise   30, 31, 33
  potential matter   44, 48, 57, 65, 83, 118, 132, 135, 137, 140, 149, 154, 156, 176, 237, *see* also archetype/ transcendence/ first cause physical/ implicit universe
  precondition/ prerequisite/ unexpressed pattern of behaviour 35, 133, 139, 147, 235, 331
  pre-dynamic absence of motion   52, 53, 84, 153
  principle/ intrinsic character/ quality   17, 19, 21, 22, 27, 34, 35, 41, 44, 47, 70, 72, 74, 79, 93, 107, 109, 129, 140, 160, 169, 222, 225, 269, 306, 347, 405, 436, *see* also principle
  source of typical action   48, 91
  unexpressed capacity   84, 106, 151

unrealised possibilities as *opp.*
impotence (none)   106, 111, 116, 147, 156, 160
   informative-metaphysical   *see* consciousness/ conceptual development/ archetype/ source/ code
codified arrangement/ blueprint stored/ memorised for later expression/
archetypal record   99, 266, 371, 405
metaphysical egg/ idea   29, 96, 371
potential mind   *see* Potential
purposeful scheme/ plan   35, 44, 97, 103, 112, 306
Potential   *see* Essence/ Neutrality/ Consciousness/ Source/ Archetype/ First Cause Psychological/ Super-State/ *Logos*
potential matter *see* potential, informative + energetic, *see* potential/ energetic-physical
preordination*see* informative potential/ law/ order/ archetype
Prigogine Ilya   326
Primal Motion   *see Logos*
Primaries
   Axiom of Materialism   76, 121, 203, 358
   Axiom of Natural Dialectic44, 77, 121, 203
   Corollary of Materialism 203, 358, 414
   Corollary of Natural Dialectic *see* also Glossary
   Dialectic   37, 39, 134, 138
   Ovoid   *see* First Cause Psychological
principle   *see* also law/ order/ potential/ information density/ intelligence
   metaphysical entity   129
   principle > practice/ generality > individual cases   108, *see* quantum aspects/ variation/ creation/ archetypal projection/ matter-in-principle/ matter-in-practice
probability   *see* also chance/ potential/ variation-on-theme
   0 = impossibility/ 1 = complete certainty   43, 85, 106, 122, 193, 318, 340, 360
   collapse of wavefunction = 1   114, 116, 189
   degrees of quantum probability lack 0 & 1   106
   mind seeks certainty   *see* mind/ chance/ randomness
   more improbable (i.e. specific)/ more informative   77, 313, 325, 326, 340, 358
   more improbable/ less likely   77, 84, 197, 358
   nature viewed statistically 52, 109, 114, 115, 122, 125, 142
   Penrose computation   193
   scientific relativity/ degrees of doubt   125
program*see* computation/ code/ teleological algorithm
promissory faith   *see* faith
protein
   + *DNA* > chromatin   368
   20-letter alphabet   86, 319
   always codified   *see* code
   amino acid   312, 313, 314, 315, 386
   amino acid handedness   317
   bio-electrical property   256, 257, *see* also cell
   DNA > protein synthesis   322
   DNA or protein first?   323, 324, 331
   enzyme/ metabolic machine   *see*
   enzyme/ metabolic pathways as bio-machines   315, 318, 328
   folding   317, 386
   functional innovation by chance? 359, 375
   informative, functional and structural bio-factor   316, 319
   proteinaceous probabilities   122, 318, 334, 360
   synthesis   262, 322, 386
proto-cell   334, 339, *see* abiogenesis
proton   *see* sub-atomic particle/ polarity physical
pseudoscience   87, 105, 201, 300, 429
PSI/ psychosomatic interface   *see* psychosomasis
Psyche   201, *see* Soul
psycho-biologic   *see* archetype/ psychosomasis
psycho-logic   *see* also consciousness/ sub-consciousness/ information/ mind/ subjectivity/ immateriality
   association   *see* unity/ brain/ memory/ psychosomasis/ resonant association
   emotional life/ psychodrama   286
   human norm   221
   informative qualities   35
   mind = immaterial entity/ informative component/ psychological pole   *see* also information/ mind/ mnemone/ archetype/ Archetype
   psychiatry   228
   psychological development
   development of idea/ plan *see* conceptual development
   educational/ moral   19, 125, 228, 417, 430, 432, 434
   spiritual evolution 75, 90, 221, 225, 228, 286, 438
   psychological entropy/ information loss

232, *see* sleep/ ignorance/ randomness/ sub-consciousness
psychological force/ volitio-attraction   44, 47, 51, 56, 73, 90, 96, 97, 125, 163, 185, 222, 224, 225, 227, 254, 273, 278, 283, 289, 303, 431, 442
   psychological impotence   *see* sub-consciousness
   psychological negentropy > understanding/ learning/ lightness of being   75, 90, 222, 228, 439, *see* also negentropy/ informative negentropy/ knowledge/ understanding
   psychological record/ database *see* sub-consciousness/ memory
   psycho-space   137, 138, 152
psychology   *see* psycho-logic/consciousness/ subjectivity/ information/ mind/ mnemone
psychosomasis   *see* also resonance/ harmonic oscillation/ vibration
   archetypal memory/ program 256, 257
   border/ interface/ medium   16, 243
   caduceus   268, 272
   informative transduction   68, 257, 260, 269
   interface/ linkage   16, 27, 57, 63, 67, 81, 84, 100, 160, 240, 244, 254, 257, 261, 275, 276
   mind-matter exchange   243, 254, 260, 276
   morphogene   250, 262, 273
   psycho-biological exchange 63, 67, 84, 226, 273, 275, *see* signal translation
   psychosomatic control   274
   resonant association   260, 276
   sub-conscious/ psycho-biological mechanism   63, 84, 203, 218, 238, 239, 240, 254, 256, 260, 383
punctuated equilibrium   *see* fossil
purpose   *opp.* physical chance/ *see* force psychological/ cause/ reason/ teleology/ metaphysic
pyramid   see cosmic model
Pythagoras   83, 120, 259

Q

quality   *see* also cosmic fundamentals
   aesthetic criterion   101, 228
   expression of cosmic fundamentals 34, 58
   metaphysical value/ yardstick 35, 59, 71, 72, 73, 93, 121, 201, 227, 228, 286, 288, 434
quality of information   227
quantum aspects   *see* also matter-in-principle/ subatomic particles/ ZPE

(*raj*) kinetic phase of material phenomena   61, 63, 83, 170, 257
collapse of wave function > resolution of probabilities/ precise certainty/ physical actuality   *see* sub-atomic particle
cosmic ocean/ quantum water/ bulk-bergs   137
creation fields   145, 170, 196
determined quantum characters > natural law   108, 117, 119, 125, 195
HUP (uncertainty principle) 113, 116, *see* also chance/ probability/ randomness
informant agency   63, 125, 177, 254, 257, 258, 259
is uncertainty certain?   114
letters/ punctuation of cosmic language   81, 100, 177, *see* also alphabet/ cosmo-logical language/ score
paradoxical wave-particle duality *see* also wave/ energy
psychosomatic agency?   118, 238, 239, 255, 260
quantum biology   256, 261, 332, *see* also *H. electromagneticus*/ psychosomasis/ resonance
quantum theory   105, 113, 114, 115, 116, 127, 132, 141, 142, 196
statistical quantification   *see* chance/ probability/ potential
string theory/ quantum extravaganzas   142, 144, 196
sub-strate of potential matter   57, 61, 118, 119
super-strate of condensed/ classical matter   57, 61, 83, 154
vacuum's plenitude   156, 159
whence quantum characters? 65, 110, 119, 178
quark   *see* subatomic particle/ polarity physical

R

radiometric dating   *see* dating methods
raj ……… *see* cosmic fundamental/ positive influence
   vector of levity   35
Ramayan   276
randomness   78, 99, 124, *see* also chance/ probability
   anti-reason   121, 429
   noise/ bug/ interference   323
   random variation   299, 342, 363
   reason-in-reverse   78
rationalism   22, *see* also reason
reality   *see* also truth/ objectivity/ subjectivity

common delusions/ illusions 23, 50, 195, *see* also delusion/ illusion
exclusively materialistic/ scientific version 19, 23, 50, 68, 105, 116, 137, 202, 435
filtrate of brain 68, 148
hierarchical/ lesser realities/ relative appearances 48, 69, 71, 137, 186
inclusive holistic (materialistic + immaterialistic) version 19, 69, 209, 238, 435
material = energy/ immaterial = information 137, 203, 215
personal/ cosmological *see* cosmological axis/ eye-centre
psychological reality 117, 215
quantum reality 113, 116, 146, *see* also quantum aspects
special delusion
informative/ matter generates information 24, 78, 117, 203, 204, 209, 307
informative/ matter innovates 78, 111, 209, 308, 325, see also innovation/ code/ creativity
neuro-scientific reductionism/ mind = matter 24, 59, 68, 209, 215, *see* also brain/ mind/ neuro-hormonal system
neuro-scientific/ mind = matter 202
physical/ starry cosmos represents the whole truth 70
Reality *see* Essence/ Absolute/ Super-Nature/ Consciousness
reason *see* also logic/ psychological force/ cause/ archetype
   anti-reason/ inconsistency/ irrationality 90, 124
   cosmos for no reason? 22, 41, 108
   fossil rationality 300, 352
   gives meaning/ informs 97, 275, 368, 440
   Greek word λόγος (*logos*) 93
   inward/ subjective/ insightful 20
   is the origin of reason irrational? 122, 124
   metaphysical exercise 59, 74, 77, 118, 120, 304, 442
   no reason = irrationality 107, 111, 122, 191, 209, 275, 358, 399, 413
   opp. randomness/ chance 78, 124, 306
   opp. sensation 19
   orderly/ intellectual exercise 90, 93, 120, 140, 222, 437
   outward/ objective/ analytical 20, 25, 88
   rational/ intentional/ teleological exercise 87, 102, 120, 124, 173, 209, 302, 401
   reason = cause 90, 96
   reason/ logic/ mathematics are metaphysical *see* metaphysic/ teleology
   sub-reason/ whimsy/ thoughtless reflex/ instinct 90, 222
   supra-reason/ accord with *Logos* i.e. Logical 90, 93
Reason *see* First Cause/ *Logos*/ Archetype
recapitulation theory of 387
red shift 187
Rees Martin 183
reflective asymmetry *see* symmetry/ inversion/ mirror image/ chirality
reincarnation
   ante-natal psychology 282, 288
reincarnation/ metempsychosis 276, 279, 282, 283, 289, *see* also homeostasis/ cyclical equilibration/ *karma*
relativity *see* also motion/ change
   < motion/ change 31
   = mental relationship/ perspective 140, 228, 418, 435
   = motion/ change 52, 64, 112, 156, 177
   = relationship 16, 58, 72, 129
   bio-logical relationships 332, 345, 348, 352, 396, 429
   communal 418, 432
   dialectical relativity 28, 29, 36, 417
   Einstein's physical theories of 105, 127, 140, 141, 182
   existential truth/ reality is relative 31, 37, 48, 69, 70, 71, 112, 134
   fundamental relativity 36, 69, 258
   mind unifies/ orders relativities 120, 125, 140, 228, 245
   moral 73, 75, 228, 430, 439
   opp. absolution 28, 36, 51, 93, 134, 283
   personal axis of 418, *see* cosmological axis/ eye-centre
   physical theories of 42, 187
   relationship with absolution 439
   relative probability/ quantum uncertainty 114, 142
religion/ formal canon *see* also faith/ world-view/ delusion/ truth
   anti-religion = atheism 430
   conscientious devotees 25
   delusions/ imperfections 25
   religion/ world-view unavoidable 435
   science and religion 27, 90, 416
   social/ life-style 432, 435
   state religion/ ideology 89, 419, 430, 433
   sub-religion 73
   supra- or nuclear religion 221
   supra-religion 73, 90, 93, 435, 438, 439, 444, *see* also Essence/ Super-Nature/ Transcendence/

Consciousness
REM sleep　　　　　　　　　　233
replication　　　　　*see DNA* molecule
replisome　see molecular/ cellular machinery
reproduction　　*see* also polarity/ sex/ teleology/ survival/ conceptual & bio-logical developments
　> end-product/ reproductively enabled form　247, 254, 293, 371, 387, 395, 411
　alternation of generations/ metagenesis　398
　anticipatory/ teleological process　247, 254, 262, 293, 295, 323, 386, 396, 400, 403
　archetypal stability/ genetic variation　353, 387, 397, 400
　archetypally informed　　　387, 399
　asex/ mitotic sub-routine　368, 387, 388, 395
　asexual cellular/ cell cycle's reproductive phase　　　　　309
　asexual/ invariance/ cloning/ two < one
　complex algorithms/ programs　328, 386, 388
　evolution of?　　　　　　　　388
　one > two/ dispersal　　　　　397
　prokaryotic fissive sub-routines　309, 368, 387
　reproductive archetype　*see* archetype/ sex/ polarity
　sexual　　　　　　　　　*see* sex
resistance (expression of (*tam* ↓) cosmic fundamental/ concretizing influence *see* gravity/ materialisation/ negative power
resonance　*see* also harmonic oscillation
　attunement　　　　　　258, 260
　electromagnetic　　　　258, 274
　harmonic interaction　　258, 274
　psychosomatic attunement　263, 274
　quartz crystal　　　　　　　258
　resonant association　243, 245, 258, 260, 263, 266
　synergetic　　　　　　　　　260
　vibratory transfer of energy　258, 263
respiration　*see* metabolism/ polarity bio-energetic
resurrection kits　　　　　　　280
retrotransposon　*see* gene and Glossary, transposon
ribosome　　*see* molecular/ cellular machinery
ribozyme　　　　*see RNA* molecule
RNA **molecule**
　information-carrier　　　　　301
　micro-*RNA* 347, 373, *see* also Glossary
　m-*RNA*　　　　　　　　　　319

non-protein-coding　　　　　　407
ribosomal r-*RNA*　　　　　　　319
ribozyme　　　　　　　　　　　324
RNA gene　　　　　　　　366, 373
t-*RNA*　　　　　　　　　　　319
Ryle Gilbert　　　　　　　　　414

S

saint　19, 27, 71, 93, 101, 149, 228, 437
Salvation　　　　　　*see* Communion
samadhi　　　　　*see* Consciousness
sanskara　　　　　　　*see karma*
Sat see cosmic fundamental/ tendency to balance/ neutralising influence/ potential
　anti-vector/ quality of balance　35
Schrödinger Erwin　　　　114, 115
science of the soul　　　75, 95, 228
science/ natural science - philosophical self-definition　*see* materialism/ naturalistic methodology/ scientism/ faith
scientism　　　*see* faith and Glossary
secondaries
　dialectic　　　　　　38, 39, 138
　ovoid　　　*see* first cause physical
seed　33, 387, *see* also egg/ zygote/ idea
Seed　　*see* tree of life/ world tree.
self-organisation　*see* order/ innovation/ code/ creativity/ circuitry
sensation　*see* also neuro-hormonal system/ objectivity
　opp. contemplation　　　　　219
　outward diffusion of attention　19, 20, 69, 222, 226
　physical connection　　　　　132
separation/ isolation/ differentiation　*see* (dis-)unity
SETI　　　　　　　　　　　　　300
sex　　　*see* also polarity/ symmetry/ reproduction
　> variation-on-theme　*see* variation
　anticipates/ is teleological　　400
　cerebral sexuality　　　　　　381
　complex algorithm/ program　389, 398, *see* egg/ bio-logical development/ conceptual development
　deliberate (meiotic and genomic) lottery　　　　　　　　　　389
　evolution of?　262, 392, 396, 399
　genetic recombination　*see* variation
　hermaphroditic
　neutrality/ one > two/ > polarity 265, 267, 381, 389, 395, 396, 397
　physicality　　　　　　　　　398
　potential > dimorphic male/ female expression　265, 267, 381, 389, 395, 396

511

sexual potential   *see* neutral archetype
   information < archetype   396, 403
   meiotic sub-routine 266, 368, 387, 389, 395
   outward (*tam*) externalising/ dispersive vector   380
   polar unification/ two > one/ sexual union   380, 389
   pre-programmed act of materialisation   380
   reflective asymmetry   *see* inversion/ polarity/ switch
   sexual archetype   266, *see* also neutrality/ polarity
   sexual selection   341
   sexual stereotype   392, 394, 395, see archetype

Shabda 66, 93, *see* also *Logos*/ Sound and Light

Shannon Claude   77
Sherrington Charles   238
sickle-cell anaemia   *see* mutation

signal translation
   mind <> body exchange   18, 68, 260
   sub-routine of typical mnemone   240, 242, 244, 251, 256, 260

sin/ swollen passion *opp*. beneficence 222, 428, 431, *see* also negative power/ immorality/ evil

sink *see* also oblivion/ expression of (*tam*)
   cosmic fundamental/ subtendence/ sink
      conscio-material base   32
      cosmic periphery/ solidity   153
      extreme subtendency   54
      impotence/ exhaustion/ end   61, 147
      physical sink/ matter   22, 104, 147
      psychological sink/ sub-conscious mind   22, 155
      subjective annihilation   33, 58, 155
      sub-state/ exhausted phase 33, 36, 61, 64, 151

sleep 35, 63, 147, 150, 167, 230, *see* also ignorance/ oblivion/ sub-consciousness/ switch
   evolution of?   231, 232
   switches *SCN*, *VLPO*, *RAS* and orexin   231

SMC (standard model of cosmology)   *see* also big-bang/ inflation/ matter/ cosmos

SMC (standard model of cosmology) 105

SMPP 105, 141, *see* also standard model of particle physics

snowflake   95
social Darwinism   429
society   *see* community
solar plexus   271
somato-sensory arc   383, *see* neuro-hormonal *sys*tem

soul   *see* also neuroscientific imperative/ materialism/ atheism
   materialism's figment of imagination   103, 202
   naturalistic non-entity 23, 76, 103, 126, 201, 202

Soul/ True Self   *see* Psyche/ Essence/ Consciousness

source   *see* also creativity/ alpha/ expression of (*sat*) cosmic fundamental/ transcendence
   centrality   31, 32, 59, 108, 155
   cosmic/ existential   *see* Source/ First Cause
   dot/ point/ point of origin/ first hint of form/ primal motion   29, 35, 65, 71, 137, 156, 159, 161, 340, 408, *see* also quantum physics
   formless cause/ potential 29, 33, 37, 41, 71, 106, 137, 151, 161, 340
   informative component 81, 93, 96, 104, 109, 116, 121, 129, 132, 155, 156, 159, 259, 269, 335, 355, *see* also archetype/ information/ conceptual development
   inverted opposite of sink 33, 36, 38, 71, 151, *see* symmetry/ inversion
   is matter source of mind? 103, 132, 154, 215, 299, 408, 413
   latent or incipient expression/ egg 121, 132, 192, 399, *see* egg/ metaphysical egg/ idea
   of creativity   21, 35, 104, 222, 378
   of mind   *see* Source
   of purpose/ intent/ design   *see* idea
   precedes sink   121
   principal/ source of order   44, 96, 102, 106, 116, 129, 155, 160, 259, 291, 355, 362, 443, *see* also order
   projector   32, 33, 41, 48, 61, 65, 91, 104, 107, 119, 144, 171, 180, 186, 190, 290, 401
   super-state/ potent start/ first phase   61, 71, 116

space 178, *see* also apparent absolutes/ void
   3-D/ dimensional trinity   29
   apparent absolute   197
   archetypal property of pre-space   118, 120, 156, 157, 160
   as frame of reference   158
   Calabi-Yau geometry/ topology   196
   combinatorial   359, 360
   continuity
   classical continuity of unbroken nothingness   144
   quantum discontinuity of broken/ granular/ pixellated nothingness 108, 144

coordinate of motion/ change 176
dialectical trinity 160
ether 157, 158
formless page for matter's geometric
  script 29, 98, 136, 157
impotent/ exhausted sink 42, 43, 154, 156, 161
inner space *see* psycho-space/ subjectivity
intra-atomic space/ vacuum 69, 146, 151, 170
levitational/ expansive 187, 191, 192
outer space/ range of extra-atomic vacuum densities 146, 154, 157, 158, 159, 162, 167, 170, 186, *see* also void/ physical nothingness
physical nothingness/ emptiness/ abstraction 154, 157, 161, 179
polarised vacuum 145
potential source/ vacuous quantum plenitude 41, 44, 48, 156, 159, 160
pre-vacuous absence 42, 48, 160, 186
vacuum catastrophe 142
vacuum energy 146, 157, 161, 179, *see* ZPE

space-time *see* also space/ time
  cosmic envelope 41, 157
  dynamic nothingness 151
  flexi-world 141
  materialism's immateriality/ apparent nothingness/ intangible physic 163
  phenomenal extension/ basic physical coordinates 41, 42, 141, 147, 163
  quantum blur 115, 143

special cases
  for materialism mind = special case of matter 151
  of consciousness/ non-consciousness = matter 22
  of metaphysic/ materialism denies metaphysic 21

species *see* also variety/ bio-classification
  artificial/ breeder speciation 307, 336, 345, 373, 414
  binomial system 343
  cross-species cloning 371
  Darwin's plastic species 343, 429
  hybridisation 345, 375
  lowest unit of standard bio-classification 343
  speciation 332, 344, 351, 361, *see* also variation/ plasticity
  species problem/ hard to fix 343, 429
  standard definition 343
  sub-divisions 266, 343, 429
  taxonomic rank 86, 266, 343
spectrum *see* cosmic models

sperm/ male informative agent 264, 371, 389, 398, *see* also sex/ polarity
Sperry Roger 209
splicing/ spliceosome *see* molecular/ cellular machinery
split-brain model 384
spontaneous generation *see* abiogenesis
Sraosha *see Logos*/ Sound and Light
SRM (self-replicating molecule) 323
stack *see* also dialectical stack
  primary 37
  secondary 38
Stalin Josef 430
standard model of particle physics 105, 141, 142, 143, 170, 171, 182
STE (special theory of evolution) *see* micro-evolution/ variation-on-theme)/ plasticity
steady-state theory 180, 187
stereo-computation *see* computation and Glossary
stickleback 373, 410
stratigraphy 172
string theory 142
stromatolite 422
sub-atomic particle *see* also cosmo-logical language/ Glossary - elementary particle
  collapse of wave function 116, 189, 196, *see* also quantum aspects
  electron/ electrical charge 28, 63, 69, 84, 112, 115, 144, 145, 146, 160, 167, 170, 178, 194, 195, 209, 239, 257, 259, 284
  enduring characters 119, 167
  force-carrier/ boson 142, 143, 178
  harmonic modulation 177
  Higgs boson/ mass-donor? 178
  mass-carrier/ fermion 143
  neutrino 170, 194
  neutron 146, 170, 178, *see* also neutrality/ polarity cosmic
  nucleon 69
  photon
  least material levitational entity (*opp.* quark) 160
  quantum of radiant energy 28, 65, 115, 142, 143, 144, 145, 166, 178, 195, 239, 255, 257
    positron 28, 145, 160
    proton 49, 115, 146, 148, 159, 160, 167, 170, 178, 184, 194, 195, 284
  quark
    least material gravitational entity (*opp.* photon) 160
    quantum of contractive energy 142, 160, 170, 209, 229, 326
sub-conscious mind *see* sub-consciousness
sub-consciousness *see* also mnemone/

archetype
- = memories/ mnemone/ psychological database 50, *see* also Chapters 15 and 16 *passim*
- = psychological experience/ condition of sleep 50, 232
- archetypal memory 50, 63, 99, 118, 153, 235, 267, 268, 272, 273, 302
- archetypal program 104, 235, 239, 249
- archetypal/ psychosomatic control 247, 250, 268, 273
- automatic informant/ sub-conscious instructor 63, 254, 257, 268, 269, 273
- dormant/ comatose conditions 32, 63, 154, 217, 218, 232, 233, 236, 250, 251
- dreaming & (semi-dormant) day-dreaming states 35, 104, 217, 232, 236
- evolution of memory inc. sub-conscious mnemone & corresponding bio-physical mechanisms? 253
- evolution of sub-consciousness? 230
- finished state/ 'solid' mind 16, 153, 218
- fixed image/ frozen time 84, 104, 155, 166, 234, 235, 248, 254
- habit 215, 248
- immaterial element/ metaphysical entity/ form/ architecture 104, 240, 241, 254, 306
- memories personal & typical *see* mnemone
- non-waking zone 35, 57, 232
- passive mind 98, 104, 235, 257
- psychological context/ environment 16, 244
- psychological files/ 'photographic' records 50, 80, 82, 235, 248, 249, 302, 306
- psychosomatic 'sandwiched' zone 63, 67, 155, 222, 229, 239, 273, 275, 302
- rationally inaccessible 230, 237
- sub-state/ subtendent mind 63, 66, 132, 218

subjectivity
- experiential being/ awakened life 18, 125, 174, 204, 209
- formful = mind 51, 125
- immaterial element 18, 54, 84, 104, 121, 200, 201, 209, 229, 276, 289, 443, *see* also subjectivity
- inner subject as *opp.* outer object 18, 20, 51, 54, 74, 121, 126, 137, 176, 210, 279, 431
- inward focus as *opp.* sensation/ physical action 19, 20, 96, 222, 225
- materialistic nervousness 23, 91, 201, 203, 209, 215, 229, 276
- perspective involves informative meaning 18, 21, 35, 72, 78, 100, 104
- range from 1 (Subjectivity) through mind to special case (0) matter 61, 67, 68, 165
- special case (zero) = matter 33, 51, 58, 100, 155
- subjective recycling 277
- subjective science 20, 225, 228

Subjectivity *see* Consciousness/ Essence/ Absolute

sub-routine *see* code/ computation; *top-down* programming

Substance *see* Essence (Glossary)/ *Logos*/ Absolute

sub-state *see* impotence/ subtendence
- (*tam*) position 61, 63
- base pole/ physical universe 53, 61, 63, 83, 286
- impotence *opp.* super-state potential 30, 33, 36, 61
- informative/ energetic void 43
- sink/ phase below 33, 61, 62, 63, 218, 426

subtendence *see* also negative power/ sink
- 'dark/ nether pole' as *opp.* transcendence 61
- inferior effect/ next level down/ sub-state 62
- non-consciousness *see* non-consciousness
- *opp.* transcendence 54
- psychological hell *see* hell/ negative power, *see* hell/ negative power

sun *see* astronomical bodies/ time/ ecology
sunlight *see* light/ photosynthesis
super-code/ data compression *see* code/ epigenetics

Super-Consciousness *see* Consciousness
Super-Nature *see* also Essence
- (N)One 151, 435
- Consciousness 72, 93, *see* also Supreme Being/ Enlightenment
- Most Natural/ Heart of Nature 73, 137, 153, 228, 282, 435
- Natural Essence/ Nucleus/ Substance 72, 93, 137
- super-matter *see* potential matter/ first cause physical/ archetype
- super-natural = immaterial 62

super-state *see* also potential/ transcendence
- (*sat*) position 60, 65
- informative/ energetic poise 60, 65
- source/ potential/ phase above 60, 65, *see* also potential/ informative component/ archetype

Super-State *see* Transcendent Information/ Source/ Potential

super-symmetry *see* symmetry
Supreme Being *see* Essence
Supreme Court, USA 88
survival *see* also reproduction/ reincarnation/ immortality/ Consciousness
  'selfish genes' 285, 386
  > more life 245, 262, 275, 302, 397
  cyclical equilibration/ homeostatic regulation 302, *see* also vibration/ homeostasis
  economically comfortable 72, 245, 423, 424
  entropic destruction of complex (bio-)molecules 321
  genetic/ reproductive strategy *see* reproduction
  little deaths - conscious pain/ unconscious periods 233, 425, *see* also sleep/ pain/ negative power
  metaphysical recycling *see* reincarnation
  non-survival/ death/ extinction 201, 279, 281, 283, 285, 296, 303, 308, 341, 353
  self-seeking schemes 74, 222
  temporary reprieve/ remission from death 334, *see* also gravity/ impotence/ exhaustion/ death
  Ultimate Survivor 286
  what survives death? 250, 280, 281, 289
survival of the fittest 341, 342, 428
  battle 430, 431
  molecular? 84, 314, 315, 317, 321, 323, 325
  survival not arrival 342
  the fittest survive (tautology) 86
switch *see* also inversion/ polarity
  binary 52, 77, 87
  biochemical switches/ signal messengers 231, 274, 281, 299, 394, 404, *see* also information/ bio-logic/ homeostasis/ neuro-hormonal system
  chiral 316, 382, *see* also mirror image
  computational 304, 307, 370, 381, *see* also hierarchy/ computation/ circuitry/ teleology
  electronic 197, *see* circuitry
  epigenetic 366, 369, 370, 373, 396, *see* also epigenetics
  gender 395, 396, 398, *see* sex
  genetic 250, 262, 299, 304, 307, 366, 372, 404, 407, *see* circuitry/ computation
  hierarchical switching system 263, 307, 369, 374, 399, 404, 409, *see* also hierarchy/ genetic linguistic hierarchy/ epigenetics
  homeostatic switches *see* homeostasis
  inter-modular/ between genetic routines 304, *see* also computation
  inversion 54, 382, *see* also inversion/ reflective asymmetry
  mind-body decussation 212, 269, 273, 382
  sleep/ waking toggle 208, 232
symbol/ symbolic information *see* language/ cosmo- and bio-logical languages/ reason/ mathematics
symmetry *see* also balance/ equilibrium/ polarity/ mirror image
  (*raj*)/ (*tam*) asymmetry-generators 36, 156, 213
  (*sat*) quality of balance/ equilibrator/ symmetry generator 35
  archetypal/ pre-actively latent 104, 108, 109, 160
  balanced bilateral/ radial and other biological symmetries 213, 272, 273, 346, 382
  breakage > variation-on-theme 108, 435, *see* also variation
  central = neutral/ undifferentiated potential 104, 108, 144, 156, 258, 435
  electromagnetic oscillation 136
  external/ large-scale/ variation-on-principle 177
  geometry of balance/ proportion 120, 158, 213, 272, 320, 385, *see* also golden mean
  internal quantum/ gauge/ invariant/ local/ space-time independent 109, 143, 177
  invariance of natural law 107, 108, 177, 435
  physics - study of symmetrical transformations 156
  polar bifurcation 104, 108, 144, 258, 329, 381, 395
  reflective/ polar asymmetry of complementary opposites 36, 50, 52, 53, 54, 104, 108, 109, 137, 143, 145, 156, 160, 177, 205, 212, 213, 272, 273, 329, 377, 378, 381, 383, 385, 395, 396, 435, *see* also inversion
  scale symmetry 326, *see* also mathematics/ Mandelbrot
  super-symmetry/ *SUSY* 105, 141, 143
Symmetry *see* Potential/ Unity/ Essence
synchromesh 1
  cognizance ◇ mnemone 244
  conscious ◇ sub-conscious gearing 244
  present experience ◇ memory 244
  psychological ◇ psycho-biological

gearing 238
synchromesh 2
  memory ◇ oblivion/ physical agency 257
  mnemonic ◇ body 257
  psycho-biological ◇ physical
    gearing 238, 257
  sub-conscious ◇ non-conscious
    gearing 257
Synthia 336
system/ systematic integrity *see* cosmo- & bio-logics/ irreducible complexity/ minimal functionality/ computation/ neuro-hormonal system/ brain/ genetic circuitry/ machine/ molecular & cellular machinery/ teleology
systematics *see* bio-classification/ homology

T

T. rex *see* dinosaur
tam *see* cosmic fundamental/ negative influence
  vector of gravity 36
Tao *see* dialectical operator
Tao (or *Dao*) 35, 38, 93
tautology 86, 122, 172, 352, 376, *see* also logic
taxonomy *see* bio-classification/ homology
tectonic plates and cycles 422
Teleological High-Spot *see* First Cause/ Logos/ Natural Teleology
teleology *see* also cause/ first cause/ reason/ information/ mind/ force psychological/ nature/ machine
  algorithm to specific target 355, 395, 405, *see* also program/ computation
  bio-logical development 247, 295, 407, *see* also bio-logic/ bio-logical development
  causal/ provident phase 96
  conceptual development 94, 254, 307, 337, *see* also conceptual development/ idea/ innovation/ creativity
  doctrine of first-or-final causes 124
  fulfilment of expectation/ anticipation 49, 103, 227, 327, 386
  goal-oriented program 304, 306, 350, 411, *see* also reason/ computation
  immaterial/ metaphysical/ psychological 102, 227
  informative intent 77, 247, 403, *see* also force psychological
  matter cannot (self-)engineer 111, 247, 294, 335, 413, *see* also matter
  mind > matter 263
  outside naturalistic/ scientific methodology 102, 124, 412, *see* also naturalistic methodology/ force psychological/ mind
  rational creativity 86, 263, 323, 405
  study of purposive design in nature 49, 102, 173
  teleological *opp.* accidental 49, 323, 413
theme *see* law/ archetype/ invariance
Theodora 282
theories of relativity *see* relativity, Einstein's theories
Theory of Intelligence 125, 198, 367, *opp.* Theory of No Intelligence *see* also intelligence/ reason/ innovation/ code
Theory of No Intelligence 91, 353, 367, 379, 399, *opp.* Theory of Intelligence *see* also evolutionary theory/ chance/ entropy/ random pointlessness
thermodynamics first law 186
thermodynamics second law 186
Thermoplasma acidiphilum 333
third eye *see* eye-centre
Thompson D'Arcy 262
thought *see* active information/ consciousness-in-motion/ mind
throat/ brainstem plexus 273
time *see* also apparent absolutes
  3-D/ dimensional trinity 168
  as frame of reference 158
  coordinate of motion/ change 165, 176
  dating methods 172, see also fossil
  dynamic nothingness 42
  era/ grade 169
  fixed image/ frozen time *see* sub-consciousness/ memory/ archetype
  fundamental trinity 169
  is creatively impotent 25, 85, 107, 173, 311, 336, 358, 372
  motion's ghost 162
  physical/ sub-state timelessness 154, 167
  relativities of time 164
  species of time
archetypal time 166
arrow/ straight-line/ DC-time 48, 168
bio-time 166
cycle/ period/ AC-time 166, 167, 168, 181, 220, 421, *see* also wave/ vibration
psycho-time 165
super-time 165, *see* archetypal time/ timelessness
  time relative 164
  timelessness/ no-time 164, *see* super-time/ sub-state oblivion
  unseen space for motion's tale 29, 162
tonoscope 258
TOP (Theory of Potential) *see* GUT, TOE, GUE, archetype, conscio-material

dipole, potential, hierarchy, *fig.* 5.1 and Glossary: unification, *see CUT, GUT, TOE, GUE*, archetype, conscio-material dipole, potential, hierarchy, *fig.* 5.1 and Glossary: unification
top-down   21, 27
transcendence   *see* also super-state/ potential
    'light/ upper pole' as *opp.* subtendence   65
    archetypal informant   58, 107
    informative source   91, *see* also source/ information
    mind transcends matter   204
    next level up/ level above/ super-state   65
    of logic/ mathematics   120
    psychological heaven   282
    source/ superior cause of inferior effect   36, 61, 65
transcendent
matter   *see* potential matter/ physical archetype/ physical absolute
projection of biological form   290
projection of physical phenomena 41, 48, 107, 144, 153, 171, 180, 187, 190, 290, 401, *see* also first cause physical/ archetype/ source/ informative potential
Transcendence   *see* Super-State/ Illumination/ Consciousness
    metaphysical step from existence to Essence   66
transcription   322, *see* also protein synthesis
transformism   *see* macro-evolution
translation   322, *see* also protein synthesis
transmigration   *see* reincarnation
transmutation 123, 143, 311, 348, 425, *see* also macro-evolution/ alchemy
transposon   *see* gene and Glossary
transposon/ retrotransposon   *see* gene:jumping gene
tree of life
    developmental stereo-tree   *see* bio-logical development discordance
    developmental   348
    genetic/ phyletic   346, 348
    family tree
    literal, personal hereditary tree   345
    metaphorical/ Darwinian icon   290, 309, 339, 345, 347, 363, 391
    inverted
    immaterial, informative root > material, energetic outcome   104, 209, 435
    neuro-hormonal/ brain > body   209, 346, *see* also neuro-hormonal system/ hierarchy/ information

    mythical world tree   104
    one historical/ many hypothetical   346
    single tree or living forest?   346, 347, 354
Triangle of Truth   38
trinity   *see* cosmic fundamentals/ fundamental order of creation/ hierarchy/ cosmos/ Natural Dialectic
    bio-logical   *see figs.* 19.1 and 19.2
    codified potential/ physiology/ anatomy 297
    information/ function/ structure   297
    nucleus/ cytoplasmic business/structural sub-units inc. membrane   297
    cosmo-logical
    macrocosmic   *see figs*. 1.3/ 2.3/ 2.6/ 3.1/ 13.1 also conscio-material (c-m) dipole/ cosmic models (ziggurat)
    microcosmic   *see figs*. 13.1/14.1/ 15.1/ 17.3/ 17.4 also conscio-material (c-m) dipole
    dialectical   34, *see* also cosmic fundamentals/ dialectical operators
    physical/ informed   *see figs*. 3.3/ 10.2 also energetic trinity/ potential matter/ quantum matter-in-principle/ bulk matter-in-practice
    psychological/ informant   34, *see figs*. 3.2/ 5.1 also informative trinity/ archetype/ Archetype
    psychosomatic/ psycho-biological   *see figs*. 15.2/ 15.4/ 15.5/ 16.1 also archetypal bio-classification
    reproductive   *see figs*. 24.4/ 24.5 also reproduction/ sex/ bio-logical development
    social   *see fig*. 26.4 also religion/ politics/ law
Trinity *see figs*. 2.3/ 8.2 also Metaphysical Tri-unity/ Transcendent Potential-*Logos*-human
truth   *see* also reality/ objectivity/ subjectivity
    graded/ sliding-scale/ hierarchical priorities   69, 71, 186, 228
    holism's = mind (based on consciousness) *and* matter   68
    materialism's = matter   68
    relative lack of ignorance/ illusion   68
    relative/ lesser existential truths   71
Truth   221
two pillars of faith   *see* faith
type/ typology   *see* bio-classification/ homology/ archetype
typical mnemone   *see* mnemone

U

ultra-conscious = super-conscious   66, 218
uncertainty principle   *see* chance

unconsciousness   *see* psychological sub-consciousness/ physical non-consciousness
Uncreated Axis   *see also* Essence/ Consciousness
understanding   *see also* knowledge/ illumination
     grasp of principle   128, 129, 222, 273, 379, 380, 435, *see also* intelligence
     perception of cause/ how it works   14, 31, 69, 74, 94, 112, 127, 129, 157, 225, 381, 425, *see also* intelligence
uniformitarianism   172
unity
     (dis-)unity/ isolation (↓)/ apparently separate forms 69, 134, 138, 144, 147
     initial unity/ archetype   142
     opp. duality/ polarity/ multiplicity   40
     opp. nothingness   134
     social community   417
     unification/ (↑) *raj* tendency   69, 127, 138, 145
     unit/ integral whole e.g. proton/ atom/ protein/ cell/ multicellular body   69, 352
     unitification/ (↓) *tam* tendency 69, 134, 138, 144, 147, 395
     unity/ duality
     ◇ duality/ polarity   28, 35, 145, *see also* Natural Dialectic
     asex - one > two   388
     duplication/ division (↓) one-into-two (or many)   38, 145, 395
     of complementary opposites   45, 51
     polar bifurcation   *see* symmetry
     sex - two > one(s)   389
     union (↑)/ mergeance/ two-(or many)-into-one   69
     Unity   *see also* Essence/ Communion/ Super-Nature
         Cosmic Monopole   45, 53, 426
         GUE (Grand Unified Experience)   139
         Pre-existential Potential   138, 139
         term of the Essential Equation   134, 140
         Transcendent (N)One 45, 66, 130, 138, 149, 154
universal
     body   *see* cosmos/ matter
     idea   *see* archetype
     memory   *see* archetypal memory
     mind   *see* mind/ archetype/ implicit universe
universal *modus operandi*
     systematic action/ planned dynamism   *see* archetype/ creation
     unplanned 'progress'   *see* evolution theory
uranium   422

Urey Harold   313
Ussher James   171
utopia/ idealised society   431, 433

V

vacuum   *see* space/ physical void/ nothingness/ ZPE
vacuum energy   *see* vacuum/ levity
variation   *see also* plasticity/ order/ randomness
     bio-logical
     evolutionary transformism   307, 310, 344, *see also* macro-evolution/ innovation
     intra-typical change   387, *see also* micro-evolution/ limited plasticity
     invariance = reason for asexual reproduction   388
     low-level, local variation within high-level, general invariance   264, 351, 373, 389, 397, 405, 409
     macro-boundary 343, 350, 351, 355, 363, 414
     micro-evolution/ variation-on-theme   267, 307, 308, 340, 342, 343, 344, 345, 353, 363, 365, 405, 413, *see also* plasticity limited/ species
     variance = reason for sexual reproduction 389
     change/ relativity   95, 112, 114, 177
     cosmo-logical
     low-level, local variation within high-level, general invariance   108, 109, 112, 113, 115, 116, 118, 119, 137, 142, 158, 350, 351
     order of the finite   137
     variation-on-theme/ -principle/ -law 50, 85, 86, 87, 95, 99, 107, 108, 112, 115, 116, 119, 122, 142, 266, 306, 307, 310, 345, 347, 351, 353, 365, 389, 404, 405, 409, *see also* principle
     hidden variable   115, 116, 186, 203
     non-random   *see* adaptive potential
     on principle/ designed   96, 122, 306, 349, 389
     physical invariant/ non-consciousness   204
     psychological invariant/ mind   204
     random   *see* randomness/ chance/ probability
     special case - fixity/ invariance/ archetypal law   107, 108, 109, 110, 142, 350
     the equivocation   86, 310, 363
     without-theme = biological/ cosmological concepts of evolution 343
variation-on-theme   355
variety
     Darwinian species   *see* species
     incipient species   343, 344

On the Tendency of Varieties.... 341, 343
    sub-species/ race   345
vector   *see* cosmic fundamentals/ anti-parallels/ polarity informative/ levity/ gravity
Venter Craig   335
vestigial organs/ bio-logical junk   88
vibration   *see* also wave/ (*raj*) cosmic fundamental/ cosmos/ resonance/ homeostasis/ energy
    energetic oscillation   36, 70, 127, 144, 177, 258, 274, *see* also light/ atom/ sub-atomic particle
    harmonic   146, 177, 257, 259, *see* also oscillation/ music
    resonance/ resonant association   101, 177, 258, 260
    vibratory sustenance > cosmos/ appearances   69
    vibratory time   168
    wireless communication   101, 110, 144, 177, 239, 257, 258, 260, 293, *see* also information/ light/ resonance
virtuality   *see* also ZPE
    archetype   98, 108, 109, 157, 177, 246, 272, *see* also archetype/ potential matter
    subliminal actuality 100, 145, 159, 160, 177, 246
vis medicatrix naturae   *see* homeostasis/ archetype/ caduceus/ health
void   *see* nothingness/ space
    subjective suction   *see* death, type-2 perspective
Void   *see* Nothingness/ Essence
volitio-attractive/ mental push-pull   *see* psycho-logic

W

Wallace Alfred Russel   341
wave/ cyclical carriage of energy   *see* cosmic models/ vibration/ light/ harmonic oscillation
    energetic order/ ring-bound creation 136
    quantum wave-particle duality   *see* quantum aspects
    radiates from central source/ radiance 136
    sound/ light/ wave-borne information 25, 38, 66, 79, 93, 100, 109, 135, 257, 258, 259, 263, 367, *see* also harmonic oscillation
Weaver Warren   77
Weiner Jonathan   345
wheel of time   *see* time's cycles
Wheeler John   160
Wiener Norbert   79
wild-type   *see* gene/ archetype/ typology
    typological norm   267, 342, 348, 364, 389, 392
Wilson Robert   188
WMAP   188
world tree   104, *see* also Seed/ tree of life
world-view   *see* also faith
    anti-parallels/ *top-down - bottom-up*   20, 21, 68, 378, 379
    bottom-up   *see* materialism
    definition   19
    needs account for metaphysic   15, 88, 200, 230, 281, 311
    top-down   *see* holism
    unavoidable   435
Wright Sewall Green   360

Y

yantra   *see mandala*
yeast   361, 390
yin/ yang/ yin-yang 35, 396, *see* dialectical operator/ polarity

Z

Zend Avesta   276
zero   *see* Glossary/ zero-point/ non-existence/ nothingness
zero-point   *see* also balance/ nothingness/ apparent absolute/ ZPE
    complete exhaustion/ sink   43, 64, 156, 159, 188
    material absence   77, 79, 157
    non-existence   77, 134, 140, 157, 190, 233
    nothing everywhere/ big bang singularity/ cosmic ignition   188
    nothing *opp.* something   41, 77
    number zero/ symbolic absence 77, 134, *see* also Glossary
    pivot/ point of balance   35, 185
    positive void   *see* void/ source/ archetype
    zero degrees (0K)   *see* absolute physical/ apparent absolute
    zero probability = impossibility   123, 193, 318, 334, 340, 358, 361, 363
ZPE *see* also space/ vacuum/ Glossary
    quantum energy of a vacuum   145, 146, 154, 156, 159, 161
    ZPR (zero-point radiation)   *see* ZPE
Zero-Point
    Nothingness/ Source/ Super-Nature   134, 150, 155, *see* Essence
ziggurat   *see* cosmic models
zoopharmacognosy   252

Bibliography

God, Science and Evolution	Andrews E.	1980
Who Made God?	Andrews E.	2009
New Biology	Augros R. & Stanciu N.	1988
New Story of Science	Augros R. & Stanciu N.	1986
Undeniable	Axe D.	2016
Chemical Evolution	Aw S.	1976
Dawkins' Proof for the Existence of God	Barns R.	2009
Book of Nothing	Barrow J.	2000
Constants of Nature	Barrow J.	2002
Infinite Book	Barrow J.	2005
New Theories of Everything	Barrow J.	2007
Darwin's Black Box	Behe M	1996
Edge of Evolution	Behe M.	2007
Science & Evidence for Design in the Universe	Behe, Dembski, Meyer	1999
Devil's Delusion	Berlinski D.	2009
Origin of Species Revisited	Bird W	1989
Consciousness	Blackmore S.	2003
Hallmarks of Design	Burgess S.	2000
Tao of Physics	Capra F.	1976
Web of Life	Capra F.	1996
Endless Forms Most Beautiful	Carroll S.	2011
Evolution's Achilles Heels	ed. Carter R.	2014
God beyond Nature	Clark R.E.D	1982
Universe, Plan or Accident?	Clark R.E.D	1961
Void	Close F.	2007
Language of God	Collins F.	2007
Life's Solution	Conway Morris S.	2003
Runes of Evolution	Conway Morris S.	2015
Wonders of the Universe	Cox B.	2011
Life Itself	Crick F.	1972
How Life Began	Croft L.	1988
Origin of Species	Darwin C.	1859
Secret of the Creative Vacuum	Davidson J.	1989
Web of Life	Davidson J.	1988
Accidental Universe	Davies P.	1982
Edge of Infinity	Davies P.	1981
God and the New Physics	Davies P.	1983
Goldilocks Enigma	Davies P.	2007
Mind of God	Davies P.	1992
Blind Watchmaker	Dawkins R.	1986
God Delusion	Dawkins R.	2006
Intelligent Design Uncensored	Dembski W. and Witt J.	2010
Evolution, A Theory in Crisis	Denton M.	1985
Nature's Destiny	Denton M.	1998
Transformist Illusion	Dewar D.	1957
Thermodynamics & the Development of Order	ed. Williams E.	2002
Information and the Nature of Reality	eds. Davies & Gregersen	2010
Mind, Body & Electromagnetism	Evans J.	1992
What Darwin Got Wrong	Fodor, Piatelli-Palmarini	2010
Science and Human Origins	Gauger, Axe, Luskin	2012
Mysterious Epigenome	Gills, Woodward	2012

Title	Author	Year
In the Beginning was Information	Gitt W.	1997
Without Excuse	Gitt W.	2011
Cheating Time	Gosden R.	1996
Ever Since Darwin	Gould S.	1977
Panda's Thumb	Gould S.	1980
Wonderful Life	Gould S.	1989
Evolution of Living Organisms	Grassé P-P.	1978
Elegant Universe	Greene B.	1999
Brief History of Time	Hawking S.	1988
Creation and Evolution	Hayward A.	1985
Neck of the Giraffe	Hitching F.	1982
Dreaming	Hobson J. Allan	2002
Fallacies of Evolution	Hoover A.	1977
Intelligent Universe	Hoyle F.	1983
Darwin on Trial	Johnson P.	1991
Reason in the Balance	Johnson P.	1995
Y, The Descent of Men	Jones S.	2002
Abusing Science	Kitcher P.	1982
Ghost in the Machine	Koestler A.	1975
Roots of Coincidence	Koestler A.	1976
On Guard	Lane Craig William	2010
Naked Emperor: Darwinism Exposed	Latham A.	2005
God and Stephen Hawking	Lennox C	2011
God's Undertaker	Lennox C.	2009
There is a God	Flew A.	2007
Gaia: Practical Science of Planetary Medicine	Lovelock J.	1991
Gaia	Lovelock J.	1979
Darwinism: Refutation of a Myth	Lovtrup S.	1987
Darwin Retried	MacBeth N.	1971
Evolution of Sex	Maynard-Smith J.	1978
Master and his Emissary	McGilchrist I.	2009
Dawkins' God	McGrath A.	2005
Darwin's Doubt	Meyer S.	2013
Signature in the Cell	Meyer S.	2009
Facts of Life	Milton R.	1992
Chance and Necessity	Monod J.	1970
Mind and Cosmos	Nagel T.	2012
Brain Science & Biology of Belief	Neuberg A. et alii	2001
Natural Theology	Paley W.	1802
Shadows of the Mind	Penrose R.	1995
How the Mind Works	Pinker S.	1997
Quantum World	Polkinghorne J.	1984
Science and Creation	Polkinghorne J.	1988
Origins of Life	Rana F, Ross H.	2004
Cell's Design	Rana F.	2008
Creating Life in the Lab	Rana F.	2011
Great Evolution Mystery	Rattray Taylor G.	1983
Rocks Aren't Clocks	Reed J.	2013
Just Six Numbers	Rees M.	1999
Our Cosmic Habitat	Rees M.	2001
Time, Space and Things	Ridley B.	1976
Dawkins Letters	Robertson D.	2007
Alas, Poor Darwin	eds. Rose H. & S.	2000
Darwin and Design	Ruse M.	2003

Title	Author	Year
Genetic Entropy & the Mystery of the Genome	Sandford J.	2005
Refuting Evolution 1 and 2	Sarfati J.	2007
Seven Sins of Memory	Schacter D.	2001
What is Life?	Schrödinger E.	1944
When is a Fly not a Horse?	Sermonti G.	2005
Evolution, A View from the 21st Century	Shapiro J.	2011
New Science of Life	Sheldrake R	1981
Presence of the Past	Sheldrake R.	1988
Flaws in the Theory of Evolution	Shute E.	1962
Billions of Missing Links	Simmons G.	2007
Not By Chance	Spetner Lee	1997
Macro-evolution	Stanley S.	1979
Darwinian Fairytales	Stove D.	1995
Mystery of Life's Origin	Thaxton, Bradley, Olsen.	1985
Double Helix	Watson J.	1970
Icons of Evolution	Wells J.	2000
Myth of Junk DNA	Wells J.	2011
Natural Sciences Know Nothing of Evolution	Wilder Smith A.E.	1981
Basis for a New Biology	Wilder Smith A. E.	1976
Creation of Life	Wilder Smith A. E.	1974
God: To Be or Not To Be	Wilder Smith A. E.	1975
Deluded by Dawkins?	Wilson A. J.	2007
Charles Darwin: Victorian Mythmaker	Wilson A. N.	2017

The author has recently written a few more books (available from Amazon, Foyles, Waterstones, Barnes & Noble etc. and see website addresses on p.2):

 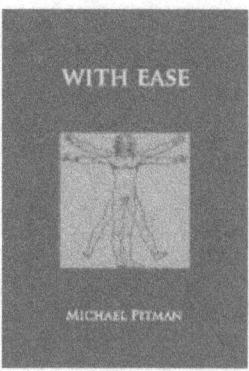 

www.ingramcontent.com/pod-product-compliance
Lightning Source LLC
Chambersburg PA
CBHW071850290426
44110CB00013B/1089